maya 2 character animation

nathan vogel
sherri sheridan
tim coleman

Contributions by Alex Alvarez, Jesse Andrewartha, Kevin Cain, Matt Ontiveros, and David Tart

Cover design: Nick Phillips and Jeff Taylor, Alloy Design, San Francisco, CA, USA

Image of Lanker created by and used with permission of Alex Alvarez

New Riders Publishing, 201 W. 103rd St., Indianapolis, Indiana 46290

Publisher
David Dwyer

Executive Editor
Steve Weiss

Acquisitions Editor
Laura Frey

Development Editors
Audrey Doyle
Barb Terry

Managing Editor
Jennifer Eberhardt

Project Editor
Caroline Wise

Copy Editor
Daryl Kessler

Indexer
Angie Bess

Technical Editors
Adam Holmes
Mike Laubach
Will McCullough
Bert van Brande

Media Development Specialist
Craig Atkins

Compositors
Wil Cruz
Steve Gifford
Amy Parker
Carl Peters
Gina Rexrode

Maya 2 Character Animation

International Standard Book Number: 0-7357-0866-5

Library of Congress Catalog Card Number: 99-64954

Printed in the United States of America

First Printing: December 1999

03 02 01 00 99   7 6 5 4 3 2 1

Interpretation of the printing code: The rightmost double-digit number is the year of the book's printing; the rightmost single-digit number is the number of the book's printing. For example, the printing code 99-1 shows that the first printing of the book occurred in 1999.

Trademarks

All terms mentioned in this book that are known to be trademarks or service marks have been appropriately capitalized. New Riders Publishing cannot attest to the accuracy of this information. Use of a term in this book should not be regarded as affecting the validity of any trademark or service mark.

Warning and Disclaimer

Every effort has been made to make this book as complete and as accurate as possible, but no warranty or fitness is implied. The information provided is on an "as is" basis. The authors and the publisher shall have neither liability nor responsibility to any person or entity with respect to any loss or damages arising from the information contained in this book or from the use of the CD or programs accompanying it.

Contents at a Glance

Appendixes

You can find Appendix A, "MEL Scripting," and Appendix B, "Maya Plug-Ins and Third-Party Solutions," on the accompanying CD-ROM. Both appendixes can be found in PDF format in the CD's Appendixes folder.

Table of Contents

Appendixes

You can find Appendix A, "MEL Scripting," and Appendix B, "Maya Plug-Ins and Third-Party Solutions," on the accompanying CD-ROM. Both appendixes can be found in PDF format in the CD's Appendixes folder.

About the Authors

Nathan Vogel, 3D Special Effects Director at Minds Eye Media in San Francisco (`www.mindseyemedia.com`), is a certified Alias | Wavefront Maya character animation instructor. His latest projects include releasing a set of Maya training videos, producing a 3D-fantasy game using Maya with RenderMan, and creating several original 3D-animated TV and film projects. An award-winning animator, Nathan recently received the 1998 World Animation Celebration Best Animated Music Video award.

Sherri Sheridan is the creative director and co-founder of Minds Eye Media in San Francisco. She is currently working on a series of original 2D and 3D animations for the Web, film, and TV. Her most recent projects include the animated Web show "Goth Grrls", the independent feature film *The Sundial Solution*, the upcoming children's HDTV show *Sassafrass the Psychic Warrior*, and a series of 3D meditation videos due out soon. She is an award-winning animator who received the "Best Music Video" award at the 1998 World Animation Celebration for "Beyond," by the recording artist Young American Primitive (Geffen Records). Sherri teaches part-time at the Academy of Art College and conducts international Animated Story Concept Development workshops between her other projects.

Before forming Minds Eye Media in 1995, Sherri worked in the multimedia industry for several years on various CD-ROM, video, and Web projects. This included some time at Macromedia, where she helped develop Shockwave animation and created the first Shockwave movies on the Web. She has a BA from U.C. Berkeley and went to graduate school at San Francisco State to study animation and interactive design.

Tim Coleman started out exploring 3D space in the fields of architecture and environmental design in college. After moving to San Francisco in 1992, Tim was exposed to the emergence of computer graphics and animation. Over the last seven years, Tim has been involved with numerous TV commercial, video game, and software development projects, including spots for the California Lottery, the video game *Gauntlet Legends*, and the development of real-time compositing software for TV broadcast use. In addition, Tim has taught and trained computer artists in 3D-computer animation across the country. Tim has worked as a freelance certified Maya instructor for Alias | Wavefront over the last two years, where he teaches a wide range of Maya classes to corporate customers. In addition, Tim teaches at the undergraduate and graduate levels at the Academy of Art College in San Francisco.

Recently, Tim, along with his partner Olivier Wolfson, started a small graphics/animation studio called Supergenius Animation, in San Francisco. At Supergenius, Tim finished a short film titled "Bowling Fer Souls," which can be seen at Spike and

Mike's Festival of Animation 1999. The three-minute animation was completed using Maya.

In the future, Tim would like to continue to pursue his interests in animating fun and unique 3D characters. He also wishes to take his teaching and training experience with 3D graphics to artists around the world.

Alex Alvarez, Director of Gnomon, Inc., School of Visual Effects for Film, Television, and Games (Los Angeles), is an alumnus of the Art Center College of Design and the University of Pennsylvania. He has a background in illustration using traditional media and computer graphics, having worked for companies such as Malibu Comics and Activision. Prior to Gnomon, Alex worked as an applications engineer for Alias | Wavefront, where he was a consultant and trainer for studios in the Los Angeles area. He is a contributor to *3D Magazine* and has spoken at several digital conferences.

Jesse Andrewartha is an expatriate Australian who now lives in Berkeley, California. For one reason or another, his dream of becoming a nuclear technician gave way to a fascination with technical imagery. After graduating from RMIT with a degree in scientific photography, he worked for a year with the Australian Federal Police, earned an honors degree, and wound up entering the world of computer graphics. An Alias | Wavefront certified instructor, Jesse has taught classes in animation and visual effects for the Ex'pressions Center for New Media. He is also an instructor and UNIX systems administrator for Mesmer Animation Labs. An aspiring technical director, Jesse has worked in this capacity for Minds Eye Media, San Francisco. Jesse writes regularly for *SIGGRAPH* and *3D Magazine* on high-end 3D products and techniques, including MEL programming and RenderMan.

Kevin Cain is the creative director at the Pelleas Design Studios and director of the SGI/PC Computer Education Center at the Academy of Art in San Francisco, where he supervises 1,400 undergraduate and graduate students. He is also the founding director of the New Media Colloquia, which presents yearly conferences linking technology and the arts. Kevin has participated in numerous computer projects around the world, including the Urban Design Workshop in Massa Marritima, Italy, and an ongoing project in Egypt to document endangered sites of cultural heritage.

As a lighting designer, Kevin has worked on more than 100 theatrical and film projects over the past decade. Kevin has also designed the lighting for numerous short films, including the 1999 AIA award-winner "Maybeck's Palace."

Matt Ontiveros is the lab manager and principal Alias | Wavefront Maya instructor at Mesmer Animation Labs. Matt teaches all three Maya certified courses. One of only a handful of Alias | Wavefront certified instructors in the country, his background includes an M.F.A. in sculpture from the University of Washington and formal training in photography, printmaking, and drawing. Matt's teaching experience includes courses in Alias | Wavefront Maya, Softimage 3D | Extreme, Composer, Alias | Wavefront PowerAnimator, traditional 3D design courses, sculpture, and drawing. In addition to his knowledge of UNIX-based platforms, Matt has graphics and production experience on Macintosh-based programs, such as Quark, Illustrator, and Photoshop.

David Tart is a native Californian who spent the majority of his formative years in Berkeley. For reasons unknown, his early influences induced him to pursue a career that allowed him to make little objects move around in funny ways on the screens of televisions and movie theaters. He is now in his sixth year of actually earning a living doing just that. After graduating from San Francisco State University (no, you *don't* have to go to Cal Arts to be a good animator), he worked for a year doing stop-motion animation for an ABC Saturday morning television show called *Bump in the Night.* While working there, he was mentored by the fine crew who'd just wrapped from *Nightmare Before Christmas.* Since then, he's been working at Pixar Animation Studios on a variety of projects, most notably *Toy Story* and *A Bug's Life.*

Dedications

This book is dedicated to the memory of Mary Jane Sheridan, who we know would have been very proud to see this book in print, since she was always bugging her daughter to publish something along the way.

It is also for Theresa Vogel, who bought Nathan his first Apple LC computer, knowing one day her grandson would be a 3D super maven.

Acknowledgments

Nathan Vogel:
I would like to thank my grandmother for supporting my computer art career; my mother and Tom and the Llewellyn clan for being such a positive influence; Sherri for making this all happen with me; Apollo the Lab Cat for all his excellentness and for keeping me sane during the rough times; the development staff and tech editors, both at New Riders and independent, for sticking through on this huge project; and Anthony Rossano and Mesmer Animation Labs, where I tested my madness on the student body.

Special thanks to the software and hardware vendors who made this book possible with their generous and appreciated contributions: Carla and Jeff at Lambsoft for MoveTools; Nancy and Gordon at Geometrix for their 3D scanners; Ali at ID8 with the Gypsy motion capture system; the cool people at Artiface; our friends at SGI with the new 320 Visual Computers; Tom Paterson at AMD with the new Athlon K7 750mhz processor; Intergraph and their WildCat graphics; Tracey Hawken at Alias | Wavefront and all of their wonderful engineers who wrote Maya; Stuart McSherry at Green Works for XFROG; Eric Hernandez and Reid Nicol at Arete for DNT; Lola Gill and Ray Davis at Pixar for RenderMan and MTOR;

Igor and Alex at Animatek and their awesome software World Builder; and all of the rest of the very important contributors who helped in some important way to get this project completed—you are never forgotten.

I would also like to thank some of the key inspirations in my life who helped allow this book to be created: George Maestri for writing *[digital] Character Animation*; George Lucas for making *Star Wars* (I saw it when I was five and it left a lasting impression); Chris Landreth for being such a hardcore independent for so long; my *Dungeons and Dragons* gaming posse for the endless creative inspiration; and Donnie and Ivan for drawing such brilliant source imagery.

Special thanks to TAO/PlaneTrance and the Global Trance Music collectives, CCC, Koinanea, Blue Room, Eye Phunk, Frequency 8, UV/99, Dimension 7, Visionary Designer Eugene Tsui, Alex Alvarez, Alex Lindsey, Brennan Doyle at Tippet Studios, Weta Studio and the Lord of the Rings project, Joel Hornsby, Seanchan Owens, Jason Halverson, Troy Sutton, David Tart, the Superior Olive, Jane Veeder, Stephen Wilson, George Legradi, Chris Robbins, the SFSU Conceptual Design department, the SFSU Downtown Center, the Academy of Art College SF CGI staff and lab techs (Joaq Psycho and crew), CAI, CEA, Media Bytes, the Neuromancer project, William Gibson, Neal Stephenson, Duncan at A|W, 3D Gear, Night Tribe, Glenn Grillo, Ken Solomon, Patrick and Stephanie, 3D Design, *DV* and Miller Freeman, Viewpoint for the killer geometry, PDI, ILM and Manix, Pierre and Peter, Jonathan Quentin's Sacred Geometry, Bob and the Dose Hermanos, Joe and Marsha and the Burning Man project, Radio V and the Verbum crew, Michael and Carla, Mediaman, Spookbot, Donna Matrix, and the Core.

And finally, thank you to Sherri Sheridan, Tim Coleman, Matt Ontiveros (Appendix A), Jesse Andrewartha (Appendix A and Appendix B), Kevin Cain (Chapter 8), Tadao M. (help with Chapter 7), David Tart (character animation insert), Alex Alvarez (quotes from Chapter 5), Alex Lindsay (quotes and concepts in Chapter 7), Donnie Bruce (dragon illustrator in Chapter 6; dinosaur, dragon fly, and dragon textures in Chapter 7), Ivan Wachter (hairy man illustrator in Chapter 5), Apollo the Lab Cat (for editing liberally with his shredding style), and Viewpoint for lending us the Dinosaur model.

Sherri Sheridan:
I would like to extend my heartfelt appreciation to everyone who helped write this book, including Tim Coleman, Matt Ontiveros, Kevin Cain, David Tart, Carlos Petros, Jesse Andrewartha, all the folks at Mesmer Animation, and the Minds Eye Media late-night hackers for testing out all our ideas.

Many thanks to all the wonderful people at New Riders, including Chris Nelson, Laura Frey, Steve Weiss, Barb Terry, Audrey Doyle, Caroline Wise, and all the other people who helped make this such an incredible book. We especially thank this group for believing in our vision of the book and supporting the direction we wanted to take the information at every turn to make this a uniquely useful computer graphics book.

Also, thanks to David Biedny for helping blaze new trails into publishing and for giving us deep wisdom tidbits along the way, such as, "It's going to take longer than you think."

Thanks also to Maia Sanders, Sam Miceli, Donnie Bruce, Chris Hatala, Debbie Rich, Jeff Vacanti, Ryan Nishimoto, and all the other artists who contributed their work as inspiring examples in these pages.

Special thanks to SGI for providing incredibly fast computers that made Maya hum.

Tim Coleman:
I would like to thank my parents, Anong, and my brother Greg for their love and support; Olivier Wolfson for his insightful viewpoints and comic critique; everybody at Alias | Wavefront; Spike from Spike and Mike's Animation Festival for his support of independent animators worldwide; and my students and friends at the Academy of Art who have taught me so much.

A Message from New Riders

Like it says in the old Dead tune, "What a long, strange trip it's been…"

This book has been a labor of passion for everyone involved. Flip through the pages. There's nothing quite like this book in existence. And anyone playing around with Maya would likely tell you there's nothing quite like Maya, either. Appropriate.

At New Riders, we take a lot of pride in the fact that our books are borne of true partnerships between publisher and author. The author/publisher entity is unique to each book, and hopefully, it's the best partnership imaginable for that particular title. It's with that in mind that we say: No one besides this amazing mix of people could have created the book you hold in your hands right now: Nate Vogel and Sherri Sheridan and the rest of the folk at Minds Eye; Tim Coleman; Matt Ontiveros; Caroline Wise; Barb Terry; Audrey Doyle; Alex Alvarez and Lanker; Laura Frey; Jennifer Eberhardt; a remaining cast of crucial contributors who were in the right place at the right time; and the gods of kismet (with help from Alias | Wavefront). Gracias, all.

And to you, the reader, on behalf of everyone involved: Fire up your copy of Maya, crack open this book, and learn what *you* can do with this truly amazing tool. Don't forget to have a bonfire of fun while you're at it, and drop us a line to let us know how you're doing. Thanks.

How to Contact Us

As the reader of this book, *you* are our most important critic and commentator. We value your opinion and want to know what we're doing right, what we could do better, in what areas you'd like to see us publish, and any other words of wisdom you're willing to pass our way.

As the Executive Editor for the Graphics team at New Riders, I welcome your comments. You can fax, email, or write me directly to let me know what you did or didn't like about this book—as well as what we can do to make our books better. When you write, please be sure to include this book's title, ISBN, and author, as well as your name and phone or fax number. I will carefully review your comments and share them with the authors and editors who worked on the book. For any issues directly related to this or other titles:

Email: steve.weiss@newriders.com
Mail: Steve Weiss
Executive Editor
Professional Graphics & Design Publishing
New Riders Publishing
201 West 103rd Street
Indianapolis, IN 46290 USA

Visit Our Website: *www.newriders.com*

Go to www.newriders.com and click on the Contact link if you

Have comments or questions about this book

Want to report errors that you have found in this book

Have a book proposal or are otherwise interested in writing with New Riders

Would like us to send you one of our author kits

Are an expert in a computer topic or technology and are interested in being a reviewer or technical editor

Want to find a distributor for our titles in your area

Are an educator/instructor who wishes to preview New Riders book for classroom use. (Include your name, school, department, address, phone number, office days/hours, text currently in use, and enrollment in your department in the body/comments area, along with your request for desk/examination copies, or for additional information.

Call Us or Fax Us

You can reach us toll-free at (800) 571-5840 + 9+ 3567. Ask for New Riders. If outside the USA, please call 1-317-581-3500 and ask for New Riders. If you prefer, you can fax us at 1-317-581-4663, Attention: New Riders.

Technical Support/Customer Support Issues

Call 1-317-581-3833, from 10:00 a.m. to 3 p.m. US EST (CST from April through October of each year—unlike most of the rest of the United States, Indiana doesn't change to Daylight Savings Time each April.

You can also email our tech support team at userservices@macmillanusa.com and you can access our tech support website at http://www.mcp.com/product_support/mail_support.cfm.

Introduction

This book is about animating characters in Maya. One of the most amazing things about Maya is that it enables you to create complicated, organic-looking characters and animate them rather quickly. Maya has introduced a new level in 3D animation software by incorporating a very deep and powerful feature set into a groundbreaking interface along with a fast renderer to make animating 3D characters easier. Maya is a professional-level animation package that will free you to create almost anything you can imagine on a desktop computer.

Traditional character animation skills are still more important than knowing how to use the latest computer software, however. We encourage any computer animator to study 2D animation and understand the rich history of this world that has been bringing characters to life for more than 50 years. You should have a firm grounding in the foundations of animation, including anatomy, weight, timing, and motion, before attempting to become a really good 3D animator.

This book was written by a team of working Maya animators, each of whom contributed his own special areas of knowledge to bring you a very succinct and insightful book. Our goal was to write a useful computer book that we would like to read and keep beside us as we work late into the night on our dream animation projects.

We have added extra information from top animators in the industry who wanted to contribute their tips but did not want to write an entire book on their particular subject. Our hope is that *Maya 2 Character Animation* will be a handbook that you carry with you on your character animation journey and use as a valuable resource to help you create 3D animations that win many awards at prestigious festivals.

Story Concept Development

Do you have problems coming up with good characters and ideas to animate in 3D? Because Maya gives you an almost unlimited power to create any kind of animation, the challenge now seems to be coming up with fresh original characters and story ideas. As computers get faster and 3D programs become easier to use, visual storytelling skills become more valuable. We firmly believe that you should not go anywhere near a computer until you have a solid idea to animate that works well at the storyboard level and has received positive feedback from a variety of people.

The first three chapters of this book, written by Sherri Sheridan, take you through a conceptual development process that will help you become a better visual storyteller, character designer, and 3D animator. The first chapter discusses how to generate story ideas for fresh 3D characters and animations in Maya. The next chapter deals with 3D character design and how to create original characters that will leap off the screen visually and grab your audience emotionally. The third chapter introduces some basic visual storytelling concepts including storyboarding, camera shots, color maps, timing, and creating an animatic. The goal of these first three chapters is to help you think up solid short stories that can be created by one person in Maya for use on demo reels or at animation festivals, or to develop your own TV show or feature film. You may also just want to improve your ability to design and compose animated stories, characters, and shots.

George Lucas thinks the next *Star Wars* is going to be made on a desktop computer. Maya is the type of software that gives 3D artists the power to create their own TV shows and films. What kinds of stories do you have to tell that might be considered the next *Star Wars*?

Modeling and Texture Mapping

Using Maya's powerful modeling tools and revolutionary new interfaces, a single artist can accomplish in hours or days what might have taken several artists weeks or even months to accomplish.

Chapters 4, 5, and 6 provide detailed, step-by-step tutorials that cover much of the intricate process of creating organic humanoids and fantastic creatures. After completing the three modeling chapters, you will have a good foundation for creating nearly any character that you can imagine.

Chapter 7 is a detailed chapter providing in-depth information on texture-mapping NURBS characters. This chapter focuses on solving some of the more complex problems (generally left unexplained in the Maya manuals) of dealing with detailed textures that cover complex, seamless, multi-object NURBS characters. Along the way, several unique methods are introduced explaining special tools to help you add character and realism to your projects.

Maya Character Animation

Chapters 9, 10, and 11 give you insight in setting up and animating your 3D characters. You will look at concepts that give you the most flexibility in controlling your characters. In addition, you will examine different approaches to animating your 3D character. Finally, you will look at techniques for setting up and animating the facial features of your character.

Other Important Information

Several special chapters have been included to help take you through the entire process of animating characters in Maya. Chapter 8 is written by Kevin Cain, who has a long background in theater lighting and teaching Maya at the Academy of Art College in San Francisco. This chapter focuses on lighting, providing valuable tips on how lighting works in Maya and introducing the basics of lighting scenes and characters to really make your animation look professional. Traditional lighting concepts are presented because many animators have not had much experience in this area, and lighting can make or break an animation, depending on the quality of the lighting setup.

In Appendix A, Matt Oliveras and the team from Mesmer Animation School provide details on MEL scripting, introducing the awesome power of the ability to get into the code level of the architecture in Maya. This is one of Maya's strongest features, and you may find you have a flair for writing scripts that save hours of time or make the program do amazing things no one has ever seen before.

In Appendix B, Nathan Vogel and Jesse Andrewartha have gathered up as much first-hand information as possible to give you a solid introduction to some third-party plug-ins for Maya. We tried to cover our favorite plug-ins, as well as products we have used on projects, to give you the best information possible.

Things You Need to Know

This book assumes that you have some basic knowledge to get into some deeper parts of the Maya character animation process. You should be familiar with the following to feel comfortable completing the tutorials:

- **Basic understanding of Maya.** You should complete the "Salty the Seal" tutorial from the Maya training manual to have an overview of the program.
- **Good knowledge of Photoshop.** This knowledge is essential for the modeling and texture mapping sections.

Hardware Recommendations

Maya can run quite well on a $500 PC for simple animations. The more complex your scenes and characters become, however, the more processor speed and RAM you are going to need to have a good time animating. This is true with any 3D software package.

It seems that no matter how fast things become, we as animators keep making our characters and sets more complicated—ending up where we started speed-wise, but with cooler-looking results. There are lots of tips on optimizing your interface for maximum performance. A good 3D card is essential to take advantage of Maya's redraw capabilities. We use and recommend the following hardware to run Maya 2.0 for this book:

We have found that the SGI 320 NT workstation loaded with two Pentium III 500MHz processors and 512MB of RAM was a stable and reliable workhorse capable of dealing with complex scenes and detailed characters. That is not to say that a custom-built, dual P3 500 with 512MB of off-the-shelf memory with a 3D Labs GMX2000 AGP card won't give the 320 a real run for its money. But you have to be ready to give your custom creation some extra time hacking it to life and sweating it out so that it runs solidly.

The hardware clock of constant forward change continues to ring. As this book went to press, several benchmarks were published that have put the high-end Intergraph NT workstations with their own custom Wildcat graphics cards at the top in terms of graphics acceleration and rendering speed. But the built-in, uncompressed video and Firewire ports on the 320, combined with the ultra-hip and sassy LCD wide-screen panel, make it a joy to create art digitally.

About the CD

A CD-ROM is included with this book that contains some really helpful information and is worth popping into your computer as you read the book. Project files are included from the tutorials to help you follow along if you get stuck on something. Alex Alverez has graciously included his "Lanker" tutorial, and we put the Lanker character on the cover because he was one of the first 3D Maya characters that inspired us to use the program and write the book. Models and texture maps are also included on the CD-ROM for you to play with while learning the program.

About New Riders

One of the reasons we chose New Riders to publish this book is because they really try to produce useful computer software books and are supportive of doing full-color throughout. We hope you enjoy this aspect of our book because we as visual artists feel it is much more interesting to read colorful software manuals, and appreciate New Riders for letting us write the kind of book we felt we would like to use.

A Final Note

Being a great 3D character animator is a lifelong process. It is one of few professions that require a person to have so many different types of talent and knowledge. You must be constantly open to learning new software and skills as they become necessary to progress along your path.

Being an animator is also one of the most exciting, entertaining, and fulfilling careers you could ever imagine. Maya is a complex program to master and will leave you frustrated at times, and then jubilant when you figure out how to make it do what you want. This is the first time in our known history that we as individuals can create whatever we can imagine on a desktop computer. Hopefully, you will find much information in these pages to help you take another step toward realizing your vision and make animating more fun.

Part I

Creating Stories for 3D Animation

Creating Story Concepts for 3D Animation

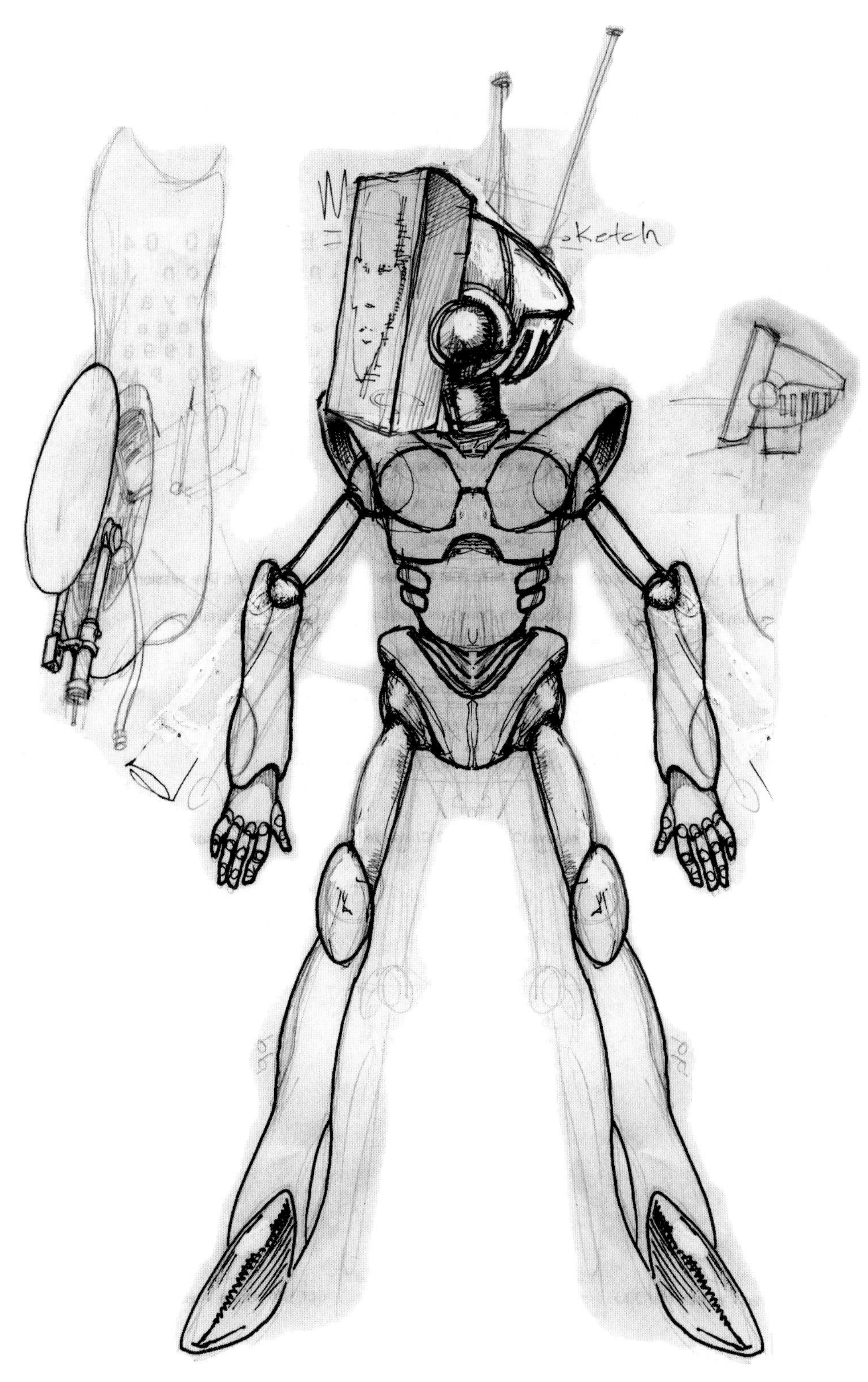

CHAPTER 1

Creating Story Concepts for 3D Animation

by Sherri Sheridan

One of the most amazing things about Maya is that it allows you to create complicated organic-looking characters and animate them rather quickly. The challenge now seems to be in coming up with fresh original characters and story ideas. As computers get faster and 3D programs become easier to use, visual storytelling skills become more valuable.

The first three chapters of this book take you though a conceptual development process that will help you become a better visual storyteller, character designer, and 3D animator. The first step is learning how to generate story ideas for fresh 3D characters and animations in Maya. The next chapter deals with 3D character design and how to create original characters who will leap off the screen visually and grab your audience emotionally. The third chapter introduces some basic visual storytelling concepts, including story boarding, camera shots, color maps, timing, and creating an animatic. The goal of these first three chapters is to help you think up solid short stories that can be created by one person in Maya for use on demo reels or animation festivals or to develop your own TV show or feature film. You may also just want to improve your ability to design and compose animated stories, characters, and shots.

George Lucas thinks the next *Star Wars* will be made on a desktop computer. Maya is the type of software that gives 3D artists the power to create their own TV shows and films. What kinds of stories do you have to tell that might be considered the next *Star Wars*?

1.1 Understanding the Importance of Planning Your Story Concept

We have all seen animations that were technically perfect or had great 3D characters but left us feeling like the story was not really there and that a great deal of time had been wasted. The most effective way to learn Maya is to come up with an idea and create it using the same type of production process most animation houses employ. However, the best design tool still is a pencil, which is what we will use to begin generating ideas. It takes only a little more time at the beginning to develop a good idea for a 3D project than it does to just sit down and start clicking away, doing what the computer wants. The time you spend planning helps you avoid the problem of wasting all those hours you spend working.

Top Ten Reasons to Become a Visual Storytelling Expert

You probably have your own reasons for using Maya to animate characters, but here are the reasons we think you should learn to become an expert in the field:

1. **To get the job you want.** Most animation houses hire animators who understand how to compose a shot and design a good character rather than animators who are more technically oriented and have limited artistic and conceptual skills.
2. **To win at an animation festival.** You may get the job you want because someone saw your piece in the festival, or you may get a chance to pitch a TV series based on your characters to the head of Comedy Central who happened to be in the festival audience and loved your animation.
3. **To have more fun animating.** It feels better to be working on a really good idea for months on end late into the night than it does to be working on an idea you suspect may have very limited viewing potential.
4. **To gain more flexibility in the job market.** The animation industry can be unpredictable for employment. The more skills you have, the better off you will be if your film company suddenly runs out of money and you need to get another job fast.
5. **To escape animating other people's ideas for rest of your life.** Some 3D people are content to model monsters for 3D game companies year in and year out. Others find that they need a change of pace or perhaps have their own ideas of projects they would like to create. The more visual storytelling skills you cultivate, the better position you are in to pull together your own dream projects.
6. **To be ready when visual storytelling skills become mandatory in the very near future.** Technical skills and knowing which button to push will become less important in the future as interfaces improve. You can teach an artist to use a computer, but you usually can't teach a computer person to be an artist as quickly.

7. **To be ready when huge content markets open up for original animations**. When video streaming hits the Internet with TV-type speed and quality, a big market for original animated shorts and series will emerge. The future offers many other new options, such as the digital TV market, animation festivals, and commercial work.
8. **To get started on a lifelong process.** It takes a long time to get good at visual storytelling, so you should start now. Being a good visual artist and animator is a lifelong process, one that you can't start too soon.
9. **To be able to offer more intelligent and useful input at production meetings, which will further your career options.** After you learn how to tell a good visual story, you will be able to spot problems and come up with solutions for making difficult shots work. This ability will also make you a more valuable animator and give you opportunities to work on the kind of projects you prefer.
10. **To become really rich and famous by making the next *Star Wars* on your $500 PC using Maya.** But first you need to know how to tell a good visual story.

1.2 Becoming a Better 3D Animator

To become a better visual storyteller and 3D animator, you need to spend some time studying some other areas between your renderings. If you want to be a really good 3D person you should be taking classes, reading books, and learning as much as you can about the following subjects:

- Figure drawing
- Filmmaking and Traditional Cinematography
- 2D Animation
- Graphic design, typography, and Sacred Geometry
- Acting
- Screen writing
- Fine arts and performing arts
- Photography

You may be wondering how you are going to have time to do any 3D animation at all if you have to know all of these other things. This is why you need to start *now*. The best animators usually have some training in several of these fields, if not all. Being a 3D animator is one of the most challenging careers around, but at least you will never have to worry about being bored!

1.2.1 Figure Drawing

If you don't know how to draw really well, you will have a rough time as a professional animator these days. You may have once been able to get by without knowing how to draw, but that is changing fast. Draw as much as possible. Most animation houses will ask to see some figure or life drawings before hiring you.

Try to fit in at least nine hours a week of figure drawing off a nude model. Concentrate on getting the energy of the pose quickly with short gesture drawings. Check local art schools for life drawing classes and open studios. You must be able to draw to communicate your ideas and to create storyboards, and you must understand light, composition, color, and values. Life drawing also keeps your eye finely tuned to the nuances of how to really see the world—its shadows, light, and so on. Most large animation studios consider life drawing important enough to offer open drawing studios for their animators during lunchtime.

1.2.2 Filmmaking and Traditional Cinematography

Visually creative people have been setting up shots, constructing interesting characters, and designing sets ever since the first story was told. Study the great directors behind the cameras of the best films ever made. It's all been done before in one visual form or another. You don't have to reinvent the wheel on every storyboard panel. Storyboard shot by shot some films or scenes from films that you think worked really well or are similar to current projects you are developing. Try to figure out what the director was thinking when he composed each shot.

Figure 1.1
Being good at figure drawing enables you to design anatomically correct exaggerated characters such as this giant carnival fat man or anything else your imagination concocts.

Credit: Illustration by Chris Hatala.

Storyboard the first five minutes of the movie classic *Citizen Kane*. Pay attention to the composition, contrast of lights and darks, lighting, camera movement, transitions from scene to scene, timing of shots, cuts, and overall look and feel creation. What are your top ten favorite movies of all time? Storyboard the best parts and try to figure out why you liked each movie so much. Borrow some of these ideas in your next animation.

1.2.3 2D Animation

Most really good 3D character animators have spent time doing 2D animations drawing moving characters at 24 frames per second. Read *The Animator's Workbook* by Tony White (Watson-Guptill Publications) and the *Illusion of Life* by Frank Thomas and Ollie Johnston (Hyperion Press). Study the "Principles of Animation" in the *Illusion of Life*. See how much squash and stretch, anticipation, staging, follow through, overlapping action, ease in/ease out, arcs, secondary action, timing, exaggeration and appeal you can add to your next 3D project (see Chapter 3, "Designing Storyboards for Animation," for brief definitions of these terms).

Set up a pencil tester with a cheap video camera and try to hand-draw animated bouncing balls, walking sacks of flour, and simple characters doing things. Develop a sense of timing for movement. Learn how to put some personality into a character's walk cycle. Then, get a VCR or DVD with a clean freeze frame and study each frame of a cartoon. Try to copy what works and learn what doesn't.

3D animators usually continue to do some 2D animation to keep their traditional skills fresh. If you are just getting into 3D animation and are serious about becoming a character animator as a career, you may want to go to a reputable school such as CalArts or Sheridan College to study 2D animation first.

1.2.4 Graphic Design, Typography, and Sacred Geometry

Good animation requires good design sense. Understanding type, layout, visual communication, and how colors work together will improve the way you design your characters and shots.

Take a graphic design course or practice copying designs that appeal to your sense of style. You will gain a greater understanding of the process involved if you try to re-create a nice layout with pictures and professional type. An increased awareness of design will also make the credit sequences of your animated projects look more professional.

Using the principles of Sacred Geometry will also help you design sets and characters by promoting an understanding of how we relate to shapes and mathematical constructs in our visual world. *A Beginner's Guide to Constructing the Universe* by Michael Schneider (Thames and Hudson Inc.) is a great book to get you started.

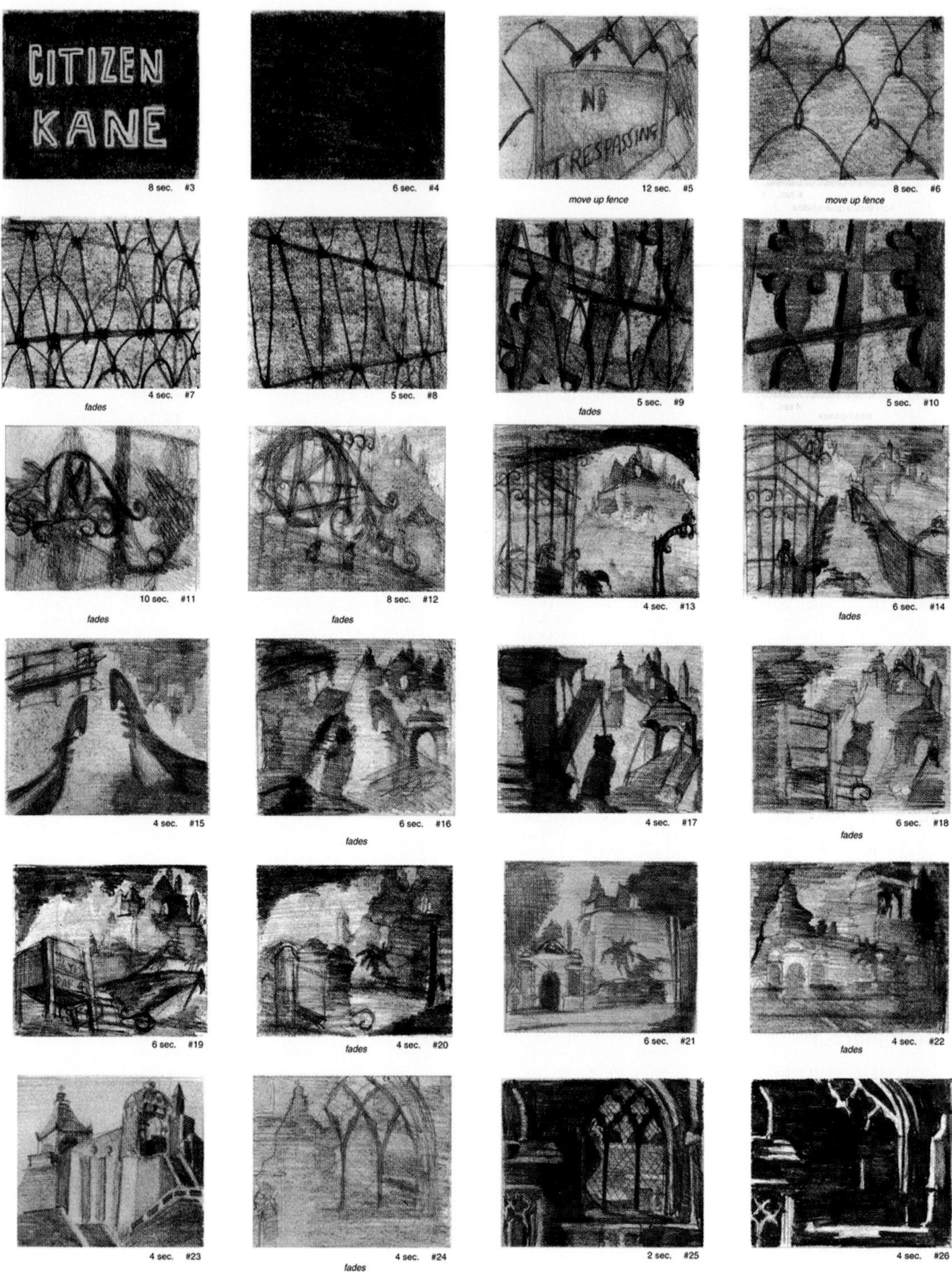

Figure 1.2

Storyboard your favorite shots from good films to learn more about how to tell a visual story. Shown here are the first few minutes of ***Citizen Kane***. Notice how beautifully composed and balanced each shot appears and how together they tell a very dramatic story visually.

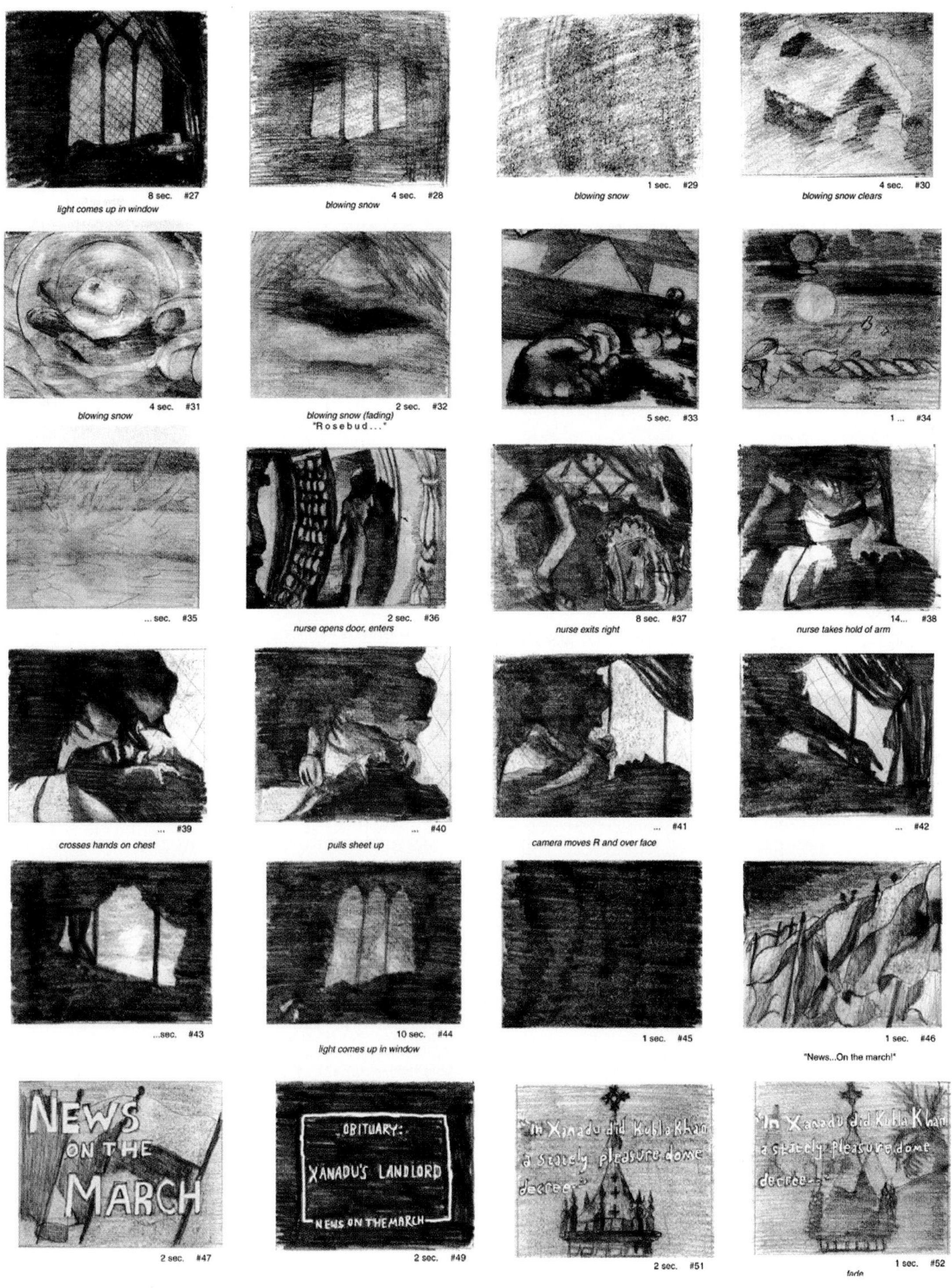

Credit: Illustrations by Maia Sanders.

1.2.5 Acting

Take an acting class for these five reasons:

- To learn about dramatic tension and three-act structure
- To understand the importance of giving characters clear motivations
- To get a better sense of how to make your characters come alive in a 3D scene
- To feel comfortable acting out your 3D characters in front of a mirror to get the motion right
- To feel more comfortable acting out your story and the voices of your 3D characters while pitching your storyboards in front of a group

1.2.6 Screen Writing

Read some books on the screen writing, such as *Writing Screenplays That Sell* by Michael Hauge (Harperperennial Library) and then write your dream screenplay. Or take a screen writing class to learn how to write a good screenplay. You will learn all about formatting, beats, scene structure, visual storytelling, and lots of other information you can use later. Typically, 3D animation houses borrow lots of ideas from movie making, so you should get familiar with the lingo.

1.2.7 Fine Arts/Performing Arts

Studying any area of the arts can help you become a better animator. Obviously, you can't master painting, sculpture, music, and theater, but, if you already have an interest in one area, you may want to further develop that talent.

Volunteer at a community theater to light their next play. You will learn a great deal about how important dramatic lighting can be to a 3D scene. Take a clay figure sculpture class, and learn how to make little models of 3D characters you may want to animate one day.

A Graphic Artist's Survival Guide

The Artist's Way by Julia Cameron (J. P. Tarcher) is a great artistic survival book for people embarking on creative careers. Cameron's book will also help you think of yourself more as an artist if you feel a little weak in this area. This book is also a great help for handling the pressures of maintaining creativity on tight deadlines.

1.2.8 Photography

Carry a good camera with you at all times to take pictures of rusted walls, strange pets, graffiti, interesting-looking people, architecture, sunsets, and so on. Start your own clip media library of 3D texture maps and reference shots for brainstorming. (Commercial clip media textures are often overused and easy to spot.) Photography helps develop your eye to compose better-moving camera shots after you master the still frame.

1.3 Beginning the Story Concept Process

You are about to embark on a story-telling journey. The ideas presented in these first three chapters were developed to help even people who have never told a story before formulate a good solid idea for a short (2 to 10 minutes) 3D animated film. This step-by-step process continues throughout the book. A sample story called "Media Man in the Net" will be partially developed in these first three chapters to give you an idea of the conceptual design process for 3D animated projects.

1.3.1 The Most Important Thing

The first question to consider is the primary goal of any film, TV show, play, or animation. Now think about this for a moment before you read on because it is a very important point. Most people say that the primary goal is to entertain or make money. These goals are important, but they won't happen unless you accomplish your primary goal:

To evoke a strong emotional response from the audience

Ideally, you want your audience to be laughing out loud, crying out in surprise, biting their fingernails while worrying about the characters, jumping out of their seats with joy, recoiling in fear, screaming out in bursts of terror, smoldering in emotional pain, feeling at one with the universe, or exhausted from all the action. These reactions are harder to elicit then you might think.

Review the emotional impacts of your favorite films. When you go out to movies, keep an eye on the audience. See if you can get an emotional read off of how much they are involved in the story. Films that don't do well usually fail to evoke any emotion at all. *The Avengers* (1998) is a painful example of a technically good film full of hot stars and special effects that failed to evoke any emotion at all and did miserably at the box office.

The first step to showing off a really great 3D animated character is coming up with a good story that you can then mold into a deeply satisfying emotional ride.

1.3.2 Formulating a Basic 3D Story

I've developed a process I use to think up 3D stories to animate and to help others who are not as familiar with storytelling think up original concepts. These steps are not always done in the same order depending on how the idea develops, but I usually follow this progression:

1. Research story ideas.
2. Brainstorm specific ideas good for 3D and Maya.
3. Create a story concept sentence.
4. Fill in three-act structure points.
5. Flush out the story to a one-page draft.
6. Create some rough thumbnail drawings. (*Thumbnails* are small rough storyboard panels used to visually block out scenes or illustrate ideas in progress.)
7. Evaluate your idea using the "Using the Story Concept Checklist" at the end of this chapter.

You may have a creative process that works better for the way you think up ideas to animate. Feel free to come up with your own 3D story-generating formula. For example, I often write a short story on paper, realize that it has 3D appeal, and then turn it into an animated idea.

1.3.3 Researching Ideas for 3D Animated Stories

I've found many places to look for 3D story ideas, characters, and settings:

- **Personal Experiences.** Write about what you know. This is the oldest rule in the book. If you choose stories about which you have some prior knowledge, you will be able to communicate more effectively the slight nuances and details that will bring the characters and situations to life. Try to pick experiences that have a high emotional charge in your life. Make a list of your top 10 emotional experiences and see whether you can turn any of them into some form of 3D animation.
- **Novels, magazines, comic books, newspapers, Internet sites, and short story books.** All contain great ideas to animate. After you become more comfortable writing stories specifically geared for 3D animations, you will be able to re-purpose practically any story for your needs. Keep a scrapbook with clippings from different places full of stories that somehow fascinate you. After reading a good novel, summarize the story on a sheet of paper and add it to your story files. You may want to start grouping stories accord-

ing to subjects such as science fiction, adventure, game ideas, fantasy, or whatever areas interest you.

- **Movies, plays, TV shows, operas, and 2D animations.** What kinds of ideas can you borrow from existing visual stories? Disney has a pretty standard formula for most of its films; it just changes characters and settings. Most myths and stories have already been told in one form or another. Strive to notice patterns in storytelling. Practice combining ideas from different stories to create new ones. Each time you see a visual story, write a summary of the plot and note any exceptional characters or sets. Be a critic and write down what worked in the story and what didn't. Storyboard your favorite shots or any clever cuts for future reference.
- **Historical Events.** History is full of epic stories packed with emotional intensity, unexpected plot twists, and larger-than-life characters. What events in the history of the world fascinate you the most? Why? How could you turn them into original 3D animations? James Cameron says he was always interested in the sinking of the *Titanic*. His passion about this subject helped him make such a successful film. You could feel his wonder and excitement for the subject in almost every frame of the movie.
- **Dreams.** The subconscious mind deals with visual symbols that can work great for animations. The trick to re-purposing dreams is to avoid being so abstract that your audience has no idea about the structure of the story. It is virtually impossible to get an emotional response from the audience with abstract story lines in 3D. There is a great danger that the audience may simply write off your whole animation as an experimental trip video. Be careful with dream information. Try to use the symbolic references or parts that have the most emotional impact and work them into a solid story line as a visual twist. Keep a dream journal beside your bed and write down any interesting ones when you wake up.
- **Myths.** If you want to have a guaranteed good story, you may want to use the classic hero myth structure. *Star Wars* is a classic hero adventure and can be broken down into the steps, as shown in the sidebar "The Hero's Journey."

Figure 1.3
These two characters would be good for a semi-historical story with a detailed culture and background. The character on the left looks Asian but has funny ears. The leaping animal-head character looks Egyptian but has a very original flair. Both of these characters would work well in 3D and have the added advantage of looking somewhat culturally historic.

Credit: Illustrations by Donnie Bruce.

Figure 1.4
This character would be good for a 3D mythic story full of legends and magic. The wings look fun to animate, and the character has a fresh visual appeal that makes you want to keep looking at him.
Credit: Illustration by Donnie Bruce.

The Hero's Journey

Read Joseph Campbell's books ***The Power of Myth*** (Anchor Books) and ***Hero Of a 1000 Faces: Mythos: Princeton/Bollingen Series in World Mythology, Vol. 17*** (Princeton University Press) to get a better understanding of how to create really solid myths and hero journeys. Another great book for hero stories is ***The Writer's Journey: Mythic Structure for Writers*** by Christopher Vogler (Michael Wiese Productions). The following is a very abbreviated summary of the classic hero journey; you may want to study these books more in depth before attempting such a story.

1. Normal person in ordinary world with limited awareness of bigger problem developing in outside world.
2. Call to adventure. Event occurs that thrusts main character into uncommon situation requiring great amounts of courage as awareness of problem increases.
3. A refusal of the call to action may occur that makes the reluctant hero doubt whether he wants to go on this adventure. Reluctance to change. May meet a mentor to help him make the decision.
4. A crossing of the first threshold, when the hero leaves the place that is familiar and goes into unknown realms either physically, emotionally, or spiritually. Commits to making a change on many levels, both with himself and outside world.
5. A series of tests that increase with risk occurs, when the hero encounters allies, mentors, and enemies. Experiment with first change.
6. A journey into the innermost cave or "belly of the whale," when the hero usually goes into the stronghold of the opposition. New perceptions and skills put to test. Conflict needs to be building and obstacles getting much more difficult. Preparing for big change.
7. The main crisis or ordeal takes place, when the hero comes face to face with death or greatest fear. Attempting big change while "seizing the sword." Hero somehow magically escapes and becomes transformed or reborn after final attempt at big change.
8. Reward/Road back home has the hero returning a new person from going through this set of events. Great celebrations from final mastery of the problem.

What other movies or stories can you think of that follow this hero journey formula?

- **Software Capabilities.** What does Maya do that you've heard about or seen that you are dying to try? Perhaps you just got the new Cloth plug-in and suddenly want to do a futuristic fashion show. Every time you get a new software capability, brainstorm at least 50 different uses for it in your story lines. This will force you to come up with ideas beyond the first few everyone thinks about. Try to do what your software does best. For example, Maya is known for character animation, so you might not want to try to do a huge city getting washed away by a tidal wave in Maya for your first project. Most animation houses use several different software packages to take advantage of what each one does best for the job. Create a table listing columns for basic functionality categories at the top, and then brainstorm ideas for each one.

In a big drawing book or on a big sheet of paper, make four columns with these main subject headings:

Dynamics Particles Cloth/Fur/Paint Effect Maya Live

Try to come up with at least 60 specific uses for each of the following four software-capability groups.

Story Ideas Based on Maya Capabilities

Dynamics	Particles	Cloth/Fur/ Paint Effect	Maya Live
Car crashes	Drool	Viking sailboat	3D character in 2D environment
Picture-type fireworks	Bee swarms	Futuristic fashion show	Shaking camera effects
Visible electrical fields	Hail	Alien fashion show	3D wings on a blue-screened angel
Stylized science experiments	Splashing water; waterfalls	Ape with mohawk	3D tail or horns on a blue-screened actor; actor; 3D dinosaurs in a video of a forest
Ball rolling through mazes	Dimensional vortex	Flaming hair	3D UFOs flying over a video of New York City
Jello people	Tiny water creatures	English sheep dog-type characters	
Mousetraps		Furry tunnel you can fly through	
Rube Goldberg contraptions		Furry sea creatures	

Add columns to the chart as new capabilities are introduced in the form of plug-ins and upgrades.

Not all the preceding ideas are going to become Oscar-winning Maya animation shorts. That's okay. Whenever you are in the brainstorming phase, try not to judge the quality of the ideas—just get a bunch down on paper for future reference.

1.3.4 Brainstorming Technique for Original 3D Ideas

Most of the ideas you think of while watching *Star Trek* or walking down the street have also been thought of by some other dedicated 3D maven who is probably working on them right now. You need to go beyond the first idea and dig deeper. The majority of screenplays in Hollywood are rewritten from scratch at least 10 times before they are considered ready to shoot. You should spend as much time as possible making sure your idea is the best and most original one you can generate before committing it to production.

Never touch a computer unless you have final storyboards that have undergone a process such as the one described in these first three chapters. If the story doesn't work on paper, you can't fix it by adding a bunch of whiz-bang special effects or even great animation. Be honest with yourself, get feedback from a variety of sources, and be prepared to make your final concept as successful as possible at the storyboard stage. A great deal of "rough" storyboard ideas end up in final animations because the animator was just so enthusiastic he couldn't wait to put that computer to work. Sometimes you do get a great idea out of the blue, but it will still need considerable work to make it a tight animated story.

In a big drawing book or on a big sheet of paper, make four columns with these main subject headings:

3D Characters	3D Settings	Motivations	Two Obstacles

Think of at least 20 to 60 ideas for each column. Add to this list as you go through your daily life. This brainstorming tool can also be tailored for specific ideas, such as an alien game world. You could then concentrate on just alien 3D characters, alien sets, and corresponding motivations and obstacles. It's even okay to refer to the other columns sometimes if doing so helps you generate better ideas. See what works best for you and your story needs. The following sections describe the focus of each column.

1.3.4.1 3D Characters Column

Think of some 3D characters you have always wanted to animate. You must be able to provide a good reason to do them in 3D or you can't put them on the list. Be specific and describe them using visual adjectives whenever possible. The goal is to be able to "see" the character and understand its essence from the description. "Robot dog" won't get you as far as "robot dog made of found junkyard objects; rusted; pit bull; sassy male punk teenager; white trash mutt with one eye." Now you can *see* the robot dog better and get a good visual sense of his personality. Don't worry about sentence structure or good grammar. Sometimes just a few words are fine. Underline main words such as robot dog to help you glance quickly at the list later if this seems helps you.

Figure 1.5
This is a 500-year-old woman with lots of wrinkles. It would be very difficult to find an actress to play this role, which makes it a great idea for 3D animation.
Credit: Illustration by Donnie Bruce.

1.3.4.2 3D Settings Column

Think of your ultimate list of 3D sets. Again, it must make sense to do it in 3D. Audiences want to see things they have never seen before. Base all settings on actual existing places to give them a style. Feel free to combine ideas. Add dates for reference to help you identify the look and feel. "A city" gets you nowhere. Try "Ice crystal city; Paris-type metropolis; 2300; future primitive cave buildings; powder blue with a silver mercury river." Once again, underline the main idea if it is hard for you to pick the idea out from the rest of the words. As you do the settings, don't look at the first character column list. You shouldn't be trying to think of where the robot dog lives at this point.

Figure 1.6
These prehistoric characters and this setting would make sense to do in 3D. You couldn't take a camera crew out to shoot this and it is something people have not seen much of before, which will make it more interesting.
Credit: Illustration by Donnie Bruce.

1.3.4.3 Motivations Column

Again, don't look at the previous two columns. Think of good motivations for characters in movies or stories you have heard. Motivations can be a bit more general or really specific if you want. "To get rich" sometimes works as well as "to get rich by inventing new happy drug." If you get stuck on thinking about good motivations, go down to your local video store and read the back of movie boxes. Video stores are also great resources for thinking up 3D characters and sets. Just add slight design twists to make them 3D worthy.

The visible motivation on the part of your main character is what will drive the story, so be sure to choose motivations you can *show* well.

1.3.4.4 Obstacles Column

You need to look at the Motivations column to complete the obstacles because these two ideas are linked. Think of at least two obstacles for each motivation. "To get rich by inventing new happy drug" could have obstacles such as "bad side effects/drug companies trying to kill you for the formula." Try to have the first obstacle be less challenging than the second one. You usually want to have an escalating series of challenges for your characters in the story to build to a strong climax toward the end.

1.3.4.5 A Sample Brainstorming Chart

Look at the following example of a brainstorming chart. Remember that the value of a brainstorming chart is that you make it your personal list of ideas; adding ideas to it regularly will increase your 3D conceptual creativity.

A Sample Brainstorming Chart

3D Characters	3D Settings	Motivations	Two Obstacles
Δ Robot dog made of junk, pit bull, sassy teenage punkrock boy (16 years old)	Δ Ice crystal city in 2300 AD future, primitive Paris	Δ To get rich by inventing new happy drug from rare animal	1. Bad side effects 2. Drug companies willing to kill for formula
Δ Obese kung-fu warrior into disco	Δ Inside the human heart	Δ To get girl by getting rid of husband	1. Evil people take her 2. She thinks hero is silly
Δ 500-yr-old Chinese hero woman, covered w/ huge wrinkles, wooden leg, 3 ft tall	Δ Magical redwood forest with talking trees	Δ To win the race	1. Vehicle always breaks down 2. People try to kill hero
	Δ Haunted castle, 1800 AD,		

3D Characters	3D Settings	Motivations	Two Obstacles
	Transylvania, demon statues that come to life	Δ To stop bomb that could blow up planet	1. Hero doesn't know where bomb is 2. Only 3 minutes left
Δ Mayan King stone statue 10-ft tall, playboy, ancient	Δ Alien planet made of fluffy clouds thick enough to live on, low gravity	Δ To have everyone worship hero as a god	1. Planet is about to explode 2. Hero lost voice from yelling
Δ 3D Avatar techno DJ, live in Internet, 007 fix-ittype	Δ Deep-ocean jellyfish city, in crevice, dome fish skin dwellings	Δ To keep evil virus from killing everything	1. Virus is smarter than hero 2. Virus kills through visuals
Δ Alien octopus that eats airplanes, 3 miles wide, lives in Bermuda Triangle	Δ 3D Internet city, Metaverse San Francisco, 2050 AD	Δ To learn how to be psychic at militant institution	1. Hero keeps seeing all the demons 2. CIA and Mafia are trying to recruit hero
Δ Atlantis crystal princess, telepathic, tones nordic	Δ Subterranean rat world made of garbage, New York City-type	Δ To win the dance contest at a galatic competition	1. Competition has more legs 2. Penalty is death for losers
Δ Ultimate frisbee pig, 500 lbs, teenager	Δ Futuristic robot factory run by cyborgs with human slaves	Δ To hunt down an evil poacher	1. Poacher is diabolical madman 2. Poacher is hunting hero
Δ Red blood cell who wants to be white, into swing music, rockabilly	Δ Ancient tomb city from lost civilization with high tech stuff	Δ To get rid of half the world's population to save the planet from destruction	1. Breeding outpaces extermination 2. Doctors keep coming up with cures for synthetic diseases
Δ Biker bumblebee hates the hive life, rebel, James Dean-type	Δ Space station vacation land 3000 miles	Δ To rule the world	1. Lots of competition 2. Partner is trying to kill hero

continues

continued

3D Characters	3D Settings	Motivations	Two Obstacles
Δ Shape-shifting, liquid alien, assumes different human forms	above earth, Balinese theme	Δ To get enlightened at kung fu yoga camp	1. Head guru is wise but insane 2. They want hero to go to war for the camp
Δ Black widow spider race with human heads, '60s Valley of the Dolls	Δ Hopi village at height of civilization	Δ To kill bad aliens invading the planet	1. Aliens have better technology 2. Aliens can look like humans
Δ Toy wookie doll, violent temper, purple fur, glowing red eyes, 9-in tall	Δ Space station on edge of universe, seaport, seaport, ancient Turkish theme	Δ To kidnap all the earth and women and sell them on Mars	1. Earth men trying to kill hero 2. Women fight back
Δ Sunflower opera singer in a field with other singing plants	Δ Huge futuristic aquarium in which fish communicate	Δ To find another planet to live on	1. Stolen spaceships hard to fix 2. Other aliens are hunting hero
Δ Race of 12-ft tall humans without mouths, ears, or hair	Δ Glass world made of blown liquid shapes, Prague-like	Δ To make the galaxy peaceful	1. Huge 300-yr war in progress 2. Hero in prison
Δ Lizard-headed humanoids, corn-worshippers from Taos	Δ Village inside the belly of a huge whale-like like creature, French seaport	Δ To find way back to home planet	1. Hero lost, no idea of location 2. Hero's eyes damaged by starlight
Δ Animated talking star constellations in the night sky	Δ Futuristic organic pirate ship that talks and has own agenda	Δ To go to the edges of the known world exploring	1. Hero can't get out of tiny home town 2. Travel is forbidden
Δ Biped alligator race, drunk Ozark outlaws, cowboys, warriors, ancient	Δ Swamp world full of quicksand, huge snakes, Venus fly traps		
	Δ Life inside beehive, 500 miles wide, metropolis Bangkok		

3D Characters

Δ Medusa-headed femme fatale, 1000 gold cobras on head, skin of sand, gemstone eyes, Courtney Love-type

Δ 3-legged humanoid with snake head, covered in rainbow scales, 14-ft tall African tribal witch doctor theme

1.4 Defining the Story Concept Sentence

Now that you have filled in your chart, you are ready to create some original stories. First, you need to understand the *story concept sentence*. Here is the basic structure:

It is a story about a character who wants something.

You should be able to define any story ideas for animation projects, film, or TV shows using this format. Animators also call it the "30-second elevator pitch"—the idea being that if you are stuck in the elevator with Steven Spielberg, you can pitch your idea before he reaches his floor and has a chance to get away from you. The story concept sentence also helps to orient people to your idea before presenting the storyboards or just to explain to other animators or people in the industry the concept of what you are doing.

Once again, the idea is to be as specific as possible to make the sentence a good synopsis of your particular story. Avoid references to popular movies or characters. Stating that a character is a Buckaroo Bonzai type in a *2001* spaceship setting may lose people who are not as familiar with these references as you are. Pretend that the person hearing your story concept sentence is from another planet and knows nothing about our culture but understands the language. It will also help you to redefine popular characters and settings in concise adjectives and understand their essence more fully.

What do you think the story concept sentence is for the movie *Titanic*? (We use *Titanic* as an example because it is one of the few movies in recent times that almost everyone has seen). Most people would say *Titanic* is about a boat that sinks. Everyone knows that the boat is going to sink before seeing the movie. Cameron had to instead come up with a way to get the audience to care about the people on the boat and the reality of ship life.

It is a story about a well-to-do, young, recently engaged woman who falls in love with poor young artist aboard a luxury ocean liner on its maiden voyage. As the ship hits an iceberg and sinks, we get to see how this event affects their relationship and lives.

What is the story concept sentence of the movie *Star Wars*? You might say it is a story about Luke. Unfortunately, some people don't know Luke, so you have to be more descriptive.

It is a story about a young adventurous man coming of age during a galactic civil war in outer space and his struggle to save a princess and become a hero by using a spiritual type of universal "force" power.

Now try generating the story concept sentence for your top 10 favorite films. If you get stuck, start by figuring out who the main character is and then what the character's main goal was throughout the film for each film. Then generate story concept sentences for TV episodes, comic books, stories you read, or personal experiences. Keep a running list to have ideas for future Maya projects.

1.4.1 Using a Brainstorming Chart to Create Story Concept Sentences

Now that you understand the story concept sentence, you can use your brainstorming chart to come up with some new ideas to animate. The characters are almost sure to be original, good to do in 3D and live in 3D environments with some unexpected motivations. By using this brainstorming technique, you will be able to generate a large number of original ideas quickly. Save your lists in a notebook to refer to later for ideas on other projects or for elements in your story.

You will be using a slightly altered story concept sentence to ensure that you get a good original 3D setting in the mix:

It is a story about (insert a 3D character) who lives in (insert a 3D setting) who wants (insert a motivation).

Abbreviate the first 3D character in your brainstorming chart, and place it in the "3D character" space of the preceding story concept sentence. Now go through the settings in your brainstorming chart until you find one that makes the most sense or sounds interesting. Add a motivation that fits with the character and setting you have already combined. Cross off each entry in the brainstorming chart as you use it so that you won't use it again.

I've created a list of 20 combinations of characters, settings, and motivations taken from the preceding brainstorming chart to give you an idea of the combination techniques. Don't worry about grammar or sentence structure—just get the necessary information in the sentence. You can also add or delete little twists or details that will make the ideas hold together better but try to stick as close to your original list as possible. Don't worry about the obstacles yet.

> It is a story about a robot dog who lives in a futuristic factory run by robots with human slaves who wants to hunt down evil poacher.
>
> It is a story about an obese kung-fu warrior who lives in a Hopi-type village who wants to get enlightened by going through intense training.
>
> It is a story about a 500-year-old Chinese woman who lives in a haunted castle who wants to get rich by inventing new happy drug.
>
> It is a story about a Mayan statue king who lives in a alien planet made of fluffy clouds who wants to get girl of his dreams.
>
> It is a story about a 3D Avatar techno DJ who lives in a Metaverse Internet world who wants to keep evil virus from killing everything.
>
> It is a story about an alien octopus who eats airplanes who lives in a city deep below the ocean floor who wants to win a dance contest at a galactic competition.
>
> It is a story about an Atlantis crystal princess who lives in a ancient tomb who wants to get rid of half the world's population.
>
> It is a story about an ultimate Frisbee pig who lives in a space station who wants to win the futuristic Frisbee trick competition.
>
> It is a story about a red blood cell who lives in a human heart who wants to go to beyond the edges of his known world.
>
> It is a story about a biker bumblebee who lives in a sub-terrain world run by rats-who wants to rob the ultimate honey bank.
>
> It is a story about a shape-shifting humanoid alien who lives in an organic pirate ship who wants to kidnap all earth women and sell them to other planets.
>
> It is a story about a spider with a human head who lives in a swamp world who wants to rule the world.
>
> It is a story about an evil toy wookie doll who lives in a space station who wants to kill bad aliens invading the ships through the children.
>
> It is a story about a sunflower opera singer who lives in a magical forest with talking Redwood trees who wants to make the whole galaxy a peaceful place by singing her beautiful songs.
>
> It is a story about a race of people without ears or mouths who lives in a sand world who wants to find another planet to live on to survive.
>
> It is a story about a lizard-headed humanoids who live in a huge beehive who wants to fly the beehive back to their own planet far away.

It is a story about an animated talking constellations who lives in a glass world who wants to make everyone worship them as gods.

It is a story about an alligator biped race of drunks who lives in a futuristic aquarium and wants to win the annual speed race around the big tank.

It is a story about a medusa-headed woman made of dirt who lives in a crystal city who wants to learn to be psychic at a militant institute.

It is a story about a snake-headed, two-legged humanoid who lives in a belly of a beast who wants to stop the bomb it just swallowed from exploding.

1.4.2 Creating a Brainstorming Chart for Particular Themes

Not all the ideas you come up with will work well enough to develop into 3D animations. If you created a good chart of 60 rows for each of the columns, you may have only two or three good ideas at the end. If you feel you have only a very few good ideas, try limiting your brainstorming charts to themes that interest you. Some example themes include alien worlds, gothic fantasies, medieval samurai, or dog stories, as in the brief example of the dog stories theme:

Dog Stories Brainstorming Chart

3D Characters	3D Settings	Motivations	Two Obstacles
geometric stylized dog made out of primitives	vast bone yard wonderland	to be king/ queen of the local dog park	1. current top dog is trying to kill you 2. your owner has abandoned you
ghost pet Great Dane dog	doggie amusement park	to win the dog show	1. you just broke out out in a hot spot rash 2. the announcer's voice hurts your ears
Irish Setter movie starlet that walks like a woman in heels	big dog town with little dog houses	to break into the butcher shop	1. you are trapped inside your own house 2. Bulldog guard dog twice your size & mean

From each theme chart, create as many story concept sentences as you can. The brief Dog Stories Chart led to these three sentences:

It is a story about a geometric style dog who lives in a dog town and wants to break into the butcher shop.

It is a story about a ghost Great Dane who lives in a vast bone yard wonderland who wants to win the ghost pet dog show.

It is a story about an Irish Setter movie starlet dog who walks like a woman who lives in a doggie amusement park who wants to be queen.

1.5 Using Three-Act Structures

After you get the hang of creating 3D story concept sentences, you can use them to generate *three-act structure* animations. Some people work best by first writing a short story based on their story concept sentence on paper to see what types of plot points appear by surprise. You should try both ways and see what works best for you. If you are not used to writing stories, using the three-act structure formula might help you construct a beginning, middle, and end much more easily. Start by choosing your favorite story concept sentence from the brainstorming list you created.

Keep adding to your brainstorming lists and making up new ones. Exper-iment with the theme lists to help tailor the stories for specific genres. The strange thing about ideas is that you always seem to have a plethora of them until you really need one. Dedicate one drawing book to brainstorming story concept lists, sentences, and three-act structure points.

1.5.1 Defining Three-Act Structure

The idea of three-act structure has been around since ancient oral storytelling traditions. Greek playwrights created the term, and it is still used today in most of our modern stories. Three-act structure permeates almost every aspect of commercial entertainment. TV shows, feature films, animated shows, books, operas, and plays all follow the three-act structure formula.

Stories are broken into three parts called *acts*. The first act is where the setup for the story occurs, main characters are introduced, and the look and feel is established. In Act Two, there is buildup of story tension with a lot of action, conflict, obstacles, and difficult situations that the main character must confront and overcome. Act Three contains the climax of the story, ties up loose plot ends, and resolves the situations presented.

In the graphical representation of three-act structure, the upward curve represents the buildup of dramatic tension in the story, resulting in the climax at the top of the curve. The points of the curve represent timing of particular events or turning points in the story line. Notice that Act One and Act Three are the same length, each being about one quarter of the story. Act Two is twice as long as either of the other acts and comprises half of the story in the middle section. You can use these plot points to determine the timing of the events in your story.

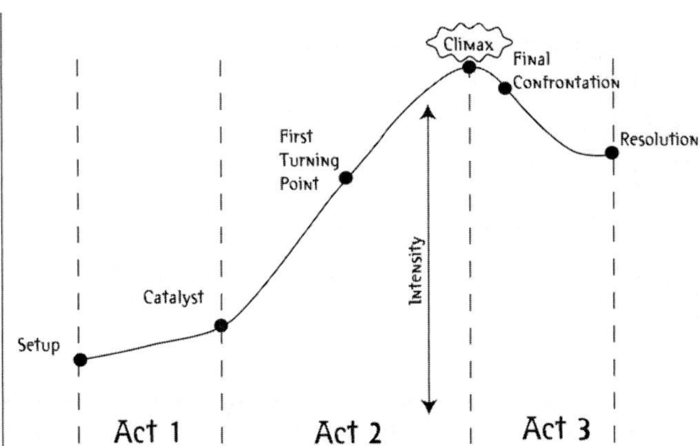

Figure 1.7
This diagram graphically illustrates the three-act structure.

Most Hollywood screenplays are 120 pages or minutes. We will be presenting a two-minute story in the following sections, which breaks down to 120 *seconds* if we use the same formula. For a four-minute animation, just double the seconds for each part. You can use the two-minute timing as the basis for estimating other lengths.

1.5.1.1 The Setup (0-30 Seconds)

Use the setup to define a status quo. In it, you tell what is normal, introduce a main character, foreshadow subplots, and introduce mood and setting.

1.5.1.2 The Catalyst (25-30 Seconds)

The *catalyst* is an event that changes a status quo. It forces a character to act and introduces conflict or problems. Without this event, there would be no story.

1.5.1.3 The First Turning Point (55-65 Seconds)

The first turning point provides a fresh turn of action for a story. The plot twists, adding surprise changes in the story direction and motivation for the main character. There is an escalation of obstacles, with new and bigger obstacles introduced.

1.5.1.4 The Climax (90-95 Seconds)

The *climax* is the point of the greatest intensity. Its goal is to get the audience sitting at the edge of their seats; they should be fully engaged in this moment of the story. This will be the hardest plot point to successfully execute. If the audience is not emotionally involved in the story, they won't experience a point of intensity that feels satisfying. This buildup of incredible tension also needs to be timed perfectly with the other plot points. If the story climaxes too early, the audience feels flat at the end. Most people fail to make their story climaxes strong enough in general. This sequence is often the most important and expensive series of shots in a 3D animation or film with special effects.

1.5.1.5 The Final Confrontation (100-110 Seconds)

Confrontation between two main characters that has been building up throughout the story finally happens. Hero and villain battle to the death, love interests have a point of reckoning, or groups of opposing characters meet head on and battle it out over a brewing issue.

1.5.1.6 The Resolution (105-120 Seconds)

The *resolution* ties up loose ends of the story. Does the boy finally get the girl after saving the world? Do the characters live happily ever? How does going through this experience change the characters or situations?

1.5.2 Exploring Examples of Three-Act Structure

Now we will break up some popular films with these plot points. Let's use the movies *Titanic* and *Star Wars* as examples since most people have seen these movies.

The three-act structure for the movie *Titanic* is as follows:

Setup	We meet Rose, a wealthy young woman, and Jack, a poor young artist, as they board the ship for its maiden voyage. Subplot starts contrasting the rich vs. poor passengers and how life aboard reflects this idea through accommodations, dining experiences, and entertainment.
Catalyst	Jack and Rose meet when Rose tries to kill herself by jumping off the edge of the ship. They start to fall in love. If Jack and Rose had never met, there wouldn't have been much of a story. You wouldn't have cared who survived unless you were emotionally involved with one or two main characters on a deeper level.
First Turning Point	The ship hits the iceberg. Jack and Rose then quickly discover that they have two hours to figure out how to survive and stay together. Fresh direction for the story.
Climax	Jack yells to Rose, "This is it!" as the ship's bow is erect in the water, ready to finally sink into the icy sea. In the most expensive shot in the film, the ship plunges into water with people falling all over the place. The audience has been anticipating this event throughout the entire film.
Final Confrontation	*Titanic* has two final confrontations. The first happens before the climax with Jack and Rose confronting Rose's fiancé about their true feelings. The second happens after the climax, when the lifeboat full of rich passengers who argue about whether to try to save people screaming in the water. Wraps up the subplot of class differences presented throughout the movie.
Resolution	Who lives and who dies is resolved. Rose as an old woman reveals to the audience that she has had the diamond all along and now throws it overboard. We also see how the sinking of this ship and the subsequent search for the necklace has affected all the character's lives.

The three-act structure for the movie *Star Wars* is as follows:

Setup	We meet the princess whom Darth Vader captures, and we get a taste of the light and dark sides of this galactic war. We then meet Luke on his planet as a restless young man ripe for an adventure.
Catalyst	Luke decides to leave his home and help Obi-Wan Kenobi get the plans to the rebel planet after his family is killed. If Luke had decided not to go, there wouldn't have been much of a story.
First Turning Point	Luke and Obi-Wan Kenobi find out that the rebel planet has been destroyed and discover the princess is being held captive on the Death Star, also a big new threat.
Final Confrontation	Obi-Wan Kenobi fights Darth Vader to the death to settle a long dispute.
Climax	"Use the force, Luke" is the height of tension in the film, as Luke flies in for the final chance to destroy the Death Star and save the rebel planet.
Resolution	Luke returns to the princess as a hero.

Notice that the final confrontation in *Star Wars* occurs before the climax. This happens occasionally, so keep an eye on where it may work to your advantage to make the story stronger and the climax more powerful.

1.5.3 Deciding Whether to Use Three-Act Structure

Many conceptual artists and experimental abstract filmmakers argue about the use of three-act structure. The danger in *not* using three-act structure is that people are familiar with it, expect it, and get a little frustrated and confused with an animation that doesn't give them a satisfactory story experience. It is a big risk to embark on a story without basic three-act structure, especially when utilizing something as expensive and time-consuming as 3D animation. You should master three-act structure before diving into any sort of abstract free form story lines or experimental ideas.

If you're still questioning why you should use this structure, study these four reasons:

- Three-act structure guarantees good solid story structure in a proven formula that works.
- Hollywood usually won't touch stories that don't use three-act structure because the stories tend to lose money, are viewed as experimental projects, and audiences don't respond in consistently positive ways.

- Knowledge of three-act structure will help you to fix stories that don't work or that have problems.
- Three-act structure helps you to pace the events in your story more effectively.

1.5.4 Applying Three-Act Structure to Your Ideas

Now choose your favorite story concept sentence from the list you have generated. Fill in three-act structure plot points for your idea, as demonstrated in the following example story "Media Man in the Net."

This is a story about a 3D Avatar techno DJ who lives in a Metaverse Internet world in 2020 who wants to keep an evil virus from killing everything during the launch of a new software that induces emotional states via graphics.

Notice that, to fit together better as a story, some of the information has changed—such as switching out the bad guy to for an evil virus, including a software launch, and changing the date to 2020. This story concept will be particularly difficult since it has lots of cliché elements in it that will have to be worked out, such as the overdone virus in the net theme. We chose this one to show how to fix problem stories along the way. You should choose one from your list that makes you feel excited enough to want to animate it over the course of several months or even years.

The three-act structure for the 3D animation "Media Man in the Net" is as follows:

Setup	Meet Media Man driving through a 3D Internet node on his way to pick up new anti-virus software. Introduce subplot of love interest and big software Internet launch event coming up that night.
Catalyst	Anti-virus software appears to be contaminated as a Media Man installs it to system prior to product launch.
First Turning Point	Evil virus penetrates internal security systems.
Climax	Media Man defeats virus and saves live Internet product launch party with his smart thinking.
Final Confrontation	Media Man kills the virus and Media Man and The Core finally say they love each other.
Resolution	Everyone on the Internet tuned into the broadcast feels good about new software and experiences the peak moment-induced frequency.

1.6 Writing a One-Page Summary of a Story

Using your three-act structure points as a skeleton, fill in events that happened between. You can approach this one of three ways (or you may want to do all three):

1. **Write a short story that is one to two pages.** Use one paragraph for Act 1, two paragraphs for Act 2, and one paragraph for Act 3. This technique often is the best place to start to see how the story holds together in an informal approach. It is also easier to make changes in the basic story structure with this approach.
2. **Write a short screenplay.** If you know how to write in traditional screenwriting format, you may want to rough it out on paper in the official format to get a sense of the timing (one page equals about a minute) and scenes.
3. **Write a list of scenes.** You may also just want to use a brief sentence or two to describe each scene in your animation. These sentences will help you develop the thumbnails. Start each numbered item with the location of the scene or description of the set such as "3D Metaverse" to help you understand where the event is taking place in the story. For the "Media Man in the Net" example, the rough scenes were listed in this manner to flesh out the spaces between the plot points.

The goal at this point is to get an idea of the events that need to occur to bridge the three-act structure points. This is not intended to be a final script, but more of a working tool to help you develop a set of rough thumbnails to see how much potential the idea might have for animation. If you find it hard to think of visual events that go between these plot points, you may want to choose a better story idea. The visible motivation on the part of your main character must be the driving force behind your animation.

The following Rough Scene List is an example of how to bridge the gaps for the three-act structure plot points using our sample story "Media Man in the Net." This particular story takes place mostly in a 3D Internet-type Metaverse world but also has some real video footage of actual people. It is helpful to clearly indicate whether the shot is 2D (firewire video) or 3D at this point to get an idea of how the shots will hold together.

A Rough Scene List

1. 3D Metaverse: Media Man (MM) streaks through the Metaverse (3D Internet city) in his pod car. We meet Vax, his real world human, who's talking to him from outside this 3D Internet Metaverse about his duties that day. MM is holding a Sacred Geometry sculpture-spinning hologram and talking about

giving it to his current love interest, The Core. He also mentions how lucky he was to find the new anti-virus software needed to protect his project from a new strain of AI viruses rumored to be currently close to completion.

2. 3D Diabolical Lab: We see a diabolical lab room inside the 3D Metaverse with a crazy scientist Avatar pushing buttons that make an ominous shadow behind thick glass twist and writhe. The scientist talks about how this evil virus will destroy Cyberdome, which he believes was stolen from him.

3. 3D Metaverse: MM pulls up into a side alley in 3D Chinatown and drives through a secret door after being scanned.

4. Inside 3D Pet Store: MM gets out of pod car inside virtual scary underground pet store. Mole-type bald-headed Avatar enters and gives him new virus software after brief cryptic conversation.

5. 3D Metaverse: MM heads back out through the Metaverse toward the entrance to the Cyberdome. We see this as a Megaplex-type Internet private community with lots of security gates he needs to go through as he talks more about the launch of the new emotion-inducer software. MM has a conversation with Vax, whom we see on his com screen asking if everything is ready.

6. 3D Diabolical Lab: We see the evil scientist again in his lab, working away as the virus behind the glass starts to come alive. The side door opens and the mole man comes in and says he has done what he was told and wants his family back. Evil doctor points to the door inside the glass room and tells him his family is in there, where the mole goes and is torn apart by the shadow-lit virus.

7. Firewire Video of Actual Rave Party: We see Vax as a real person at a real warehouse party-type techno-trance party scene testing the real-time graphic software. Lots of crew-type people are running around in the background as we see the conversation continue with MM on VAX's laptop.

8. 3D Cyberdome Core Central: MM is now inside the main nerve center of Cyberdome and greets The Core, who is a shining white Avatar woman hooked into a bunch of Geiger-like cables as a neural net link-up for the emotional induced states they generate. MM and The Core talk about this as he prepares to upload the anti virus software. Vax breaks in and says two minutes to launch.

9. 3D Diabolical Lab: We see the evil virus lord break out of the glass cage and kill the scientist who made him, then flee down the Metaverse streets heading toward Cyberdome.

10. 3D Cyberdome Core Central: The Core, during a routine test with the new anti-virus software, starts to jerk spastically, telling MM something is wrong.

Figure 1.8
This rough design is of The Core character, who is hooked into a machine that generates emotions via 3D computer graphic visuals.
Credit: Illustration by Sam Miceli.

MM checks and says it is corrupted and that he doesn't understand since the Mole has always given him the best and most honorable code before. MM also announces that all security systems are off-line.

11. 3D Metaverse: We see evil virus waiting outside Cyberdome gate as it suddenly flies open and he passes through the security points easily cloaked.

12. Firewire Video of Actual Rave Party: Vax, back at the party, is trying to raise MM on his laptop while on stage in front of thousands of people for the big live Internet launch. MM says something is wrong. Vax tells him to fix it fast and continue with launch since nothing can stop it from happening at this point.

13. 3D Cyberdome Core Central: The Core comes back on-line after MM goes to back up system. The Core starts to broadcast new emotion frequency of Sacred Geometry patterns onto party screen behind Vax at the event. Vax is explaining how this new software will be like a drug to induce emotional states.

14. Firewire Video of Actual Rave Party: Cut to audience, which is staring at screen and changes from mildly interesting to radiantly happy as we cut back and forth from The Core to crowd and screen.

15. 3D Cyberdome Core Central: MM is working hard to figure out what is wrong with security as evil virus comes up behind him and starts to spin, radiating these slimy tentacles that splash a rain swarm of virus worms over across MM, who immediately becomes infected. The Core sees this from the corner of her eye and starts to scream out in fear as the virus heads toward her.

16. Firewire Video of Actual Rave Party: We watch as crowd's faces go from happy to screaming scary as The Core is attacked by evil virus lord. Vax is now trance-locked on the screen too and screaming.

17. 3D Cyberdome Core Central: MM pulls himself up and inserts a backup experimental AI anti-virus dose into security system, which starts to bring it back on-line. Virus sees him and they fight it out with The Core helping and the audience going through all the emotions at the same time from watching the screen patterns generate.

18. 3D Cyberdome Core Central: MM defeats virus in some ingenious way. The Core and MM profess love for each other, and all of crowd goes into love mode, watching screen as MM and The Core stare into each other's eyes and then turn off live emotion feed.

19. Firewire Video Of Actual Rave Party: Crowd looks all loving as screen at party goes to black and Vax snaps out of phase lock into his announcer role as everyone starts to cheer enthusiastically.

1.7 Creating Rough Thumbnails

It is now time to test the visual potential of your story. If the story seems hard to draw into individual scenes, it may not be visual enough and need some reworking. You are on the right track if you can immediately think up lots of "wow" camera shots for the thumbnails.

At this point, it is very useful to know how to draw. This first pass at thumbnails is just to get a feel for the story and possible obvious camera shots. You may want to use some sticky notes (2" × 3") for the thumbnails or just roughly block out where you see the characters and sets in small boxes.

Notice how rough the initial thumbnails for the "Media Man in the Net" are at this stage. Your goal is to try to just block out the shot. Thumbnails can be any size you want, but it is generally better to keep them around one to two inches across. Your goal is to see how the shots are working together, which is easier if you can fit more of them on a single page—depending on how you are doing your layout.

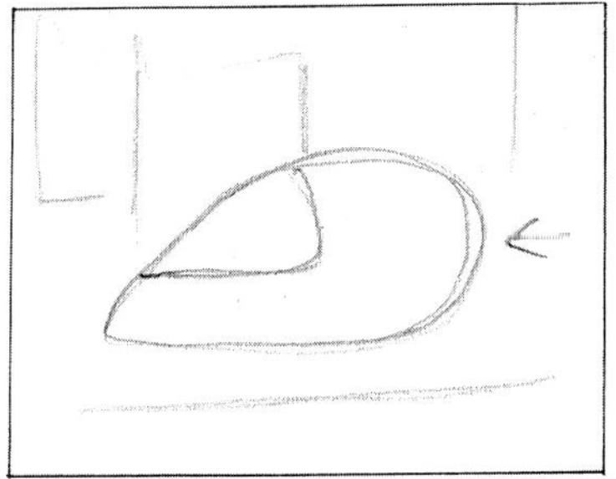

Figure 1.9
These initial thumbnails are for the first few shots in the "Media Man in the Net."
Credit: Illustration by Sam Miceli.

1.8 Using the Story Concept Checklist

Now you are ready to test your story idea against a set of questions that will help you evaluate whether you should proceed with this idea. If your rough story and thumbnails don't meet all these requirements, change them until they do or develop another idea that will work better.

- **Does the story work?**

  Was it easy to draw the thumbnails? Do you sense plot problems or areas that may confuse people? Some stories feel more problematic than others at the initial stage, and you need to get sense whether your story is going to be "problem child."

- **Is it a good visual story?**

  Could you watch it with the sound off and still understand the basic story line? If your story depends heavily on dialogue or other devices, it may not be best suited for animation or as a visual story.

- **Is it a clever and original idea?**

  Do you feel excited about the idea? Or does it seem that you've heard it somewhere before? Coming up with a great idea that is both clever and really original is the hardest thing to do for a 3D animated project. This is where the conceptual genius part enters into the formula. The originality of your 3D animation will win awards at festivals and get you dream jobs and projects if everything else is working too. The goal here is to have everyone hear your concept or see your animation and think "Wow! What a great idea for 3D! I wish I had thought of that!"

- **Do you have a solid three-act structure with good pacing and your main plot points in the right place?**

  Your biggest challenge will be making your climax strong enough.

- **Does the story evoke a good range of strong emotions in both the characters and audience?**

  Try to visualize how your audience will respond to each scene in your animation. Are they laughing, worried, frightened…or bored?

- **Are all the characters' motivation clear?**

  Do we visually know what they want from scene to scene?

- **Is there good reason to do your story in 3D?**

Are you showing the audience something that they have never seen before? Would the story work better shot with traditional actors and sets, or as 2D animation? Be honest here or you could waste a great deal of time on something people may not appreciate.

- Do you have feedback from several people or groups of people and have you incorporated the changes into your story lines?

  Make a short list of general questions for people giving you feedback to answer: Did you understand the story? If not, where did you get confused? How did you feel about the main characters? What were your favorite parts? What parts seemed uninteresting or boring? Did you feel any emotions when going through the story? If so, which ones and at what points? Any ideas on making the story better?

  It generally takes about 20 minutes to pitch and receive feedback for every four minutes of final storyboards. Offer to take people out for coffee or lunch to really listen to your idea. Cultivate a good group of people who know that their feedback is appreciated so you can train them to give you useful feedback; they'll become a valuable resource. Be very nice to these people and honor their time. Give them good feedback on their projects, too, if possible, to keep the relationship balanced.

  Don't just ask animators for feedback. Ask your parents, friends, or anyone who will listen. Post your completed storyboards on a secret page of your Web site and direct people you know to go over them and email back their questions and suggestions when they have the time.

Figure 1.10
This interesting character may work better in 2D since it looks similar to Japanamation-style characters. Sometimes you can make these look interesting in 3D, but you need to be careful when introducing established 2D styles into 3D. There are visual expectations associated with some styles, which may be difficult to overcome in the 3D realm. You may also get into trouble with regard to originality.

Credit: Illustration by Donnie Bruce.

Let's run "Media Man in the Net" through this checklist to how well it's working at this stage.

- Does the story work?

  The story seems to be working, but we still have some cliché elements to work out. There is also the question of whether the computer graphics for the software launch can visually convince the audience that the characters are transmitting an emotional state of some kind. The story feels too long and may need to be scaled back to under ten minutes.

- Is it a good visual story? Could you watch with the sound off and still understand the basic story line?

  The main motivations of the characters are pretty clear. We could use more visual motivation for Vax—perhaps a series of screen graphics behind him, a pre-party test phase that tells about the emotion-inducer software product with a countdown to add timing tension in background. We could add subtitles for the Metaverse intro so people immediately know where the story takes place, "Metaverse 3D Internet Node July 2020," for example.

Δ Is it a clever and original idea?

The originality is the weakest point in this story. Sounds derivative of *Reboot*, *Tron*, *The Lawnmower Man*, *The Matrix*, and other take-offs. Hopefully the Sacred Geometry graphics and 3D Internet mixed with live action will save the idea. Needs more feedback from neutral third parties.

Δ Do you have a solid three-act structure with good pacing and your main plot points in the right place?

The three-act structure seems to be working fine since we did plot points ahead of time and stuck to pacing between events. The climax is very tension-filled, with lots of high-stake emotions from different places riding on the outcome. Live feed, product launch, security problems, virus attack, love interest threatened, death to main characters threatened, and audience fed into emotions of violent-struggle-gone-bad builds tension visually.

Δ Does the story evoke a good range of strong emotions in both the characters and audience?

Audience should key into every aspect of emotion since the main story line is based on a product that generates emotions visually. Nice range of emotions too.

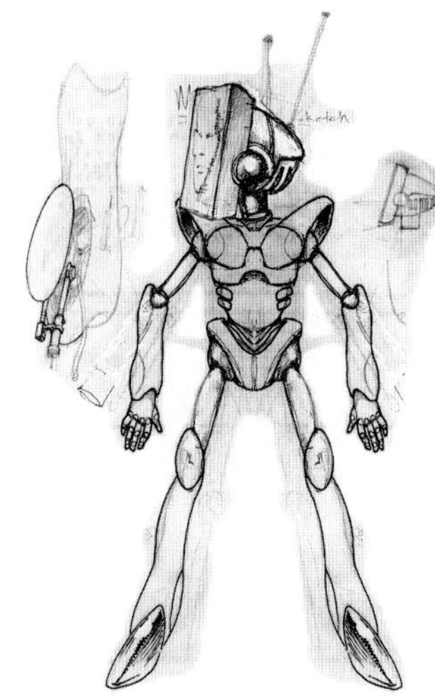

Figure 1.11
This rough sketch is of the first Media Man character. A TV takes the place of a head to display animations based on the character's thoughts and emotions.
Credit: Illustration by Nathan Vogel.

Δ Are all the characters' motivations clear?

Yes, MM wants to protect Cyberdome from virus and win The Core's heart. Vax wants a good software product launch. Virus wants to kill Cyberdome systems.

Δ Is there good reason to do your story in 3D?

It's a 3D Internet world so, yes, this makes sense because all the technology is headed this way and people would probably like to see what the Internet might look like in 20 years. Adding the live footage of the techno-party people helps break up the visual styles and adds a sense of realism, too.

Δ Get feedback and incorporate the changes into your story lines.

The feedback said to make it less cheesy and more edgy. Need to figure out whether the tone of this animation is supposed to feel more like *The Matrix* or *Reboot*.

1.9 Avoiding Common 3D Animated Plot Pitfalls

Most 3D story ideas share similar types of problems. Try to add to the following list of problems as you get better at creating original stories and characters.

- A linear progression of events
- Animate/inanimate objects in human world/situations
- Inappropriate use of 3D
- Dream sequences or flashbacks
- The four-minute movie trailer/TV show pitch problem
- Fear of dialogue

The following sections discuss these common pitfalls. Make sure your 3D story does not contain any of them.

1.9.1 A Linear Progression of Events

Your story will sound flat and boring if events happen too casually, predictably, or easily for the characters. You need some unexpected twists and turns in your story that keeps the audience guessing as to what will happen next. Don't make things too obvious or convenient for your characters, or the story will seem contrived and fake. This is the most common problem in 3D animations conceived by people unfamiliar with telling dramatic visual stories.

1.9.2 Animals/Inanimate Objects in Human World/Situations

If you are doing a biker rat story, don't put your rats in a human biker bar. Put them in a sewer rat hole with settings that feel authentic to what rats would use to decorate a biker bar, such as bottle cap stools and used tin can tables. Have them talk about what rats are concerned about, such as some new sticky foot trap or poison pellets that look like cheese.

project

Study movies like *Antz* and *A Bug's Life*. These were excellent examples of showing how the animals use things people felt were familiar to those animals as props and sets. They use leaves to build a fake bird monster and harvested seeds to eat or built ant colonies for work. The ants in these movies didn't eat hamburgers or drive Ford pickup trucks.

Remain true to the essence and desires of the types of characters you design, or the story will seem false and contrived. The same thing goes for any inanimate object that you bring to life through the wonders of 3D animation. If you are doing a dancing blender, make sure the audience believes that the character is really a blender and has blender wants and needs, not those of a typical housewife or baseball coach—even though that personality may be part of your dancing blender character.

1.9.3 Inappropriate Use of 3D

You can argue stylistic choice for a long time. 3D is expensive and time consuming. People expect you to show them something they couldn't see otherwise, or they may feel somehow cheated. Animations that use 3D in clever ways get much more respect and attention than those that are not as well thought out. You will just have to work much harder to make the animation perfect in every way if the 3D conceptual base is absent. Why make it hard on yourself? Start with a good 3D idea, and your feedback will be more encouraging through out the process and subsequent screenings.

Figure 1.12
This cute little character feels more 2D than 3D but still might work depending on the story and sets.
Credit: Illustration by Jeff Vacanti.

2D animation can do anything. All 3D stories could just as well be done in 2D—except when photo-realism is essential. The more you write animated stories, you will get a feel for which ones would work better in 2D rather then 3D. 3D tends to feel more serious. Try to get a sense of 2D versus 3D as you watch TV and go to films and animation festivals. Ask yourself how the piece would feel in 2D rather than 3D, or visa versa. Why did the director choose that particular medium? It's a blurry line that's changing constantly, so try to get a sense of it.

1.9.4 Dream Sequences or Flashbacks

Dream sequences and flashbacks are usually used to get in and out of problem plot areas or to justify 3D animation in some fantastic way. Audiences are getting pretty weary of the old "character falls asleep at table, all this crazy stuff happens, he wakes up, and it was all a dream" story. If you use these types of story-telling devices in really fresh and original ways, you may be able to get away with it, but it is always a gamble and one we don't recommend unless you are a very good and experienced visual story teller.

1.9.5 The Four-Minute Movie Trailer/TV Show Pitch Problem

It's great to have an idea you would like to one day develop into a feature film or TV series. Unfortunately, trying to shove that whole idea into a few minutes will most likely come off as a movie trailer type of animation in which the events are happening too fast to digest emotionally and huge gaps in the story seem to be appear every few seconds.

Try to take a small, interesting piece of the script and make it strong enough to stand alone as a festival piece. This approach will give you a greater opportunity to fully develop the characters in more subtle ways. The story will become deeper and have good look-and-feel potential that will reflect more of what you had in mind for the original piece as a whole.

Festivals are littered with animation-type trailers that show lots of hard work on the screen but leave audiences feeling dazed and confused as to what the story is about—which means no emotion and no points for professional storytelling discretion. Better to tell a short, slow, simple story full of emotions than a fast, choppy, confusing, action-packed trailer whirlwind no one will understand or remember much about later. "Media Man in the Net" suffers from this problem because it was already developed as a full-length screenplay. The part chosen for the short animation is a version of the beginning sequence for the longer movie.

1.9.6 Fear of Dialogue

The phrase *lip synch* can send some 3D animators into a panic attack. Fortunately, this book covers several techniques for making your 3D characters talk easily and quickly. Use dialogue in your stories when appropriate. People glean a great deal of information about characters from their voices and how they talk. Back in the old days (a few years ago), it was much harder to do character animation of any kind, and you didn't see many talking 3D characters. With programs such as Maya, you can now realistically do dialogue and enrich your story lines in the process. Audiences in general respond better and identify more with characters who talk. You don't want to have your characters talk all the time since you are doing an animation, but a little dialogue goes a long way in making your story stronger.

Some animators admit to having a limited ear when it comes to writing good dialogue. This is an honest concern. Try to pay attention to how people talk in conversations to get an idea of the differences between written dialogue and stiff sentences that some actors and characters speak. Use a tape recorder and try to mimic the characters in your story with a friend—talking out the ideas that need to be communicated and sample dialogue tracks. When you listen to your recording later, you will most likely be able to tell which lines or phrases work and which ones aren't spoken naturally. After you have the general idea of your entire story, it is a good idea to tape record it all—much like a book on tape—to hear how the story sounds and to hear whether it is working. You will also be able to get a sense of the timing from doing a tape version of your story.

1.10 In Summary

If you have followed all the steps in this chapter and incorporated all the feedback, you should have a good idea for your Maya 3D animation. You are now ready to concentrate on the character design and development aspects of your rough story idea. Remember that coming up with a good 3D idea to animate takes practice and lots of creative thinking. It is challenging to create really good short form 3D animations, and you need to be patient with yourself as you learn how to tell visual stories.

CHAPTER 2

Designing Characters

by Sherri Sheridan

A good main character is essential to any successful 3D animated story. Maya excels at animating simple 3D characters or complex, organic-looking creatures. As computers get faster, the complexity of 3D characters and the audience's viewing expectations rise. Maya gives you the power to build and animate almost any character you can imagine. The big challenge now is coming up with original characters who are alive with personality and have amazing 3D visual appeal.

Conceptual 3D character designers work in many different ways to come up with their ideas. This chapter takes you through my step-by-step character design process. You will learn how to create a main character for the story you developed in Chapter 1, "Creating Story Concepts for 3D Animation," or for any other projects you may encounter.

The process presented in this chapter is designed to help you think of your 3D character as if it were an actor you create who is perfect for the lead role in your animation. The more you know your character, the better you can animate it for the most believability and emotional impact. You can then use these same steps to create a whole cast of characters who visually relate to one another and to their environments.

2.1 Creating Character Types

Lots of different types of characters are well suited for 3D. In the sections that follow, you will find some ideas to get you started. Add to this list as you see more and more 3D characters so that you'll have a broad selection of brainstorming types to refer to for fresh approaches.

Try designing several versions of the same character but using different types, such as a biped, quadruped, and wing creature. You might first think that the character should be a biped and then later see something coming out in the winged creature with possibly four legs that may work better for a more interesting animation.

Character Types to Mix and Match Randomly

Body Type	Head	Legs/Feet	Arms/Hands	Unique Features
alien	humanoid	biped	human	tail
humanoid	alien head	quadruped	monkey	trunk
sea creature	feline	fins	claws	horn
demon	reptilian	webbed feet	feline	huge warts
animal	borg	triped	hooves	tentacles
mythical	dragon	octoped	lady	big hair
spirit	elephant	bird	vampiric	mohawk
object	dog	objects	boney	brain helmet
light	goat	ape	gloved	plates
hybrid	bloated	bugs	hairy	tubes
plant	shrunken	stilty	beefy	parasites
dinosaur	mutant	obese	long fingers	ball and chain
toy	voodoo	wooden	alien	implants
bacterium	Draconian	peg leg	dismembered	beard
subatomic	insectoid	amputated	glowing	skull plates
element	arachnid	cyborg	cut up	spots
food	bull	elephant	fingernails	holes
symbol	devilish	mermaid	metal	drawers
robot	troll	cowboy	scaly	TV screens
liquid	bird	sumo	child's	hooks
spaceship	grassy	mechanical	ancient	widgets
vehicle	squid	elf	mummy	organic
bird				weapons

Figure 2.1
This character looks fun to animate. He has an interesting trunk and an alien zoo-creature type of feel.
Credit: Illustration by Sam Miceli.

Body Type	Head	Legs/Feet	Arms/Hands	Unique Features
superheroes				
shape-shifters				
prehistoric				
insect				
reptile				

If you randomly chose one characteristic from each column, for example, you could come up with this creature:

An alien with a reptilian head, three legs, claws, and skull plates.

2.1.1 Start with a Rough Sketch

You should have a good idea of your story by now, which will help you create some rough sketches for your character designs. This is where being able to draw fast and accurately comes in handy. Even if you cannot draw really well, you should try to get the basic concept or energy of the character down on paper. These types of fast sketches are commonly referred to as *thumbnails*.

In a sketchbook, keep a visual library of different body part types. Use pictures of real animals, insects, fish, plants, or people for a variety of ideas.

Decide who is the main character in your story. Using a pencil, draw at least ten different gesture-type versions (two to three inches tall) of your character. Spend about a minute or two on each sketch. Experiment with proportions of body parts, head shapes, clothing, and postures. This will prevent your imagination from being stuck on one idea at this point. Try not to fall in love with any of your ideas during the conceptual development stages of any project; you don't want to limit your willingness to try new things that may work better than the initial concept.

Look at the drawings of Media Man, the main character for our example story. The initial idea was that he have a TV set as a head and be a superhero, techno DJ, 3D, Avatar-type character.

Notice the variety of versions and what changes between each one. The shape of the TV set was of particular concern. Nathan Vogel created the character over six years ago and now wanted it to have a 16:9 aspect ratio for film or HDTV projections. Some felt a horizontal rectangular head would present several problems:

- It would make the character seem less intelligent.
- The audience wouldn't identify with it because it was so anti-human.
- The character needed to be updated from the old designs to add more visual appeal.

This character is an example of a time when the story line was generated from an existing character design that didn't have a solid script yet. Some animators first come up with a strong 3D character they want to animate and then develop a story to showcase that character. We experimented with more oval shapes for the head and tried not to get stuck with any one idea at this point. Quantity, not quality, was our goal at this stage.

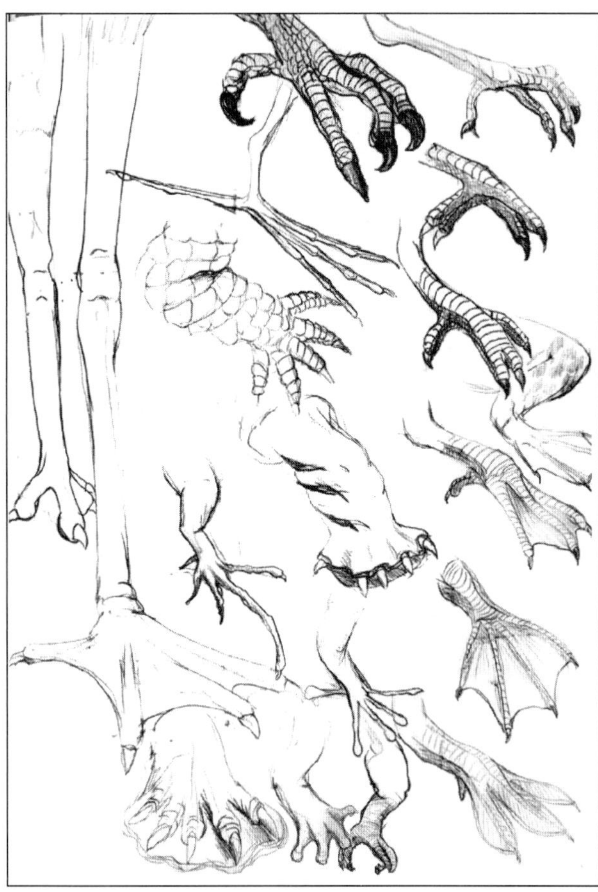

Figure 2.2
Draw lots of different thumbnails of basic body shapes or multiple versions of each body part to examine all the possibilities.

Credit: Illustrations by Maia Sanders.

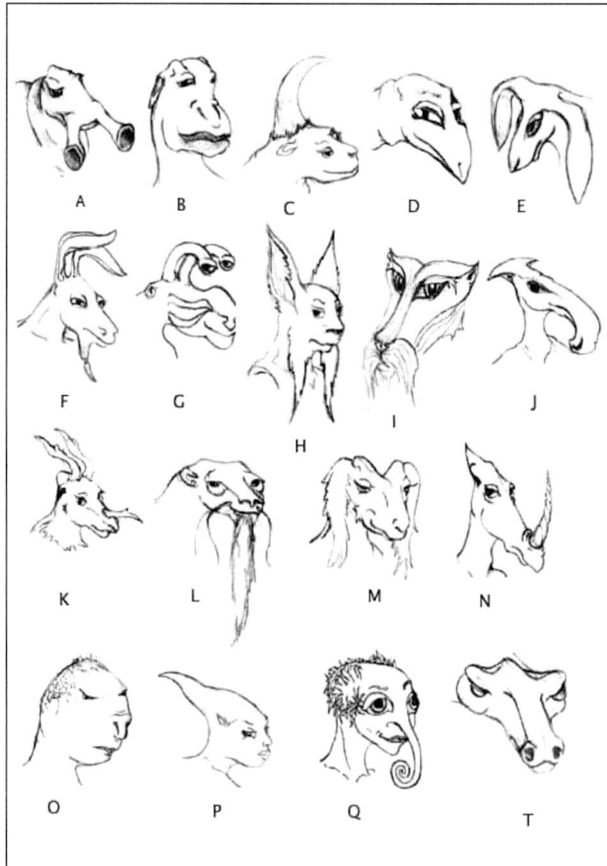

Figure 2.3
Maia Sanders came up with the idea of using the alphabet to design character head shapes. Try numbers, symbols, foreign alphabets, or other shapes to help you design beyond the basic ideas.

Credit: Illustrations by Maia Sanders.

Figure 2.4
Notice the different versions of the head as this 3D character starts to take shape.
Credit: Illustration by Donnie Bruce.

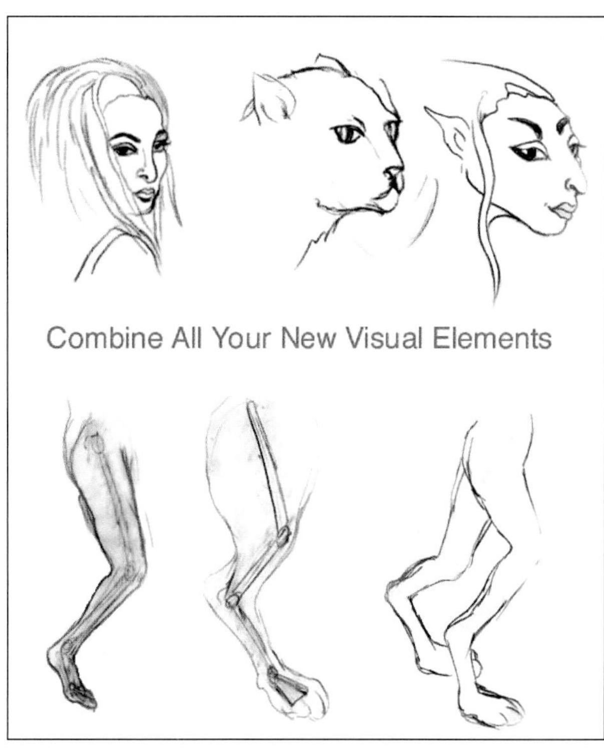

Figure 2.5
These sketches show the progression between incorporating two ideas for a character into one design.
Credit: Illustration by Maia Sanders.

2.1.2 Create Strong Silhouttes

Strong silhouettes make for good character design. You should do at least one big drawing-book size page (9"×12") of 10 to 12 rough sketches to get a feel for all the possibilities toward which your character design might veer. Pay particular attention to strong silhouettes at this stage. The design must have a very distinct solid black shape to ensure that the audience can pick the character out of backgrounds and easily read whatever motion the character is attempting.

2.1.3 Use Familiar Visual Themes

A *visual theme* is a design thread that has some familiar attributes running through everything you create in your animation. You don't want to create 3D characters who seem disconnected from their environments or the other characters in the story. Often you are creating your own visual reality in a 3D animation; it will look more professional if you pick a visual theme or familiar touchstone to design all the elements around for each particular set of characters or locations. This theme helps the audience relate to your characters.

Figure 2.6
Here are some pictures of a wax model photographed from several different angles. This character has a strong profile from some directions, but it gets a little confusing from the sides due to the low chin.

Credit: Wax model and character created by Ryan Nishimura.

If you were to pick the visual theme of a giant beehive for a futuristic NYC design, you would incorporate aspects of traditional beehives with your buildings, city layout, and spaceships. If you were to choose a New age, hippie, utopian visual theme for a futuristic NYC design, it would look quite different from the beehive one. In general, you would base your design partly on the common NYC elements we all know (dense buildings, taxis, subways, endless layers of life) and partly on the particular theme that says something about how that city has changed or is special in some visual way. Sometimes you may want your design to have the look and feel of NYC but in another place or time.

Visual themes, when designed consciously, tell the audience what is important to the characters in this world you create and how they look at life on a symbolic level. Think of how you feel you know a person better after seeing where they live, what kind of pictures are on their walls, and their furniture choices.

2.1.3.1 Studying Existing Uses of Visual Themes

Star Trek is a good example of characters and sets with strong specific visual themes. The main *Star Trek* crewmembers are much like the stereotypic good Americans who are out to explore the universe and follow their "prime directive." The Romulans seem to be based on French people who have stylish blue outfits and classy haircuts with a rather cool demeanor. The Klingons come off with a heavy metal biker gang appeal with hot tempers, brown colors, lots of barbaric weapons, and big hair. Vulcans may remind you of Czechoslovakian intellectuals who are reserved and wise from a long rich history of civilization.

The animators we know spend long hours debating these types of theme issues. If the creators of *Star Trek* designed a bunch of really cool-looking characters with no themes, the audience wouldn't be able to get a handle on each character and would think the show was silly. You may not agree with these generalizations of *Star Trek* species. That's okay. Please feel free to come up with classifications of your own. The point is that the species' ships, uniforms, and character designs all match in general look and feel. Even if you have never seen that particular ship, the design and colors tell you are on a *Starfleet* ship as opposed to a Klingon battle cruiser. Visual themes are particularly important for creating believable alien races that the audience can identify with on some common reality ground.

Keep lists and visual notes of how other conceptual designers have classified their characters or sets in any science fiction material you encounter. Studying others' classifications will help you grow accustomed to designing your own visual themes. (*Babylon 5*, for example, is another TV show with good design themes to study.)

If you are a 3D animator, immerse yourself in any and all science fiction and special effects TV shows or movies. This background information will help you understand what other animators are talking about when they say things like "an organic Vorlon-type ship," or it will keep you informed on what has been done before so that you don't duplicate common ideas. You can also "borrow" ideas for shots that work really well. You may even start to recognize particular Maya-based dynamics, plug-ins, or features!

Figure 2.7
With this character, we use the visual theme of a pirate to give it a well-traveled, exotic look.
Credit: Illustration by Maia Sanders.

Figure 2.8
Another pirate theme creature that easily relates to the previous one since the visual theme is executed so strongly.
Credit: Illustration by Maia Sanders.

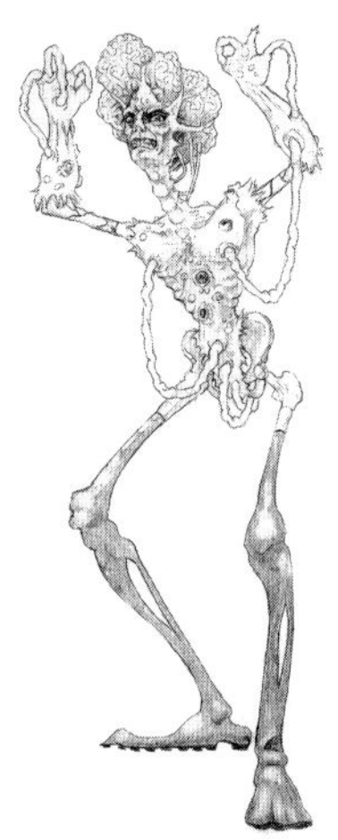

Figure 2.9
The drawings for the general visual theme "scary virus" in "Media Man in the Net."

Credits: First two illustrations by Donnie Bruce. Third illustration by Sam Miceli.

2.1.3.2 Developing Your Own Strong Visual Themes

A visual theme consists of any group of people or things that have specific styles associated with them with regard to body shape, architecture, clothing, skin type, jewelry, music, transportation, weapons, demeanor/attitude, colors, time periods,or cultural history. You can tell whether a visual theme is a strong one in your own mind by seeing how many of specific ideas you can come up for each of these categories.

Notice the progression of the general visual theme "scary virus," the bad guy for the Media Man story, throughout these drawings. This theme didn't have a strong culture associated with it, but the idea of creepy, infectious, diseased organic tissue is woven into the design. The tentacles were voted down as being too time-consuming to animate.

Try to make a list of around 60 visual themes you'd like to work on at some level—complete with a few descriptive words to help you define the theme. You might want to keep a separate "visual themes" drawing book to refer to when designing characters. You can tell a lot about people by the types of themes that they enjoy using. Choose themes you are familiar with so that you can provide a deeper level of details for animation ideas.

Examine the following visual theme list to get a feel for what sorts of things constitute strong visual themes. Try designing the same character, using each of these themes, to see how different it becomes.

Visual Themes

Mayan	rave scene	Day of the Dead
Egyptian	surreal Dali	SPCA kennel
French Court 1600	Haitian voodoo	racial stereotypes
Hell's Angels	American Old West	pirate ship 1650
modern Gothic	Eskimo 1800	LA rockabilly bar
New Age hippie	tacky cruise ship	tribal African
coral reef	beehive	transvestites
'80s metal bands	kiddy play school	Viking sailors
rainforest tribal	Arabian Nights	Atlantis
Viva Las Vegas	disco Detroit 1980	vampire

Hawaiian luau	Mafia 1960 NYC	Hell
psycho ward	Catholic school	Ancient Greece
medieval prison	southern hog farm	Hopi 1600
Swan Lake	1950 beauty parlor	trashy Texas trailer
Frank Lloyd Wright	World	park
Australian outback	Aztec Empire	Hare Krishna
Quaker	Rasta Jamaica	Chinese Ming dynasty
Hasidic Jews	Charleston flappers	Planet of the (animals)
punk rock	blue-collar factory	Disneyland
Italian Renaissance	circus freak show 1940	Nazi Germany
primeval forest	Florida swamp	genetic lab 2050
organic high-tech	Mardi Gras	Druid village
futuristic NYC 3012	Texas rodeo	World War II

2.1.3.2.1 Resources for Visual Themes One of the best resources for visual themes is an old *National Geographic* magazine. The pictures and stories in each issue are gold mines for conceptual designers. You can now purchase the last 100 years of National Geographic on CD or DVD for around $140, which includes a search engine to help you find particular themes.

You can type in the word *Mayan,* for example, and all the issues with stories on the Mayan people come up. You can use the pictures to make a collage in Photoshop and then print color versions of the collage to have beside you while you design your characters. A good color printer is an essential tool even if it costs only a couple hundred dollars.

If you don't want to purchase the CD, thrift stores sometimes have big bags of old *National Geographic* magazines for about 25 cents each. You can thumb through them and then choose which themes appeal to you. Having the actual paper versions is nice, but clipping and cataloging all that paper can be rather time-consuming.

Subscribe to magazines with good pictures, put them in a big pile at the end of the year and clip out the best images that inspire good designs. European fashion magazines are a great resource for costume designs. *Architectural Digest* has some wonderful palettes for sets and furniture designs. What magazines do you like that

Figure 2.10
This wonderful character uses the theme of a Yemen woman from Saudi Arabia.

Credit: Illustration by Ann Mikulka Beckman.

have visual themes just waiting for you to explore? Start your own visual theme library today and organize them into binders according to civilizations or groups that make sense to you. It's a great way to spend time while rendering.

Group themes into categories such as clothing, bodies/faces, architecture, art, jewelry, transportation, religion, landscape, and any other categories that seem relevant. Create collages for each of these sections for every specific visual theme. Place all the faces on one page so you can refer to it while designing your character's face. Then you won't have to go through page after page of collages with only one or two faces on a page. You will also notice patterns and themes more easily when grouping the visual areas.

Perhaps all the Mayan people have strange crooks on their noses or strangely shaped ears you can exaggerate in some way on your 3D character. Maybe when you're looking at the clothing pages, you notice that they wear a very specific shade of cobalt blue made from the dye of some local plant, or something. The details you spot as a cultural detective will weave a history (also called the *background* or *back story*) for your 3D character—even if it's an alien from another planet who walks on four feet and has a nose like an elephant's.

To create the character, a visual theme collage was first made from a *National Geographic* article. Then the designer went through several versions of the character, finally arriving at this soulful interpretation of an ancient alien-type Yemen.

2.1.3.2.2 Elements to Include in Your Visual Theme After you have a preliminary list of 60 or so themes that interest you (with a few key descriptive words if you find that helps you), it is time to choose one for the main character in your 3D story. Make several collages for the different visual areas (people, architecture, art, and so on). Then look at these pictures and ask yourself which are key visual elements. Set a goal of listing at least 20 for each section. Include a separate column for colors and create a five-color RGB palette.

Start your own personal clip media library with your top 10 visual themes, using photographs from old magazines. Make collages for each of the main groups (faces, jewelry, clothing, and so on) and see what types of design patterns you start to notice that might work well for your 3D character.

2.1.3.2.3 Moods in Visual Themes Every visual theme has a particular mood associated with it; you can discover it by examining your collages and visual element lists. The mood is important to help you ensure that your character or

group of characters is going to fit this particular theme's essence. You can also use the opposite of the essence if you do this consciously.
For example, say you want to use a Hell Angels, acid rock, biker visual theme for a gang of chickens you are animating. You would have an interesting juxtaposition with chickens perceived as lacking courage and making your chickens a very tough biker poultry gang.

Or you could make your chickens really rough and tough—perhaps a gang of renegade ex-cock fighters. You could also approach the mood or essence of your chicken gang as being spineless, wimpy chickens dressed up as bikers that get flustered every time anything even slightly scary happens.

The point is that each of these examples has a very specific mood attached to it that you need to establish in order to decide which design path and attitude you want your chicken gang to convey. You may also want to have two chicken biker gangs with similar themes but completely different moods to show contrast and to build conflict. Play with these ideas and see how many versions you can come up with for each character design, as in the following example:

Figure 2.11
This spider character was designed with a Mayan visual theme.
Credit: Illustration by Debbie Rich.

Elements List

Visual theme	Hell's Angel biker chicken
Visual elements	Harleys, tattoos, leather vests, beards, rock t-shirts, scars, bad boys, rebels
Visual mood #1	Tough, rebel, wanderer, criminal, free, dangerous, fun loving, daredevil, redneck, macho
Visual mood #2	spineless, dainty, wimpy, delicate, easily frightened, urban wanna-be, poser
Color palette	White, black, red, tan

Notice the different moods of the two biker chickens. Although they have the same biker chicken visual theme, the biker chicken on the right is very macho, whereas the one on the left is rather silly looking and softer.

Figure 2.12
Two biker chickens with entirely different attitudes.

Credits: Illustration on left by Sam Miceli. Illustration on right by Maia Sanders.

2.1.3.3 Creating Rough Sketches Based on Visual Themes

When you have a clear idea of your character's basic shape and silhouette, you can combine it with your new visual theme ideas into something original. You may change your character dramatically at this stage but try to use as many of the rough sketches you did earlier as guidelines if possible.

Do another full page of rough thumbnail sketches, incorporating your new visual theme elements and mood. The elements will come out in the clothing, jewelry, facial features, and body builds. It is better to layer on as many of your elements from your list as possible and scale back later than it is to keep it too bland or simple at this point. The mood you have developed will come into play with the character's posture, facial expressions, and pose. Remember to keep the idea of strong silhouette in mind while you design at this stage.

You should also being thinking of the types of existing visual elements you can exaggerate using 3D. Perhaps your biker chicken has an animated tattoo on his arm that talks to him or something else that would make the animation angle more visually entertaining.

2.2 Adding Essence to Your Characters

Figure 2.13
These rough sketches show different visual themes and moods for an evil serpent-type bad guy.

Credit: Illustration by Maia Sanders.

One of the most challenging parts of animating a 3D character is giving it a powerful screen presence, or *essence*. Great actors and actresses have the ability to "hold the screen" with their charisma and soulful qualities. You can make your 3D character more interesting by making it as real as possible. There are several ways to approach this, such as adding a background history and little visual and personality details that will make your 3D character really come alive.

2.2.1 Make the Connection Between the Audience and Your Character

What do you think people mean when they say, "I don't identify with your characters?" This is important question, so think about for a moment.

Character identification is how well the audience relates emotionally to the characters in a story.

If the audience doesn't relate to or care about your characters, they won't become involved in your story or enjoy it very much.

Character identification is the one of the hardest things to achieve when creating a 3D character. For the last 50 or so years, we've become so used to seeing 2D animated characters that they get away with a lot; they are so stylized that the audience doesn't care whether they look real.

People generally have higher expectations of 3D characters. Studios such as Industrial Light & Magic have trained audiences to expect very high levels of design and animation from the 3D world. I think expectations will level off in the coming years as audiences learn to enjoy and forgive the more stylized versions of 3D characters coming out more and more each day. Most stylized 3D characters fall flat against the traditional 2D characters unless the design, story, and animation are absolutely superb.

Keep in mind the following 10 points when you're trying to get the audience to identify with your characters. Try to incorporate as many of these as possible in your designs, character history, story, and visual shots as possible.

- Your character is likable
- The visual theme and setting are in some way familiar
- Your character has visual appeal
- Your character has imperfections
- Your character has superhero qualities
- Your character elicits sympathy
- You introduce your main character ASAP
- You place your character in jeopardy
- You empower your character
- You use the eyes of the audience

2.2.1.1 Make Sure Your Character Is Likable

Making your characters likable doesn't mean that they would win the high school popularity contest by any means. Likability refers to the idea that your character should have some redeeming personality attribute that the members of the audience see in themselves or in someone they know. Never make a character's personality "black" or "white"—or all good or all evil; make "gray" characters who have a mixture of good and bad traits that reflect the traits of most people you know. For even the most diabolical madman, you should allude to what event in his life made him so mean. Add a love of something sensitive, such as cats, or a funny sense of humor to a really evil character to make the audience care about the character on some level.

Figure 2.14
Wouldn't you enjoy animating this character?
Credit: Illustration by Jeff Vacanti.

This extremely likable character looks fun to animate. The visual theme is a forest bacchanal rave, and the design is based around the actress Gwyneth Paltrow. Try to draw your characters so you can almost see them move with lots of energy and tension in the pose like this one.

2.2.1.2 Make Sure the Visual Theme and Setting Are Familiar

Design your 3D character with a strong visual theme to give it some history and personality. The audience will subconsciously relate to the character more easily if there is something already familiar in its design. Abstract 3D characters without visual themes can come off as contrived and flat to audiences; the audience has nothing visually or emotionally familiar to latch onto to help them understand the characters' orientation.

2.2.1.3 Make Sure Your Character Has Visual Appeal

Make your characters interesting to look at. She doesn't have to be cute or pretty—just designed well. Sometimes the simplest character makes the best one, but 3D characters tend to have more detail than 2D animated characters, and you need to add details that make them interesting and develop the character's personality and cultural orientation in the process. A strong silhouette once again is very important to help make characters visually appealing.

2.2.1.4 Make Sure Your Character Has Imperfections

All the characters you design must have imperfections of some kind. Do you know any perfect people? Then don't design perfect characters; they're annoying and unrealistic. Add as many imperfections as possible—drawing from both physical disabilities and character flaws (such as personality problems or mental or emotional

Figure 2.15
This character uses the idea of a macho-type Robert DeNiro as a ballerina with shotgun barrels for legs. These familiar ideas put together in a fresh way create an original 3D character.
Credit: Illustration by Sam Miceli.

Figure 2.16
Although she isn't beautiful or cute, this elderly alien female has strong visual appeal; she is interesting to look at and could hold a screen. Lots of personality comes across in this design.
Credit: Illustration by Sam Miceli.

weaknesses). Audiences like for main characters to have imperfections that they themselves possess or notice in the people around them.

Make a list of imperfections you've noticed in the following:

- Your own life
- Your family members
- A few people who you know really well
- Interesting characters you've seen in TV shows, animations, or films

Notice how imperfections go together with certain personality types. Try designing a character using as many of these imperfections as possible. You can begin with this list:

Imperfections for Characters

eye patch	missing teeth	clumsy
lots of scars	skin rash	bad breath clouds
fresh wounds	exaggerated limb/parts	dumb
missing limbs	hairy clumps	stutters
handicapped	wrinkles	poor eye contact
thick glasses	cellulite	seizures
stitches	obesity	ridiculous laugh
limpy legs	acne	

2.2.1.5 Make Sure Your Character Has Superhero Qualities

3D characters have an added advantage over human actors for conceptual designers because 3D characters can do anything. Try to really understand the essence of your character and then apply some superhero qualities. These may be as simple as detachable wings that can make the character fly like a UFO or weapons that spring out of body parts in a jam.

Remember to always balance any superhero qualities with imperfections equally as strong. Superman would be very boring if it weren't for his helplessness in the presence of Kryptonite. Audiences like to see characters who can do things they wish they could in their everyday lives despite whatever imperfections they possess. What types of original superhero qualities do wish you had and could bestow upon a 3D character? Try to fit superhero traits with visual themes as much as possible. An animated Mayan stone statue may have the ability to lift huge blocks of rock or talk to star constellations, for example.

The next five character identification points are ideas that you need to work into the story line. Try to see some design elements coming through for each point if possible, although these ideas aren't as visual as the first five.

2.2.1.6 Make Sure Your Character Elicits Sympathy

Place the character in difficult situations that are beyond his control, or create sympathy for him from the audience. You don't want the character to have too much of a victim complex, but he must be able to endure situations that the audience perceives as being hard to deal with emotionally. Pay attention to how good films use, to some extent, varieties of plot devices and character designs to elicit sympathy from the audience. A common visual sympathy device is to portray the character with a limp, an arm in a cast, one eye missing, scars, or sad eyes. You can also elicit sympathy by something unexpected happening, such as an accident that creates challenges for the character to overcome.

2.2.1.7 Introduce Your Main Character ASAP

This point is especially important if you are doing a short animation. Audiences think of a main character as a guide to lead them emotionally through the story. You should have only one main character and make it very clear who the main character is in the very beginning. Just remember how you feel when you watch a movie and get attached to who you think is the main character, see her get shot 10 minutes into the story, and find out that another character is actually the main character. You may feel cheated, tricked, and ripped off. You may have a hard time getting into the actual main character and may harbor resentment emotionally for the whole story after that point, which will make it hard to identify with any of the characters.

Figure 2.17
This character has all sorts of superhero possibilities. He looks very strong and fierce.

Credit: Illustration by Donnie Bruce.

2.2.1.8 Place Your Character in Jeopardy

The main story line of most movies is as follows: You meet a likable character with some imperfections and in a sympathetic situation, and get attached to him emotionally; then a whole stream difficult things start to happen to the character. An escalating array of obstacles starts to pile up, placing the main character in extreme emotional or physical jeopardy. Audiences love to worry about characters. The more obstacles, the better. Get your audience attached to the character before placing him or her in too much jeopardy, or they may not even care.

Titanic is a classic example of how to develop strong attachment to characters before presenting major obstacles. Cameron had his audience care about and identify with the main characters before placing them in great jeopardy by sinking the ship. If he'd started sinking the ship right away, he'd have had a difficult time trying to blend all the subtle personality quirks and sympathetic situations into the chaos of survival.

2.2.1.9 Empower Your Character

Give your character the ability to save herself or get out of really difficult situations in clever ways. People like to see stories about a character who changes her life by making decisions and using her strengths to affect her situations. Don't just have linear events happen to the character. Play with cause and effect in which the character's actions create struggles or resolve situations, weaving a more involving story.

2.2.1.10 Use the Eyes of the Audience

The phrase "eyes of the audience" means that the audience finds out and experiences things when the character does. You may not always want to use this approach, but you should try to fit it in as much as possible.

Figure 2.18
This powerful and beautiful-looking 3D character who can fly and runs around in cool boots look as though she can handle anything.
Credit: Illustration by Sam Miceli.

Sometimes you can build tension, for example, if you cut back and forth between two situations, moving the situations toward an intersection without the characters' knowledge.

There is also a great deal of emotional charge in having your audience walk through a dark hallway, as in the movie *Alien*, and discover where the monster is along with the characters who are soon going to be devoured. You will start to get a feel for this the more you create visual stories. Keep a list of how filmmakers use this plot device effectively and try to use it in your 3D animations in a similar way.

2.2.2 Give Your Character a History

All 3D characters need to have a history behind them to come off as real and believable. The more you know about your character, the easier it will be to design, animate, and have it say the right things if it talks.

Think of how flat movie characters are who wake up with complete amnesia. Many 3D characters have this same problem because their animators didn't take the time to think deeply about who their characters really are from a life experience and genetic level. Cast your 3D characters as if they are the perfect thespians for the roles. Give them backgrounds that will make the audience fall in love them because they are so full of personality and depth.

You should write character-driven stories as much as possible—meaning that your concepts don't rely on funny situations or superficial gags but rather on interesting and engaging the characters. Creating a good character starts with your deep reflection on who the character is by writing his complete history. Many screenwriters use this trick for the characters in their films before writing any of the script. To keep your characters feeling genuine, try to base them on people you know or on information you know about people in general.

Character History List

Full birth name:	Natural defenses:
Birthplace/date:	Attitude toward work:
Relationship w/ mother:	Education/training:

Relationship w/ father:	Occupation:
First memory:	Occupations of parents:
Favorite food:	Social status:
Talents:	Pets:
Flaws:	Travels:
Perceived purpose in life:	Sense of humor:
Astrological sign:	Favorite band:
Emotional makeup:	Favorite restaurant:
Superhero qualities:	Family backgrounds:
Loves most:	Most frightening memory:
Best moment in life:	Main goals in life:
Worst fear:	Favorite color:
Biggest enemy:	Favorite known designer:
Love interest:	Type of vehicle:
Best friend:	3D world limits/possibilities:
Favorite hobby:	Favorite drink:
Spiritual orientation:	Most prized possession:
Political orientation:	Secret insecurity:
Three key turning points in life:	Siblings/relationships:

Figure 2.19
Sloxan is full of personality and energy and quite fun to watch. She would have no problem holding her own during a Barbara Walters interview.
Credit: Illustration by Sam Miceli.

Write your own version of a character history list for your main 3D character. Sometimes you need to tailor your list to reflect the types of 3D characters you are focusing on. If you are creating a cast of sea creatures, for example, you might add questions such as the following: Do they breathe air through their mouths or through gills? How fast can they swim? Can creatures live inside their bellies if swallowed whole?

The Barbara Walters 3D Character Interview Test

Listen to the types of questions journalists like Barbara Walters ask people they interview to try to get the greatest emotional response. Then pretend that your 3D character has become so successful that Barbara Walters has requested a 20-minute interview with her—as if she were a human being. What questions would Barbara ask your character? Use questions that will help you define the essence of your character. You may even want to write a brief script of Barbara Walters interviewing your character to see what would make the best questions and answers for a TV interview.

How would you animate your character during this interview to show what she is about visually? Pretend that your character is famous in some way that is connected with your animation idea and see what happens. You may come up with unusual questions and interview behaviors that you hadn't thought of when writing a straight character history. Any great 3D character should be able to hold her own in a Barbara Walters interview for 20 minutes on prime time. If you feel your character isn't interesting enough, then change her until she is. Study good TV interviews to see what makes a character or person fascinating enough to watch for this long in an interview format. Pay particular attention to the setting of the interview since it should be one familiar to the character's everyday life.

2.2.3 Attach an Actor's Voice to Your 3D Character

Each 3D characters should have a movie star-type persona attached to it in some way to help the audience understand its essence more clearly. Even if you are suffering from fear of "lip synch" and have opted not to have your character speak, you still need to attach an actor's voice to your character. Choose well-known voices from famous film stars. Try to stay away from the more obscure TV and foreign actors unless it is just for your own information.

When you go into a Hollywood studio to pitch an idea for an animation, you'll hear, "Who would play the voice of this character?" before you can even pitch your story. If you answer, "Robin Williams," it
doesn't mean that Robin Williams is going to be available to play your biker chicken character. It means that a biker chicken played by a Robin Williams-type actor is completely different than a biker chicken played by a voice like Arnold Schwarzenegger, Mel Gibson, or Nicholas Cage. The name you use merely helps orient people and you to how this character sounds and what type of essence and personality he has beyond the basic storyboard drawing and descriptions.

When trying to cast an actor to play the essence of your 3D character, it is helpful to determine the basic character type. Use the following list to narrow down the possibilities.

Basic Cliché Character Types

Actors	
Leading man	handsome, intelligent, educated, suave, hero potential, cool
Sensitive artist	beautiful, vulnerable, moody, edgy, (talented), (struggling)
Funny guy	goofy, prankster, silly, jovial, (idiotic), (neurotic), (cynical)
Diabolical madman	genius, evil, warped, deranged, vengeful, (amusing), (creepy)
Working-class	takes orders, hard
blue-collar man	worker,uneducated, (beer drinker)
Authority figure	powerful, educated, respected, wise, aloof, conservative
Mercenary soldier of fortune	aggressive, combative, hard, good shape, unemotional
White trash	poor, uneducated, survival-oriented, uncouth, crude, unkempt
Wimpy	frail, passive, afraid, sensitive
Sleazy bachelor	macho, womanizer, superficial, sports fan
Actresses	
Leading lady	Beautiful, smart, funny, sexy
Femme fatale	Beautiful, dangerous, indecisive, moody
Funny girl	Goofy, prankster, silly, jovial, (idiotic), (neurotic), (cynical)
Classy broad	Rich, pretty, well-educated, well traveled, cosmopolitan
Girl next door	Happy, cute, perky, smiling, athletic, light-hearted
Crazy woman	edgy, dangerous, willful
Mother	nurturing, mature, compassionate, wise
Artistic quirky chick	weird, unpredictable, creative, pretty in offbeat way
Innocent virginal	pure, small town,
Church-going girl	sheltered, religious, feminine
Sleazy barfly/ party girl	worn out, cynical, tired, drinker, social, wild

Figure 2.20
A Jeremy Irons type could be the voice for this creature.
Credit: Illustration by Donnie Bruce.

How would your perception of the Jeremy Irons character change if the voice were played by Robin Williams? You would probably have to redesign the character to make it more compatible with a voice like Robin's.

2.2.4 Base Your Character Design on an Actor

You might also want to design your character's facial features or body type to somehow resemble the actor you have chosen for the voice essence. This will make the character easier for the audience to identify with. Choose actors who, of course, would fit the history, visual theme, and mood of your animation.

When you hire a voice actor to play a character, he will appreciate being told that his futuristic valley girl ape is played by a Courtney Love-type voice rather then being told she sounds "cute but rebellious."

Try designing variations of the same character, using each of the actors in the following list to see how different the design becomes.

The Essence of the Actor List

Actors	Actresses
Jack Nicholson	Liz Taylor
Jeff Goldblum	Drew Barrymore
Robert DeNiro	Sharon Stone
Val Kilmer	Jodie Foster
Wesley Snipes	Courtney Love
Woody Harrelson	Alicia Silverstone
Christopher Walken	Madonna
Bruce Lee	Sigourney Weaver
Nicholas Cage	Gwyneth Paltrow
Matt Dillon	Michelle Pfeiffer
Pee Wee Herman	Shirley MacLaine
Clint Eastwood	Cher
Robin Williams	Rosie Perez
Keanu Reeves	Angela Bassett

The Essence of the Actor List

Bruce Willis	Liv Tyler
Arnold Schwarzenegger	Roseanne
Bill Murray	Mia Farrow
James Earl Jones	Lucy Lawless
Dennis Hopper	Demi Moore
Leonardo DeCaprio	Uma Thurman
Johnny Depp	Whoopie Goldberg
James Woods	Meryl Streep
Tom Hanks	Zsa Zsa Gabor
Willem Dafoe	Oprah Winfrey
Jackie Chan	Pamela Sue Anderson
John Travolta	Winona Ryder
Samuel L. Jackson	Elizabeth Shue
Ben Affleck	Anjelica Huston
Billy Crystal	Lucille Ball
Anthony Hopkins	Marilyn Monroe
Sean Penn	Angela Lansbury
Harrison Ford	Dolly Parton

Figure 2.21
This character, designed around Sigourney Weaver, is made of flesh and bone but moves like an organic motorcycle.
Credit: Illustration by Sam Miceli.

2.3 Animating to Create Screen Presence

When people say that an actor has a great screen presence, they mean that when Liz Taylor, for example, walks into the frame you cannot take your eyes off of her. Actors are either born with or spend a long time developing that screen presence, and it's also one of the more challenging aspects of animating a 3D character. Your goal is to get the character to the point where she holds a screen as well as Elizabeth Taylor does. Screen presence is hard to define for those who haven't thought about it much.

Figure 2.22
This pirate-theme character was designed around Samuel L. Jackson. When animating him, you mightwant to include some little visual trick Samuel does to make the character come alive.

Credit: Illustration by Maia Sanders.

2.3.1 Study Actors' Mannerisms

Study actors in films to see what makes some more interesting to watch then others. A great deal of screen presence has to do with the energy of the actor. How do you translate this magical energy to your 3D character? Great animation and design are a good start, but you also need to be studying the little nuances, such as those Jack Nicholson projects onscreen with his eyes, eyebrows, mouth, and swaggering walk. They are part of what makes him so electric. Choose one actor each month and rent his top four films. As you watch each film, write down what little ticks he gives to his characters to make them so appealing and original.

2.3.2 Consider the Anatomy

One dangerous design pitfall in 3D is creating characters who look great on paper but may be anatomically difficult or impossible to animate well. Obvious problems such as short legs with really big feet are going to make that walk cycle a very challenging set of keyframes. Study anatomy books on animals and humans to understand how they are built to move. Take an anatomy class for artists to get a good solid understanding of muscle and bone structure to apply to your 3D characters.

A great example of questionable 3D character design for animation was in a recent science fiction movie. The 3D alien in this film had these really funky inverted knees, and the center of gravity seemed completely off as the alien ran around. During the screening I saw, the audience laughed when this character started to move in what was supposed to a very scary chase scene. After that scene, the alien seemed pretty unbelievable, and the whole movie started to unravel quickly. Once you show people a 3D element that looks a little off, they get pickier with everything else you try to make them believe.

Design using pictures of real creatures for reference. All the bug ideas were taken from actual photographs of real bugs zoomed way up. By designing from real creatures, you also run less of a risk in having anatomy problems when animating.

2.3.3 Act Out the Animations

You should be able to act out a walk cycle for your character. You may want to try acting out the motion for each design. Try to get something original that looks good and moves really nicely. Most animators work with long, floor-length mirrors near their stations, close-up mirrors for facial animations next to their computers, and stopwatches to time the movements of each sequence.

Remember this suggestion presented in Chapter 1: Take an acting class to learn how to act so that you can make your characters leap off the screen. Give your 3D characters some activity to do during dialog that not only says something about what is important to that character but also gives you a chance to show off your great animation abilities. You could have your biker chicken fixing a delicate music box, building a bomb, or pulling the wings off of butterflies during a conversation, for example. Any one of those activities will help the audience get a better visual read of what this character is about.

2.4 Adding Clothing and Accessories

Clothing and accessories are treasure troves of information about what your character is like inside. Watch the movies *Antz* and *A Bug's Life.* Then write down the name of each main character with a rough sketch and a list of what visual elements make that character different from all the other ants or bugs. How do these different accessories make the character more interesting? Ask yourself why the designers made each decision and what they wanted you to feel about that particular element.

Figure 2.23
This character looks great on paper but may be challenging to animate due to the knees joined at the spoke.

Credit: Illustration by Sam Miceli.

Figure 2.24
These drawings of bugs were taken from actual photographs of real bugs.

Figure 2.25
How would you act out a walk cycle for this particular character?
Credit: Illustration by Sam Miceli.

Use accessories with your 3D character to help develop the history of the character and situations. If your character is a lizard warrior who comes from a long clan of sword fighters, for example, embellish him with all kinds of symbols from the clan. You don't need to explain the story behind each piece of jewelry or clothing or items in the animation. All the visual details will add to the character's essence simply by being well thought out and supported by a good solid history. Following is a list of possible accessories to get you started. Which specific ones could you add to your 3D character to make it more interesting?

Clothing and Accessories List

Clothing	Jewelry	Weapons	Misc.	Misc.
cape	rings	knife	helmet	animal hides
coat	necklaces	sword	sporting equipment	eyeglasses
hat	chokers	guns	briefcase	pets
uniform	bracelets	bow and arrow	laptop computer	painted skin
veils	bangles	utility belt	wheelchair	sun glasses
sari	pendants	lasso	skateboard	visor
socks	hair ties	whip	surfboard	goggles
stockings	ankle chains	stick	backpack	electronic implants
shoes	arm bangles	kung fu weapons	dead animals	headset
feathers	bones	shield	purse	nail polish
t-shirts w/ pictures	piercings	high-tech weapons	medicine satchel	nail extensions
LCD cloth	headbands	laser gun	books	facial hair
ties	head dresses	hand cannon	flowers	mask
suit	rope ties	harpoon	cups	bird
swim suit	belly chains	dagger	trophy	bicycle

Clothing	Jewelry	Weapons	Misc.	Misc.
parka	toe rings	garden tools	bottle	tattoos
apron	foot bracelets	spear	roller skates	eye patch
fur coat	chains	axe	juggling balls	ski mask
belt	ball and chains	horns	cigarettes	Lone Ranger mask
scarves	chastity belts	medieval weapons	cigar	primitive mask
space suit	S&M stuff	club	pipe	cast
costumes	brooch	trident	doll	bandages
vest	hair clips	boomerang	binoculars	
gloves	watch	flame thrower	camera	butterfly net
turban	crown	bomb	can	umbrella
lace	beads	explosives	walking stick	
sashes	bindi	spikes	miner's head light	
ribbons	safety pins	fingernail weapons	air tanks	
corset	nails		radio	
team shirts	primitive		musical instruments	
muff	religious icons		fan	
armor			spurs	

Visual themes are filled with ideas for accessories and clothing options. Collect costume picture books to get more ideas on different outfits for your 3D characters. Just remember: You are the wardrobe department and costume designer for your 3D animation, so spend some time thinking about this aspect and you will get lots of points for originality.

Figure 2.26
This soldier has a very stylish-looking uniform. And that cape would be fun to animate using Maya Cloth.
Credit: Illustration by Donnie Bruce.

Figure 2.27
This pirate theme lady was based on Angela Lansbury. Great clothes make the woman.
Credit: Illustration by Maia Sanders.

With the advent of Maya cloth and hair, we're going to see some fashionably dressed characters in the near future. This aspect of character design will be taking on a role of great importance in the coming years.

The fashion industry is really into Maya right now, so make friends with an up-and-coming clothing designer who can dress your cast of 3D actors. We will see more of this synergy very soon in the 3D industry—much the same way designers compete to dress the stars for the Academy Awards each year. Already clothing companies such as Diesel are sponsoring 3D action games that have characters wearing their signature line of clothes. What types of deals can you swing in this area to fund your next animation project?

2.5 Sketching Your Main Character

After completing your visual theme, character history, identification checklist, actor's voice, and costume design you should once again draw more small quick sketches incorporating all these ideas. Don't rush through this step. If you go straight toward a clean sheet of paper for the final design, you may come up with a stiff character full of eraser marks.

2.5.1 Draw Several Versions

Try drawing at least 10 different versions of body types, feet, hands, head shapes, clothing, facial features, and 3D elements. You don't have to draw the entire character each time—just focus on parts first and then put together the pieces that work the best.

2.5.2 Select a Color Palette

Select a color palette of no more than five colors, basing it on your visual theme. Make copies of the final black-and-white pencil designs to use as a test template to experiment with different color combinations.

2.5.3 Narrow the Field

For your main character, choose three to five versions drawn at least 6 to 10 inches tall on good paper. You may find that a direction you want to go with the main character that will somehow effect the design of the secondary ones.

2.5.4 Get Quick, Informal Feedback

When you have a main character and a palette you think looks good, get some feedback. Ask anyone who will look at the drawings what they think about them.

2.6 Sketching the Other 3D Characters as a Group

It is much easier to create the secondary characters when you have a really solid main character who, after multiple rounds of revisions, has received positive feedback. Go through the same sketching process, using the same visual theme but different moods, types, essences, histories, and so on. You should end up with a great group of 3D actors at the end of the process.

Figure 2.28
This nearly final version of our 3D Avatar techno DJ Media Man has changed a lot from the initial sketches after we worked on his essence.
Credit: Illustration by Donnie Bruce.

2.6.1 Design Around Four Basic Types

Try to design character sets based around these four types: main character, love interest, sidekick, and enemy. You can add or subtract characters later as the story requires.

This complete set of characters is based on the visual theme of a futuristic traveling pirate acrobatic troupe. Notice how well these creatures all go together and how the limited color palette works. The enemy in a long coat was designed around a family member where lots of personality seems to just be oozing through visually. Family members make great 3D essences for strong characters since you know them so well.

2.6.2 Avoid Stereotypical Characters

If you are creating a stylized 3D cheerleader, give her a different twist, such as a love for kung fu, an obsession with frogs, or a fear of sunlight. Anything you can add to a stereotypical character that will surprise the audience will make the character deeper and more interesting.

2.6.3 Create Contrast and Conflict

Your sidekick needs to be different enough from your main character to stand out as original in itself but not so far out that it either overshadows the main character or makes the relationship between them hard to believe. Your enemy needs to have some connection to the main character that comes through in either the design or the back story. The love interest shouldn't be too stereotypical as a leading lady or man and should have a distinct personality. Study secondary characters in movies, TV shows, and animations to see what works best.

2.6.4 Get Detailed Feedback

When you have rough sketches of all your characters, you're ready to get more detailed feedback. Even include some of your preliminary

Figure 2.29a
Main Character
Credit: Illustration by Maia Sanders.

Figure 2.29b
Love Interest
Credit: Illustration by Maia Sanders.

Figure 2.29c
Sidekick
Credit: Illustration by Maia Sanders.

Figure 2.29d
Enemy
Credit: Illustration by Maia Sanders.

sketches; a rough design you came up with earlier may have more appeal than the one you settled on. The following two sections provide questions that should help you get feedback that is valuable.

2.6.4.1 Good Questions to Ask Yourself Before You Ask Others for Feedback

Laying your hard work in front of others can be intimidating. You'll feel much better if you know that you're as prepared as possible before you ask others' opinion of your work. The easiest way to do that is to take some time to get your own feedback. Following is a list of questions you should ask *yourself*. Be as objective as possible.

Questions To Ask Yourself

- Do I like this character?
- Am I excited about this character and want to spend months of my life working on it, deep into the night?
- Does this character have a good silhouette?
- Will this character be fun, easy, and interesting to animate?

- Is the mood and visual theme clear in design?
- Does this character leap off the page visually?
- Is there a strong essence and actor's voice attached to this 3D character that makes sense visually?
- Do the accessories tie into the character's history to help develop the back story?
- Would the character be interesting enough to hold its own on a 20-minute Barbara Walters-type TV interview?
- Is this character clever and original?
- Is there a good technical reason to do this character in 3D, or would it be cheaper to use a real actor?
- Does this character avoid stereotypes and cliches?

2.4.6.2 Good Questions to Ask Other People for Feedback

If you've done your homework, getting feedback from others is exciting. Following is a list of questions that should elicit valuable feedback. Just make certain that you value the opinions of those evaluating your work.

Questions To Ask Others

- Which version do you like best and why?
- What kind of emotion do you feel when you look at this character?
- What would you change if you could to make it more interesting?
- Does this character look like Jack Nicholson or Robin Williams to you? (Don't trick the audience. Design around one or the other.)
- What would you say is the personality of this character?
- What can you tell me about this character just by looking at him?

2.7 In Summary

As you may be starting to suspect by now, the conceptual design process can be deep and time consuming, but the rewards are worth all the effort. This process isn't a linear one; you may have to go back and change your story several times to fit new character histories, visual themes, or design decisions. After you get more of a handle on all the things you need to be aware of, you will do them automatically and the entire process will go a lot faster.

Chapter 3

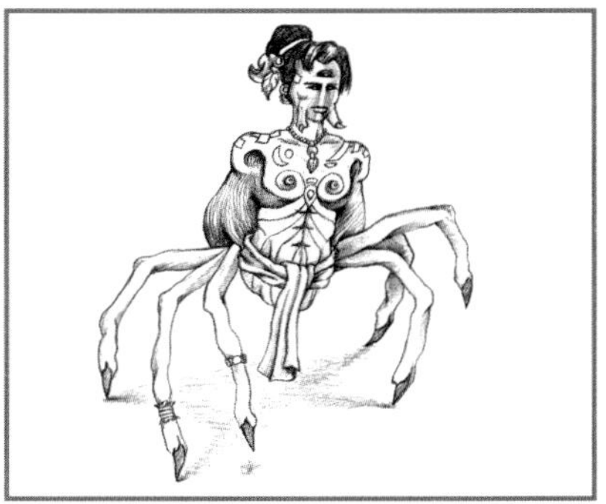

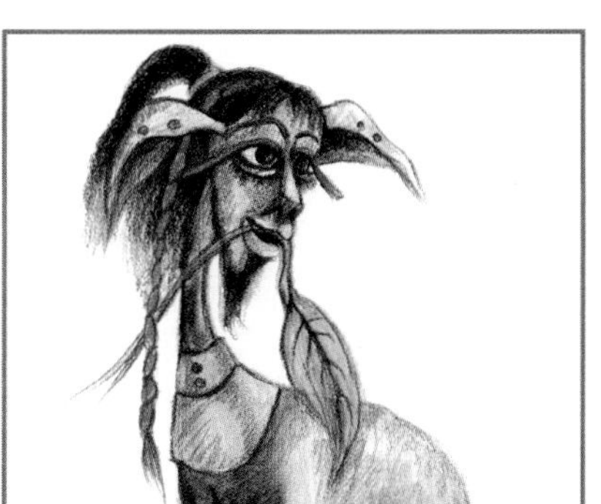

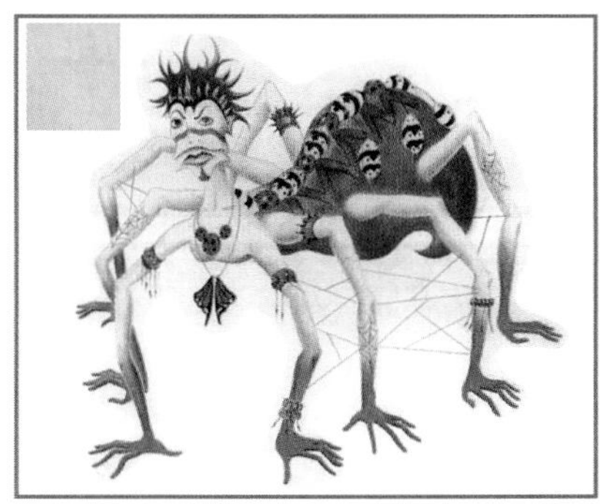

Designing Storyboards for Animation

by Sherri Sheridan

It is now time to design the visual road maps for your 3D animated story. You should have a solid understanding of your story structure and main characters before proceeding to this stage. You may have to go back and rewrite scenes at the story level or redesign parts of your characters based on problems you encounter from feedback after pitching the storyboards.

Understand that this is part of the process and know that this is the time to make those kinds of changes. If every story were perfect from start to finish the first time you thought of it, creating original 3D animations would be a breeze. For most visual stories, at least one-third to one-half of the total production time is spent getting to the final animatic stage before any 3D has been created on the computer.

3.1 Set Design

Great 3D characters can be enhanced by well-thought-out set designs that reflect character essence and personality. When you meet a person, you can gain a greater knowledge about who he or she is after seeing where the person lives. The types of pictures people hang on their walls and whether they live in a dusty gray warehouse or a spacious Mediterranean villa overlooking a sculpture garden they created expands your perception of who they are as people and what is important to them. Therefore, before you ever begin designing the set, have this important issue resolved:

Know where your 3d character sleeps, works, plays, and eats.

Figure 3.1
This production sketch shows where the evil virus who attacks Media Man lives. The image has a retro-futuristic Frankenstein visual theme that is in line with the funky feel of the whole story.

Credit: Illustration by Sam Miceli.

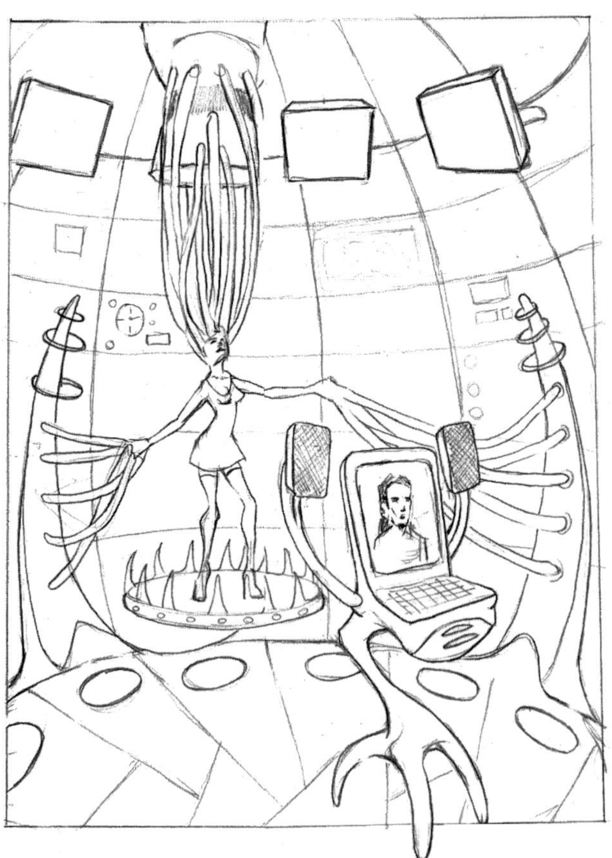

Figure 3.2
This production sketch shows where The Core lives in "Media Man in the Net." She is wired into the 3D net and has a techno retro-future visual theme. We can see that she is an integrated part of the 3D world by the way she is inserted into the Internet architecture of the Cyberdome.

Credit: Illustration by Sam Miceli.

Try designing a rough bedroom or living environment for each 3D character you develop, even if the audience will never see this particular location. By doing this, you may gain many visual clues about who the character is and possibly even how to animate it. If your 500-year-old Chinese woman lives in a cave, she may always be picking spider webs out of her hair, or perhaps bright light forces her to shield her eyes during environmental changes throughout the script. Remember the screen power around all the subtle actions and nuances of a brilliant real-life character actor. How can you make your 3D creatures come alive by having the set design interact as much as possible with their personalities and animated movements? The following sections suggest several tips to increase your design awareness.

3.1.1 Study the Masters

Study existing films for set design tips. *The Usual Suspects* (1998) is a film that did some incredible character development through set design. Ask yourself questions about every background detail, such as "Why is a stuffed white ferret posed to strike behind this character? How does this make me feel about him? How would my feelings change about this character if the background object were a beat-up, old, stuffed Raggedy Ann doll instead of the ferret? Why are the walls covered with brown wood paneling?" Old film noir movies do lots with background set design; you should watch them for ideas.

In designing virtual characters and sets, keep in mind that someone in the audience will be more familiar than you are with the environment you are trying to portray. They will love you if you are authentic and hate you if you are fake or clichéd. Think about the really cheesy "gothic club scenes" in mainstream movies where the set dresser and costumer have obviously never been to such a place. The more authentic, small details you put into a set, the better it will read. Always try to reach that set specialist in the audience.

Audiences need things to look at in the background that tell them more about what the reality is like in these often strange and unfamiliar 3D worlds. Professional set designers spend a great deal of time on each detail of a big film.

Study good architectural design magazines and books to expand the possibilities of what you can do with your 3D sets. Even if you create a matte painting, it still needs to convey some visual connection to the characters and story. Use the same visual themes you developed for your characters (refer to Chapter 2, "Designing Characters") when designing the sets they live in.

The latest *Star Wars* film *The Phantom Menace* has some incredible conceptual design with regard to characters, sets, and visual themes. See if you can figure out the basis for each location and character theme. For example, Jar Jar Binks has some Jamaican qualities, whereas Queen Amidala when dressed up looks similar to a Mongolian princess.

Figure 3.3
This character has an American Hopi Indian visual theme and fits nicely into the set using the same theme. Notice that the accessories on the creature match the ones on the hut.
Credit: Illustration by Debbie Rich.

3.1.2 Use Your Imagination

Try to avoid stereotypical bedrooms, offices, or houses. This is 3D! Embrace it! For the first time in history, you now have the power to create anything you can imagine—thanks to a desktop computer. Defy physics, create new planets, make an 800-foot-tall solid gold temple or a miniature world inside a dirty shag carpet. Most stories have been told before in some form: Only the sets and characters change.

3.1.3 Add Background Looping Animations

Try to loop animations in the background to add movement and make the place feel alive. Add old clocks ticking, clouds moving through windows, weird birds flying off in the distance, spaceships or vehicles dashing by, fans whirling, candles burning, fountains, aquariums, flashing signs, TV monitors, strange creatures walking around, or pets curled up on a rug and breathing hard. Make a list of possible looping animations that fit each environment you plan to have in your animation and use the ones that work best. Try to keep it simple in regard to modeling and texture mapping, especially if they are far off in the distance.

3.1.4 Resources for 3D Set Ideas

Keep a clip art file of set design ideas for reference. A great place to get new ideas is the 100-year *National Geographic* CD-ROM set mentioned in Chapter 1. You might be able to pull off a fresh 3D rock-city design by looking at a microscopic picture of granite or a hillside in Bangladesh. Try picking five pictures you think fit the mood and feel for your animation and design five different rock cities based on each picture's form and shape. Make sure to include your visual theme as strongly as possible in each of the five designs. Pick the one that works best, and try to exaggerate even more the visual parts that are working, such as the curves in the rock wall or the number of crevices. People like to see things they have never seen before in 3D, and set design is great way to get more "wow" shots.

Another fun exercise to get to know your 3D characters is to try to design their work environments, neighborhoods, backyards, familiar landscapes, grocery stores, gardens, favorite restauranst or bars, and vehicles. Even though you may never show your characters in all these environments, at least you'll have a great feel for who your character is, and you will be more likely to pass on such information to your viewers.

3.1.5 Using the Set Design Checklist

When creating your 3D sets, make sure you do the following:

- Show the audience something they haven't seen before.

  A bedroom made of organic breathing furniture is more interesting to look at than one made from a clip art set of the basic wooden designs we have all seen a thousand times before.

- Provide amazing visual appeal that makes you just want to stare endlessly in complete awe.
- Have a clever and original angle.
- Design sets to match the environment your characters inhabit.

  Suppose that you're doing an animation about rats in a bar. Instead of making your bar set look like Al's Bar down the street, transform your set into a far-out rat's biker bar and incorporate into the environment rat-type objects such as bottle cap tables, matchbox stools, and found sewer objects. Be true to your character's essence in some way. Study the films *A Bug's Life*, *Antz*, and *Toy Story* to get a feel for how this all works. A good rule of thumb for creating animate or inanimate object animations is to make sure that half of the objects in the set are from the animal or object's world and the other half is from a human's world. This will help the audience identify with the story and the characters. Creating a totally realistic ant hole is a bit dry for an entertainment show, and if all the ants look identical, it's hard to tell which ant is which.

Figure 3.4
This character looks well suited to his environment with a good sense of balance and feet that can walk on jagged rocks.
Credit: Illustration by Chris Hatala.

- Develop your character and story in a deep way by adding background visual details that give viewers information about the world your characters live in and what is important to them.

  An alien living in a frilly, lacy cocoon is perceived completely differently from one dwelling inside a cocoon made of sharp metal knives.

- Match your visual themes to your characters and their environments. Don't create them separately, throw them together, and expect everything to work out.

  We have all seen 3D animations in which the characters look completely different from each other and the worlds they live in, resulting in confusion and a lack of belief in the reality the animator is trying to create. Think of how you would feel if you went to India and all the surroundings looked the way they do today, but the people were wearing day-glo shiny silver glam rock outfits. You would feel as though something was very off.

- Be careful not to overwhelm your foreground characters with backgrounds that are too busy.

  You need to create strong silhouettes for your 3D characters. Make sure backgrounds are of contrasting colors or are lit differently and don't have so much detail that your characters gets lost. Looping animations are great to make your set come alive, but too many can make the audience dizzy. Just use them economically and wisely.

3.2 Production Design

Collect "wow" shots from existing films that may work in some way in your animation. You can grab frames from a video tape, or you can storyboard the shots. Keep a separate drawing book just for clever camera work. Each time you see a film, make note of at least three great shots that affected you emotionally or made you visually appreciative.

Collect photographs that have the look and feel for what you think might work with your animation. Pay attention to the composition and lighting and see what ideas you can borrow. Paintings are another great source for good conceptual design ideas. See if you can capture the essence of a painting inside a shot, character, or set in your animation.

After you have researched your visual theme and created strong 3D characters and sets, you are ready to pull together the visual look and feel for the entire animation. Gather all your thumbnails, character designs, illustrations, pictures, and photographs and lay them out next to each other on the floor or on a big table. See whether the pieces look like they fit with the rest of the mood and theme you are developing.

If you have multiple sets of characters or environments, lay them out in separate areas and see whether they still hold together in some way. Begin weaving together ideas for your own camera shots.

Start to see the animation taking shape in your mind. Make changes to any images or characters that don't fit or feel weak visually. Try mixing together different set designs and photographs for fresh lighting or composition ideas. This is a great time to try different combinations before you start locking down the storyboards.

3.2.1 Timing Considerations

You are almost ready to start storyboarding. One problem animators run into is deciding how many shots to give to each section of a story. Another problem is deciding which events occur in each scene. Even if you have a rough shot list as we did in Chapter 1, "Creating Story Concepts for 3D Animation," you may still run into some confusion about the individual scenes and how many panels to allow for each part. These decisions all involve aspects of timing.

People can mean a great deal of things when they make comments such as "The timing in your animation feels off." Consider the points discussed in the following sections when working on the various aspects of timing in your 3D animation.

3.2.1.1 One Thing at a Time

When storyboarding your animation, try to have only one thing happening or moving at a time (not including background loops). Some 3D animators are tempted to move everything all at once, and it usually comes off as a general lack of cinematic elegance. We have all seen the 360-degree spinning 3D camera with 50 looping animations going on in the scene; our eyes get lost and confused as to what to look at next, losing the subject completely.

Take your finger and follow the movement within the frame of movies you like, such as *Citizen Kane*. Notice the timing of how each movement occurs and the directional rhythm. Try to keep your 3D cameras locked down as much as possible unless you have a really good reason to move it.

3.2.1.2 Time Your Shots According to the Three-Act Structure Formula

The average length of a shot for most films or TV shows is about four seconds. Generally you never see shots longer than 10 seconds unless they are establishing shots or creating dramatic effects. For example, you might have a montage of 10 one-second shots flashing to build a mood or tell a story. Or you may hold a shot for 15 seconds if it's some really interesting dramatic moment—but this needs to be done really well to work.

When I know my setup is supposed to take 30 seconds, I plan to include six to eight panels or shots (the average shot being three to four seconds long). Use the formula in the section "Using Three-Act Structures" of Chapter 1 to figure out how long each part of your animation should be. Table 3.1 provides a quick review of the elements, their purposes, and the number of seconds and shots of each in a two-minute animation.

Table 3.1

Three-Act Structure in a Two-Minute Animation

Element	Number of Seconds	Number of Shots	Purpose
Setup	0–30	6–8	Defines the status quo (what is normal)
Catalyst	25–30	1–2	Event that changes the status quo
First Turning Point	55–65	3–5	Fresh turn acion
Climax	90–95	1–2	Point of the greatest intensity
Final Confrontation	100–110	3–5	Confrontation between two main characters
Resolution	105–120	5–10	Ties up loose ends of the story
Totals	**120**	**20–32**	

With 3D animated short films, you might want to keep your resolution short. You could then spread these extra three to eight shots out somewhere else in the animation, depending on your story.

For a four-minute animation, simply double the time and number of shots. For other lengths, estimate based on multiples of two minutes.

3.2.1.3 Act Out Those Shots

Get a stopwatch (which you should already have if you are a 3D animator) and act out each scene of your animation. Act out each scene three times and average them according to the number of seconds you find it takes to get through each movement and dialog piece.

Better yet, videotape yourself with a few friends acting out the entire animation, much like an experimental short play without sets, and see how the timing and story work in general. This is especially important if the story is supposed to be funny. If your idea doesn't get many laughs as a rough cut, with your friends acting it out on videotape, you may want to change your idea until it's perceived as funny. No amount of good animation or design can save a bad script. It's better to find out now rather than after you've spent all those hours staring into a computer screen creating an animation idea that just doesn't work.

Figure 3.5
How would you go about acting out these two types of different characters for timing purposes? Would you have the flying one be really fast and zippy and the walking one, sort of slow and clumsy for a good contrast?
Credit: Illustration by Sam Miceli.

3.2.1.4 Practice Comedic Timing

Funny stories are some are the most difficult to execute successfully—particularly in 3D. A great sense of comedic timing is essential to pace gags and character animations effectively. Study animations such as the Looney Tunes *Road Runner*. It isn't very funny if Wile E. Coyote chases the roadrunner off a cliff, and the coyote simply falls to the ground. But the shot is funny when the coyote stops to pause in midair before falling, looks down, and then looks at the audience with a panicked expression on his face as he realizes his impending situation. Good comedic timing is perhaps more natural ability than something you can learn, but with lots of practice anything is possible.

Be careful with funny animation ideas and test them thoroughly on friends and neutral third parties first. There is a gag story-type of animation that requires lots of little funny things that happen throughout the story building to a more and more absurd climax. One mistake is to create an animated story that rests solely on one funny gag at the end; one gag usually won't be enough to hold your audience throughout the piece, and it will probably bomb at the end. The creators of popular shows such as *The Simpsons* have a great pace and feel for comedic timing; you should examine these types of funny shows thoroughly before attempting comedic stories.

3.2.1.5 Pace Your Animation

The pace of your story should generally accelerate as it nears the climax, but this isn't always the case. Pay attention to when your story starts to get slow in the wrong places and speed up the pace, change it, or cut parts out. You need to think of your animation as a carefully choreographed ballet that juggles literally thousands of components that need to all fall into one groove and rise out to the end of your story. Try slowing down or speeding up your pacing to get different feels for your story. Use pace to elicit more emotion. Study suspense or thriller films to see how pace is used to increase the tension and emotional release.

3.2.1.6 Act Out Your Characters' Movements

Once again, use a stopwatch or video camera and act out each movement of your characters to get a sense for how long things take to occur. Timing is by far one of the most difficult aspects of good character animation. Take a stopwatch everywhere you go and compare how long it takes for a big woman versus a skinny woman to walk across the street—or for a truck versus a sports car to drive by. You will start to develop a feel for the ratio between seconds and movement if you study the world around you.

A good trick is to put together a cheap motion-capture system by using two video cameras to record you at the same time from different angles. Act out your scene, map the video frames on image planes around your 3D character to the correct scale, and use the video as keyframe guides. This simple trick will help you to get the hang of character animation, and you'll learn more about timing for keyframing.

3.2.2 Cinematography

You've probably seen the work of many different types of illustrators. Some are great at drawing photo-realistically but hit a brick wall when trying to design amazing camera shots. Every illustrator creating storyboards for 3D animation must have a solid background in cinematography and editing. *Cinematography* is the art of filmmaking, particularly with regard to framing camera shots and telling a powerful visual story.

The best way to learn about cinematography is to do the following:

- Take classes in filmmaking, editing, lighting, and production design.
- Study great films and then storyboard shot-by-shot the sequences you find particularly appealing; or storyboard the entire film to get a feel for how it was put together.
- Make little movies with an inexpensive digital video camera and edit them on your computer, using programs such as Adobe Premiere and After Effects. This takes less time than building everything in 3D and trying to develop your skills around the constraints of a 3D computer world.
- Write short stories that might be re-purposed for a set of 3D characters later, or different versions of stories based on an existing animation idea. Focus on creating tension and emotion. Pay attention to making the dialog sound natural and interesting while you and your friends act out the characters in exaggerated, articulated gestures. Don't worry about props, sets, or costumes. Use your imagination and experiment with the essence of the story's emotional curve. Play with real lighting if you have time. This is a great way to spend time while rendering. It's a good form of research, too.
- Read books and magazines on filmmaking and special effects to see how to handle tricky shots, gather valuable tips, and gain awareness about specific 3D animation issues with regard to 2D elements and special effects.
- Study photography and take lots of photographs. Shooting large numbers of black-and-white photos is particularly useful for developing a good eye for composition as well as lights and darks. Increase your understanding of what ingredients make an interesting visual picture or a "wow" shot.

- Storyboard animated 3D films and compare camera shots and techniques to traditional cinematography to see how they differ.
- Do lots of life drawings off of nude models. Using a real model to do life drawings will help you keep your eyes in shape and keep you aware of the subtleties of value, shading, shadows, light, composition, and color.
- You may want to play with 3×5 index cards and sticky notes (letterbox-type size) to experiment with different versions of shot ideas and combinations of shots to change meaning. Add photographs and illustrations into the mix and try for a rough visual timeline from a look-and-feel perspective to see how all the elements are pulling together. Sketch any obviously great camera shots you may see right off the bat. This initial design process will simplify storyboarding later.
- Refer to films in similar genres for shots and blocking ideas. Suppose that you are creating an alien battle animation between two clans, and it seems a little flat. You may want to rent *Braveheart* and borrow from that film some ideas that created tension on the battlefield. Change your weapons from primitive to futuristic, alter your battlefield landscape, and incorporate any other design themes that can give your battle the correct flavor. Time the camera shots in *Braveheart* and see where you can chop off seconds or scenes and still preserve the excitement. Try to figure out what *Braveheart* did to make the scenes so successful, and take it one step further. Rent several epic battle movies, and you may even see where the filmmakers borrowed from each other.
- Go to independent international film festivals and see 40 or so films in two weeks. After you see each film, write down the story line, create a three-act structure, storyboard the best camera shots, note the cultural slant each director took on each story, and pay attention to what worked and what didn't. Try to see how you could take some or all these plots and ideas and repurpose them for a 3D film. You should do this in the back of your mind as a fun 3D conceptual exercise for every film or TV show you watch.

Independent films haven't been through the Hollywood "machine" and often have problems and successes that are easy to spot. The story lines are often more experimental, which is good to study for ideas. For example, the lighting in independent films is sometimes off, or the director finds a way to make things work on a small budget that would also save you time in 3D. Maybe it's scarier to see the monster in the shadows throughout most of the film and keep the lights low. Cable stations are currently featuring more international independent films, which you can study.

Short films are of particular interest to a 3D animator attempting to make 3D shorts, so keep an eye out for these as well and use the same evaluation process. Never "borrow" an idea outright; always change it enough to make it yours or combine it with another idea to make it fresh and original for 3D animation.

Figure 3.6
This character has good visual appeal. What do you think her world would look like? She also looks fun to animate with those hands for walking around.

Credit: Illustration by Debbie Rich.

3.2.3 Storyboarding

Storyboards are the visual blueprints of your 3D animation and are the most useful tool for any visual art form. After you start to build pieces of your story on the computer, you will be focused on just making the scene happen and shouldn't have to worry about whether the shot is going to work. That's why we encourage you to never touch a computer until you have a final set of storyboards to keep you on track. In the real world, you can't always follow that recommendation, but it is a goal you should shoot for if at all possible, especially when something as time-consuming and expensive as 3D is involved.

Top 10 Reasons Storyboards Are Essential to Any 3D Animation

You may be tempted to skip storyboarding and go straight to the animation, but we can give you several reasons why we think storyboarding is critical.

1. **To save time and money.** It is cheaper and less time-consuming to make changes at the storyboard level than it is in the middle of a whirlwind 3D production.
2. **To see whether your story works.** Never depend on whiz-bang special effects or on the coolness of the 3D world to save your story. If it doesn't work on paper, it won't work after putting in lots of hard work into creating your 3D elements. Pay particular attention to how the continuity between the shots is working. Are there any visual gaps in terms of getting from shot to shot, or are people confused when you pitch the story?
3. **To provide a visual road map for the team.** Usually you will work with other people to produce a 3D animation. Storyboards are about the only really effective way to describe an animation so that everyone can understand it clearly.
4. **To make it easier for people to provide feedback.** Without storyboards, most people's eyes will merely glaze over after your first few descriptive sentences, and people will have very little to say after you are finished explaining everything about your 3D story. With the physical boards in front of everyone, they can point to particular shots and make comments or suggestions much more easily. You will also have a hard time trying to describe each shot without a visual guide to help keep you on track as you go through each shot.
5. **To help plan the production schedule.** Storyboards spawn production bibles that list each shot and what parts need to be shot, built, painted, or created to make the scene work. Without a good set of storyboards, you may have a hard time planning your production schedule.
6. **To facilitate the client-approval process.** You draw up the idea, and the client signs off on each panel, agreeing on the animation. Any changes are extra. Never create an animation without having the client sign off on a detailed set of storyboards. They are visual contracts that protect both you and client from misunderstandings and the need for major revisions later.
7. **To make setting up your 3D world easier.** Set up is easier to accomplish if you know your shots beforehand. First, you need to know the location of

the camera that affects your lighting, texture mapping, blocking, and range of animation. Second, if you know that one of the characters will be shown in the distance, you can get by with more basic texture maps on that character. If the character is to be in an extreme close-up shot, your texture maps have to be much more detailed. It is very important to think smart and economically in 3D so that you spend your valuable time making other important things look better.

8. **To visualize multiple ideas and choose the best ones.** You could create several sets of storyboards for one animation and play with the parts that work best. Don't always go with the first idea or shot that pops into your head. Ask yourself how else you could handle the shot. How would Hitchcock have framed that scene, or what would Fellini have tried to make it more interesting?
9. **To create visual appeal with regard to look and feel as well as color.** Does the story look good as a whole? Will the colors work, or do your lizard men clash with the bumblebee people? Is the mood strong enough? Is each shot a "wow" shot? What can you change in each shot to make it hold up as its own visual masterpiece? Is something interesting going on with the animation as an art form in itself?
10. **To create an animatic.** *Animatics* are created from storyboards that are scanned into the computer and transformed into a slideshow with timing and audio to get a feel for the story as a whole piece. Some animatics have 3D elements. We provide more information on animatics later in this chapter.

It is now time to create some rough storyboards for your 3D animation. Include all the production design and research you have gathered, along with your strong visual themes. There is no established standard iconography for the correct way to execute a set of storyboards. The most important thing is that the boards convey the idea you are thinking about animating in a clear and easy-to-understand way.

Figure 3.7
Pan arrow shot to show an arching path for camera pan.

When creating your storyboards, keep in mind the following tips:

- Draw boxes that are the same aspect ratio as your output size. The aspect ratio for video is 4:3; HDTV and film aspect ratios are around 16:9. Don't plan shots in perfect squares.
- Use arrows to indicate camera movement or motion in the frame.
- Use arrows along with shot names to convey specific camera movements that may not be understood by arrows alone.
- Use a frame within a frame to indicate zoom areas.
- Use differently shaped, extra-long, or extra-wide boxes to show pans.
- Draw multiple pictures for moving objects or characters within a frame if it seems important to the story.
- Write dialog beneath storyboard panels so you can see how the audio track will help tell the story. Keep dialog to a minimum to avoid talking heads throughout your animation. Try to give the character some activity to do

Figure 3.8
Zig zag moving subject shot.

Figure 3.9
Moving subject behind another object shot.

while speaking to make it seem more full. A talking-head ostrich isn't as entertaining as a talking ostrich carving pictures in a tree with his foot.

- Include shot numbers next to each panel to indicate order. Number the shots horizontally, not vertically.
- You shouldn't need to provide a caption under each panel of what is occurring unless the shot is confusing, which may mean you need to fix the shot. However, if you plan to leave your storyboards with someone and won't be available to explain it, it may be a good idea to provide some textual explanation under each panel.
- Place major transitions between panels.
- Vary your camera angles to get more interesting shots. Try to use as much perspective as possible so that the scene has more of a 3D feel, and indicate clearly what the camera is pointing at.
- Draw with pencil or charcoal for these first drafts to get a sense of lights and darks. Use dry markers, acrylic paint, or colored pencils for final versions. You may also want to color them on the computer using programs such as Photoshop and Painter.

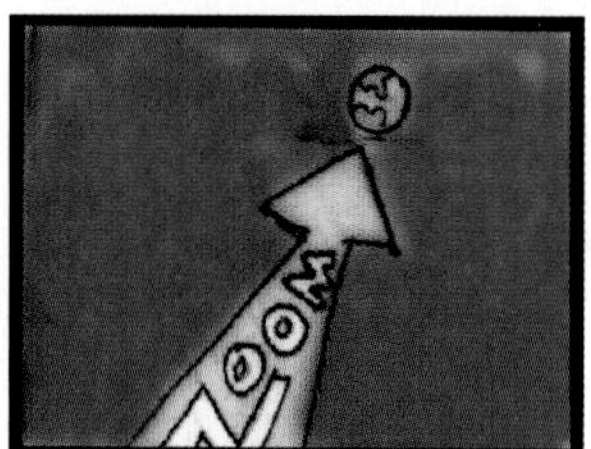

Figure 3.10
Zoom up to object.

Figure 3.11
Zoom into area.

Figure 3.12
Pan across scene.

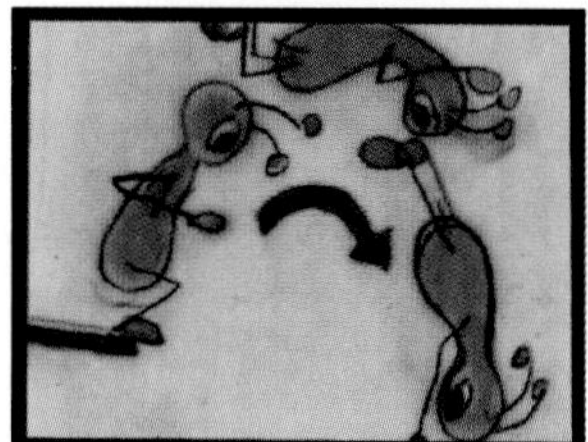

Figure 3.13
Multiple pictures of subject moving in frame to show action inside shot.

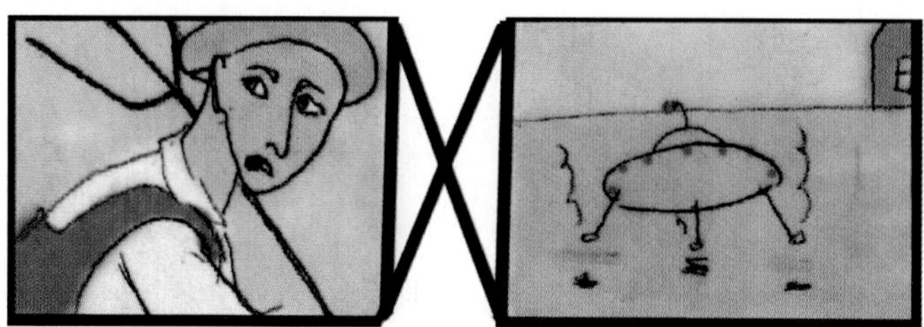

Figure 3.14
"X" used to show a dissolve between two shots.

3.2.4 Lights

Lighting can make or break a 3D animation. Here's why: Good lighting looks expensive and dramatic. Bad lighting looks cheap and amateurish. Not only that—good lighting helps tell a story and establish mood and emotion. Study the masters of cinematic lighting. Read *Painting with Light* by John Alton (UC Press). Practice lighting real objects at home with a variety of light sources or even a couple of desk lamps.

Watch movies such as *Citizen Kane* and pay attention just to where the lights are and why each scene is lit that particular way. The directors of old black-and-white classic films did some amazing things with light, and such films are worth checking out. Chapter 8, "Digital Lighting in Maya," contains good information to help you light scenes more realistically and dramatically. Keep dramatic lightening in mind while planning your camera shots and storyboards.

You need to balance your lights and darks so your animation will have good composition. Render out all final animations with a grayscale pass in a video-editing program to see if the contrast is working. It should look fine in black-and-white if you've been paying attention to your lighting and scene compositions. Using too many mid-tones in the background and foreground may make your silhouettes and animation hard to read visually.

3.2.5 Camera

Use a variety of camera shots, movements, and transitions to add a rhythm to your 3D animation. People who don't realize how many types of great camera shots there are to choose from usually use lots of monotonous medium shots. Learn the names of these shots and what they mean because animators often refer to them during production, and they are referred to in scripts. You can combine shots to make more complicated ones; just make sure you always have a good reason before moving your camera.

3.2.5.1 Camera Shots

Table 3.2 lists different types of camera shots to use in your 3D animations. Try to vary your shots as much as possible without drawing too much attention to the camera work. Remember that your goal is to evoke emotion by using the best type of camera shot possible.

Table 3.2

Forty-Seven Different Camera Shots

	Aerial shot	A shot from above, usually high up. Often used at the beginning and end of films to zoom in or out.
	Point-of-view (POV)	What the character is seeing. Usually this character isn't in the shot; it's the shot seen from the character's point of view.
	Reverse POV	Reverse POV, usually used to show character's reaction to POV.
	Over the shoulder	Looking over the shoulder of a character. Used to show where things are in relation to the character.
	Reaction shot	Shot of characters reacting emotionally to something that just occurred.
	Zoom	Moving shot where the focal length and field of view change over time. Used to bridge shots.
	Tracking shot	Shot that follows a character or an object moving through a scene. Can also be used to move away from or toward a stationary subject.
	Follow shot	Like a tracking shot, but keeps focus only on subject as it moves.
	Into view	The camera moves and reveals a new subject.
	Into frame	A new subject moves into the frame from off screen without changing the shot.
	Insert	A close-up shot of an object that implies great importance, such as a lost key under a table.
	Montage	A series of shots used to build moods or provide information. Typical ones include passage-of-time shots with seasons changing around one tree quickly, or calendar pages flipping. Try to use different angles or similar shapes to create more dramatic montages.

	Split screen	A frame with two shots occurring simultaneously, often divided by a line. Two people talking on the telephone is common for a split screen. Screen can be split diagonally, horizontally, vertically, or with an alpha channel shape mask for interesting effects.
	Pause or beat	Script term to control timing. You might use a reflective shot of a babbling brook and incorporate a long pause to let the audience think about things.
	Freeze frame	Single frame repeated or held to imply freezing of action. Good for funny take-offs and freeze-frame endings.
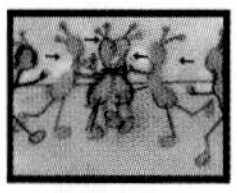	Slow motion (slow-mo)	Camera speed increased above normal frame rate to give the impression of the action being slowed down.
	Super (or extreme) slow motion	Exaggerated slow motion that is even slower Good for drawn-out, dramatic moments such as death scenes.
 Bend, Oregon 1956	Super	Text or images superimposed over the shot. Often used for credits, subtitles, or animated graphics.
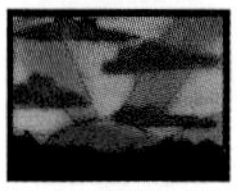	Stock shot	Footage you buy from stock shops of shots you need but don't want to create or are unable to shoot yourself. Shots can consist of newsreel footage, aerial city scenes, establishing shots of far-away locations, microscopic scientific footage, or any other previously shot footage.
	Two shot	Two people or characters in one shot.
	Angle on	Object shot at an angle rather than straight on. You may want to specify degrees or direction of angle for 3D shots.
	Another angle	Shot from another angle different from the previous angle.
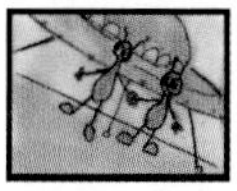	Bank shot	Camera moving on a crane-type motion in a circular orbit.

continues

continuted

Forty-Seven Different Camera Shots

	Shot	Description
	Close-up (CU)	A shot from the neck up of a person or character.
	Extreme close-up (XCU)	Just the lips or eyes or any really tight close shot. Hard to do in 3D because they usually require retexture-mapping your model with lots of extra detail to hold up to this type of close-up.
	Group shot	A group of people or characters in one shot.
	Interior/exterior	Script term used to specify whether the location is inside or outside.
	Long shot	A subject with lots of background and some far-off scenery.
	Extreme long shot	A subject with some background and lots of far-off scenery.
	Medium shot	A shot taken from the waist up on a person or character.
	Reverse angle	The reverse of the previous shot angle.
	Take	Exaggerated animated look in which the character does an extreme reaction shot.
	Three shot	Three people or characters in one shot.

3.2.5.2 Camera Movement

Good camera movement, described in Table 3.3, has a nice flow with a carefully planned visual rhythm. Study existing films, TV shows, animations, and even TV commercials to get a better sense of how to use these together.

Table 3.3

Types of Camera Movement

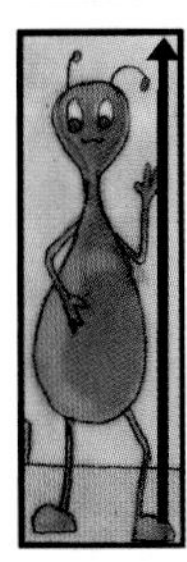 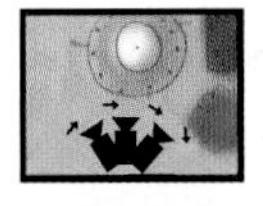 	Pan left/right	Camera rotates horizontally around its base without any dolly movement.
	Truck left/right	Horizontal movement of the entire camera with a dolly-type device.
	Dolly in/out	Horizontal movement along the lens axis of the entire camera with a dolly-type device. Also referred to as camera up/back, dolly up/back.
	Tilt up/down	Vertical rotation of the camera about the lens axis parallel to the scene.
	Pedestal up/down	Camera going up and down along the vertical axis.

3.2.5.3 Transitions

Transitions, described in Table 3.4, add a subtle bridge between shots that convey moods and emotions all by themselves. Pay attention to good uses of these types of transitions to get a better feel for how to use them with your camera shots.

Table 3.4

Transitions from One Scene to Another

	Cut	Abrupt change from one scene to another in the course of one frame.
	Fade-in	Scene appears slowly from a black or colored screen. Most movies begin with a fade-in and end with a fade-out in the script.
	Crossfade	Slow fade from one scene to another that takes place over a range of frames but always involves at least one black frame between the fades.

continues

continued

Transitions from One Scene to Another

	Dissolve	Slow fade from one scene to the next without any black frames between.
	Wipe	One scene pushes another scene off the screen either horizontally, vertically, in a spiral, or in any other manner.
	Intercut	Two scenes occur simultaneously to form one sequence cut together, which shows lines of actions where parallels or contrasts are apparent. A good example is seen at the end of the film *The Godfather*, when part of the family is in a church at the christening of a new baby while the rest of the family is gunning down a rival Mafia clan at a toll booth. This shows contrast in how extreme family life can be for this group.
	Match cut	A cut between two scenes that have the same object, setting, or person in the same position. General "wow" shot if you can work it in. May be used to show passage of time or to point out identity. An example would be cutting from the shot of a ring on a girl's finger to the same ring worn around the neck of an old woman to show they are the same person in a different time.
	Alpha mask	Using digital video editing programs often gives you the option of building your own transition shapes using alpha channel masks. Adobe Premiere has lots of shapes pre-built in its transition library. Many of these work like your basic wipe but may have a star shape associated with it. Adobe After Effects lets you hand-build any transition that you can imagine. Even simple circles with feathered edges may work better than a horizontal wipe, depending on the composition of the shot before and after. Try to match the basic shapes for transitions like these, but be careful not to overuse them because they may draw too much attention to the idea itself.

3.2.6 Interesting Visual Rhythm

Go through your rough storyboards and try to use at least one of each of the shots, camera moves, and transitions listed above to see if the camera work becomes more visually interesting. There is a fine line between great cinematography and going overboard with too many interesting elements that become distracting to the audience. Get a feel for this balance by studying your favorite cinematic films. The point is to mix up your shots so they don't feel stale and boring. Try to make four different camera shots for each shot and see how the meaning changes simply by using different shots. Pick the most emotional shot from the four. Of the following images, the outlined one has the best composition to explain the staging of the shot.

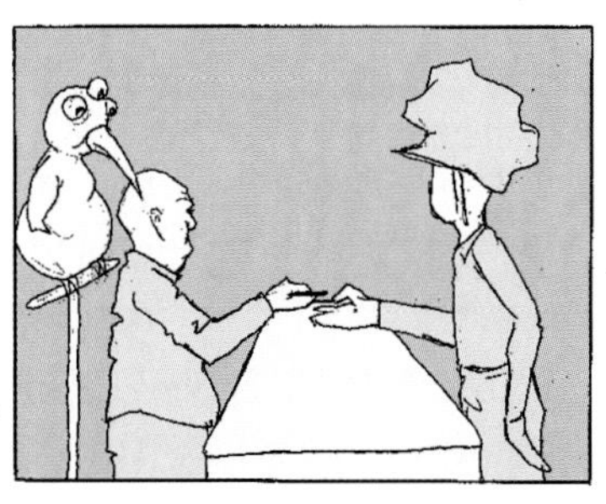

Figure 3.15
Here is the same shot with four different camera shots that change the focus. Notice how the outlined box conveys the best story.

Credit: Illustration by Sam Miceli.

3.2.6.1 Rule of Thirds

Using the Rule of Thirds is a great help for composing basic shots. The idea behind this is to break up the screen into thirds both horizontally and vertically. You then place your elements along the lines and center of interest at one of the four points where the lines cross.

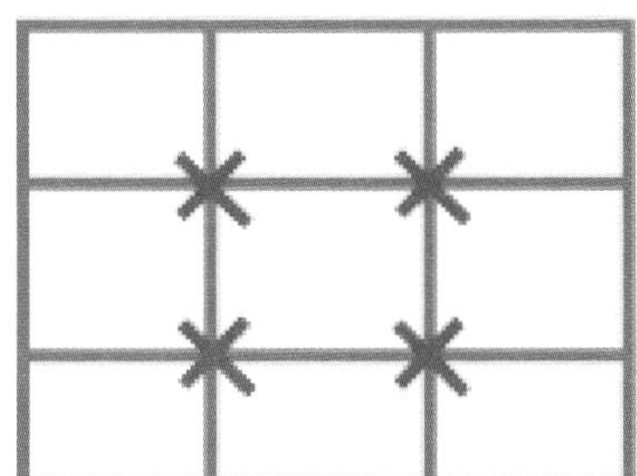

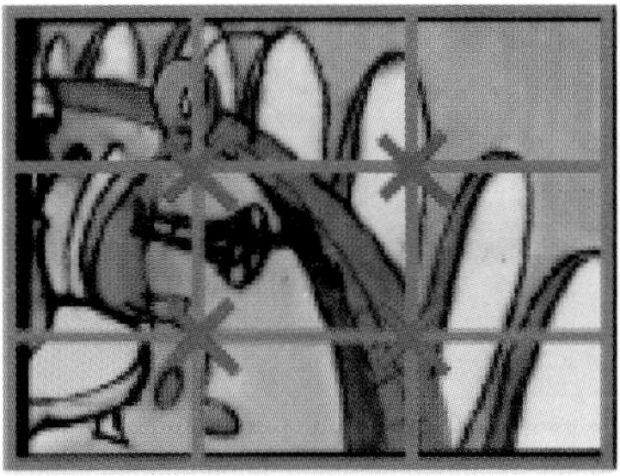

Figure 3.16
Try to place subjects and focal point on the X's where the lines cross or on the lines themselves to form stronger compositions. Placing the subject in the center all the time can become visually tedious.

3.2.6.2 Angles

3D animation can look very 2D if you don't shoot your scenes from good angles. Try to avoid flat, straight-ahead shots. Most 3D animators do this automatically when in 3D but sometimes forget to draw their storyboards with angles. The advantage of drawing with the illusion of depth is that you will then know where to put your camera in your 3D world before you build it, and whether the combination of angles you chose to use are working together visually. Pay attention to films, TV, and advertising to see how they use camera angles to feel a certain way about the subject in the shot. The basic ideas behind camera angles and the impression given to the viewer are as follows:

Figure 3.17
Try to draw your storyboard panels at the angle you want to see. This is especially important for 3D when it comes to setting up shots. By using more angled shots, you also add more dimension to your focal ranges.

Figure 3.18
This shot is nicely balanced with a character on one side and a window on another. It might not work as well if you placed the character in the middle since it would appear unbalanced behind the character.

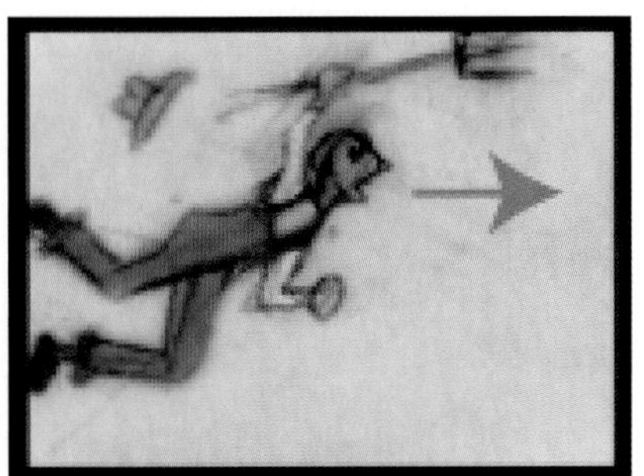

Figure 3.19
Give your character plenty of space to look in the frame. The audience likes to see what the character is looking at.

- When the camera angle is equal in height to the subject, you get the feeling of equality with the subject.
- When the camera angle is looking *down* on the subject, you get the feeling of dominance from the viewpoint of the camera, or that the subject is inferior, smaller, and less important.
- When the camera angle is looking *up* on the subject, you get the feeling of inferiority from the viewpoint of the camera, or that the subject is more important, larger, and superior.

3.2.6.3 Balance

Each shot must be scaled and balanced appropriately. If you have a big, heavy object on one end, add a smaller object on the other side to balance the shot. Objects that are closer appear larger and heavier in the shot. It is fine to use objects that are farther away and therefore lighter in the background to balance heavy object in the foreground.

Another aspect of balance is *leading looks*, meaning that you have allowed for the compositional weight of the look. If you have a side view of a person looking to the left, put this person on the right end of the frame to allow room for the look to occur. Objects such as TV sets, vehicles in motion, and subjects with slanted compositions also need to have room ahead of them for good balance.

Color affects balance because our eyes naturally go toward bright or white colors in the frame. Keep this in mind as you compose your color maps and shots. Try to arrange your frame so that the viewer's eye goes to the brightest area first. Brightness gives an object extra weight in a composition. An object with greater mass may be used to balance out one with brighter colors.

3.2.7 Color and Your 3D Animation

Color is another area that can make or break your animation. Study color theory (if you haven't done so before) to get an idea of how deep this area can go. You need to develop a limited palette of color for your animation and stick to it throughout the animation. This will also make it much easier for you when you are texture mapping late into the night and have to once again choose a color for something.

Try these color tips to help make your 3D animations look more professional:

- Stay away from really bright, highly saturated colors. These won't look good on NTSC and generally will give your animation a less expensive look and feel. If you do need a bright palette, try using one from a TV show like *The Simpsons*. These colors are proven safe bets.
- Use lots of neutral tones such as browns, grays, golds, blacks, and whites with small touches of brighter colors. Study movies such as *Dune*, *Dark City*, *Alien*, *The Matrix*, and *Blade Runner* to get a better sense of monochromatic color schemes that have high emotional impacts.
- Scan a painting using a palette you think fits the mood of your animation and has a limited color scheme. Use the eyedropper from Photoshop and pull colors off the painting for your 3D animation palette range. Artists with some good palettes to start with are H. R. Giger, Salvador Dali, or Bosche. A good reference book to look for traditional palettes is Gardner's *Art through The Ages* by Richard G. Tansey and Fred Kleiner (Harcourt Brace).
- Grab a frame from a favorite movie that uses color powerfully and use a palette from it as a guide. You may want to grab two or three different frames with slightly different palettes to cover multiple sets in your animation. Choose films with similar moods, looks and feels.

Figure 3.20
This example has a character with a palette in which the colors are limited to similar hues.

Credit: Illustration by Donnie Bruce.

Figure 3.21
This painting is similar to the one we will use for our Media Man story. The painting has a nice palette but isn't our first choice. Due to copyright issues, we cannot show you the H. R. Giger painting we are actually using for palettes on this project.

Credit: Illustration by Maia Sanders.

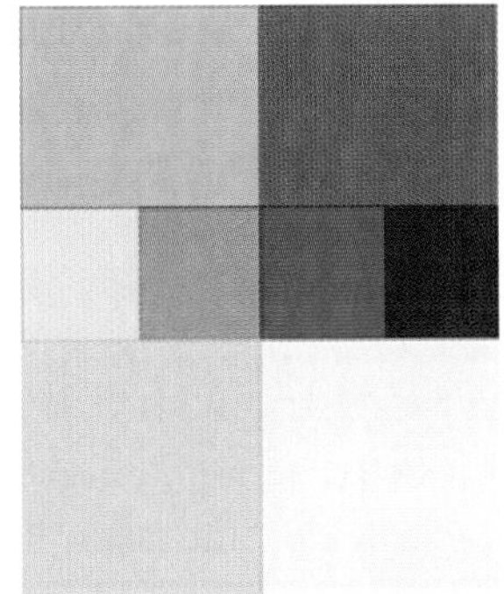

Figure 3.22
We pulled this bright palette from the painting after it was scanned and imported into Photoshop. Use the color picker to pull out a limited number of colors to use for your animation.

Figure 3.23
Notice how all the colors automatically go together really well.

Most of our favorite palettes are pulled from famous paintings and films that we cannot show you here because of copyright reasons. Some movies we are currently using for pulling palettes include *Phantom Menace*, *Dune*, *Blade Runner*, *The Fifth Element*, *The Matrix*, *Dark City*, *City of the Lost Children*, and *Baraka* (great outdoor and visual theme palettes from exotic places).

Don't grab palettes from a movie such as *Blade* for a lighthearted children's animation. You would grab palettes off of movies such as *The Lion King* or from TV shows such as *The Simpsons* to appeal to that lighter emotional level.

- Collect film design books and scan palettes from the color pictures of actual shots if you are unable to do high-quality video grabs. A number of production picture books have been released to coincide with the new *Star Wars* series, and these books have wonderful color schemes from which you can choose for your 3D animations. These color schemes are tried and tested colors guaranteed to look good on film, TV, and computer monitors. Keep a notebook full of colored shot copies, accompanied by 5 to 10 color RGB-coded palettes, to choose from for different 3D projects.
- Choose a four- or five-color palette with colors that go together in the same hue range and use only these as much as possible. Collect color palette books with RGB values and four to five color-swatch combinations used by graphic designers to make sure your colors go together.
- All monitors have slight color variations. Get your system vectrascoped to make sure it is calibrated correctly; you will avoid unpleasant surprises when you go to a professional post house to output to tape, such as discovering that your purples are actually more of a pink! If you are working on different computers or with a team, use the RGB color value numbers with your palettes to make sure everything will be the same color in the end.

3.2.7.1 Color Maps

2D animators have used color maps for a long time. Disney started the trend, and you can find many versions of color-map formats floating around today. A *color map* is a grid of colored bars for the length of your animations that tells you exactly how the colors are going to change from scene to scene and throughout the length of the piece. These helpful visual road maps give you a great understanding of how well the color scheme will hold together as a whole.

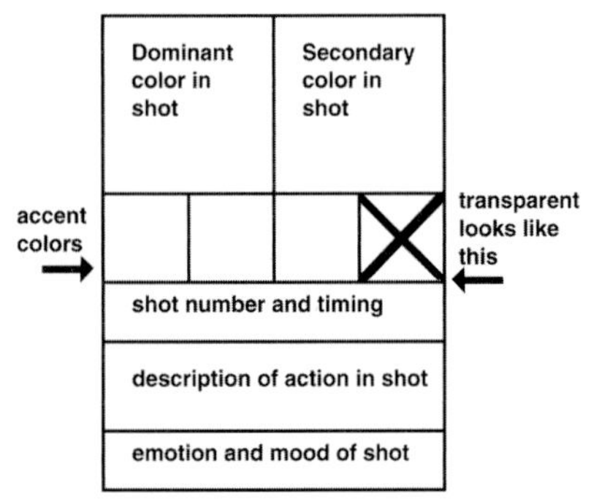

Figure 3.24
This example shows one way to create a color map for an animation.

One way to create a color map for an animation is for each scene to have a set of boxes to describe the colors used in the shot. The first box is the color used most, the second box is the color used nearly as often, and the series of little boxes show accent colors. You should be able to glance at a color map and see where the major plot points occur or where big emotional changes occur.

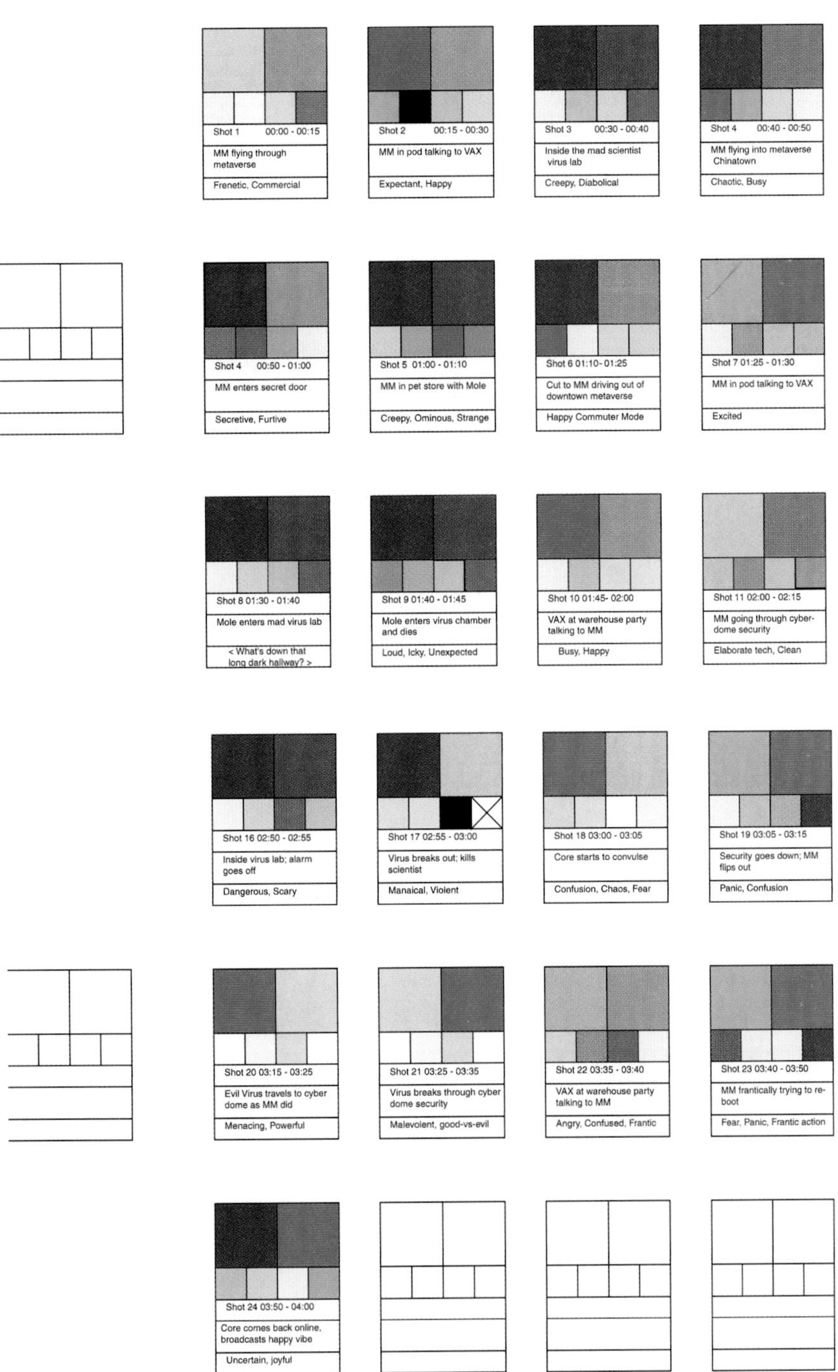

Figure 3.25
This color map is for the Media Man story. See the storyboards near the end of this chapter to see how these colors correspond with this color map.

3.2.7.2 Emotion Maps

Add a line in your color maps for the emotion of the scene. The color map can get a great deal more complicated if you want to go a step further and map emotions of each character, the scene, and what you want the audience to feel because the emotions can all be slightly different. Evoking emotion is the main goal in 3D animation, and having even a simple emotional game plan will help you determine everything from lighting choices, to music, to character animation.

3.2.8 Sound

Sound is at least 50 percent of the emotional impact of your 3D animation. Try watching *Star Wars* with the sound off and see what you think. Good sound can sometimes save bad animation. Bad sound can kill good animation. Great sound and great animation are magical.

Make friends with professional sound people or amateur audiophiles who work for cheap. You will need some help creating a good soundtrack for your animation if you haven't done much sound editing before. Sound is one job that animators shouldn't feel they should have to handle alone. It is too important and takes people who live and breath kilohertz and decibels to do it justice. Many professional sound people enjoy doing soundtracks for animations to get exposure and do something different for a change. They will be more inclined to help you out if they like the story and feel it has potential to be successful.

To create a rough sound track to animate, you can use a program such as Macromedia's Sound Edit, but try to find experienced sound people to create the final cut. Proper sound effects and background music will greatly enhance the viewing pleasure of your 3D animations.

3.3 Principles of Animation

The following "Disney 12" principles of animation are good things to keep in mind when storyboarding and planning your shots. Try to use as many of these as possible to keep your animation more of an animation art form rather than a technical execution without personality or soul. Read *The Illusion of Life* (mentioned in Chapter 1) for a deeper explanation of these ideas.

1. **Squash and Stretch.** Fill a flour sack with flour and drop it on the floor. The bottom stretches out while the top squashes. Take this same sack of flour and hurl it through the air in slow motion, watching as the flour inside changes shape, depending on velocity and impact. A bouncing rubber ball changes shape in much the same way: When it hits the floor hard, it flattens out. The most important thing is to maintain the same mass. If one part gets bigger by being stretched, another part gets smaller by being squashed.

2. **Anticipation.** Plan sequences of actions from one event to the next. Precede each major action with one that introduces it to the audience so that they don't become confused or wonder what a character is doing. Before your alien reaches for a small critter to eat, have him first look at the critter for a moment and reach out his arm.

3. **Staging.** Every visual detail in the frame needs to help convey the story line and emotion. If you are creating a futuristic haunted house, use dark colors, shadowy creatures, howling sounds, and other scary things. A bright flowerbed would be out of place.

   The same goes for staging actions; you need to direct the audience's attention by showing them precise shots using close-ups and clear actions whenever possible. Your character animation should always look good in silhouette to help the viewer see clearly what the character is doing. Render out an alpha channel movie of each scene where there is heavy character animation with just the character. If you can't tell what the character is doing and what the attitude is of the shot, you need to change it until you can to improve your staging. Study Charlie Chaplin films. He also believed that if an actor really knew his emotion, he could show it in silhouette.

4. **Ease In, Ease Out.** Also called *slow in, slow out,* the timing of the motion of the shots changes between keyframes to give more of a realistic movement and spirited result.

5. **Arcs.** Avoid mechanical precision in your character's movement. Movements of most creatures follow circular paths. Move in arcs between positions rather than straight lines. Think of how your arm is in a socket, which causes it to move in arcs. Most creatures have this type of arcing biological construction. Keep this in mind when setting keyframes for characters. Straight in-betweens run the risk of killing the essence of the action.

6. **Secondary Action.** This action accompanies the main action for added impact. If your 3D dragon is going to cry, form a tear in its eye, and then have his hand come up and wipe away the tear while his face gets sad and his lips start to quiver. If the dragon were to just sit there with tears streaming down its face, it wouldn't be as emotional.

7. **Timing.** In the context of the Principles of Animation, this term refers to the number of in-betweens to time action with character animation drawn frame by frame. The fewer in-betweens, the faster and more abrupt the action.

8. **Exaggeration.** Take a normal set of actions and make them extreme. This is done without distortion but with more of a hyper-realistic exaggeration.

   With the spider, for example, you could take a normal walk cycle and make every like move slightly exaggerated to give it more impact. Try to exaggerate

Figure 3.26
This spider would be fun to animate with some extreme finger-crawling exaggeration.
Credit: Illustation by Debbie Rich.

Figure 3.27
This colt character would have lots of overlapping action opportunities with his big floppy ears and long hair.
Illustation by Debbie Rich.

in ways that fit with your character's essence. For example, a restless, rebellious spider would move faster than a sad lonely one.

9. **Solid Drawing.** Does every frame in your animation have weight, depth, and balance? Avoid *twins*, in which the hands or legs are parallel and in the same position. Beware of plastic, perfect-looking creatures in your 3D animations. Make sure each pose has the feeling of potential activity. Try to make each shape of your character slightly off-balance with an animated mass, such as a flour sack of some kind. This way, all the parts will move and bend with the main frame of the character. Avoid symmetrical shapes. Add contrast in form and shape to achieve an active type of balance. Working in 3D makes this harder to avoid than in 2D, where each frame is redrawn.
10. **Appeal.** Live actors have a screen presence full of charisma. Animated characters need to have the same thing, which is often referred to as *appeal*. Appeal doesn't mean cuddly and cute. It refers to anything that a person likes to see: quality of charm, exceptional design elements, simplicity, communication, and magnetism. All villains should have their own type of appeal, or you won't want to watch what they are doing.
11. **Follow through and Overlapping Action.** The main idea here is that all things don't stop moving at the same time. When a character such a Jar Jar in *The Phantom Menace* comes to a stop, his long flapping ears continue to sway into place. Try to design your characters with some overlapping action if possible. Although it will be more complicated to animate, these details will make your characters look more interesting on the screen. If Jar Jar had short, stiff ears, his lack of overlapping action would have made him less fun to watch. You can add overlapping action by using folded rolls of flesh, tails, sagging body parts, or hair. Clothing such as capes, long dresses or flapping scarves is another area that can take advantage of this principle.
12. **Pose to Pose and Straight Ahead.** These are terms for the drawing process of 2D animated keyframes. In computer animation, *pose to pose* action involves keyframing each main pose and tweaking all the in-between action, which is great for keeping a shot within the timing. *Straight ahead* action is starting with the first pose and setting keyframes along the way, which can be less accurate and may take longer to perfect. These terms may sound confusing if you haven't done much animation but will become clearer when you start setting lots of keyframes for your 3D animated characters.

3.4 Pitching Storyboards for Feedback

Pitching storyboards to an audience is an art form in itself. You will need to pitch your semi-final storyboards to a variety of people to get valuable feedback. Keep in mind that making changes is much easier at the storyboard level than in the middle or end of production on a 3D animation. If the story doesn't work on paper, it will most likely not work when done as a 3D animation.

Look at the rough, colored storyboards for "Media Man in the Net." Although these aren't all of the shots, we included most of the main ones in the first half of the story to give you an idea of how storyboards might look. It is better at this stage to do rough ones and pitch for feedback rather than to spend time creating really detailed final ones that you may have to change a great deal in response to feedback.

Try to keep the following tips in mind when pitching storyboards for feedback. The more engaging your presentation, the better your chance of getting good feedback for changes that will make your 3D story better.

- Pretend whoever is listening is a five-year-old child with a short attention span.
- Don't talk about why you chose to create the animation, or about its technical aspects, camera-shot decisions, or any other extraneous information during the story pitch. People have a hard enough time getting around a new story; it isn't helpful to load them down with extraneous information before they even understand the basic premise. After your pitch, you can answer questions.
- If your audience is a small group of fewer than three people, encourage them to ask questions during your pitch. Otherwise, they may forget their questions or suggestions if they have to wait until the end. They also may not understand a key point and be lost during the rest of the story pitch. Odds are that if they are stuck on something, other people are, too, or at least some percentage of your audience will be confused later. Remember that your goal is to get feedback.
- Start with your story concept sentence to orient the audience to the basic idea first without giving away the climax or ending. If you just jump right into the first panel, they may get too confused by the setting or characters to understand the plot points you are presenting visually.
- Because storyboards don't easily convey the essence of characters, briefly introduce the main characters before starting the story pitch to the audience. Tell them which actor would play that character's voice to give them an idea of what type of essence the character is going to have in the final version. This explanation should come after the story concept sentence and right before you start to take the audience through each panel.

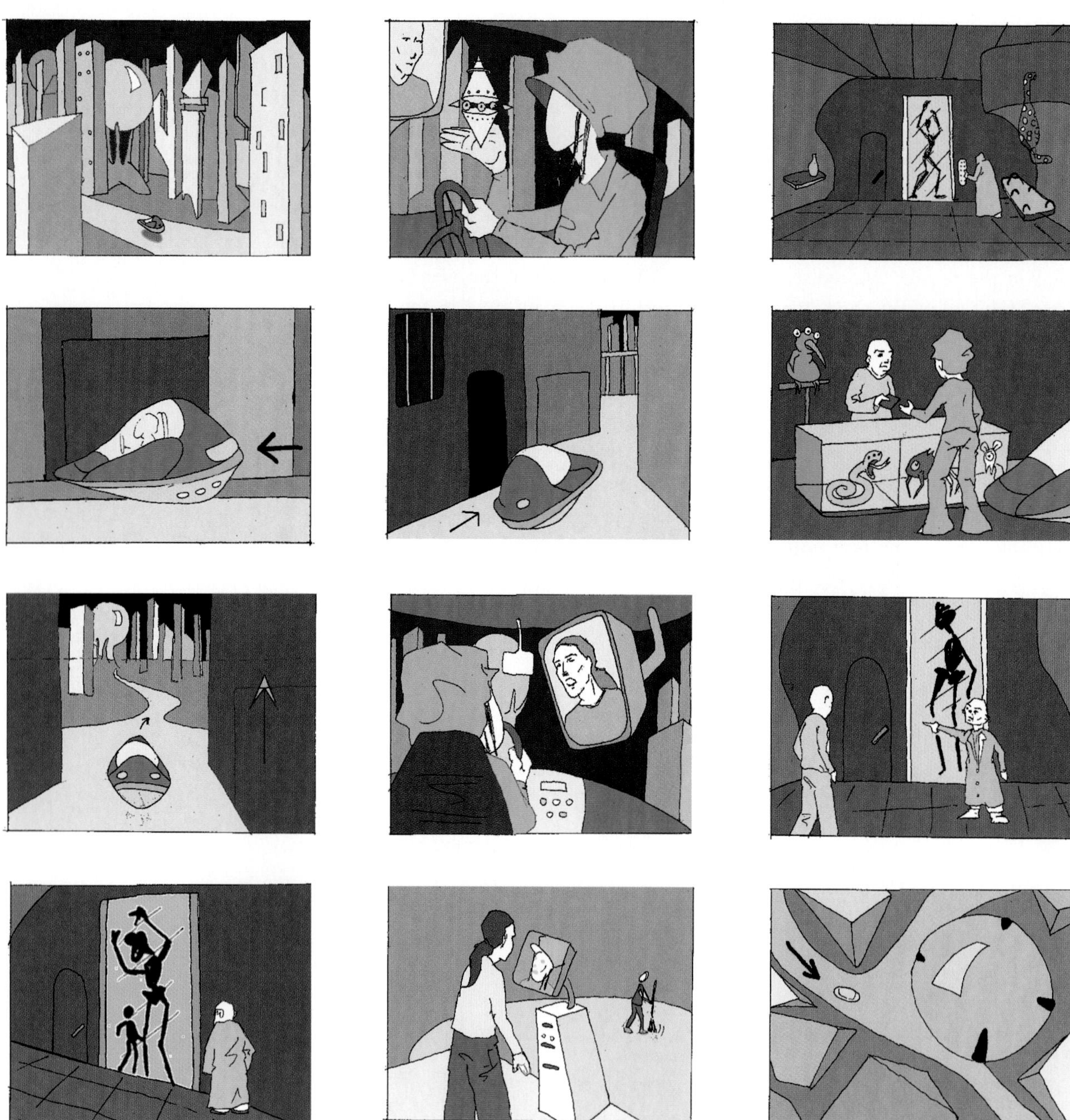

Figure 3.28
Rough storyboards for "Media Man in the Net."

Credits: Illustrated by Sam Miceli. Colored by Donnie Bruce.

- To increase the involvement of the people hearing the pitch, act out the character's voices and movements.
- Point at each panel as you tell the story to make sure everyone knows where you are visually. Draw one panel for each important shot, even if you have to copy one. Never point back and forth through your storyboards to explain some aspect of the story. Go in a linear, easy-to-understand order.
- Try to tell your story in the same amount of time it would take to see it as an animation. This will help you check the timing and pace of your story. You will also know your story is too long if it takes you 10 minutes to pitch briskly and it's supposed to be three minutes long.
- Make sure your storyboards are easy enough for everyone to see from at least 10 feet away. Use strong, dark lines and good contrasts between lights and darks.
- Pay attention to when the people hearing your story laugh, look concerned, or show any emotion; make a mental note of it to see if what you are trying to evoke is working. If the people hearing you pitch your story seem really bored or confused, you need to find out why and make major changes.
- Ask your audience these questions after the pitch: Did you like the story? What would you do to make it better? Was there anything you didn't understand? What were your favorite and least favorite parts? Did you like the main character?

  You should also ask yourself all the questions from the "Using the Story Concept Checklist" in Chapter 1 before pitching the story to anyone. Be honest with yourself and be prepared to spend some time making the story better. Most screenplays are rewritten from scratch a bunch of times before they make it into production. This is your visual script that requires the same dedication to revisions.

3.5 Animatics

After you have pitched the storyboards to at least five different people and made changes based on their feedback, pitch it again and make changes until it's as good as you can possible get it. Then you are ready to create an *animatic,* also commonly referred to as a Leica reel or story reel. The word *animatic* can mean many things these days. Disney starting creating them to test story ideas and character animation before final drawings were created.

3.5.1 Basic 2D Animatic

The first step for creating a basic animatic is to scan your storyboards, do a rough audio track with dialog, and put the storyboard and audio together in a program such as Adobe Premiere with the correct timing. This will result in a slideshow type of movie that will tell you how well the story is working on a different visual level. It is surprisingly easy to tell how well a 3D animation is going to work from even a simple animatic. Plot problems and timing issues should be readily apparent. Make the necessary changes after showing this version to several groups of people again.

3.5.2 Moving 2D Animatic

Some animatics have layers of 2D elements that move around to show motion and test timing for movement. Usually a character will slide around the screen, or a background will pan across the screen. This type of motion requires storyboards drawn in the shape of the pan to work correctly. If you have a shot that incorporates a long horizontal pan across a city street, draw a long horizontal city street panel and animate it across the screen. You might also want to draw characters or moving objects on different layers to animate around the screen to test timing. You should make one of these to test your 3D story before deciding that the idea and timing are solid.

3.5.3 Moving 3D Animatic

Some final-looking shots are still considered animatics these days. 3D animatics used to involve simple 3D primitives as place-holders for, for example, spaceship fleets traveling across the screen, with a rough motion path to check timing and camera shots. They have become more realistic, modeled in detail, texture-mapped, and lit in larger production houses because 3D is getting faster and easier to do. It is still easier to plan out rough camera shots and make changes with simple 3D objects than it is to experiment with heavy final rendered versions.

3.5.4 Getting to the Final Animatic Stage

After you have a successful final animatic, even if it is only a sliding 2D one, you will feel more confident during the computer production stage that your story works and will have a good chance at being a success. It isn't a short or easy process to get to this stage, but it is less painful than spending months on a sketchy idea for a 3D animation that doesn't get a good audience response, even if it was a good learning experience.

After you have a tried and tested final animatic, don't attempt to make changes once you start working on the computer.

Stories are like little machines with intricate parts that can't be easily swapped out or changed without affecting the entire structure. Most of the time, changes aren't recommended unless you really know what you're doing. Your goal is to not have to think about whether the story works after you start modeling, animating, and texture mapping. You will have enough on your mind at the computer stage just getting your characters and shots to look right. At times, while animating or looking at shots in progress you may come up with new ideas. These ideas can be dangerous and time consuming if they don't work out in the end. The more experienced you are, the better judgement you will develop about making changes to the animatic during production. Most of the time changes at this stage aren't recommended.

3.6 In Summary

Production design and storyboarding are careers in themselves for many professionals. Try to learn as much as you can about the ideas presented in this chapter because understanding the big production picture will make you a better 3D character animator. You will have more fun animating well-thought-out ideas and be able to fix problems with existing ideas more quickly by understanding how to tell strong visual stories that evoke a good range of emotions.

Always remember though that the steps presented in these first three chapters don't form a linear path; travel through them will occur in a variety of progressions, depending on the animator and the idea. Feel free to experiment with your own order of story concept design for 3D projects and find what works best for you.

Part II

Creating Basic Characters

Modeling an Organic Humanoid Hand with NURBS

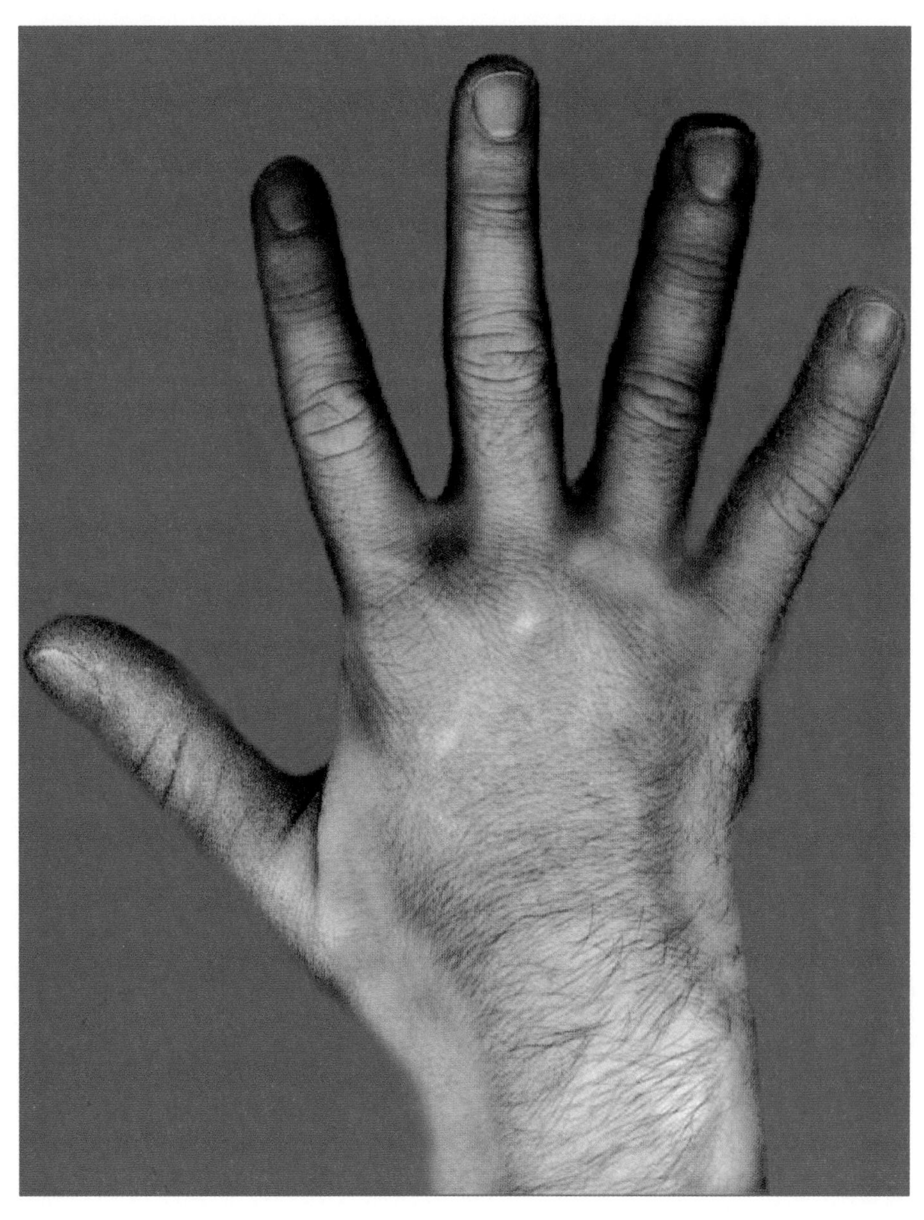

CHAPTER 4

Modeling an Organic Humanoid Hand with NURBS

by Nathan Vogel

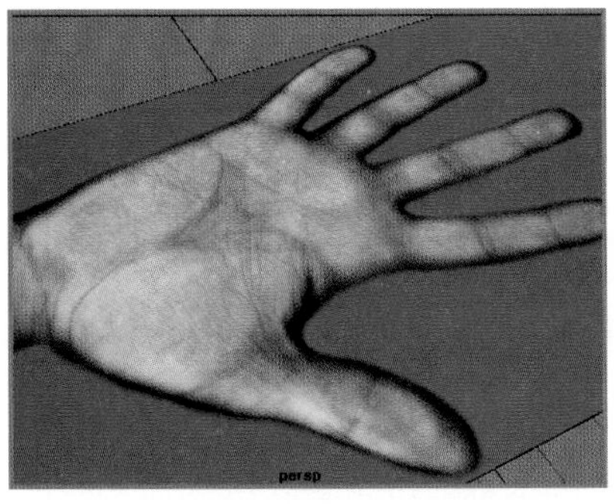

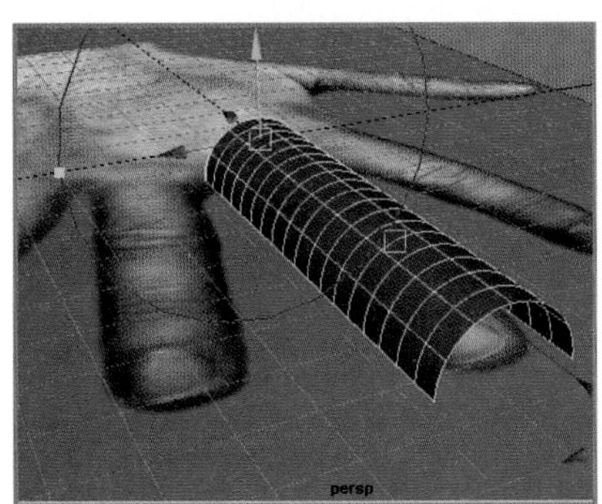

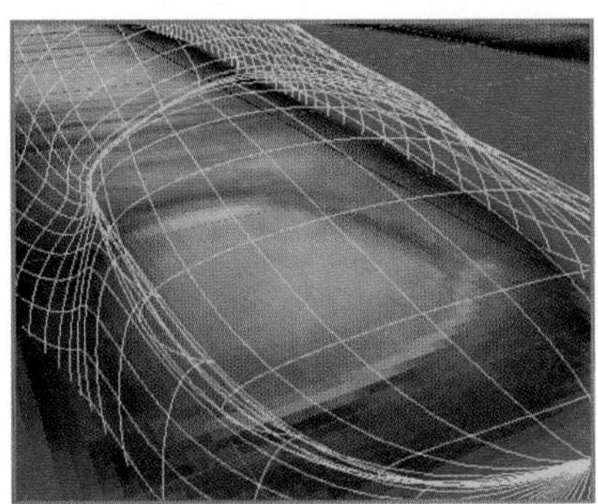

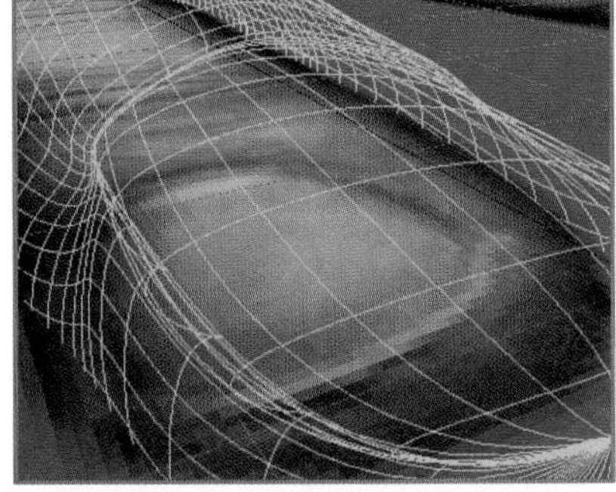

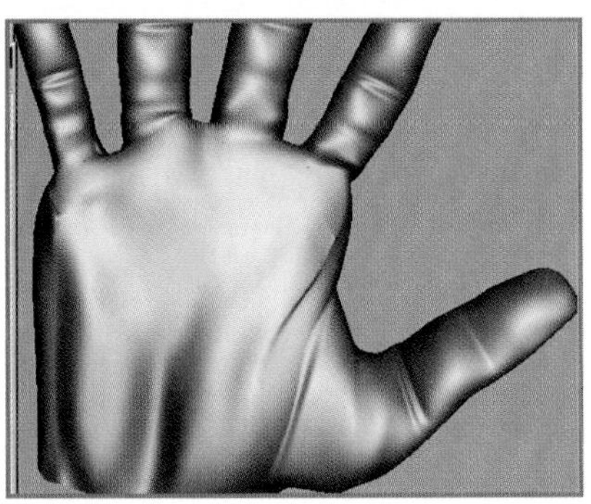

Now that we have covered in depth how to create a story and design characters, it is time to learn how to create digital characters using Maya. Building your models can be one of the most instantly rewarding steps in digital character creation. During model creation, you bring together all the visual characteristics of the character that you designed and begin to turn your ideas into a workable, digitally controlled 3D animated actor.

This book introduces you to important steps that are only the foundation to creating many possible character types. It's a good idea to familiarize yourself with existing literature and tutorials available from Alias|Wavefront. A good place to start would be books such as the *Learning Maya* resource manual, which ships with every copy of Maya.

The exercises in this book focus on creating seamless, photorealistic characters. For the first exercise, we will create an organic, humanoid hand. In creating this important character component, you will be able to use your own physical features as references for incorporating realistic characteristics on your model.

For the second project, in Chapter 5, we will use the hand as a part of the biped modeling tutorial. Another reason to start with a familiar form is that you are your own reference model. For facial and body posing, you can use your own face and body, along with a mirror or a video camera, to assess the range of motion the face and other body parts need so that you can build a model that can be animated properly.

For the third modeling exercise, in Chapter 6, we will model an organic dragon that has the characteristics of a quadruped (a feline/reptile composite), along with two additional wings (such as those of a bat) and a long, articulated tail and neck (such as those of a dinosaur). Its head will also be articulated and have as much detail as you wish to include.

These first three modeling chapters should give you the necessary experience required to solve most of the technical dilemmas that may arise when building NURBS models for your animatable characters in Maya. It is a good idea to proceed through the tutorials in order because they reference previous steps along the way.

4.1 Modeling an Organic Humanoid

This first exercise serves as a good introduction to the workflow required when building organic models. Maya offers many other modeling techniques in addition to the ones introduced in this exercise, so just think of this method as another one of the techniques available in your toolbox.

4.1.1 The Hand

First, we'll start with the human hand. To begin, you need to acquire and align the images of the hand that you will use as reference. On the CD-ROM included with this book, look for the files HANDTOP.jpg and HANDBOT.jpg. If you want to use your own images of hands, be sure to prepare them accordingly by aligning them perfectly in Photoshop. After you have the images in place, you will need to learn the initial stages of how to create a finger from a cylinder primitive, and then how to duplicate and alter it to create a whole set of four fingers and a thumb. Then we'll move on to the stump of the hand, which is created from five primitive NURBS planes that are placed on separate axes at approximately 90 degrees. Finally, we will integrate the individual segments using a combination of Stitch, Fillet Blend, Trim Surface, and Curve On Surface tools.

4.1.1.1 Creating a Finger from a Primitive

For this exercise, you will start with the middle finger and create a NURBS cylinder primitive, placing it over your image plane so that it matches your picture of a hand. By manipulating the control vertices of this cylinder, you will be able to create realistic details and definition for the finger, including a fingernail, muscles, and skin folds.

To create the finger, follow these steps:

1. Acquire an orthographic and symmetrically aligned top and bottom drawing or photograph of a hand. The image should be converted to a compatible image file format and placed into the appropriate Source Images directory of your Maya project. Maya for Windows NT and SGI have slightly differing preferred file formats. For reference imagery, I have found JPEG to be a robust and economical file type. As long as the compression setting is left on 100%, the quality is quite acceptable. Check the user manual for a complete list of compatible types.

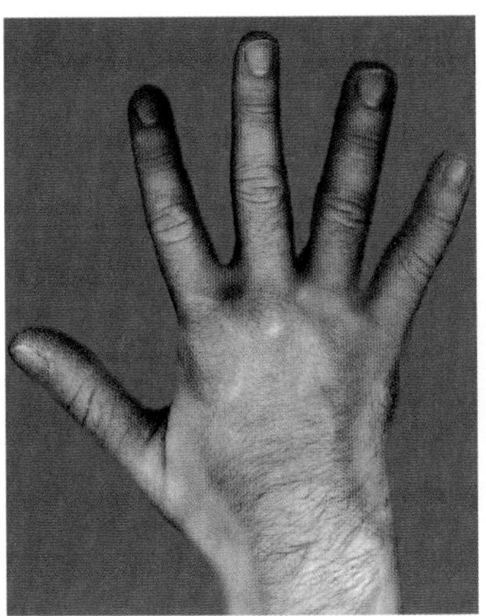
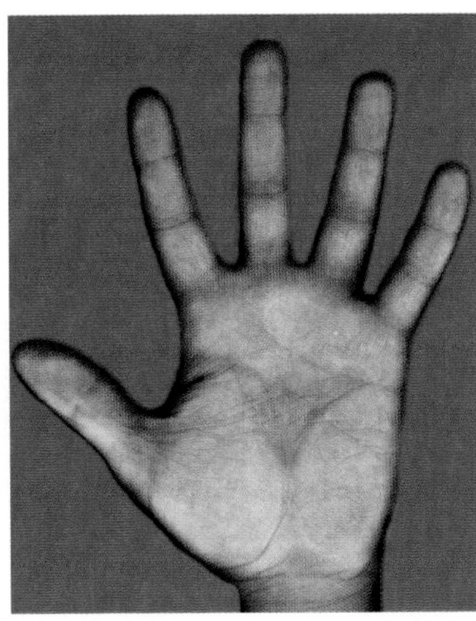

Figure 4.1
Source images for your hand templates.

2. Prepare the image so that the middle finger is in the center of the picture. This will help facilitate easy creation of all the fingers.
3. Create an image plane on the top camera by going into the Cameras tab in the Multilister and double-clicking the top camera. The Attributes Editor for the top camera appears. Scroll down the list of entries and open the Environment pop-up menu to reveal the Create Image Plane button. Click this button to create a new image plane.
4. The Attributes Editor should show the options for creating a new image plane. In the Image Name option, click the Browse icon and then load your top view of the hand.
5. Next you will make a duplicate of the first plane so that you can load the other side of the hand for a reference to the bottom half. To create another image plane for the bottom of the hand, click the Image Plane in the Multilister and, holding down the right mouse button (RMB), choose Edit, Duplicate, w/Connections to Network.
6. Double-click the new image plane and click the Select button found at the bottom of the Attributes Editor window. In the Channel box, adjust the Center X channel to read –.01, which will offset the image plane so that it is directly beneath the top view. Now you should be able to see both top and bottom views when orbiting around your scene.
7. Next, create a primitive NURBS cylinder, naming it "middleFinger," and then reposition it over the middle finger in your image, pressing the T key on the keyboard to activate the Manipulator tool. Click the makeNurbsCylinder option, highlighting it in the channel box's inputs. Set the dimensions to 12×18 sections/spans. Now press the T key to activate the Manipulator tool. With the X key pushed down, engage Grid Snapping. This makes it easier to change the cylinder's construction history.

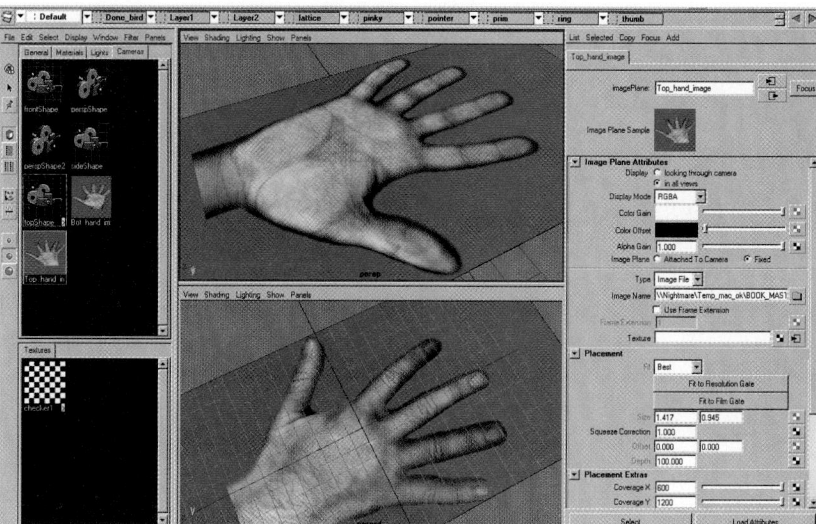

Figure 4.2
Creating image planes with the Multilister and the Attributes Editor.

Note

The abbreviation "RMB" is the accepted term for holding down the right mouse. LMB and MMB signify holding down the left and middle buttons, respectively. Maya requires a three-button mouse to function properly, as some commands cannot be re-mapped to other keys.

8. Snap the top manipulator off of its creation axis so that it matches the grid near the tip of the finger on the image plane. Now change the pivot point of the finger so that the center point is located at the base where it will connect with the palm. Be sure that the Translation tool is active, and then press the Insert key. This changes the Translation tool into the pivot point manipulator. With the X key held down, snap the pivot to the grid point near the base of the finger. Press the Insert key again to toggle out of Edit Pivot mode.

9. Press the W key to activate the Translation tool. Hold down the RMB over the cylinder and choose the Hulls option. This is a shortcut to invoking your component editing mode for NURBS Hulls. Notice how the object has a pink hull line highlighted. If your view is set to Wireframe and the model is set to Low Res (with the 1 key), the hulls and isoparms will be very easy to visualize.

10. Next, select one vertical hull at a time by clicking directly on the pink hull in an area that is not crossing another hull in the side view. It will highlight and position and scale each hull so that its silhouette and geometry distribution roughly matches the finger. Scale the fingernail tip/end hull to zero numerically, using the Channel box, and begin manipulating horizontal, ring-shaped isoparms so that they match the crease regions where the finger will bend.

11. Notice that the points in only one ring-shaped isoparm at a time are selected when using Hull mode. Adjust them so that they are spaced apart correctly over the entire image plane. For the first knuckle, position the three ring-shaped isoparms so that they cover the regions that will later have to bend. The second knuckle gets five isoparms, enough to include the oval-shaped crease in the bottom area of the knuckle.

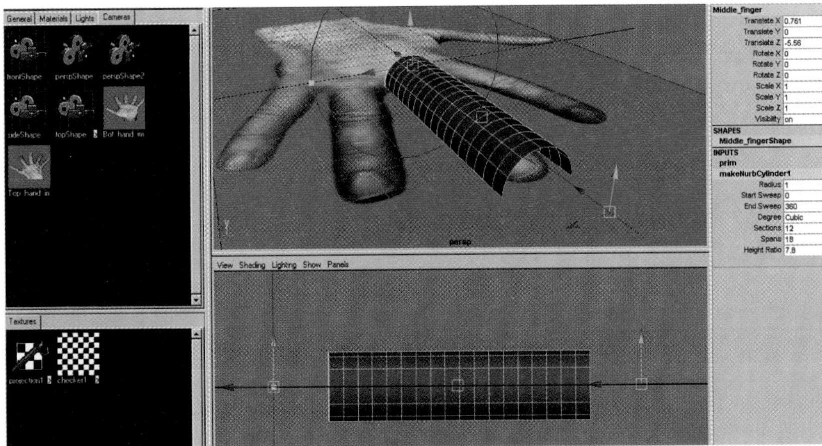

Figure 4.3a
Distributing isoparms with the Hulls mode in the Orthographic side and perspective views, respectively.

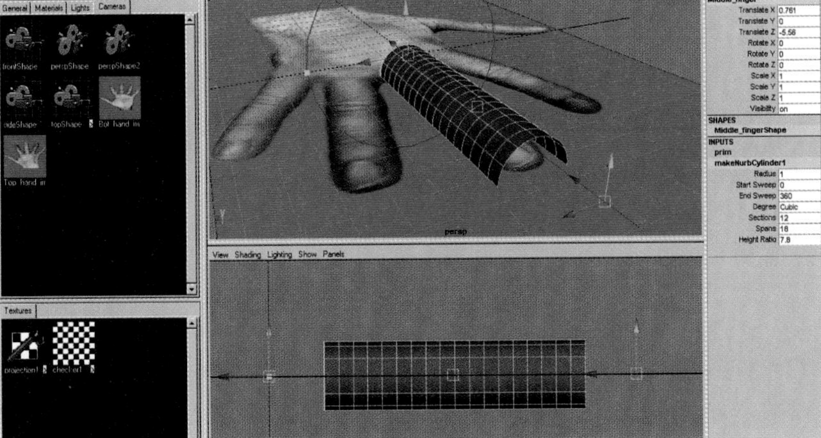

12. To create the area around the fingernail, begin by creating two isoparms and position them close together to represent the cuticle and ridge of skin above the nail. Then add three additional isoparms to form the tip of the finger. To prepare the base of the finger for joining it to the hand, flare the opening at the base of the finger upward and outward a little so that the fillet blend approximates the tangent needed to create a smooth transition. (Later, you will use NURBS planes to model the knuckles within the palm itself, so try not to over-model the base edge of the finger.)

4.1.2 Adding Fine Details to the Finger

Now it's time to begin adding fine details to your finger. You will start by building the fingernail and then adding knuckle bulges, skin creases, and muscle regions on the bottom side of the finger. (The steps necessary for creating these details, which involves careful point-by-point-based modeling, are similar to some of the work that we will complete later to create certain facial features, such as lips, eyelids, and, to some degree, nostrils.)

To create the necessary finger details, we will use a combination of CV editing, hull editing, and, for smoothing accidental creases or adding bulge areas, Artisan's brush-based modeling feature, which you can find under Modeling, Edit Surfaces, Sculpt Surfaces Tool. By opening the Settings window, you activate Artisan's custom brush and setting interface, which gives you access to a wide range of features and settings that enable you to create a variety of customized user preferences.

For basic geometry editing, click the first Settings tab labeled Sculpt. This contains the most immediately necessary tools. Some of the parameters have built-in hot-key shortcuts, as follows:

- **Radius (U).** You can interactively alter Radius (U), the general brush scale parameter, by holding down the B key and click-dragging left or right. The Artisan brush icon should visibly expand as you drag.
- **Max Displacement.** You control this hot key by pressing the up and down arrows. A small arrow should expand from the center of the stamp, indicating displacement height after the stroke's opacity reaches a cumulative 100% for each brush stroke. A *brush stroke* is defined as the duration between which the stylus or mouse button is pressed down until it is released. Opacity settings of less than 100% will cumulatively increase in value with each pass of the brush within the duration of a single stroke, but they will not exceed the Max Displacement value, which is reached in a 100% stroked region.
- **Auto Smooth.** This feature applies an additional smoothing algorithm across any stroked region equal to the strength parameter. The operations are simply the type of displacement you want the brush to have.
- **Smooth.** The Smooth operation spreads isoparms apart in a very organic way and can generally help you rid yourself of even the nastiest pinched regions of accidentally overlapping isoparms.
- **Erase.** Use this operation to remove stroke information applied to the selected model. If you delete history, you lose the stroke information associated with the model.
- **Flood.** This button applies the operation across the entire model uniformly with the value according to the Opacity and Max Displacement parameters.
- **Sculpt Variables.** This reference vector can be a useful way to determine the actual outcome of the operation's particular tool. The U and V vectors enable you to nudge isoparms across the already defined surface of the object. The Axis-Specific choices enable constrained axial displacement. The most commonly used vector is Normal, which is also the default. This option displaces based on the existing normal direction of a surface's individual faces.

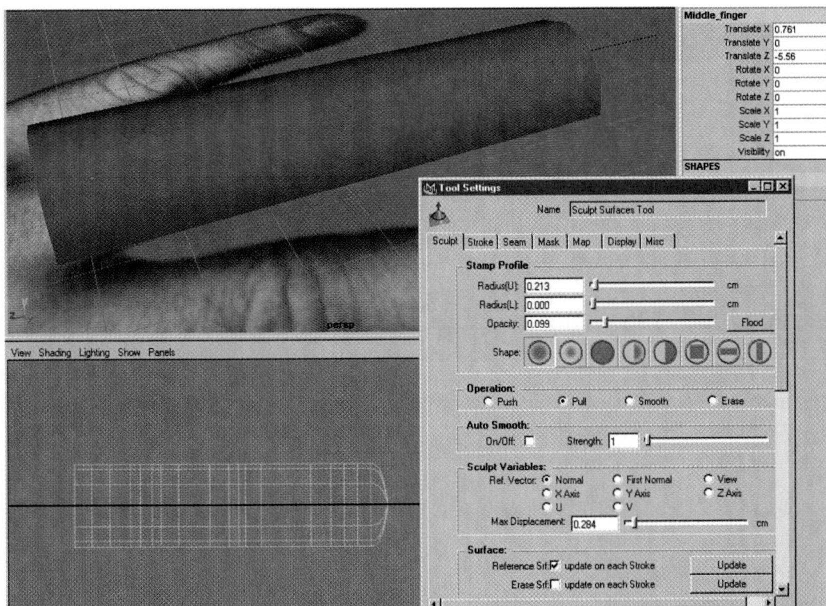

Figure 4.4
Artisan's brush interface.

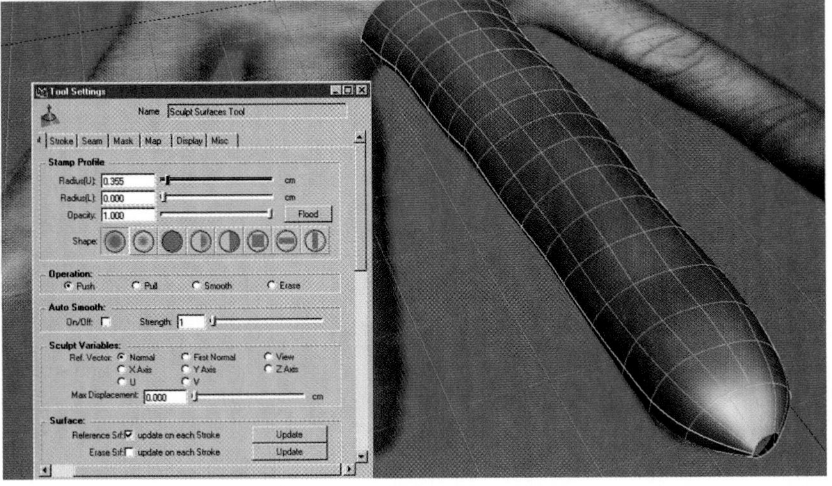

Figure 4.5
The General Prefs Packages window.

In Maya 1.0 or 1.5: If you don't see the Sculpt Surfaces tool in the menu, activate it by clicking Options, General Prefs, Packages, and then click the Maya Artisan check box. After doing that you must quit and restart Maya, so save your document, and when Maya is configured properly, move on to the next step.

4.1.2.1 Creating the Fingernail

Creating the fingernail is slightly tricky, so give yourself the time to conduct some experiments before you commit to a final version. For instance, try to get a feel for curve tension by folding the CVs around. Close-proximity CV editing is an art form that takes a while to master, so don't give up if it doesn't look right the first time. Remember that in addition to saving your work in iterative steps (such as 1a, 1b, 2a, and so on), you can configure Maya to enable infinite undos without hindering performance. You can find this capability under Options, General Prefs, General, where you then click the Queue Infinite check box. You can also duplicate the object so that you have a backup for your next, more refined, attempt after you gain some experience from the experiments.

4.1.2.1.1 Close-Proximity CV Editing Before you start building the fingernail, conduct the following experiment so that you become more familiar with close-proximity CV editing. Press F8 to go to Component mode or CV Editing mode if you have not already done so. Zoom up the fingertip and check out the isoparm's layout. Select and translate the middle CV on the 15th row from the base of the finger right below and behind the CV on the 14th row. Use the arrow keys to traverse the rows and sections, one control vertex at a time. Select the middle CV at row 14, and place it above the tip of the finger. Create the fold again to the CV near the edge of the cuticles to define the width of the fingernail.

Now continue this close-proximity folding process on the spans to both sides of the central span. This forms the completed top part of the cuticle and upper lip of the fingernail in an elegant and low-resolution, one-piece solution.

Before we move on to the next step—reproducing a version of this effect for the edge of the cuticle—you must decide on some characteristics. Does the nail stick out over the cuticle on either side, or does it tuck into the cuticle? Both phenomena are common and sometimes are found on the same hand. Although these two characteristics may seem subtle, they do influence the overall style and look of the fingertips.

After you decide how the topology of your fingers will appear, use the preceding cuticle steps to continue the technique on both sides to create a realistic-looking cuticle and fingernail. Pinch the tip of the fingernail area close together, one CV at a time. Take care to keep the fingernail ridge CVs all on the same isoparm.

Either refine your best experimental fingernail or start fresh with a duplicated backup and make the final version. By going through these basic close-proximity CV folding exercises, you will eventually learn to create detailed low-resolution one-piece fingers and nails.

This technique produces a detailed but not overly heavy model. The fingernails are not meant to be perfect when viewed close up. If the nails are to be a more central part of your character, it is a good idea to create a loft object that represents the nail itself and place it directly on top of the surface of the nail area. That way, if you need it to morph to become a claw, for example, it is more flexible.

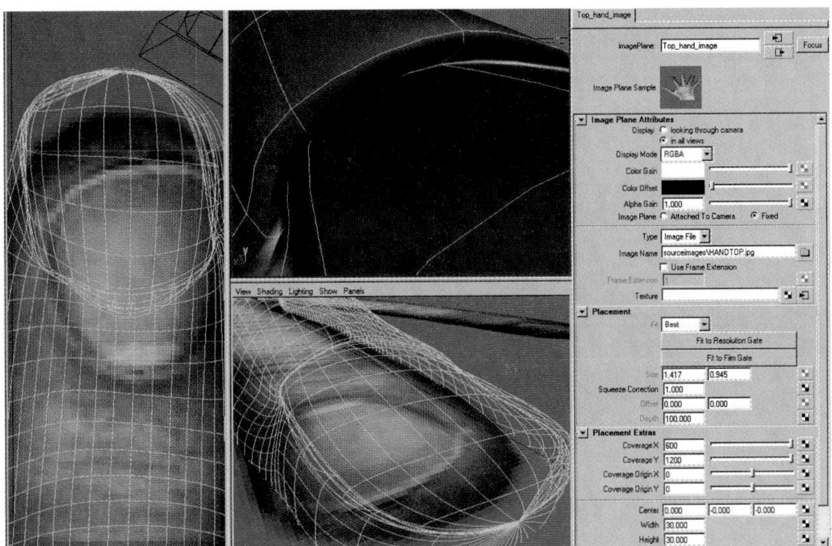

Figure 4.6
Modeling the fingernail.

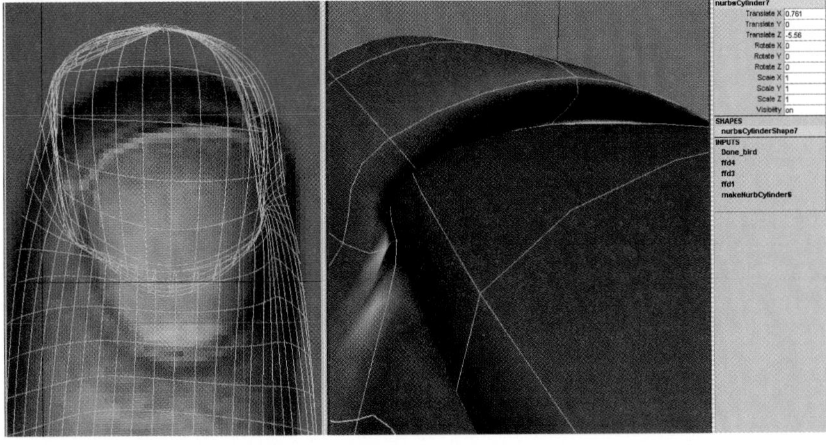

Figure 4.7
Folding the CVs.

4.1.2.2 Additional Details

Muscle, knuckles, and bones create bulges and ridges that you can achieve in a variety of ways. You can manipulate the CVs a few at a time, use the Artisan brush to mold and sculpt them, or use the deformation regions found in the Animation, Deformers menus. Generally the most useful technique is to use lattice deformers with some Artisan mesh tweaking. Sculpt deformers (different from the Sculpt Surfaces tool found in the Modeling, Edit Surfaces menus) are also quite useful for creating lumps or knuckle bulges. All these techniques can be explored and intermixed quite creatively.

Remember, you can place a deformer on a whole object or on a collection of highlighted CVs. Giving a deformer to only a section of a surface enables more specific local control. However, it also introduces the possibilities of harsh transitions between areas, so practice this technique with caution. If the deformation proves undesirable, you can always simply delete the deformer itself and try again.

To add additional detail to the finger, follow these steps:

1. Select the finger and go to Animation, Deformers, Lattice. In the Settings box choose the Group Base and Lattice Together option and click Create. It's a good idea to label each new addition to your project (and all the associated groups and base objects that come with creating deformers) with a unique and recognizable name, such as Finger_Lattice1, in case a group of animators will be working on your project with you.
2. Select the new lattice in your viewport and click the Shapes inputs found in the Channel box. Find the new ffd listed, giving it the dimensions s2×t2×u8 so that eight divisions go down the length of the finger. Set the Pick Masks to All Off, and then, holding down the RMB, choose Lattice-On under the Deformers mask.
3. Move to the side view and, with the lattice selected, hold down the RMB in the viewport and choose Lattice Point when the marking menu appears.
4. Select entire vertical rows (four lattice points per row) and scale or translate to contour the whole finger on a large overview level.
5. Now, from the perspective view, use the Modeling, Edit Surfaces, Sculpt Surfaces Tool Artisan brush and apply fine details, such as asymmetrical lumps on the knuckles.

If at any time you edit the mesh in a way that causes the geometry associated with a particular lattice deformer to go outside the area inside the lattice, you can edit the scale value of the lattice's base object. Assuming you chose the Group option when creating the lattice, open the group in the outliner and choose the Lattice Base. Then, scale it so that all the geometry associated with the lattice falls within the base's bounding box. This lets the deformer know what to include within its calculations.

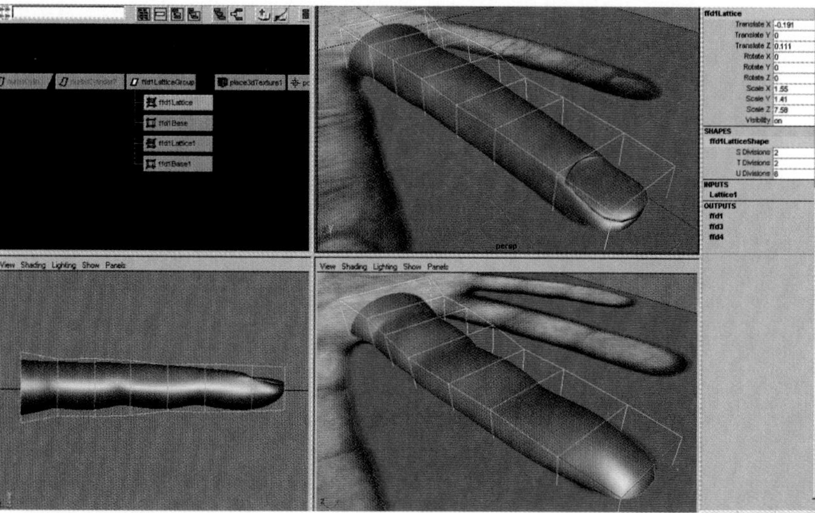

Figure 4.8
Adding a lattice and then deforming it to shape the finger.

You can create skin creases in a number of ways, both with geometry and by texture mapping a bump or displacement map. To create skin that will actually bend, fold, or uncrease, it is a safe bet that modeling the crease will provide the most realistic result. You can use textured creases to create micro folds and non-obtrusive regions of skin. You can use mesh deformations to achieve such results: Try combining a lattice with careful manipulation of the hulls from the side view using Move and Rotate on the individual rows of lattice or mesh CVs.

1. From the Outliner or Hypergraph, select the finger and go to Animation, Deformers, Lattice. With the new lattice selected, click the new ffd listed in the inputs found in the Channel box and give it the dimensions s2, t2, u20 so that 20 divisions are visible down the length of the finger. If your lattice does not look this way, then experiment with putting the 20 divisions in the "t" or "s" parameters. The most important objective is to have the lattice encompass your finger, giving convenient editability.
2. Enter the Component mode (press F8) and set Pick Masks to All Off. Then RMB over the Points mask and choose Lattice Points. From the side view select one whole vertical row at a time (four lattice points per row).
3. To determine Maya's row count, when in Component mode select a row of points and click the words "CV's (click to show)" in the Input area of the Channel box. (If determining s, t, and u values for a lattice, the third digit is the u division number.) Then rotate row 13 about –15 to –25 degrees on the X axis. Do the opposite to row 11, rotating it from 15 to 25 degrees. This should help you generate the contour of the skin creases. Remember to use the Perspective view and tumble the camera around the whole object to ascertain your progress with the finger as a whole.

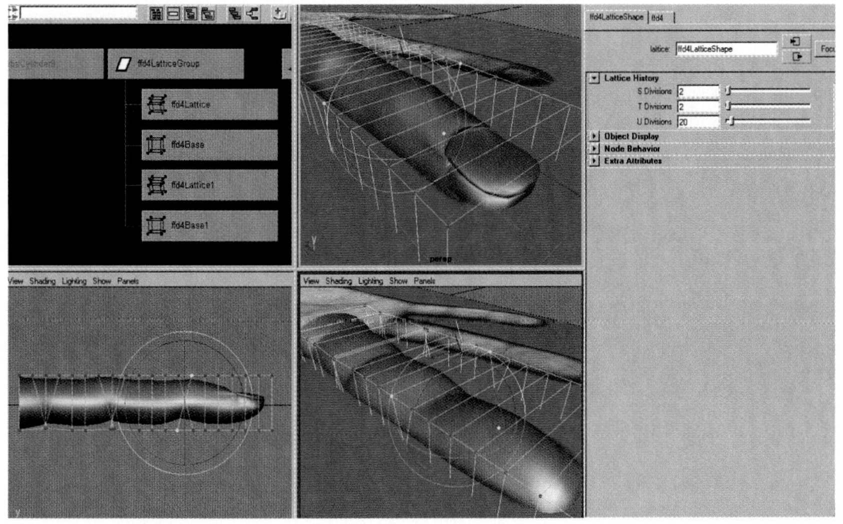

Images 4.9
Skin creases with lattices, and a way to determine row count in Maya.

4. To tweak the mesh into a fully animatable shape, add any last deformers that might be needed or any last extra isoparms. In general, it is usually better to add isoparms in the "U" direction (across the Z axis, much like rings on the finger), as this adds less overall data across the whole object. Adding isoparms in the "V" direction spans the length of the geometry and covers more area per isoparm.

5. After completing all your edits, save the file as the next iteration, select the finger surface, and choose Edit, Delete by Type, History. The goal is to get rid of any inputs to the object before integrating the finger into the rest of the hand. It is important to get rid of construction history before binding skin and duplicating because the extra information and connections tend to bog down a file. This is true especially when you are going to use this finger geometry as the source for all the other finger-like appendages that you must build.

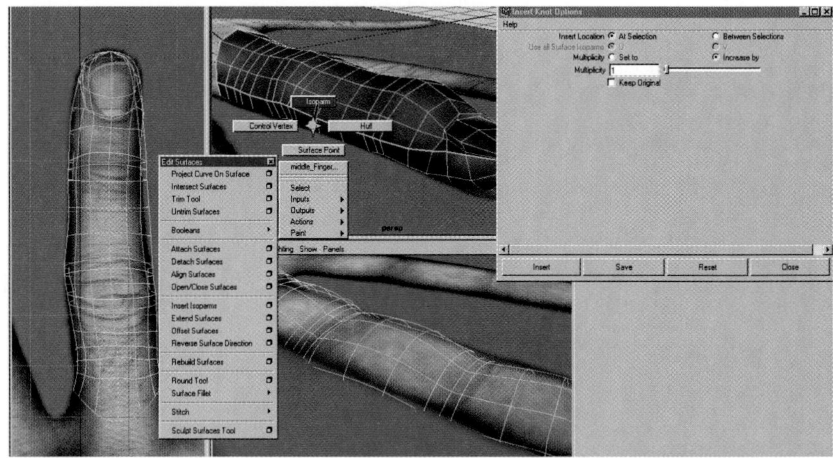

Figure 4.10
Adding more isoparms, and deleting history on the finger.

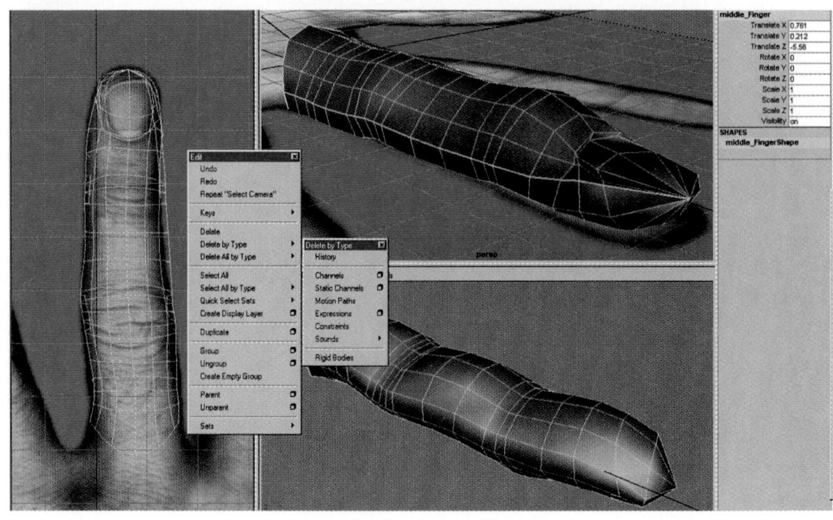

Note

Keep inputs if you are sure that you will need to alter or animate them later, or if you are creating the geometry on the fly using an active process, such as a fillet blend that uses isoparms or trim curves as source geometry. In that case, it is imperative that you keep history active for those types of functions if you want your model to react properly when you begin animating.

At the time of this writing, Maya doesn't have a truly useful History Manager, or a way of selectively deleting input histories. Therefore, you must plan the order in which you delete your input histories very carefully so that you can have access to the necessary inputs. By the end of the process, expect to have at least 22 isoparms across the length and about 12 around the circumference of the finger's surface.

4.1.2.3 Checking for Animation Characteristics

Now you're ready to test your bind skin option to check for basic animation characteristics. To do so, take the following steps:

1. First, set your object to low resolution (display mode 1) so that it is easier to figure out where to place the joint's pivot points.
2. Create a few joints in the side view that correspond to the same general proximity and characteristics of the bones in your own hand. Go to Animation, Skeleton, Create Joint Tool. Use the auto orientation set to XYZ (the default) and then click once to place each joint's pivot point where the knuckles are in the finger. (If the bones are too large or too small, you can easily alter them with the Display, Joint Size, Custom command.) If you want to learn more about creating animatable skeletons and character skin binding setups, read Chapter 10, "Animating Characters and Keyframing."
3. Select the root joint and the finger surface and choose Animation, Skinning, Bindskin, Rigid Bind. From the resulting options choose Complete Skeleton and Color Joints. (See Chapter 10 for more detail on how to bind skin.) Next, select the two middle joints and choose the Animation, Skinning, Create Flexor option. From Selected Joints choose Create Flexor and then click OK. With the tip joint selected, Shift+select each joint in the hierarchy so that it is highlighted in the Hypergraph.
4. Now, rotate the root joint on the Z axis by clicking the Rotate tool and the blue ring, and then dragging to the left with the middle mouse button. Notice that you can rotate each joint at the same time and test the range of motion. If the bones rotate a little strangely, it is because the Local Rotation Axis is out of alignment. This is a common and easy-to-fix by-product of the Auto Rotation Axis option that, by default, is "On" in the Create Joint tool (we will discuss how to fix the Local Rotation Axis later in Chapter 10).

Figure 4.11
Testing binding skin to a finger skeleton.

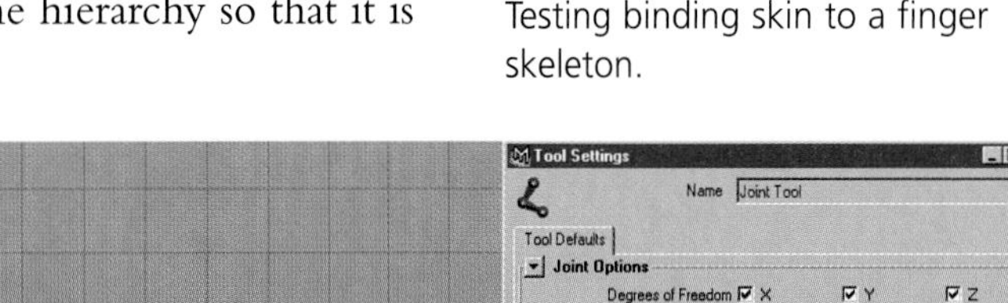

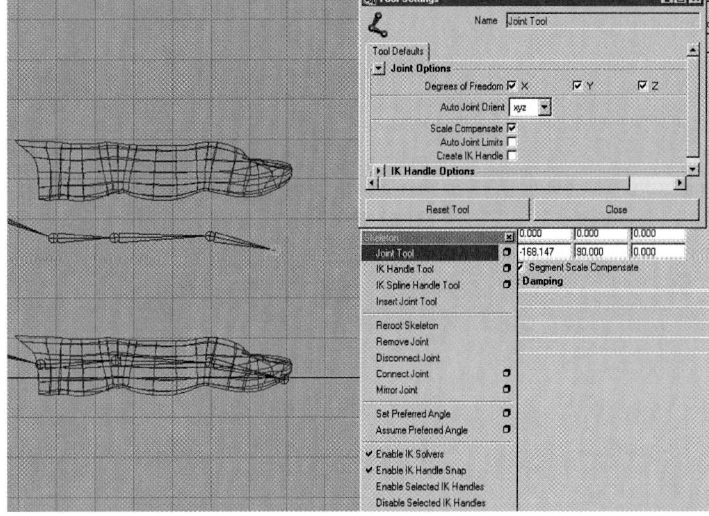

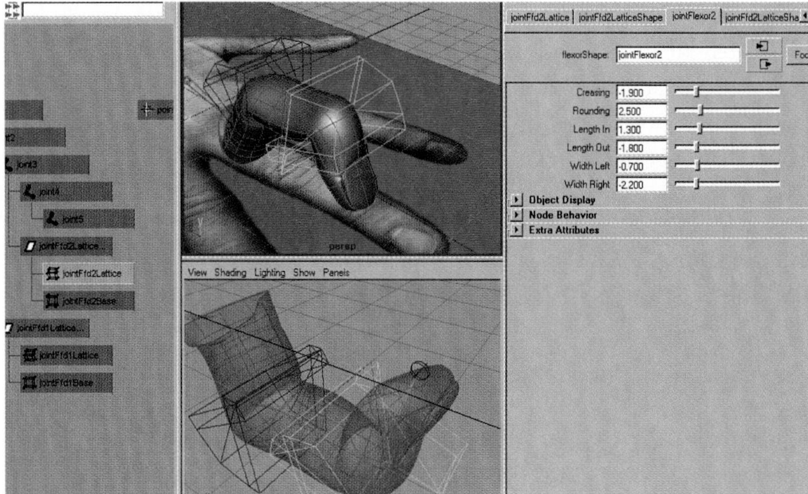

Figure 4.12
Posing the finger for deformation observation.

4.1.2.4 Duplicating the Finger

Now that you have tested the general functionality of your finger and are satisfied with the results, it is time to duplicate the finger and alter it for unique qualities. Each finger looks a little different; the thumb is the most unique of the lot. It is a good idea to modify the thumb last because of its special needs (which we will discuss later).

1. For the other three fingers, start in the top view and create three copies of the finger you already created. Then alter each one so that it matches the corresponding digit seen in the image plane.

2. For overall positioning, use the Translate tool to rotate the surface so that it matches the angle and then scale it to fit. Then, apply a new lattice and adjust the contours so that each finger uniquely resembles its matching finger on the image plane.

4.1.3 Creating the Thumb

Now that you have finished with the fingers, it's time to make some fairly radical adjustments near the thumb's base. Observe your own thumb for a moment. Notice that the thumb has just as many joints as the fingers, but its last joint is buried deep in muscle and flesh, whereas your fingers are generally free of the stump's dense, muscular base.

Using all the tools that have been introduced so far, such as hull editing, Artisan, and lattice deformers, proceed with the following steps:

1. Generate some of the connecting tissues for the thumb using an active fillet blend that spans the gap between the last isoparm on the thumb to the trimmed surface you will create later for that section of the stump.
2. Create most of the fleshy, web-like region between the fingers by combining modified lower sections of the fingers.
3. Edit the bottom five isoparms so that they become an almost "L"-shaped bend and connect with the rest of the palm.
4. Create the bottom joint from the fifth and sixth isoparms, and smooth out isoparms seven and eight.
5. Modify the oval-shaped skin crease beneath the finger so that it becomes the folding area underneath the base of the second thumb joint.
6. Fatten the thumb's nail and the first knuckle, for they are the thickest of the five digits in width.
7. Create a layer named "Fingers," add all the fingers to it, and then, with the fingers still selected, go to Modify, Freeze Transformations. This now becomes the new default for all the transformation attributes.
8. Select Edit, Delete by Type, History to remove any construction history, and go to Edit, Group, first choosing the Group Pivot Center check box, and then clicking the Group button. Name this group "Hand_Group". By this time you should have a fair approximation of an organic set of fingers that are named, grouped, and positioned properly.

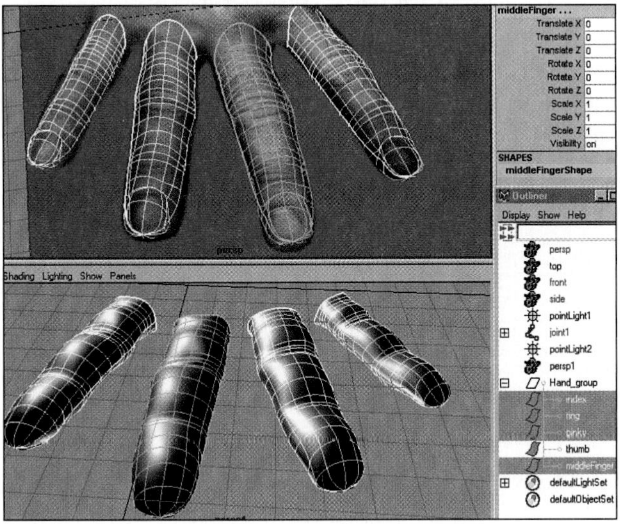

Figure 4.13
Duplicating the fingers for modification.

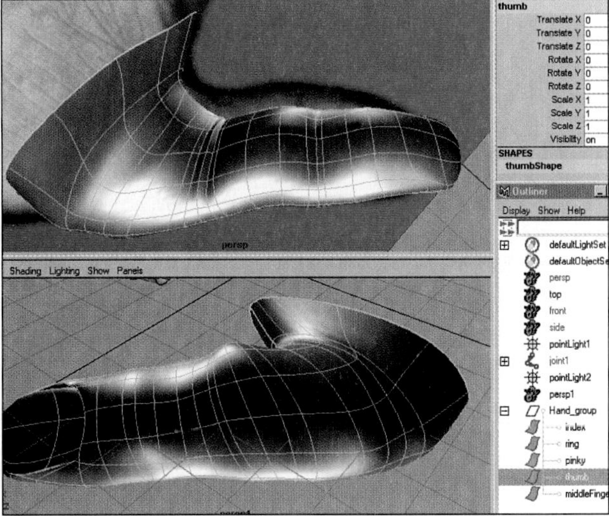

Figure 4.14
Creating a thumb.

4.1.4 Creating the Stump of the Hand from Five Planes

The next stage is to create a palm that will also serve as a foundation for the fingers, and to create the fillet blends that will join them to the base stump. Each panel is modeled separately using primitive NURBS planes, each aligned on separate axes, much like a box with an opening toward the arm. This separation will allow more flexibility and speed later, when you create a construction history lineage during the stitching and fillet-blend modeling-integration process. Keep in mind that it is important to retain the alignment of the isoparms, especially at their edges, because the Stitch tool uses each isoparm as its point of reference when attaching the two surfaces together.

To create the palm, follow these steps:

1. Go to Modeling, Create, NURBS Primitives, Plane, and then set the input values for the makeNurbsPlane to Width 5, Length Ratio 1, Patches U 9, Patches V 5, and Degree Cubic. Give it an easily recognizable name and align it over the center of the palm and up the Y axis a bit.
2. With the top of the hand selected, go to Edit, Duplicate, activating the Duplicate Upstream Graph option and then clicking Duplicate. Close the Duplicate window and, with the Translate tool selected, move the new duplicate down the Y axis until it is near the base of the thumb. Name it, and then duplicate it again using the Upstream Graph option, this time rotating it 90 degrees on the Z axis and translating on the X axis to the base of the thumb area.
3. In the inputs area, reset the makeNurbsPlane to Width 1, Length Ratio 5, Patches U 5, and Patches V 5. Name the thumb panel and duplicate it again with Upstream Graph, translating it to the area beneath the pinkie at the far right side of the palm's stump.
4. Rotate the thumb panel a little on the object's local X axis so that it approximates the angle in the hand. Duplicate the top panel and rotate it 90 degrees on the X axis, translate it to the base of the fingers, and scale its Z axis to one-fifth the original width (or .2). You should now have five named and oriented NURBS plane primitives roughly in place around the five sides of the hand. With all the hand panel pieces selected, go to Modify, Freeze Transformations, and Edit, Delete by Type, History.
5. Open an Outliner window (Maya 1.x readers must activate Drag'n'Drop Mode, Parent, under the outliner's Options menu). Drag the selected hand panels into the Hand_group that the fingers are in and save your file as an iterative version.

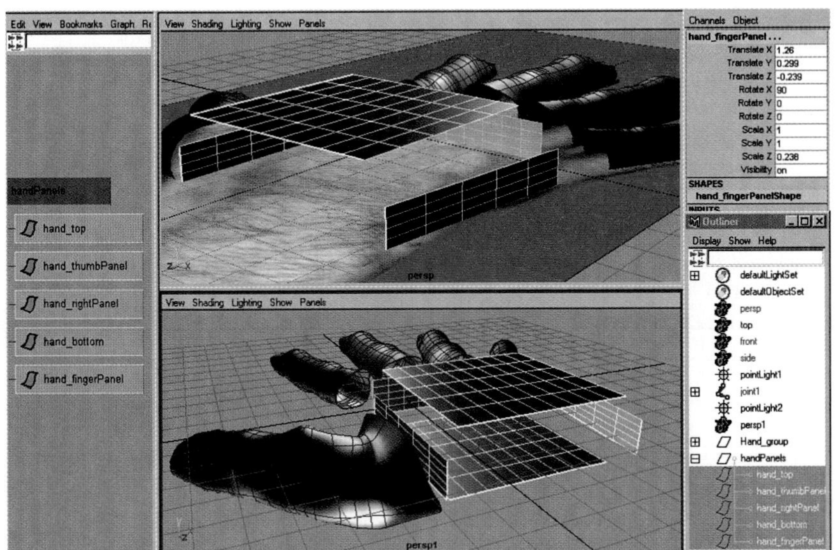

Figures 4.15
Making NURBS planes.

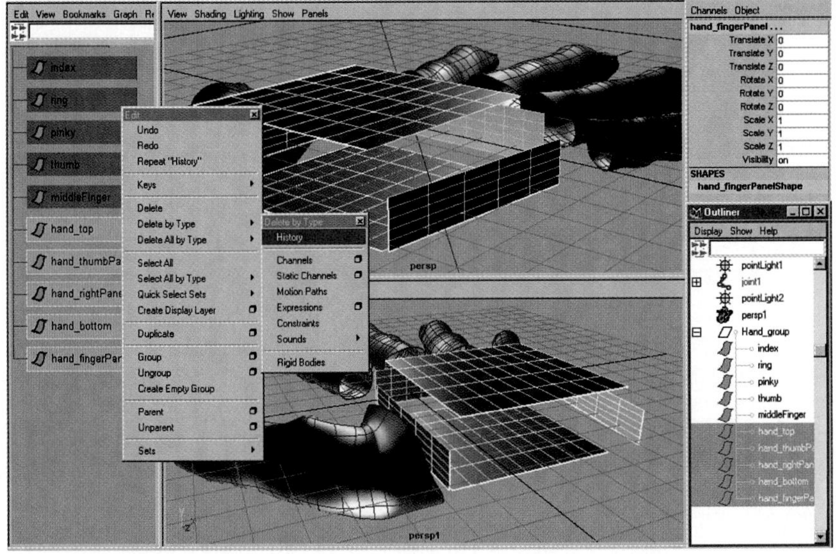

Figure 4.16
Freezing transformations with reordering in the Outliner.

4.1.4.1 Tweaking the Palm

Now we will begin the process of molding each NURBS plane primitive into place to become a seamless patch as part of the palm. To add a lattice deformer to a NURBS plane, you must translate the edge hulls toward the center of the plane a little to give the panels some depth. From here you can add a lattice and begin to sculpt the hand into place, one step at a time.

1. Select the top hand panel and go to Animation, Deformers, Lattice, choosing the Group Base and Lattice Together check box. Then click Create.

2. In the lattice's Shapes input values, set the S divisions to S 9, T 3, and U 5. Now, with the pick masks set to Lattice Only, start at the base of the hand near the wrist and begin selecting entire rows of the lattice by holding down the RMB and choosing Lattice Points. Make sure all the points in a row are selected. Translate them down the Z axis into place so that they match the hand image, and stop where the wrist ends and the arm begins.

3. Look at your hand from the side, and notice that the wrist forms a sort of ramp that goes above the palm area. To accurately model this portion of the palm, begin moving the lattice rows up the Y axis. Leave at least two or three more isoparms laterally near the knuckles to help lock everything down when the animated deformations occur later.

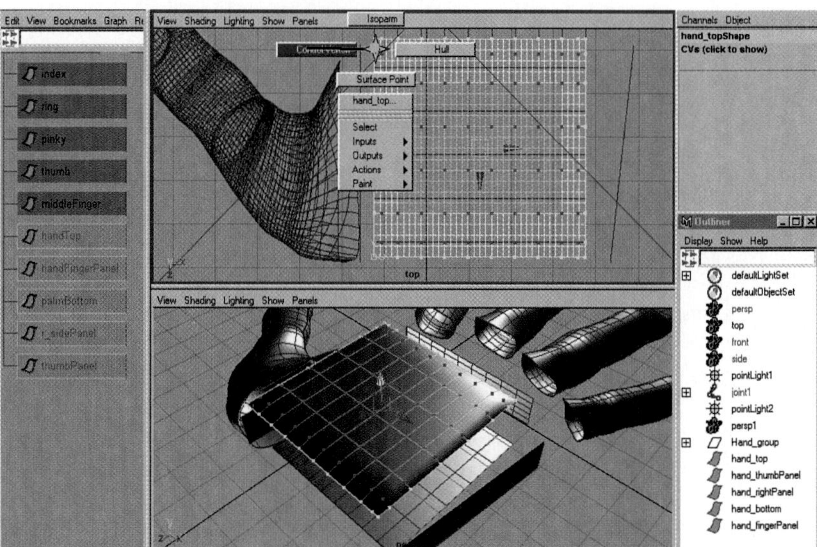

Figure 4.17
Forming hand panels.

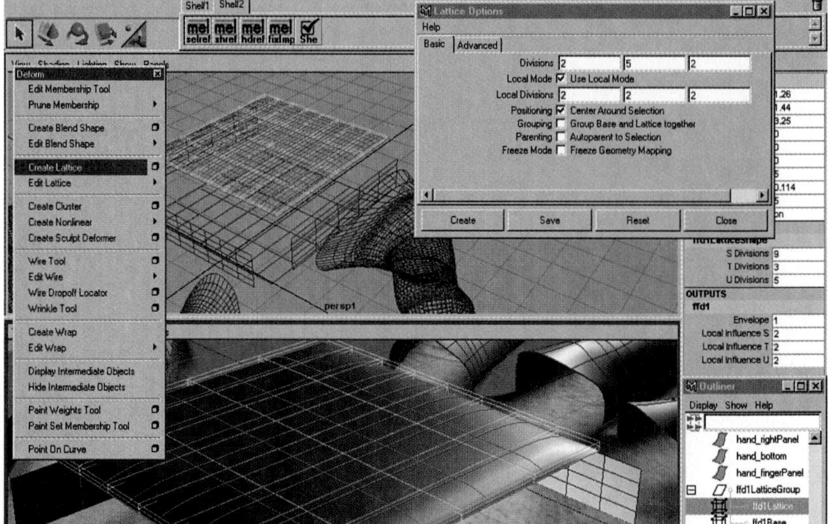

Figure 4.18
Applying lattices to the hand panels.

4. Pull the frontmost edge of the top panel slightly forward where each finger will be connected. Then use the second and third isoparms to form the slight lump that is seen above each knuckle when the hand is held out flat.

Working from the top and perspective views at the same time seems to work very well. Use the top view to select the lattice points and the perspective view to evaluate how high to displace the mesh when adding contour.

After getting a feel for using lattice editing to mold the top panel into place, continue this technique for the other panels. Try not to get too caught up in making sure the edges align perfectly, and don't spend too much time adding micro details. Now is the time to rough out the panels so that each one begins to fit together like a sort of three-dimensional jigsaw puzzle.

As a general rule, it is better to use fewer lattice divisions than more of them. However, when working with fairly low geometry, it is common to have as many divisions as the mesh has isoparms. By trying different combinations of lattice divisions for each surface, you will find the optimal balance between control and ease of use.

Remember to try and keep the isoparms generally aligned so that when it comes time to stitch them together, the alignment will dictate how and where the isoparms snap together. Edit the rest of the panels with lattice deformers so that the hand area is filled out.

When moving the lattice points of more than one lattice at a time, you might get odd results because the Translate tool is defaulted to local movement, and when you rotate the side panels into place you introduce a local rotation that will produce multiple directional movement. This is easily remedied by double-clicking the Translate tool and setting the Move tool to World Coordinates. Now, when you move points around, they will have a predictable and uniform movement.

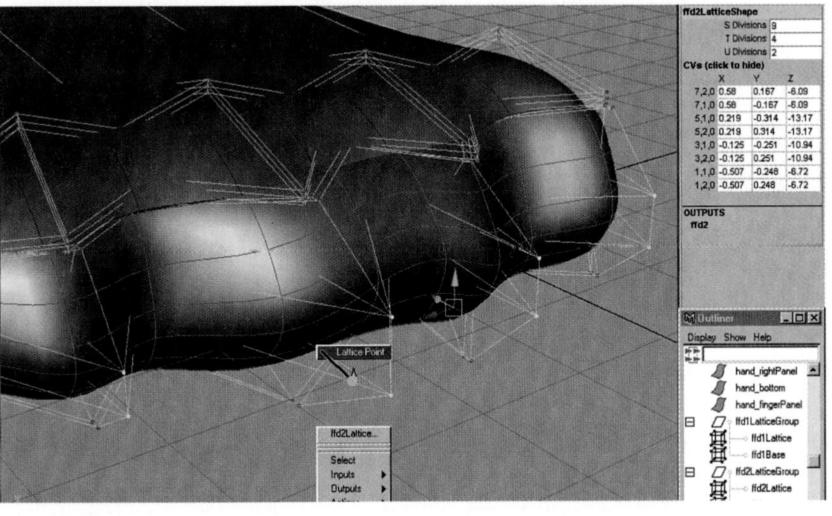

Figure 4.19
Adding lattices to the other panels.

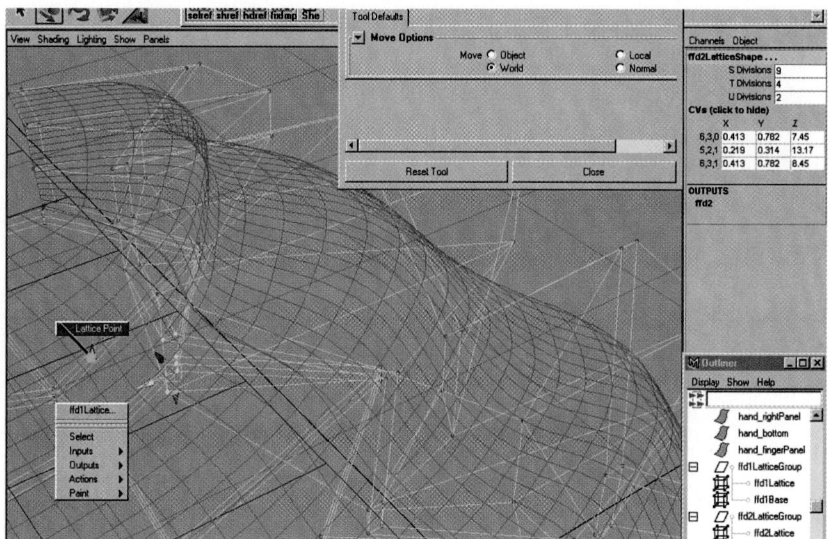

Figure 4.20
Resetting the Translate tool to World Coordinate space.

After completing the rough initial process of putting the sides together, save an iterative version and move on to the initial stitching pass.

4.1.5 Stitching the Model

The Stitch tool is a lot easier to use in Maya 2.0 than in past versions of the software, but just in case some readers have not upgraded yet, I will cover the stitching method that is compatible with versions 1.0/1.5 of Maya as well.

In earlier versions of Maya, when selecting which models you want to stitch, you must first select the edge isoparms and determine master/slave relationships. To choose which surface is to be the master, you must evaluate which surface can be pulled in regard to conforming the edge isoparms of the surfaces to each other.

To help you make such an evaluation, keep in mind that the finger panel must be the master of every other surface that touches it (that is, the other four hand panels), and the thumb panel should be the master when you're stitching the top and bottom panels. One surface can be both a master and a slave on different edges, but for seamless edge continuity and correct construction history, stitch each individual edge only once. Keep track of the master-slave order and edge relationships that you used to help avoid any stitching conflicts.

Take the following steps if you have version 1.0/1.5 of Maya:

1. First, prepare your model for stitching by removing the construction history of the models that you want to stitch. Because Maya considers all previous construction history on your model, system performance can tend to slow down, especially when every hand panel will be stitched to each other.

2. Deformers such as lattices add to the construction history list and, depending on the order in which they are added, can behave strangely when mixed with vertice level tools such as Stitch. The Stitch tool is a lower level of construction history than the lattice, and the two will sometimes argue over which has priority. So, to be safe, select the hand panels that you want to stitch and go to Edit, Delete by Type, History. Then, with the geometry still selected, reduce the display resolution to Low by pressing 1 on the keyboard.
3. Create a layer for the objects that you want to stitch, and place all the other objects into layers that you can easily hide, show, or template. This will reduce the clutter on your screen and prevent you from accidentally selecting the wrong model when performing the stitch.
4. Now, press the space bar over the window that you want zoomed to full screen. Then set your view to wireframe by pressing the 4 key. The larger workspace, combined with the less cluttered viewport and the reduced-resolution wireframes of your hand model, will enable you to easily accomplish the stitch.

The Stitch tool involves a master/slave relationship, with the first object's edge isoparm being the master and the second object's selected edge isoparm being the slave. To perform the stitching process, take the following steps:

1. Go to Modeling, Edit Surfaces, Stitch Tool, making sure that the Blending options for Position and Tangent are checked On.
2. Set the Samples Along the Edge option to at least 80 to help ensure that the seams don't tear. Occasionally, for a complex stitch, you need to set this parameter to a much higher level, but keep in mind that doing so with active history tends to slow down the file. The Cascade Stitch Nodes check box must also be on so that you can continually evaluate the edges contained in all the joining stitched pieces. This also helps you ensure that all changes correctly influence all the stitched models. Without this option turned on, the surface controls of the stitch will not correctly cascade through to the members of the stitching
3. Now, close the Tool Settings window and click the edge isoparm of the finger panel that is closest to the top hand panel. It should highlight to show you that it was correctly selected.
4. Click the edge isoparm of the top panel that is closest to the finger panel. The slave edge isoparm should now be conformed to the master's edge, with each section of the top panel connecting patch to

patch with its wireframe. This isoparm should now also be highlighted in green, indicating that the stitch was performed and can now be edited with manipulators.

5. Press Enter to complete the stitch. If you want to test the results of the stitch, select the top panel and translate it up the Y axis a few units. If everything is working correctly, the slave edge isoparm of the top panel should be welded to the master edge isoparm of the finger panel, and the object will deform more as you continue to translate it.

Note

Although you could finish now by pressing Enter, you can alternatively leave your model in the editing mode. The manipulators for this mode look like blue dots, and they are located at each far edge of the stitched region. You can use the blue manipulator dots to determine the extent of the stitch. For the hand panels we want the whole edge, but for functions such as stitching portions of segmented humanoid faces, you can use a T-Junction, in which you stitch only half of the longer side's isoparm to the other side's full isoparm that is only half the length. This way you can use the stitch tool to join two smaller objects to a third longer object of dissimilar length.

Although editing these parameters is not required for this hand model, you could customize the way the stitch is calculated by sliding these dots across the master edge isoparm. If you hold down the V key to engage point snapping, you can precisely align the stitch manipulators at an individual isoparm's intersection. Then, if you need to form a T-Junction, in which two objects are seamlessly joined to a third, point-snapping the stitch manipulator enables you to safely share an edge isoparm.

Maya 2.0's stitching system provides an alternate way to stitch multiple objects together at once. To do this, select all the models you want to stitch and then choose the Edit Surfaces, Stitch, Global Stitch Tool. By tweaking the Global Stitch parameters, such as activating the stitch smoothness tangents and the max separation slider, and then conducting some pre-stitched model refinement, you can automatically make the edges of all the selected objects seamless. Because this is an automated procedure, Maya is forced to make some assumptions about what to do. Thus, I still tend to use the Stitch tool first, one surface at a time, and then finish with the Global Stitch when all my alignment issues have been solved.

You can facilitate the stitch-creation process by previously freezing the transforms so that the hand panels and fingers are at their own zero axis. Then create a script that, when activated, simply moves all the objects outward to any location and then back again. This script will create a slightly exploded version of your hand; when you select isoparms to stitch, they will visibly jump to the location of the other edge of the isoparms to which they are being stitched.

Tip

Stitch the pieces together in the following order:

- Finger panel master on all edges
- Thumb panel master on Top and Bottom edges
- Slave on Finger Panel edge
- Right Side panel slave on all edges
- Top and Bottom panel masters on Right Side edge
- Slaves on Finger panel and Thumb panel edges

By following this order, you will distort your trim curves as little as necessary and prevent them from tearing. After you have completed the initial stitching pass, you can edit the model without having to worry about keeping the edges connected seamlessly.

To accomplish this, take the following steps:

1. Open the Script Editor and go to Windows, General Editors, Script Editor, highlight the Script Editors menu, and choose Edit, Echo All Commands toggle.
2. In the Script Editors menu, go to Edit, Clear Input. Move each of the hand objects to the desired location and observe the part of the script that appears in the Script Editor as you perform each move. (This is, in effect, a list or script of all the commands necessary to move your models to the locations desired.)
3. Assemble the important commands into a list separated by a space, a semicolon, and a carriage return. Highlight the script and drag it to an open shelf using the middle mouse button. Repeat these steps to select and zero out the position data for each of the hand objects.
4. Select and assemble the return data from the Script Editor into another script, and drag the highlighted code into the shelf once more, using the middle mouse button.
5. Rename and assign pictures to the script by moving to Options, Customize UI, Shelves, where you will see all Shelf-related preferences and settings.

By hiding all the fingers, making visible all the seamless hand panels, and setting the display smoothness to 3, you can properly evaluate your model's progress.

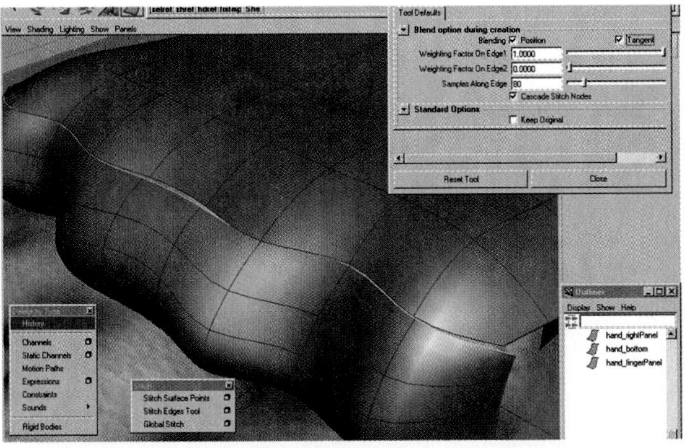

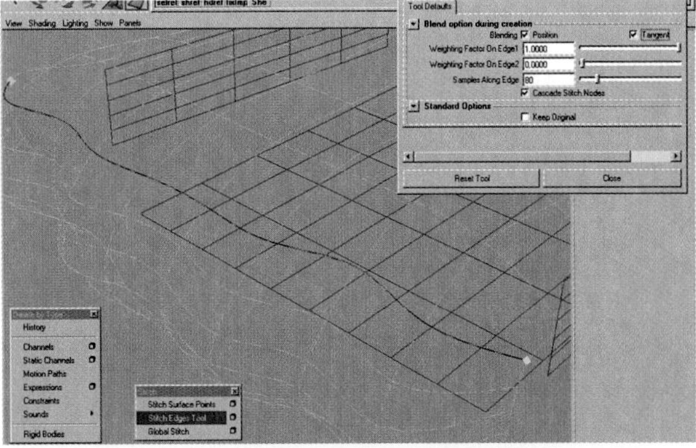

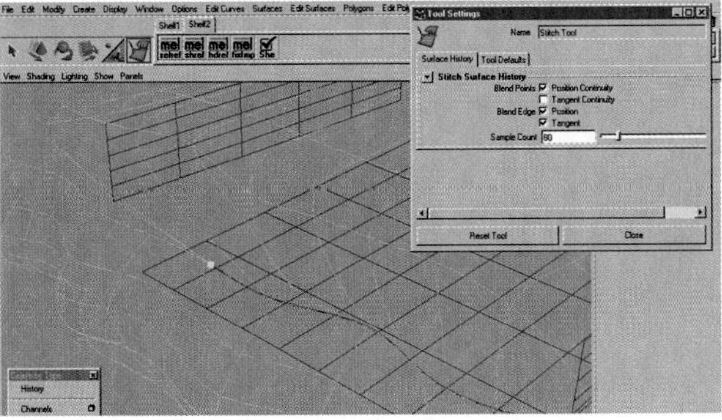

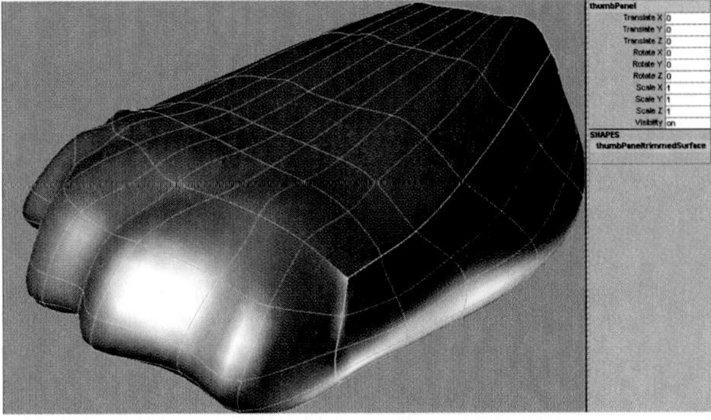

Figure 4.21
Stitching and its many stages.

MASTER SLAVE RELATIONSHIP FOR STITCHING HAND PANELS

The order in which you stitch is important for the initial editing process, although after the stitching and model sculpting is complete, you should delete history before you bind skin.

The top of the stitching master-slave relationship is the finger panel. The finger panel should control the stitch deformation of all the pieces that touch it. That means that when you start the edge-stitching process, you select the master isoparm edge first, followed by the Slave isoparm.

```
Hand_fingerPanel master of Hand_bottom
Hand_fingerPanel master of Hand_Top
Hand_fingerPanel master of Hand_thumbPanel
Hand_fingerPanel master of Hand_rightPanel
```

Next Is the Hand_thumbPanel:

```
Hand_thumbPanel master of Hand_bottom
Hand_thumbPanel master of Hand_Top
Hand_thumbPanel master of Hand_rightPanel
```

Next is the Hand_Top and Hand_bottom:

```
Hand_Top master of Hand_rightPanel
Hand_bottom master of Hand_rightPanel
```

4.1.6 Editing with or Without Construction History Retained

Now that you have prepared your object for contour editing and your model is seamless, you must decide whether to edit with or without construction history retained. Both options are possible, and each has its own unique strengths and weaknesses.

4.1.6.1 Deleting Construction History

If you decide to delete your construction history, it's important to know that Artisan has its own built-in seam correction that effectively functions like a limited but highly specialized version of the Stitch tool. Artisan's seam correction kicks in when it detects edges that fall within its threshold range, much like collision detection. If two edges touch, Artisan tries to keep them touching, even if a stroke unseams an area. The moment you lift your finger from the mouse or stylus, Artisan's seam correction kicks in, doing its best to fix the edge according to the features you activated. When you choose your next tool (other than Artisan), however, the stitching feature no longer has any effect because it is completely tool specific.

4.1.6.2 Retaining Construction History

The construction history-based Stitch tool is very handy. With the Stitch tool, when a master/slave relationship is established, an active dominance effect is exhibited when you manipulate the members of the stitching network. This ensures that no matter how you edit the individual objects, their edges remain seamlessly attached—in theory. Maya does a pretty good job of keeping the edges together, but when it comes down to it, there are some touchy aspects to making this technique work just right.

The limitations become evident when you combine stitching with deformers such as the lattice. Because the Stitch tool is very low-level, the lattice sometimes does not respect the edge correction that the stitch is trying to calculate, resulting in torn edges. This doesn't mean that the history is useless. It only means that it must be used with caution, forethought, and planning to ensure an optimized result. Don't be afraid to experiment with an alternate version of your current file.

4.1.7 Editing the Palm Model

To edit your palm model seamlessly, do the following:

1. Select all the members, being careful not to select anything else accidentally (you may want to try using the Outliner or the HyperGraph to highlight only the palm model members).
2. Activate Artisan by going to Modeling, Edit Surfaces, Sculpt Surfaces Tool. You will see the settings and tabs that allow you to customize its features. Select the Stroke tab to see the options for Reflection painting, which is useful for symmetrically modeling mirrored half-faces. Select the Seam tab to display the parameters that determine how sensitive the automatic edge detection is and to what degree the seam correction will take effect.

The Edges option box has three modes: Off, Position, and Tangent.

4.1.7.1 Off

When you set the Edges option to Off, edges are ignored. When this option is chosen, the Corners option is disabled.

Note

Stitching mode, located on the Seams tab at the bottom of the Edit Surfaces, Sculpt Surfaces Tool panel, has some significant features. The Edges, Corners, and Pole CVs option boxes all play vital roles in determining how the tool performs its solution every time the seam correction occurs. If you set them incorrectly, strange and unpredictable side effects can occur.

4.1.7.2 Position

The Position option box keeps the edges flush with one another, but it does not keep the tangency between the two selected surfaces, which often results in a pinched or creased appearance.

4.1.7.3 Tangent

The Tangent option box activates an evaluation between the angles of the isoparms on both sides of the seam. At times, this means that the surface will appear to change after the stroke is completed. This is a normal function and is required to keep a smooth, continuous tangent across the seam line.

4.1.7.4 Corners

Corners in Maya are, by definition, the place where two edge isoparms end. At the time of this writing, the Corners check box has a history of being a dangerous and unpredictable feature. Sometimes, while simply painting near an edge isoparm, the Corners option can cause the CVs being seam-corrected at the corners of two meshes to leap outside the normal area of the geometry, forming ugly tears. This, of course, leads to annoying and sometimes time-consuming fixes if you don't catch them in time. If you do end up needing this feature activated, carefully watch how the corners solve.

4.1.7.5 Pole CVs

Pole CVs is an option that refers mostly to what happens when you are working on two mirrored halves of a sphere whose CVs pinch together at the poles, forming a confusing and difficult-to-solve dilemma for the seam-correction algorithm. The Pole CVs option attempts to remedy this, but it is not 100% effective. In fact, any time you are using an Artisan brush near the poles of a sphere or any other pinched pole objects, you will need to be careful and understand why things are getting a little wacky. At times you will even see the Artisan brush become severely distorted as it passes over a polar region or some other pinched geometry.

4.1.7.6 Stitch Now

The Stitch Now button is a useful tool that applies a version of the Stitch tool with no history to all surfaces that are selected and within range for the operation to occur. Yellow lines appear on the edge isoparms of any selected objects where Artisan's internal edge-correction tool detects the surfaces it can react to, letting you know where the effect will occur. Unlike the Stitch tool, however, which defines a master-slave relationship for determining which isoparm to move, the Stitch Now tool causes all the edge isoparms to average between themselves and meet spatially in the center.

The Stitch Now button has an occasional side effect when combined with the active history of the Stitch tool. The two can, at times, have a horrendous argument over which is control, resulting in an explosion of mesh. You can avoid this by not combining these two forms of stitching.

4.1.8 Adding Additional Details

To make this hand look like that of a human, you must add the necessary details. The entire set of hand panels must be rounded and given a smooth, flowing appearance. The Finger panel needs four distinct bumps that correspond with the appropriate fingers, and a lump for the thumb panel that will act as a target for the Curve on Surfaces and Fillet Blend sequences coming up. This lump will help the fillet blend evaluate tangency, much the way the Artisan stitch tool tries to keep the curves flowing smoothly and naturally. When modeling these target bumps,

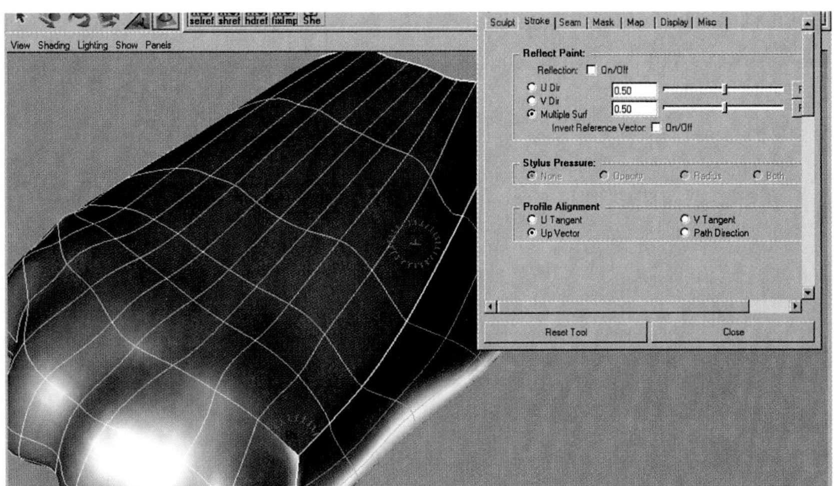

Figure 4.22
Artisan's preferences.

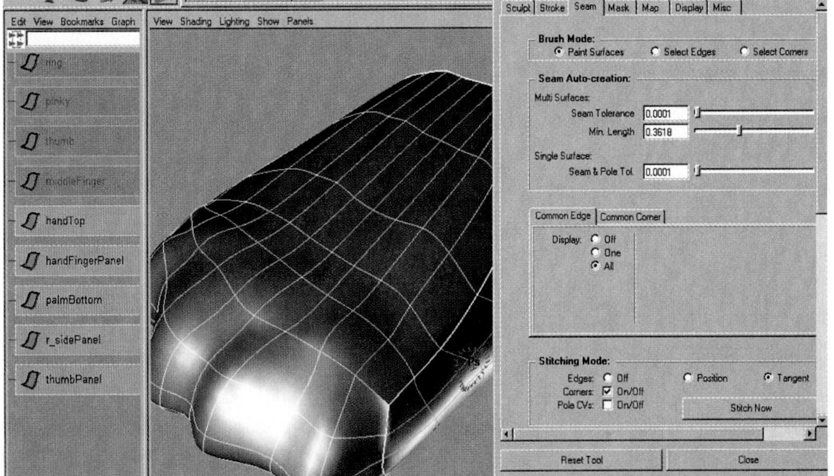

Figure 4.22
Continued.

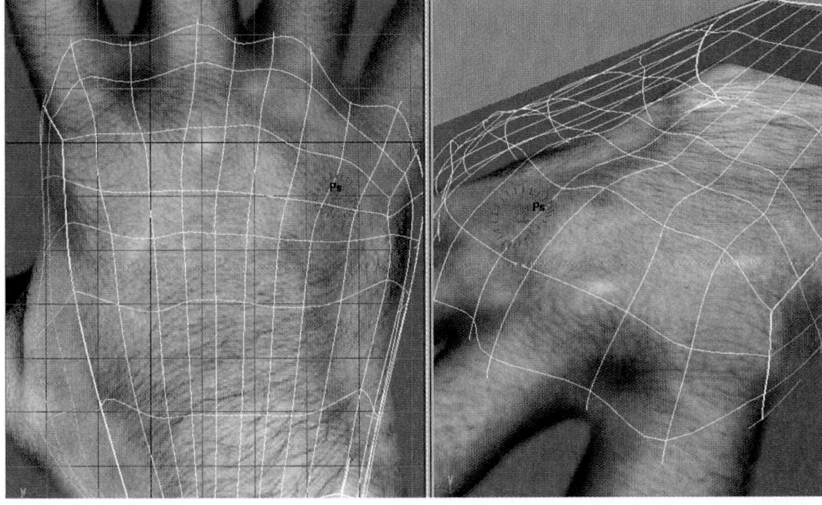

Figure 4.23
Adding details to the seamless palm models.

Note

It is imperative that the fingers and thumb panels contain the whole curve-on-surface target curves, and do not allow any to spill over to the other panels. This ensures more proper and predictable fillet blending.

keep in mind the finger-to-hand union as a whole. Try to avoid abrupt changes in form, which can result in rectangular or brick-like hands.

To add the necessary details to the hand model, do the following:

1. Make sure that the side panels are wide enough, and if necessary, select all the panels and add a low-resolution lattice to all five of them.
2. Scale the lattice and/or lattice points to match the scale. You will need to have room to attach the thumb and fingers without the fillet blends spilling to any other panels.
3. Align the thumb and finger stumps so that they match the base of the hand where they will be joined. Hold the RMB over the finger to be assigned a lattice, and choose Hulls.
4. Select the bottom five ring-shaped isoparms and add a lattice with dimensions S 3, T 3, Y 5. Create a hood-like shape as close as possible to the angle of the finger panel's foundation bump beneath it by selecting the top-middle lattice point closest to the palm and translating backward toward the palm on the Z axis.

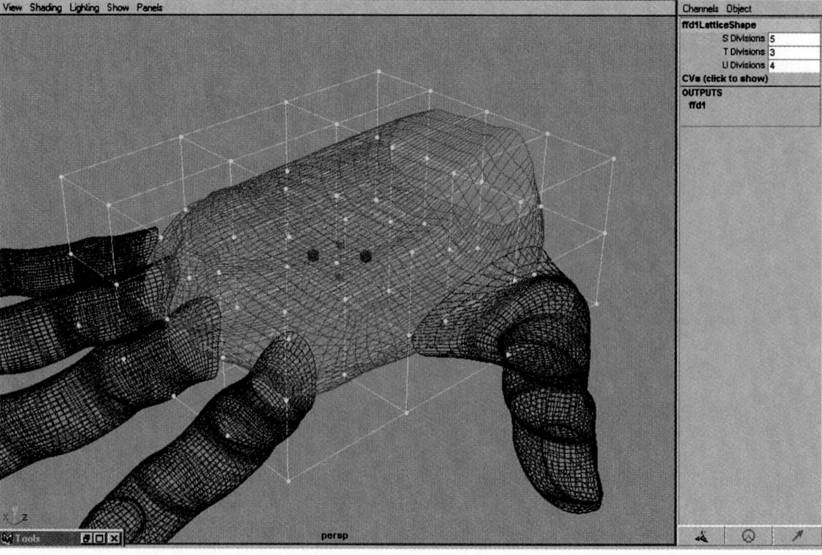

Figure 4.24
Adding lattice to scale all the hand panels simultaneously and seamlessly.

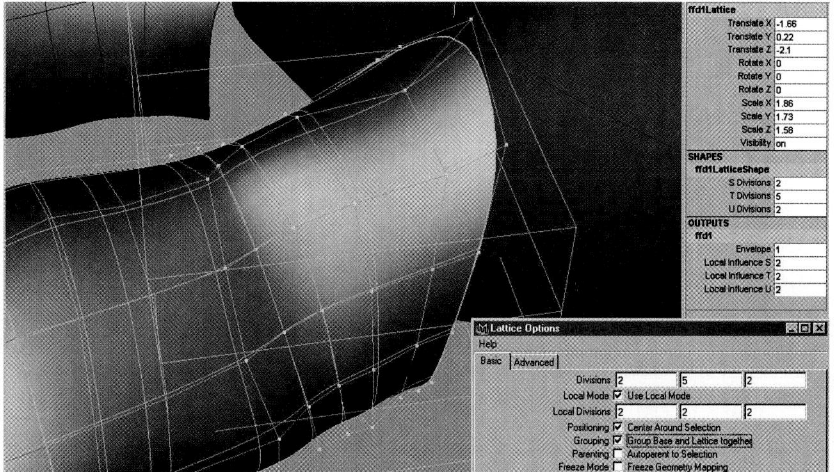

Figure 4.25
Finger stump alignment.

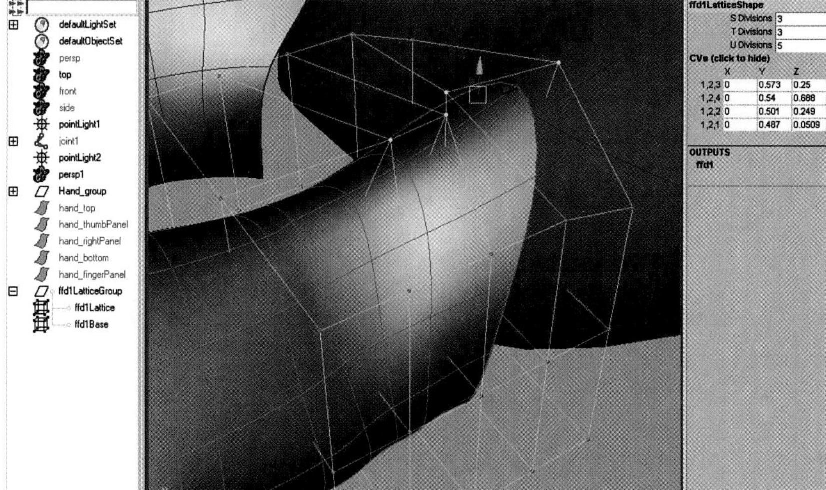

Figure 4.26
Matching the angle of the finger panel bumps.

4.1.9 Additional Modeling

When adding the micro details that make a hand look human, one must consider what must be modeled and what must be texture-mapped. At this point your hand model may still require additional modeled details, but unless the hand needs incredibly close scrutiny, texture maps are usually the preferred method of adding such details.

Any part of the hand that moves a lot will be, to some degree or another, a network of fold patterns. The back of the hand where the knuckles and hand bones pass through have lumps and bumps that you can add to subtly bring out more detail.

At the bottom section of the hand, at the base of all the fingers and thumb, are large (and somewhat subtle) muscular regions that you can pull out using a combination of lattice deformers and Artisan's displacement brushes.

If hair detail must also be considered, consider transparent texture maps or try using the new Maya 2.5 PaintEffects hair plug-in.

To align added details with the underlying hand template image, do the following:

1. Go to the Window Pane menu bar under Shading, Shade Options, X-Ray. This is often used in conjunction with the other Shade Option, Wireframe on Shaded, which helps add contour to the now semi-transparent geometry. If the X-Ray mode is too transparent, create a shader and set its transparency parameter to a more agreeable setting.
2. For a new shader, go to Window, Multilister, Edit, Create Render Node to create a new Blinn shader. Double-click the shader to edit its parameters. When using a transparent shader or X-Ray mode, it can often be easier to accurately match the contours on the bottom and top of the hand.

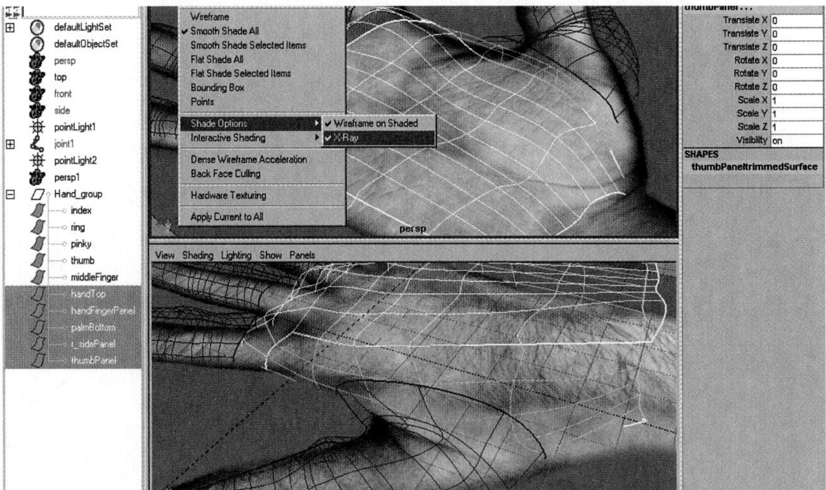

Figure 4.27
Adding details to the bottom and top of the hand.

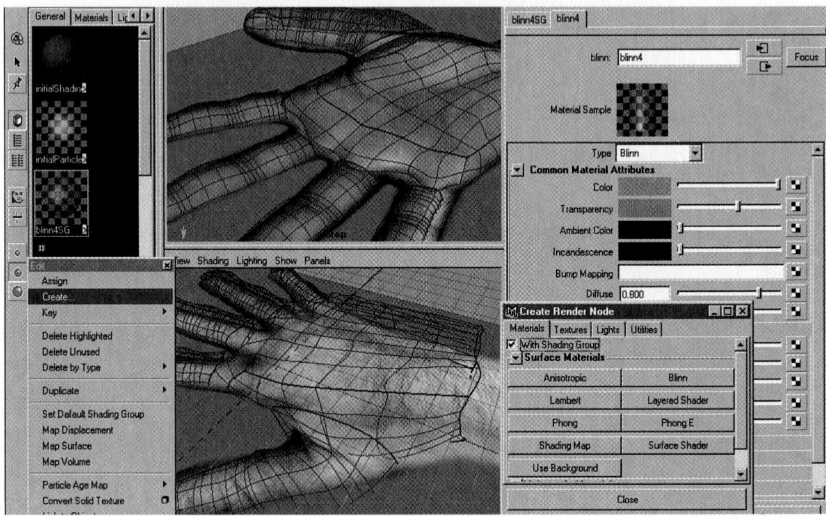

4.1.10 Turntable Animation

Before you animate your model, it is a good idea to perform a turntable animation of your model. Playblasting can be used to evaluate your geometry when the file starts becoming a little dense. By creating a movie file or a sequence of uncompressed frames, with the Playblast command you can better feel how the end result will appear, both in motion and fairly close up. This turntable can be achieved in a variety of manners. Here are some general principles to keep in mind when preparing your file for a turntable.

- The models should be given a shiny Blinn shader to emphasize any kinks, folds, or other unsightly blemishes that can often be easily recognized as the specular highlight rolls across the geometry in question. The specular highlight also gives your model a more interesting look than the default Lambert, helping one appreciate the work that has been accomplished.
- To get an interesting view of your model, try adding a couple of point lights by going to Rendering, Lights, Point Lights, with one placed on either side, the first one slightly above and the second one slightly below the model.
- Activate the Lighting Shaded mode by pressing the 7 key with the mouse over a viewport. These lights will enhance the specular effect as well as create some dynamic lighting that casts areas of light and darkness that normal shaded mode, reachable with the 5 key with the mouse over a viewport, just can't match.

Having sweetened the appearance of your display somewhat, you are now ready to keyframe a camera rotating around your model. At this point in the modeling stage, it is safer, more convenient, and instantly gratifying to not move or keyframe the hand geometry. Instead, simply rotate 360 degrees around the model. When playblasting, remember to playblast one frame less to avoid a duplicate start/end frame, which would create a hiccup as the turntable plays two of the same frame.

4.1.10.1 Creating a Locator

Some animators place a locator, scaled large enough to be easily seen and selected, in the center of the geometry to be previewed. This will serve as a target for the new camera you should now create to be spun around. To create a locator, do the following:

1. Go to Modeling, Create Locator and create a new camera by going to Modeling, Primitives, Create Camera. Parent this new camera to the locator and then translate the camera backward so that the model is in view.
2. Go the Window Panes Panels, Perspective to access the new camera you have created from one of your viewports. Align your camera to get a good view of your model, being sure that lighting shaded mode has been activated with the 7 key.

3. Set your playback range the number of frames you want to create. Select and then keyframe the locator's Y-axis rotation value in the channel box on the first frame.

4. Go to the end of the playback range and set the Y-axis rotation value in the channel box to 357 degrees. Raise or lower this number plus or minus a few degrees (always less than 360) to compensate for the fact that 0 degrees and 360 degrees are effectively the same thing when it comes to a rendered frame. Also, if the beginning and end frames are exactly the same, this will result in a brief pause that will distract the viewer slightly every time the turntable reaches the beginning of the loop.

5. Now your file is just about ready to preview. Open Windows, Playblast, and choose Reset at the bottom of the Options window. To get the most out of the turntable test, select a large display size, at least 640×480 pixels, or even the full 1.0-scale window size if you are currently zoomed to one view.

4.1.10.2 Compression Issues

When creating your keyframe animation playblast, you may encounter a compression issues option that has some very important implications. Many compression formats are not cross-platform compatible, resulting in test movies that are platform dependent.

The IRIX version of Maya uses the MVC2 compression format as its default, whereas the Windows NT version uses AVI. These codecs tend to provide fairly good results, especially when you combine them with special hardware such as a full-screen broadcast video JPEG or MPEG video output card. Use the compression and pixel sizes required by the card and its compression codec, if one is available.

Otherwise, if you have a significant amount of RAM to allow 100–300 full-screen frames loaded into memory all at once, at about 1MB per frame and up, use the Fcheck option button in the Playblast options. This option uses whatever output file format is currently set under Rendering, Render, Render Globals. The fastest file formats to preview to are Maya IFF (readable only as animations by Fcheck and the IRIX-only A|W Composer) and the native file format to the system .sgi and .bmp for each IRIX and NT, respectively.

At this point, you must determine where to save the rendered frames. It is important to note that you must activate the Save to File check box and specify a location for the frames to be saved on the fastest hard drive connected to your machine. If you do not choose the Save to File option in the Playblast settings window, the frames will be temporarily cached somewhere on your hard drive (usually your Temp directory), which must have enough space to allow for the entire sequence to be generated from beginning to end.

After evaluating the turntable animation, close the window to free up the RAM needed to display the frames or movie and decide what level of editing you want to use to create the most detailed and precise model as possible. If you must, return to any of the previous steps and fine-tune the model, if necessary. When satisfied, save your file.

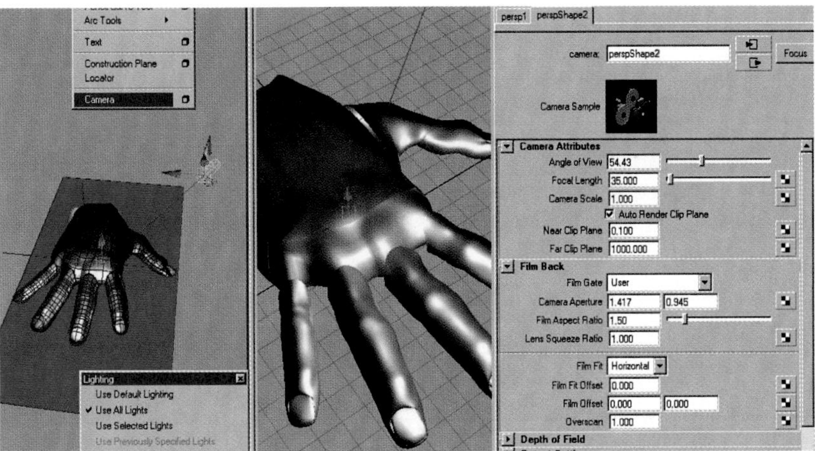

Figure 4.28
Creating Playblasts to test the shape on a lit turntable.

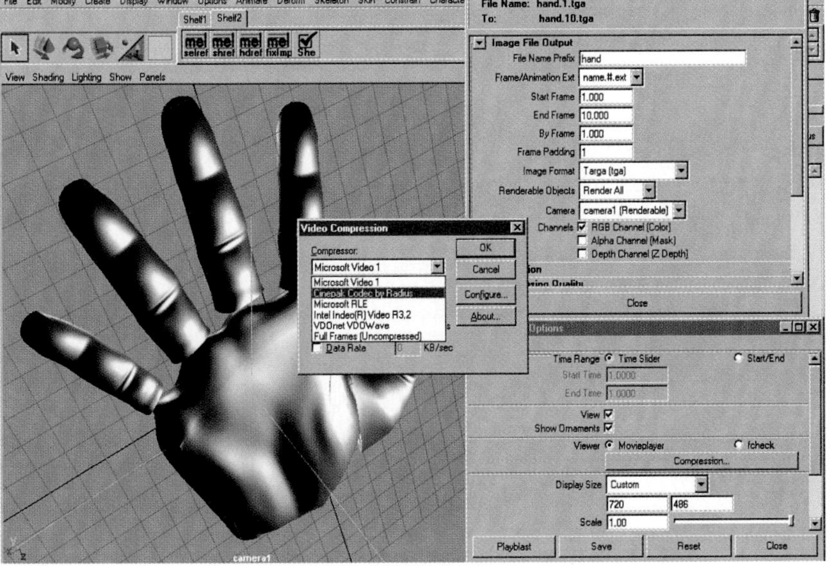

Figure 4.29
The preferences that must be set.

4.1.11 Integrating the Components

Now it is time to integrate all the components you have assembled to create a seamless NURBS model of the hand:

Note

Before moving to the final integration stage, be sure that the hand's individual components do not contain any construction history. To do this, select all the hand's pieces and go to Edit, Delete by Type, History. Now that you are organized there, you have more interactive control of your work environment because of your easy access to hiding and showing your key objects.

1. The first step is to make individual layers for each object, and to name these layers with an easy-to-recognize naming convention.
2. Go to Windows, Layer Editor. In the Layer Editor window click the New Layer button, then click the Rename button and name it Pointer_Finger. Select the Pointer finger in the outliner or Hypergraph and choose the Assign to Current button in the Layer Editor to place the finger in its new layer. Now repeat these steps for all the hand's components.
3. Hide those objects and layers that prevent you from accomplishing fine detail editing, and save the file as an iterative version.

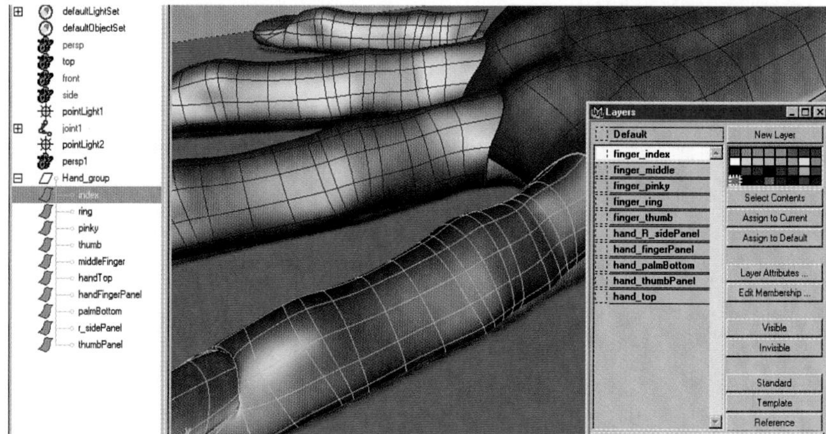

Figure 4.30
Creating layers.

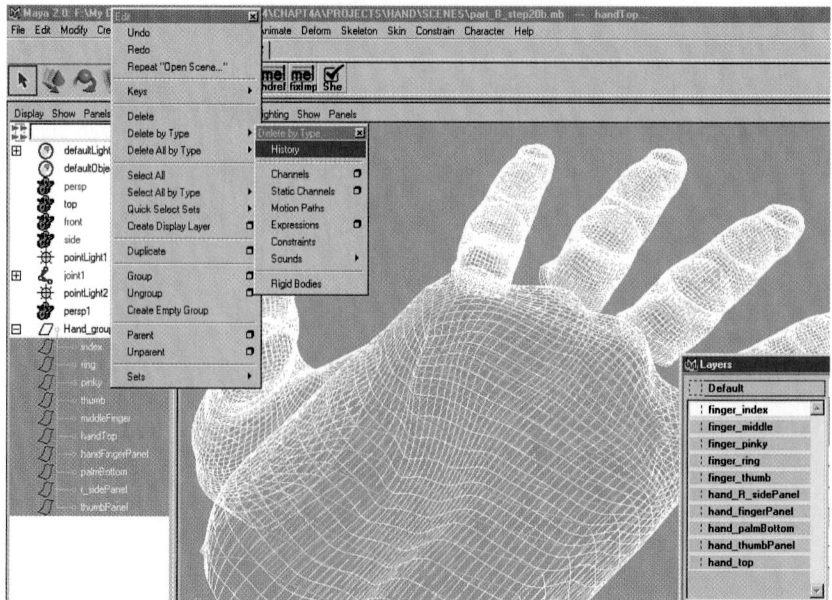

Figure 4.31
Deleting the construction history before integrating the hand pieces.

4.2 Integrating and Preparing the Hand for Animation

The first step to integrating the fingers to the palm is to create curves on surface that you will use as trim curves and fillet blend targets. You can create these curves in a variety of ways.

- You can use Make Live to draw directly on a model. By selecting a model and then choosing Modify, Make Live, you turn the surface into a temporary target for direct NURBS curve placement. You can use the feature to model a NURBS curve (using the EP, CV, or Pencil Curve tool) so that each point is placed quasi-magnetically onto the surface of a model. Do not try to close the curve while modeling. Instead, model the curve so that it finishes with some small distance between the start and end point, and then use Modeling, Curves, Open/Close Curves to close the gap.
- You can use Project Curve in conjunction with Duplicate Curve or Offset Curve on the base isoparm in a finger to prepare for projection. Projection is accomplished from the angle of the active camera view. Be aware of the construction history connected to the duplicated curve, the curve on surface, and the isoparm it was originally duplicated from because they all combine to create the final curve. Sometimes, after the duplicate or offset curve is no longer needed, it is useful to save your file with a different name, and then delete the curve, preventing it from causing any errors later.
- You can use Intersecting Surfaces, but you would use this least often because you would have to perform trims on both objects instead of just the base panel.

To add a curve on surface to your finger panel for the middle finger to connect to, do the following:

1. Make a new camera by going to Modeling, Primitives, Create Camera. Then place the camera inside the center of the middle finger, looking at the palm.
2. Set the Panels, Layouts 2 side by side. Load the new camera into one of them and the old perspective view into the other one by going to Panels, Perspective, and so on. The new camera will act as your curve projection angle and the other one will allow you to set things up from a convenient angle.
3. In your viewports set both the finger and the finger panel surfaces to low-resolution display smoothness 1 and set the viewport to wireframe by pressing the 4 key.
4. Select the base isoparm of the middle finger by holding down the RMB over the finger and choosing Isoparm. Then click the finger's base isoparm nearest the palm and go to Modeling, Edit Curves, Offset Curves. In the Channel box click in the Inputs, Offset Curve1 and set the distance somewhere between

–0.01 and –0.05. Each curve on surface should be at least slightly larger than the finger's base isoparm, or the fillet blend will not look right.

5. In the main menu, go to Modify, Center Pivot. With the offset curve selected, Shift+click the hand panel. Do you notice the blue box around the active viewport? If you move your mouse over the new projection camera and then simply press Ctrl, the blue box jumps to the new viewport. This lets you know which viewport Maya considers to be active when the projection occurs.
6. Go to Modeling, Edit Surfaces, Projection Curve, making sure that the Project Along Active View option is on. Click the Project button. The curve on surface should appear on the finger panel model.
7. With the new curve selected, press W to activate the Translate tool. Click the green arrow and translate the curve closer to the knuckles and slightly more in line with the fingers. The translation manipulator will look a little odd. It has only two arrows associated with it. This is because a curve on surface does not fully exist in three dimensions. In fact, it really is only aware of the two dimensions that make up the surface of the object upon which it exists. Thus, the two arrows represent the U and V axes.

Because of the construction history, any component of the curve on surface can be edited. The shape of the surface that the curve is on will determine the depth component.

The offset curve can be translated, rotated, and scaled, and its CVs can be edited to allow detailed tuning of the curve on surface. The curve on surface itself can be edited at component vertex level in the UV directions across the surface of the hand panel. However, the detail of the curve on surface is usually of a much higher resolution and, thus, it is more difficult to get a smooth and seamless result in this manner. You can edit the finger's base hull and corresponding CVs as well and have direct influence over the curve on surface (except for the down point that the finger itself has to modify). Sometimes this sort of editing is necessary, especially after the fillet blend is in place and the tangency of the blend needs to be fine-tuned.

Now that you are familiar with the steps involved with creating the curve on surface for one finger, build the other four for the rest of the fingers and the thumb. When you are finished, the thumb panel should have one curve on surface, and the finger panel should have four curves on surface, one for each of the fingers.

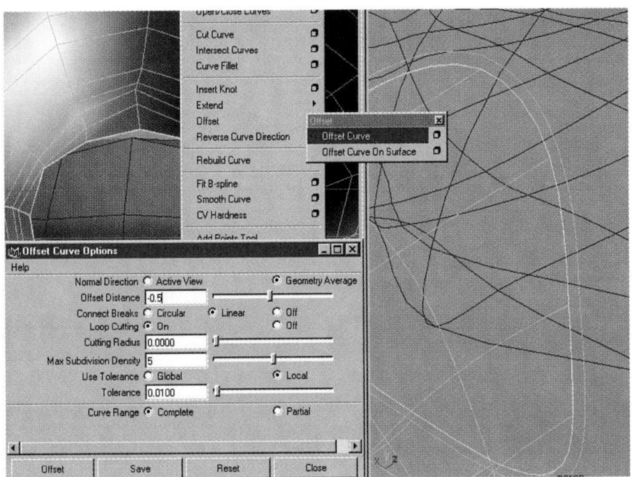

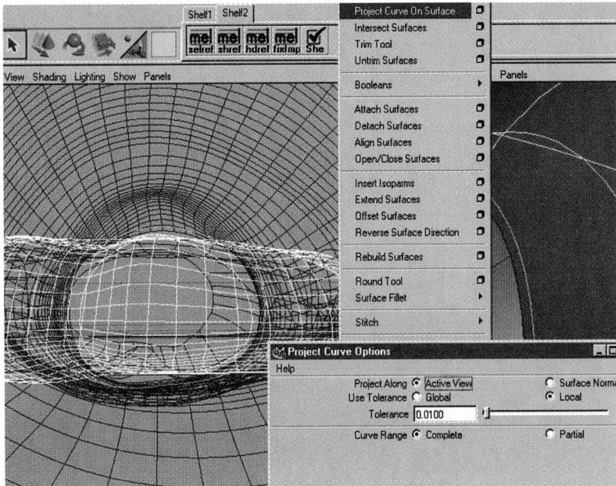

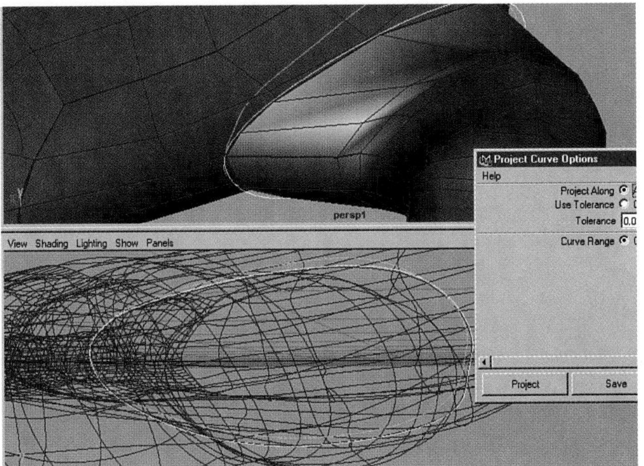

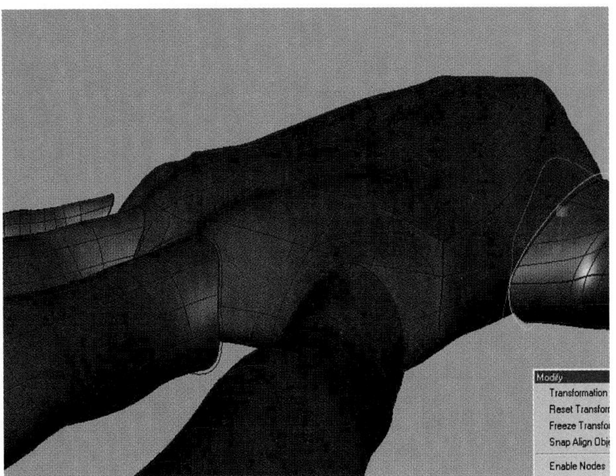

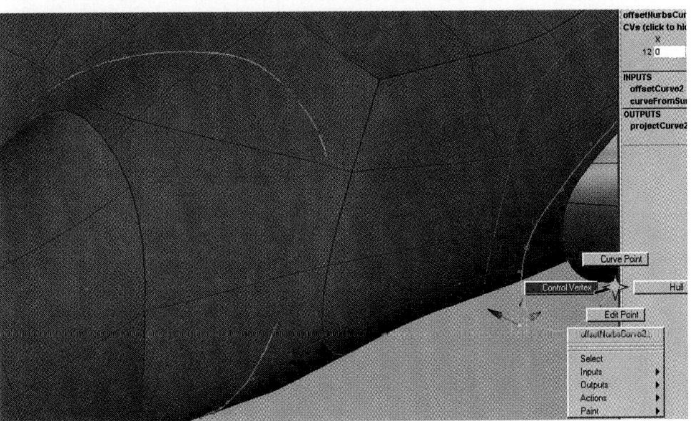

Figure 4.32
The stages involved in creating the curves on surfaces with projection curves.

4.2.1 Creating the Trim on the Finger and Thumb Panels

At this point, you are ready to create the trim on the finger panel and the thumb panel. To do this, take the following steps:

1. First select the finger panel and go to Modeling, Edit Surfaces, Trim Tool. The surface should become transparent, and a customized white wireframe should appear. Click the surface of the finger panel in an area outside of the closed circular curves on surface and then press Enter. Selecting this area will let the Trim tool know the regions of geometry that you want to keep. The trim should update and look like the finger panel, with a hole at every curve on surface.
2. Repeat these steps on the thumb panel so that there is a trim in place here also.
3. Edit the shapes of the offset curves to finesse the trim curve results. This will provide the best possible foundation for the fingers and thumb to be fillet-blended to the palm.

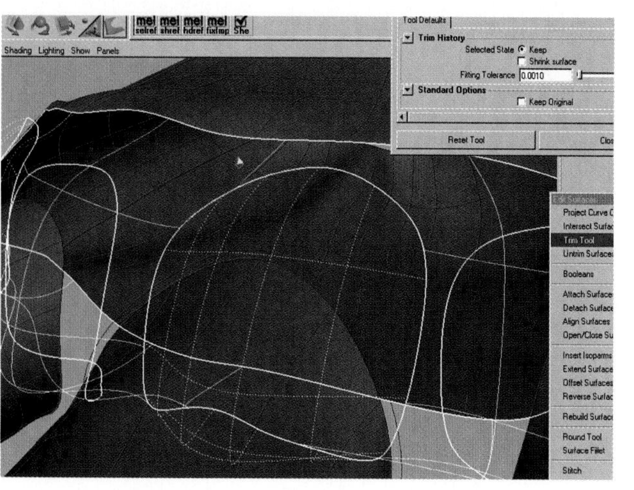

Figure 4.33
Using the Trim tool on the finger panel and on the thumb panel.

4.2.2 Creating Fillet Blends

To fillet blend the middle finger to the hand panel's middle finger trim curve, do the following:

1. Start by setting the models to low-resolution display with the 1 key and the viewport to wireframe with the 4 key. RMB on the finger panel and choose the Trim Edge option, and then click the middle finger's trim curve edge, which highlights it with a thick yellow line.

2. Now go to Modeling, Edit Surfaces, Surface Fillet, Fillet Blend Tool. The Help line on the bottom of the screen should prompt you to select a curve to blend to. With the trim edge still selected, press Enter to start the blending procedure. Now click the middle finger's base isoparm nearest the palm, highlighting it yellow, and then press Enter, creating the fillet blend.

3. The blend will look rough at first, so select it and then press the 3 key to display the highest smoothness setting. You may need to select all the connecting geometry and again press the 3 key to "upres" the model as a whole to best evaluate the progress you are making. After you have finished evaluation, return the geometry to low resolution for convenience later.

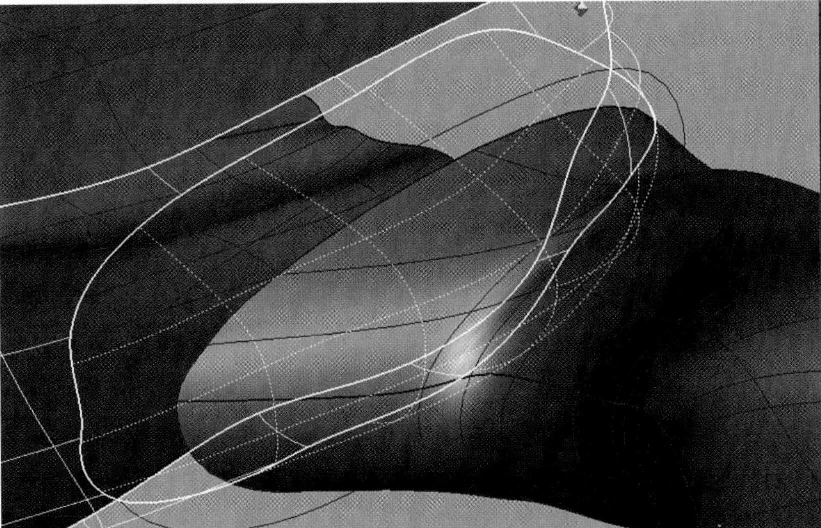

Figure 4.34
Trimming the thumb hole.

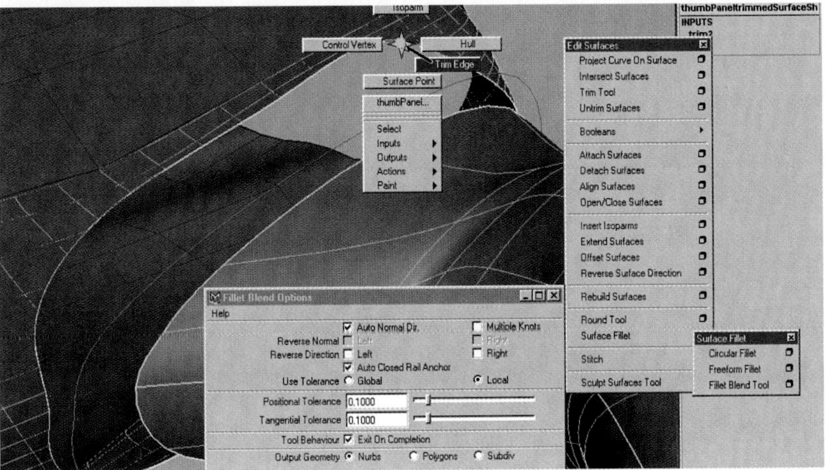

Figure 4.35
The Fillet Blend tool in action.

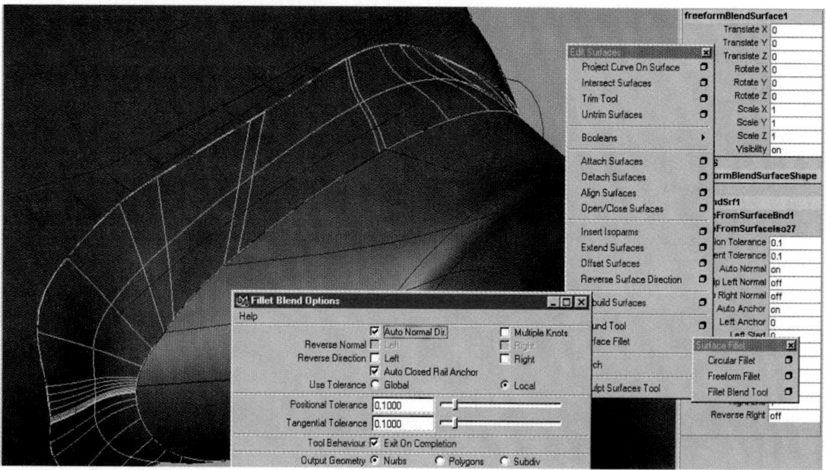

Figure 4.35
Continued.

4.3 Tweaking the Results

If the blends look twisted or otherwise unpleasing, or if the seams do not quite meet, you can edit the inputs with the Manipulator tool because all the parameters have construction history. For instance, if your blend seems to have one flipped side, you can select the blend geometry and press T to activate the Manipulator and drag the little blue dots on either curve edge of the blend. Don't hesitate to zoom way in and tumble the camera around the curve to verify that you get a pleasing result.

1. Select the Manipulator tool by pressing the T key, and then, with the Fillet Blend selected, click in the channel box under Inputs, ffBlendSrf. You should now see the little blue manipulators and the fillet blend's inputs. Use these tools to adjust the parameterization until the blend looks pleasing. Usually, most fillet blend tweaks can be fixed by moving the little blue manipulators around a little.
2. Complete the blend on the other fingers and thumb, increasing each of their smoothness settings to full resolution. Then, one by one as they are created, adjust their manipulators and parameters until they look pleasing.

Note

Remember that the fillet blends should never be bound to a skeleton if they still have their history turned on because then they would not update correctly if moved and the seams would drastically tear apart.

For organizational reasons it is important to create an order and naming convention to all your work. By placing the geometry you have created in specific layers and groups, it will be easier to hide and show as well as select and manage your models.

You can create a separate layer just for the blends. Then, while they are still selected, go to Edit, Group. Make sure that Group Pivot Center is selected, and then click Group. While the group is still selected, name it for easy identification by clicking its name in the Channel box or

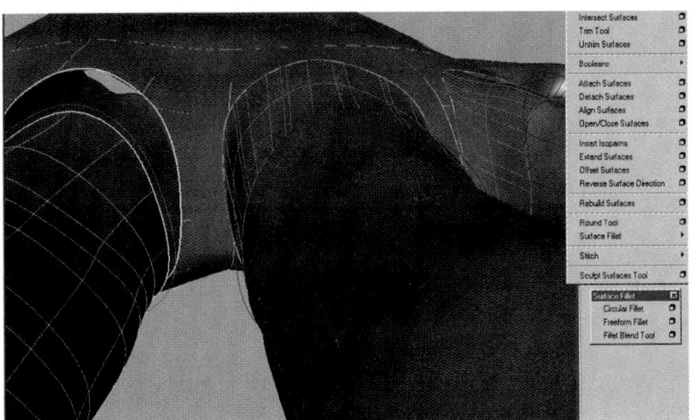

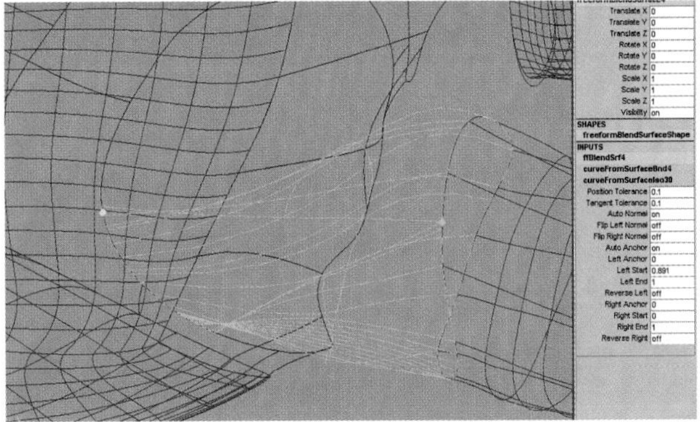

Figure 4.36
Editing the fillet blends with the Manipulator tool correctly (left) and incorrectly (right).

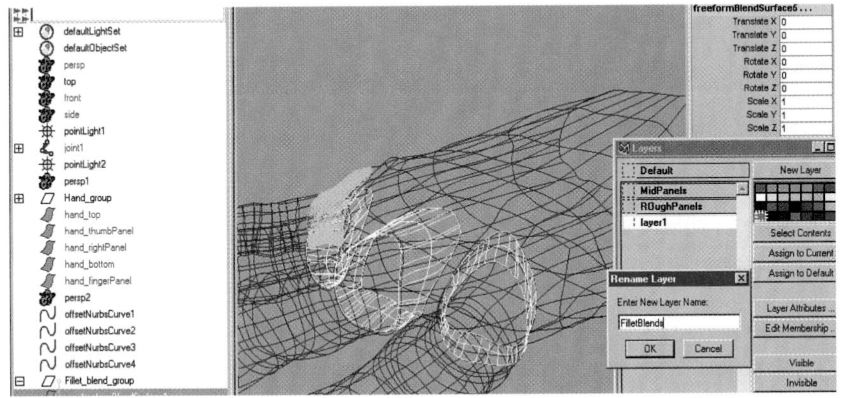

Figure 4.37
Organize the fillet blends as they are made, and place them in their own layers.

Attributes Editor. This technique goes as deep as naming every single object, shader, or construction object that is created.

Although it might seem like a hassle right now, the discipline and organization it can provide for your teams' files will come back to you down the road tenfold when your files become immense and you must coherently return to edit earlier work, which inevitably happens.

4.4 Final Fine-Tuning

After all the blends have been completed, it is time to fine-tune the overall look and feel of the integrated components. Create another Playblast and evaluate it. Add any necessary details required to make the hand look more realistic. Then, save your model.

At this point, you can delete floating offset curves and duplicate curves that might have been used to create projected curves and curves on surface. This will clean out these hanging history curves and prevent potential problems from happening that would need to be solved later on.

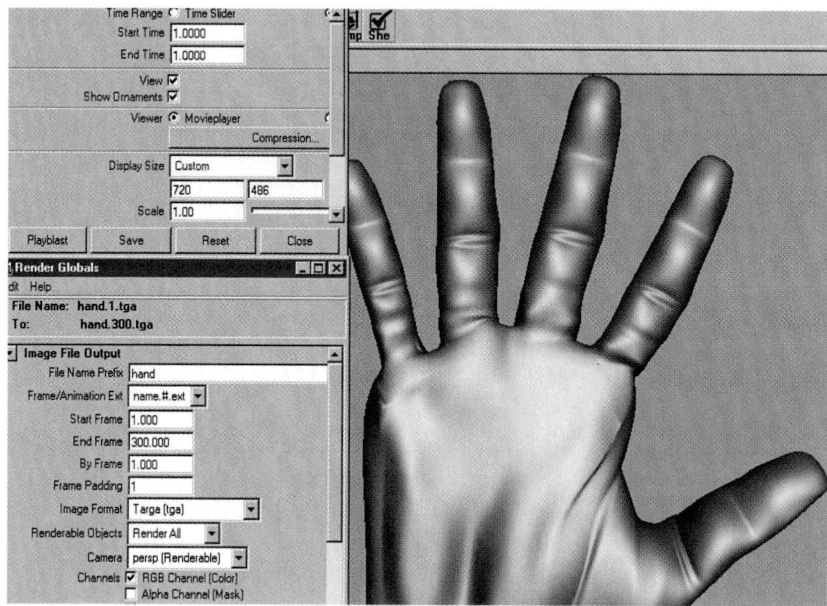

Figure 4.38
Creating another Playblast of the mostly finished hand.

4.5 Creating a Low-Resolution Version of Your Model

You will need to create a low-resolution version of your hand model that you can animate in near-real time (a heavy hand model will bog down playback). You can use this low-resolution hand as an optimized version that includes all the regions that would normally be deformed by bones removed. To do this, follow these steps:

1. The simplest way to do this is to use Modeling, Create, Polygon Cube. Use the transformation tools (translate, rotate, scale) to place one duplicated version of the cube over each area that would normally bend. Each finger gets three, and three more for the palm and base of the thumb.
2. In wireframe view, set your model to Display Mode 1 to get a good look at your model's actual parameterization. First hold down the RMB over a finger, choosing Isoparm from the pop-up menu. Select new or existing isoparms along the surface of each finger on either side of the knuckles. Then, using Modeling, Edit Surfaces, Detach Surfaces, with Keep Original toggled Off, detach the bending regions. Afterward, delete the remaining geometry, and click Edit, Delete by Type, Delete History.
3. Perform these procedures on duplicates of all the hand geometry. Fillet blends do not need to be kept for the low-resolution version. Finger and thumb panels can be deleted, and the rest of the panels can be reduced in

resolution, along with the fingers, using Modeling, Edit Surfaces, Rebuild Surfaces. Leaving the Settings window open, select one piece at a time and experiment with the parameters and geometry display resolutions to bring each object to the lowest number of isoparms possible and yet still retain full volume.

4. After the low-resolution objects are properly created and their history deleted, re-apply any shader that you had on before and proceed to place all the low-resolution objects in an easy-to-hide and -show layer.

5. Now, parent the low-resolution geometry to each bone that it will logically travel with. Select the model, Shift+select the specific bone you want to be parented to, and press the P key. Check the Hypergraph to be sure the object is correctly parented. Do this one at a time until all the objects are parented to their respective bones.

Incidentally, there are many other ways to parent in Maya. By dragging the geometry nodes in question over the chosen skeleton members so that they highlight and then releasing the mouse button (in the Hypergraph), you can also easily create hierarchical parenting structures. This metaphor of creating low-resolution geometry to animate with should become a common part of creating every CG character in your library.

Note

Another, more elegant but time-consuming method is to chop a duplicate hand model into low-resolution NURBS pieces with Detach Surfaces and Rebuild Surfaces commands.

Note

A premade skeleton is provided for your experimentation in the Hand Project's Scenes directory, labeled Hand_skeleton1.mb. Choose File, Import to integrate the premade skeleton, or experiment with adding your own.

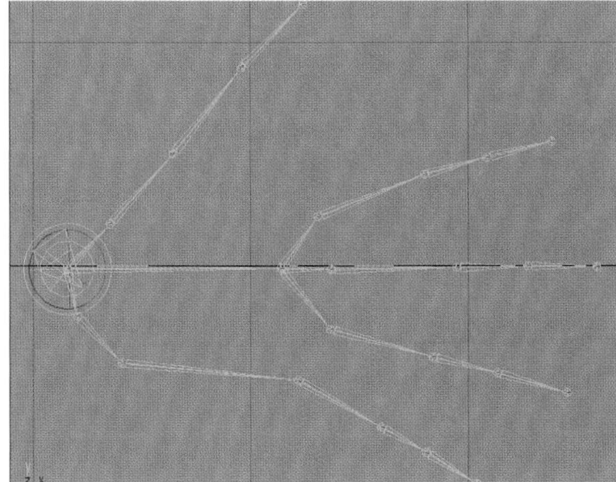

Figure 4.39
Modeling a hand skeleton.

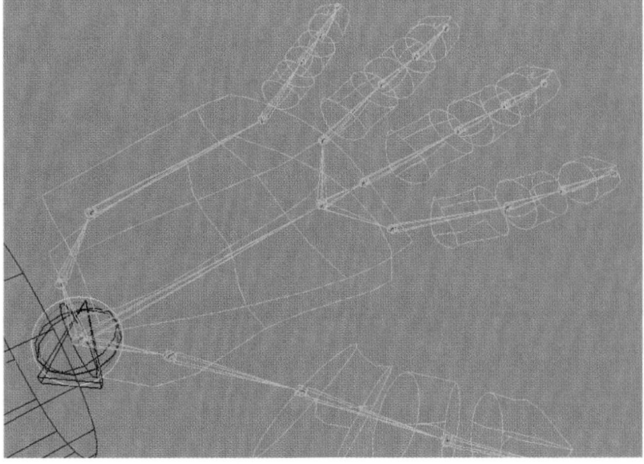

Figure 4.40
Making a low-res hand, and parenting the pieces to the skeleton.

4.5.1 Preparing the File for Animation

To prepare the file for animation, you will first model a skeleton hierarchy that correlates with the movement range that you want to exhibit with the model.

When this is done, go to Modify, Add Attribute, which opens up a window that will enable you to give this new attribute a name as well as a minimum and maximum range. Floating point number ranges between –5 and 10 are convenient, giving a finger the positive numbers to curl forward and the negative numbers to curl backward a bit, with 0 being the neutral, at-rest pose that you have modeled the hand in.

To add an automated control for posing the finger there is a command that allows you to set key frames that can be triggered with a new custom attribute called set-driven key, which you can easily add. To create a set-driven key you must do the following:

1. Select the joints that you want to curl, one finger at a time. Remember, the tip is relatively useless, so that is not needed.
2. Then go to Animation, Animate, Set Driven Key, Set. If the driven field already has the joints you want to control loaded, you are almost ready. Otherwise, click the Load Driven button at the bottom-right of the Select Driven Key window, and then select all the driven joints in the Driven field so that they are highlighted.
3. Select the base finger joint and choose the Rotation tool. Look at the color of the Rotation tool's manipulator circles and decide which axis you want to drive. Select the appropriate rotation axis in the Set Driven Key Settings window. Then select the locator and click the Load Driver button. Now select the controller locator and highlight the new attribute you just created to control the finger, and then click Key. This lays the first point for the driven key to return to.
4. Now set the new attribute to its highest positive number, and then rotate each of the finger's joints so that the finger is in a completely curled position, the maximum that you want to allow the finger to curl. Then make sure all the finger joints are selected in the Driven Key window, and click the Key button.
5. Set your rotation attribute back to the minimum negative number time, and then rotate the joints backward so the fingers look slightly hyperextended. Key the pose in the Driven Key window. Now a fairly elaborate control hierarchy can be created that exists in an easy-to-select location, ready for you to bring to life. (You'll learn more about this locator in Chapter 10.)

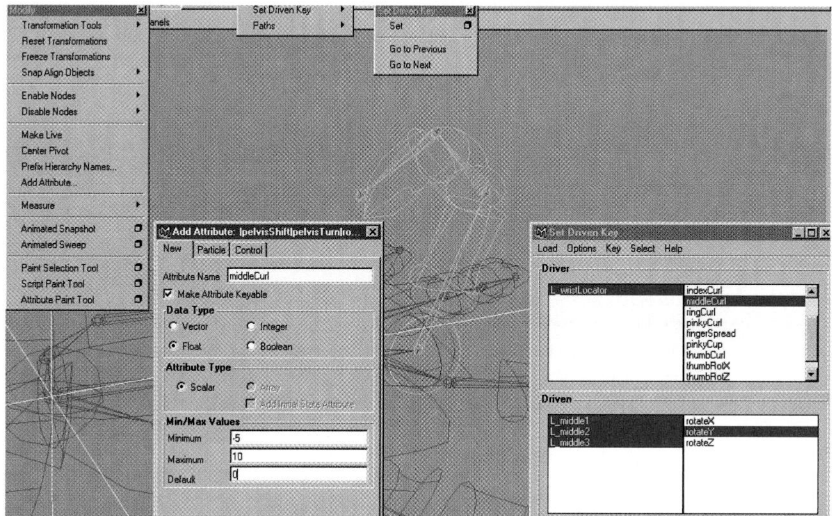

Figures 4.41
Creating set-driven keys to make forward kinematics control for the hand.

4.5.2 Binding the Hand Geometry to the Skeleton

When preparing to bind the hand geometry to the skeleton, you must decide what mixture of methodologies you want to use. Using a lattice across all the palms' hand panels to ensure seam cohesion for bone binding is usually the safest technique. The lattice welds geometry into place as long as both edges of a seam are within the lattice's boundary.

1. Select all the hand panel models and add a lattice with sufficient density to allow a human range of possible movements to bend the lattice when bound to the hand skeleton. This will, in turn, bend the mesh, causing a smoother and more skin-like effect than if you bind the panels to all the hand bones with no lattice.
2. Select the lattice and the root skeleton, and then go to Animation, Skinning, Bind Skin, (as from Maya 2.0 Rigid Bind) with the complete skeleton option bound to the hand bones. Try adding flexor lattices at certain key locations, such as where the bones pivot under the front finger panel.
3. Select the joints in question, and then go to Animation, Skinning, Create Flexor. This opens the flexor window. Using the default settings, click Create; a lattice flexor should be positioned at the selected joints. These lattice flexors help ensure that deformation occurs as smoothly as possible as well as give set-driven, key-style, cause-and-effect feedback. Flexors can be altered with their own custom muscle-related attributes in the Channel box after the joint has been bent into place, causing the surface to deform as needed.

Tip

Make sure that none of the fillet blend surfaces is highlighted when adding the lattice or when binding the lattice to the bones. It is a good idea to hide fillet blend surfaces during this whole procedure and then turn them back on when it is time to test the quality of the bind and its effects on the hand as a whole.

Also, for the fingers, it is important to use flexor lattices at the joints for all the knuckles. Otherwise, the binding the fingers to the bones is a pretty straightforward process on the selected joints of the finger. The main knuckle where the hand joins to the fingers and the tip of each finger should be left out of the selection because they do not have any direct influence on the finger geometries deforming.

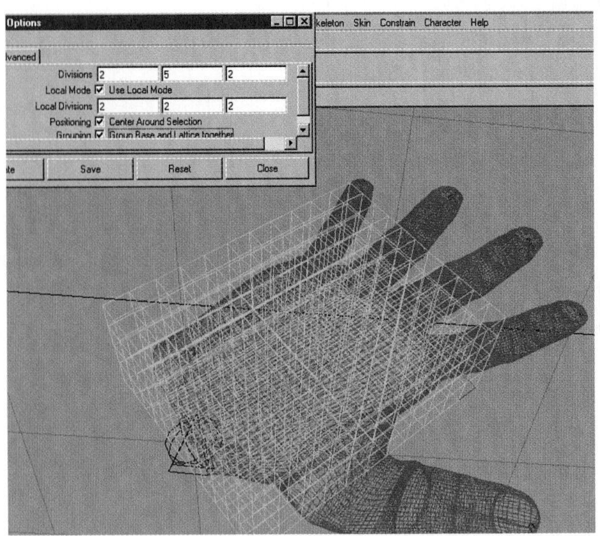

Figure 4.42
Using a lattice to do a bind skin on the hand panels.

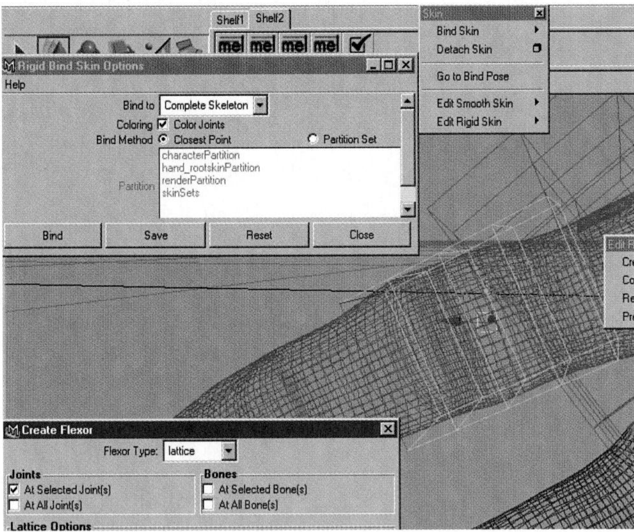

Figure 4.43
Flexor lattices on the finger skeleton after binding their geometry.

4. Prepare the file to be exported or given to another teammate by doing some simple organization. Grouping and layering all the pieces of geometry into easy-to-understand categories is very important. Group all the fillet blend geometry by itself and name it something such as DONOTMOVE_handBlends. If you were to transform these blends in any way, it would create problems because the blends are automatically re-created whenever the hand and finger geometries are repositioned. Because of their construction history, they never need to be moved and their transform nodes should be left alone. So, name them similarly to the above "warning" name.

5. Create a new locator by going to Modeling, Create, Locator. Then, with the Translation tool selected (press the W key) and holding down the V key, snap the locator to the center pivot of the hand's root joint.

6. The root joint of the skeleton can be point-constrained to the new locator. Select the Locator, and then select the root joint. Go to Animation, Constraints, Point. Select the locator and drag it around to confirm that the constraint works. Now the locator is ready to also host any custom attributes that might be necessary to create the set-driven key hand controllers. These would trigger premade keyframed motion, such as various hand gestures. With the locator as a separate entity from the other previous groups that you have made, you can create a final group that you can label Hand_complete and add all the other sub-groups to it, including the lattice's group.

Note

The palm and fingers should each be grouped separately, but they should never be moved because they are skinned to the geometry, thus handing over their position data to the skeleton deformers that now govern the geometry's CVs, overriding the Shape Transform node. The Transform node still has influence and, if moved, will cause serious harm to your hand's geometries.

7. Now, lock all the position, rotation, and scale attributes in the channel box. Do this by selecting the channels, RMB over the highlighted items, and choose Lock Selected from the marking menu. If you need to scale the hand as a group, scale the skeleton's root joint, which has influence over all the geometries and deformers that are skinned to it. This setup helps make your file as portable for you and for anyone else you may end up working with to be able to understand your database, and then most conveniently and adequately adapt the work to the next task at hand.

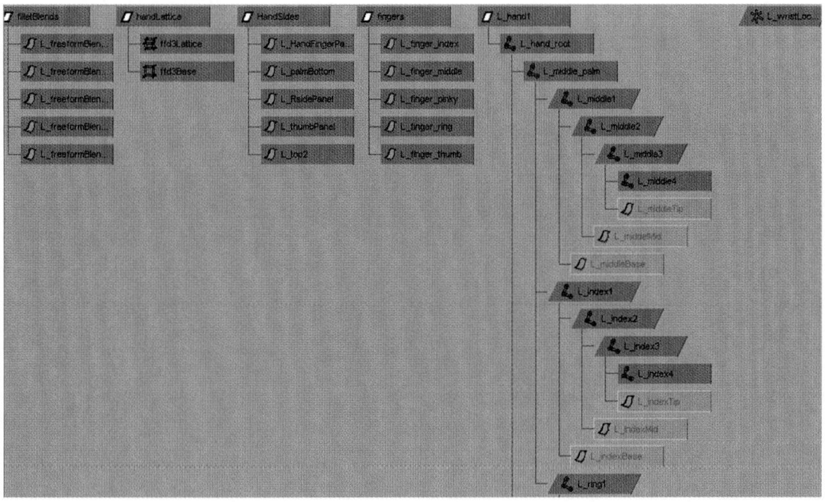

Figure 4.44
Grouping the geometry, and special naming organizational techniques.

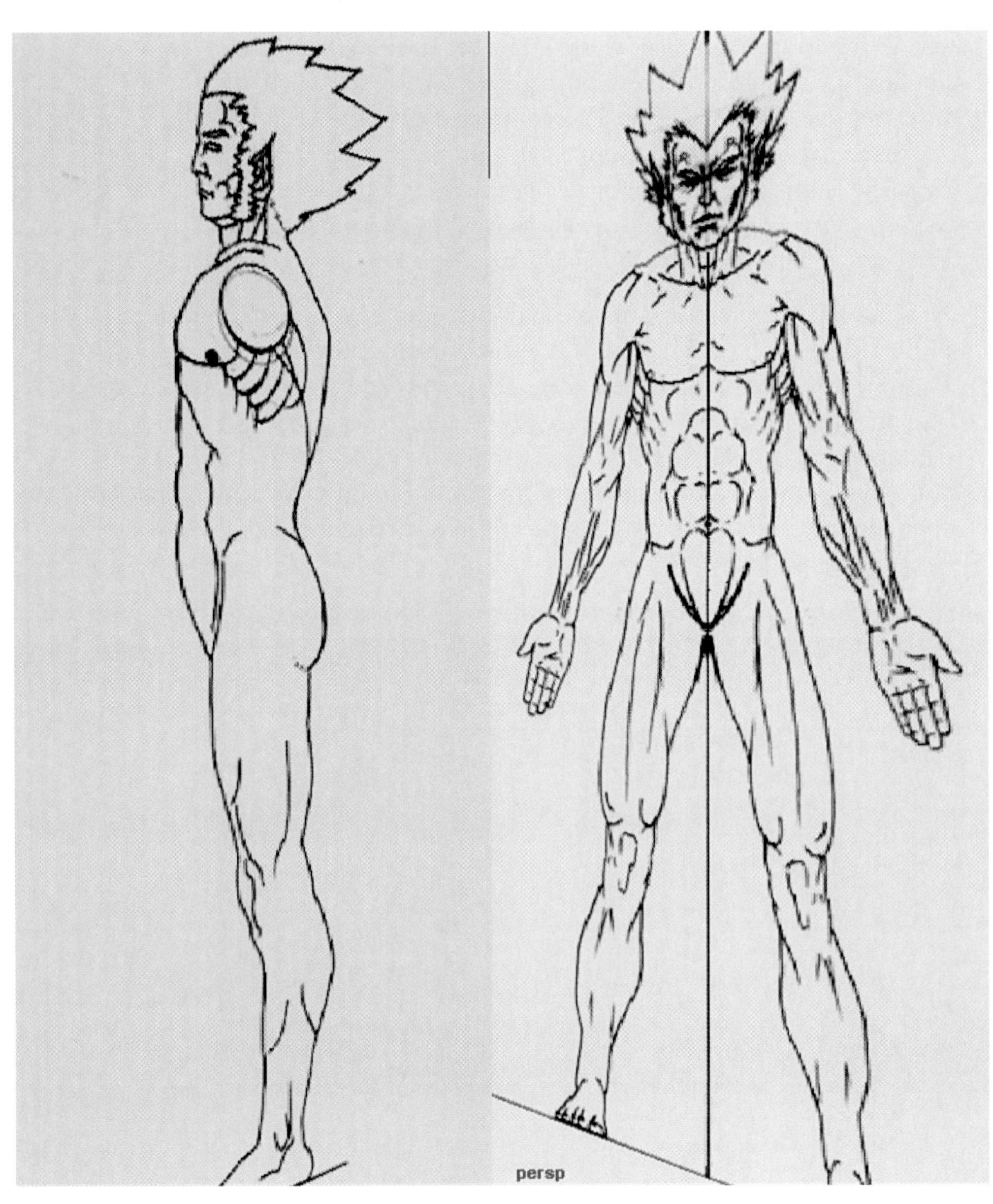
persp

CHAPTER 5

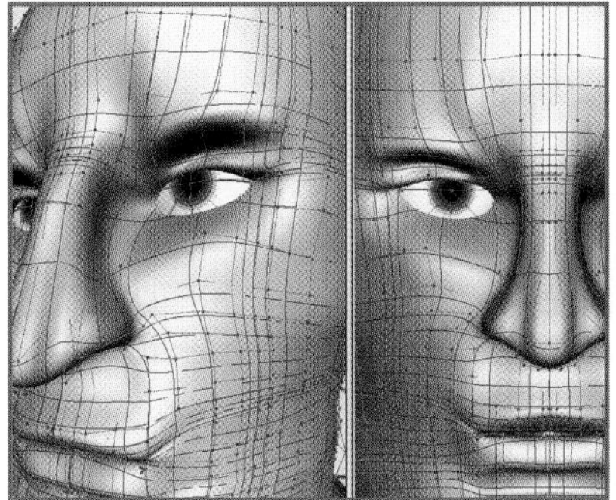

Modeling an Organic Humanoid with NURBS

by Nathan Vogel

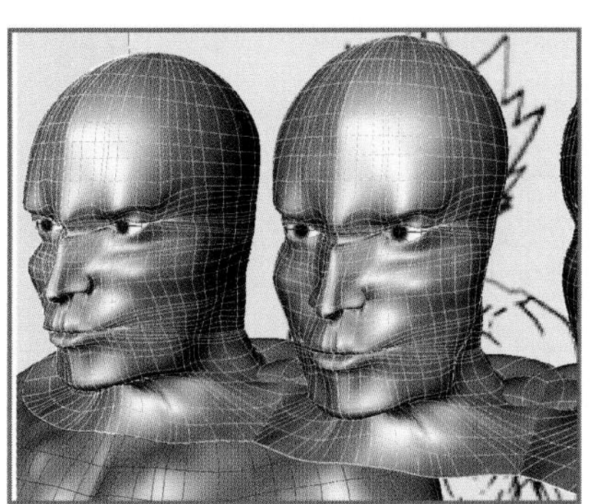

Maya offers a number of features in its state-of-the-art tool set that you can use to create realistic articulated 3D characters. In this chapter, we use these tools to model an organic biped to which you can add character animation controls for later animation. You can create articulate organic characters in Maya in a number of ways, and each method has its strengths and weaknesses depending on the requirements of the model you're creating. The following series of tutorials serves as a good introduction to many of the most important aspects of bipedal character creation in Maya.

5.1 The Lanker Tutorial

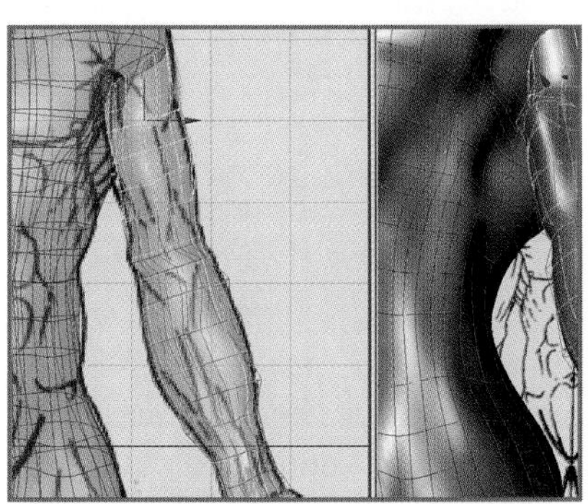

The Lanker tutorial series, authored by Alex Alvarez, provides the steps necessary for modeling an organic 3D biped character (Lanker is the name of the character on which Alex based the tutorial). This tutorial appeared in part in *3D Design* magazine, on the `www.highend3d.com` Web site under Maya Tutorials, and on the author's Gnomon 3D Film FX and Design school's Web site, `www.gnomon3D.com`. The full tutorial is included on the CD-ROM that accompanies this book.

The first part of the tutorial is referenced heavily in the next section. Because the tutorial was initially written so that it would fit in the limited space of a magazine, we added some details where it was necessary to clarify and elaborate on certain points.

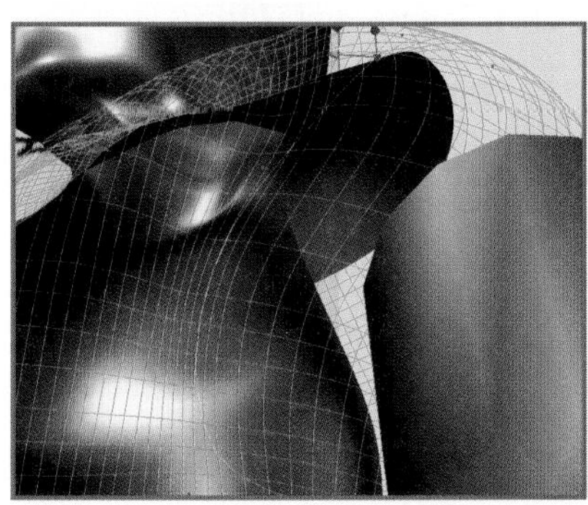

Note

When a key is referred to in a phrase such as "Using the V key...," it means push and hold the equivalent keyboard letter to activate the specified function. As long as the NumLock is on, the numeric keypad can also be used for numbers.

5.1.1 Creating a Torso and Head from a Sphere

In this section, we will learn how to create a torso and head from a sphere. As the first step in the modeling process, however, you will need to load any reference images you may have into two image planes—one orthographic front image and a matching scale orthographic side image. Next, you will create a primitive sphere and hull, edit them to match your side profile image, and add horizontal isoparms to match the contour of your source drawings. When the time comes to add vertical details that exceed the resolution of the sphere, you will detach the sphere in half to assist in symmetrical placement of new vertical isoparms. After adding the vertical details, you will mirror the half torso shape and sculpt both of the sides simultaneously with Artisan's multi-surface Brush Reflection option.

> *"The first and most important step involved in building a character is design and the preparation of orthographic drawings. The main thing to point out is that I am working off of only front and side drawings, due to the fact that the arms are at the character's side. I decided to model him with his arms down, just to make sure that his shoulder region looked exactly as I desired: a more common, neutral position."*
>
> \- A. Alvarez

You must make sure that you place the image planes so that they intersect and are aligned perfectly. Good placement helps to ensure accurate cross-sectional placement of the hulls. If you are not an illustrator, you can find many other orthographic images in classical anatomical art reference books.

Several orthographic character templates containing front- and side-view matched images of an orthographic reference humanoid are included on the CD that accompanies this book. The illustrated version of the character with the large hair-do is based on Cuthrau, a character originally designed by Ivan Wachter for a fantasy role-playing game. Wachter evolved the character into the digital realm, and an animated pilot episode is in the making. Look for the images in the directory BOOKPROJ: CHAPT4: PROJECTS: BIPED: SOURCEIMAGES: CUTSIDE.bmp and CUTFRONT.bmp, or BIPED_SD.bmp and BIPED_FT.bmp. If you want to use biped character images of your own, be sure to prepare them using the steps outlined in this section. You must decide which images you are going to use as reference. The anatomical drawings on the CD have extra contour lines that may come in useful.

5.1.1.1 Creating Image Planes

To create your image planes, follow these steps:

1. Create an image plane on the side camera by going into the Cameras tab in the Multilister window and double-clicking the side camera. The Attributes Editor for the side camera should appear. Scroll down the list of entries until you reach the Environment pop-up menu, and click the Create Image Plane button.
2. At this point, the Attributes Editor should reveal the options that you can use to create a new image plane. In the Image Name option, click Browse and then load your side view of your biped. Double-click the new image plane and click the Select button found at the bottom of the Attributes Editor window. In the Channel box, adjust the Center X,Y, and Z channels to adjust their positions in space.
3. To create the front image plane, follow the procedures outlined in step 1.
4. Next, create a reference chamber around your character that will give you instant feedback on the modeling process. To do this, align the imagery so that the side view is still centered on the Z,Y axis but is pulled back down the negative X axis so that it meets the far side of the front image plane. Then move the front image plane down the negative Z axis, away from the camera, until it meets the far right edge of the character's side view. You can use the resulting box around your character as a reference to refer to as you continue modeling your character.

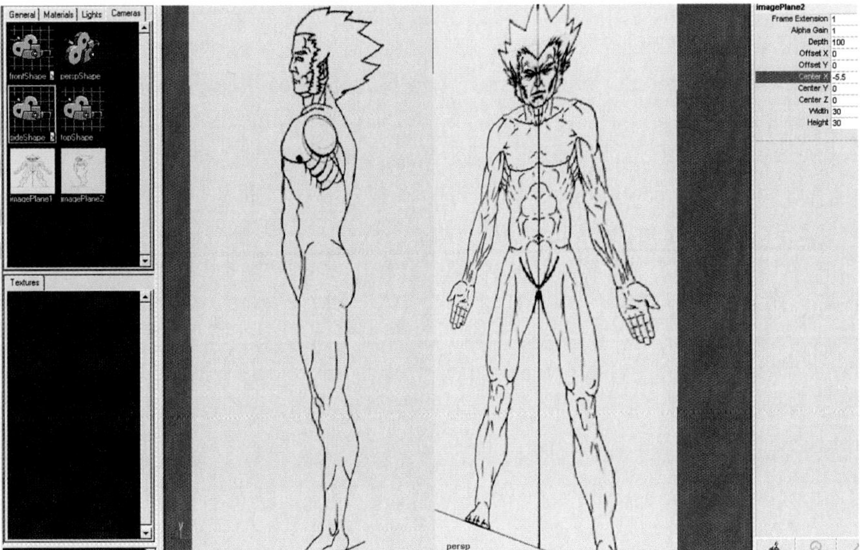

Figure 5.1
Aligning the front and side image planes assists in accurate cross-section modeling.

5.1.1.2 Beginning the Modeling Process

"With the orthos, or image planes, loaded into my front and side views, it is time to begin modeling. The first step is to place half a sphere into the side view with the poles at the top of the head and bottom of the pelvis. What we are going to do is model the entire head (including eyelids, nose, ears, and mouth), neck, torso, and pelvis out of a single half sphere. When modeling any humanoid form, this is the way to go, especially for the head. Any other technique can seriously slow things down, maybe not while modeling, but definitely down the road when setting up deformations or sculpting morph targets for facial animation. When this geometry is complete, we will then attach arms and legs with the Fillet Blend tool."

- A. Alvarez

Follow these steps to get your spheres in place:

1. Using a full sphere shape, go to Modeling, Create NURBS Sphere. (Although Alex's tutorial mentions creating a half sphere, we will create a full 360-degree sphere and split it later, after we add certain symmetrical details.) Set the sphere's parameters to 16 × 4 sections and spans so that you will have enough geometry to begin with. Set the radius to match the waist's profile in the side view.
2. Later, we will want to use symmetrical hull scaling from the front view. To prepare for this, create a full spherical object. Go to the side view, and then, from the menus or hotbox, choose Modeling, Create Surface, Create Primitive, NURBS SPHERE. Don't edit the settings in the Create Primitives settings palette. Instead, edit the model interactively from the inputs for MakeNurbsSphere in the Channel Box, or from the Attributes Editor window.
3. To display or select components for a single object, hold the RMB over the sphere and choose an edit mode. Hulls and CVs options enable direct editing, whereas isoparms and surface points enable you to mark the surface for other tools so that you know where to perform the desired tasks. Try the Hulls mode and click the pink horizontal hull floating above the surface's isoparms. Using the R key, choose the Scale tool. Notice that the Scale tool's manipulator is in the center of the hull's ring of points. If you were working with the hull of a half sphere, the manipulator would be in the center of the hull's points. Thus, the Scale tool's manipulator would not scale toward the grid, unless you created a custom MEL script that reset the center point of the hull each time you selected a point.

Note

When you select a hull object component in Maya, the default central point of manipulation is at the calculated average center of all the CVs in the hull (or of any other group of selected components). Thus, if you use the half-sphere technique, you might become frustrated when you try to scale the hull symmetrically or toward the zero point on the X axis.

4. Later, you will complete certain basic shaping steps on your model so that it bisects down the center, between the eyes. This will enable you to conduct multi-surface reflection editing in Artisan. In preparation for that step, delete the history of the object so as to lighten the geometry and prevent construction history build-up. To do this, go to Edit, Delete by Type, Delete History.

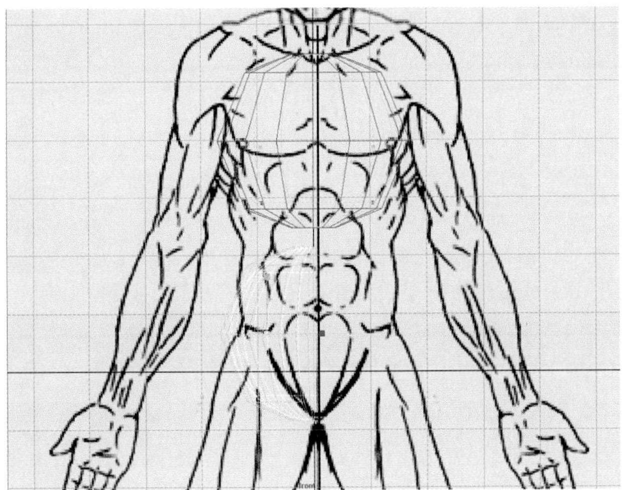

Figure 5.2
The primitive sphere is a useful and straightforward modeling foundation.

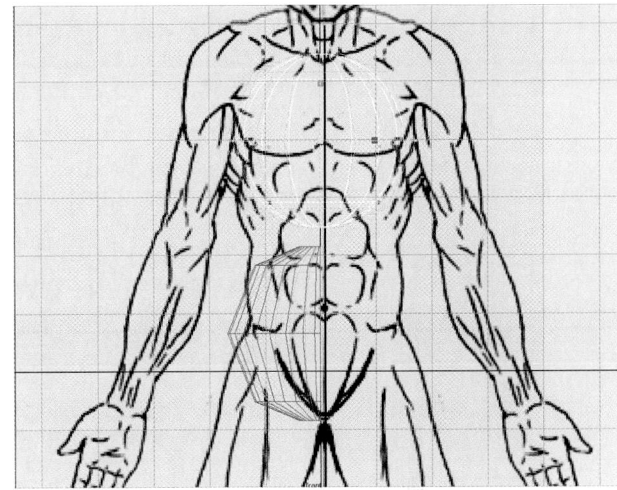

Figure 5.3
Here we see a hull of a half sphere and a full sphere, with the Scale tool active. Note the center point for scaling.

5.1.1.3 Manipulating the Hulls

> *"With the sphere in place, we now want to shape it to match the general forms of the Image Planes, starting first with the side view. A feature of Maya that makes this and many other common tasks more intuitive are the new manipulators for modifying objects. When an object is selected with a transformation tool—either translate, rotate, or scale—a manipulator appears which allows for intuitive freeform or axis-constrained editing. For objects or surfaces, this manipulator appears at their Pivot Points, but for Components, the manipulator appears at the center of the selecting vertices or hulls. This allows for editing in the Perspective window, which was before possible only in orthographic views."*
>
> \- A. Alvarez

When manipulating the hulls, you must move the top third of the hulls to the skull area to form the dome of the head. The bottom third of the hulls should be translated to the base of the groin; they will form the foundation for the hips, pelvis, and crotch region, as well as the upper buttocks.

In Maya, you can prepare a sphere for editing in a number of ways. You can either go to the Component Edit mode by pressing F8, or use the Object Components marking menu by holding down the RMB over the torso geometry. Use either CV or Hull mode to select the portions you want to edit. If you choose to edit in Component mode, set the pick masks to All Off, and individually activate either the CV or Hulls masks. If you want to use the marking menu with the RMB, select CV or Hulls to enable the editing process to transpire.

1. Activate the Translation tool using the W key and, in the side view, move each complete vertical ring of eight CVs to the designated region so that they match the image plane. If you want to move groups of CVs or Hulls, in CV mode simply drag the mouse over a larger region, making sure all points are highlighted; in Hulls mode Shift+click each corresponding hull to move them all simultaneously.
2. Remember that the topmost and bottommost tips of the sphere actually contain a ring of 16 CVs, just like the rest of the isoparm's hulls. This means that to be precise, you must be very careful to remember to pick it when moving the rest of the vertices in the region. Otherwise, spiking or tearing can occur.

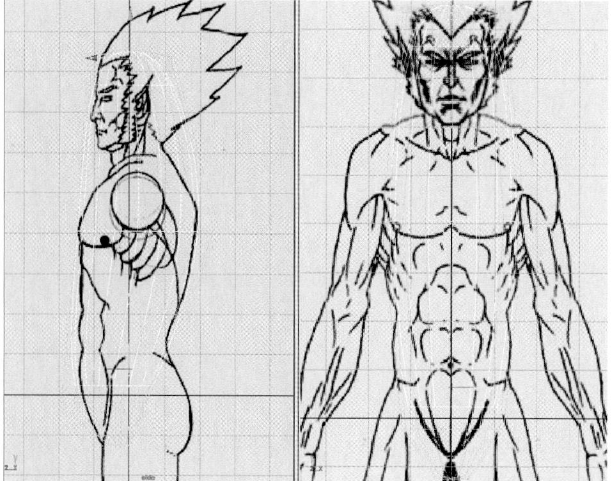

Figure 5.4
Select and place the hulls by translating or Z-axis-scaling them to fit from the side view. Match the image plane for precision.

Figure 5.5
Be sure not to forget the topmost hull.

5.1.1.4 Inserting Horizontal Isoparms

"The default primitive sphere we placed over the image plane was created with 16 vertical and four horizontal spans. The next step involves inserting enough horizontal isoparms to match the drawing, without including any areas of tight sudden changes in curvature. If the entire figure were, for example, 6 feet tall, you wouldn't want to get any horizontal isoparms that are closer than 1 inch apart for the head and 3 inches for the torso. Using the Insert Isoparm tool and the Move tool, in the side view place the hulls (rows of vertices) so that the Move manipulator is centered horizontally with the corresponding area of the drawing. Then, switching to the Scale Manipulator, scale the hull to match the design. An important note with the Scale Manip is to constrain the scaling to the Z direction. Not doing this will cause the selected vertices to also scale across the YZ plane, which will cause problems in mirror copying. Once the side view has been filled out, the same process is done in the front view. It is important for

future ease of manipulation to try and keep the hulls as 'vertical' and 'horizontal' as possible; waviness will make the wireframe very difficult to read and edit later."

- A. Alvarez

Continue adding horizontal isoparms from the side view until the geometry approaches the details needed to form the torso and basic head shape silhouettes. Attempt to add only the most essential isoparms at this time, making sure to edit them into place as you go along. (Although Alvarez uses half spheres in his example, I'm using whole spheres here.)

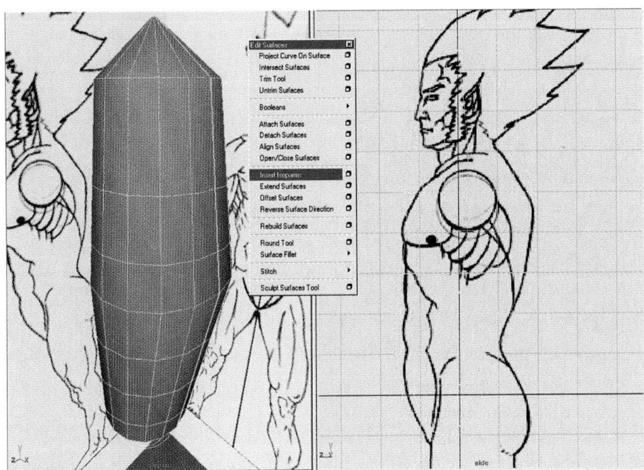

Figure 5.6
You must insert more horizontal isoparms to achieve the profile of the biped.

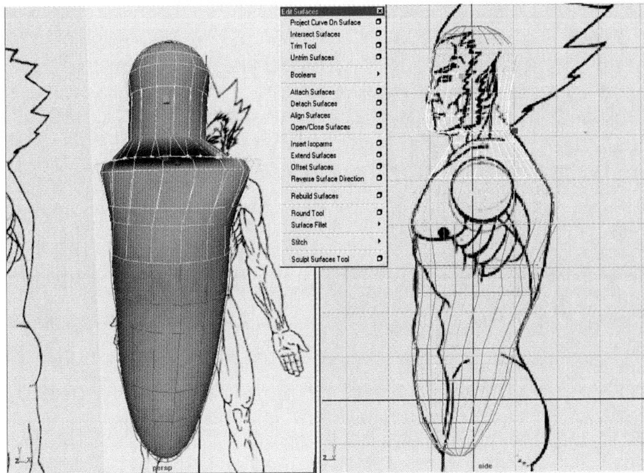

Figure 5.7
Be careful not to bunch up too many isoparms in one place quite yet, or it might cause creases or lumps.

5.1.1.5 Matching the Image Plane's Curvature

"With the basic forms filled in, it is now important to decide exactly how we want the curvature of the surface to flow. Trying to follow anatomical muscle flow can be difficult, but remember that many details can be added via bump and displacement maps later. The main problem is trying to get surfaces to deform diagonal to their parameterization. Thus, at this point it is good to try to sculpt the still-simple surface so that its isoparms, or surface curves, follow the flow of form in the design. For example, in the neck/shoulder area, muscles move from the back of the skull to the end of the collarbone, and from behind the ear to the beginning of the collarbone. The Artisan module of Maya can make this process substantially quicker, especially later on, when the geometry becomes more dense."

- A. Alvarez

At this point in the modeling process, you will be completing the torso shape.

First, select the front and back center isoparms running up and down the length of your model. You can write a MEL script to automate part of this procedure, if you want. Open your Script Editor and place it in a convenient and out-of-the-way, yet visible area of the screen. As you proceed to detach surfaces of your torso model, select and copy the MEL commands that react to your specific moves.

Here is an example of a script copied from the Script Editor by simply recording the moves on a simple default NURBS sphere primitive that had its history deleted prior to performing the process.

```
detachSurface -ch 1 -rpo 1;
select -r nurbsSphere1detachedSurface2 ;
delete;
select -r nurbsSphere2 ;
duplicate;
setAttr "nurbsSphere3.scaleX" -1;
```

The first line instructs Maya to use the Detach Surfaces function on the selected isoparms. The second line tells Maya to select the right side. The third line tells Maya to delete the selected piece. The fourth line selects the remaining half. The fifth line duplicates the half object. The sixth line scales it –1 in the X axis, effectively mirroring it.

Drag the script from the Script Editor window and into your shelf, making a convenient button with which to activate the script. You will need to make this script adapt to the names of the object that you end up choosing; MEL scripts are very name-picky.

Do not perform the Freeze Transformations command on the mirrored duplicate because Artisan needs a negative scaled object to successfully execute this technique.

Later, as you need more isoparm detail, you will again need to delete the right side, add isoparms, and once more mirror the left side to facilitate easy Artisan sculpting.

To complete the torso shape, do the following:

1. Select the torso geometry. Using the 1 key, set the display smoothness to Low, and using the 4 key, set the shading mode to Wireframe. Write a MEL script to handle this often-executed procedure by opening the Script Editor window and turning on Echo All Commands. If you have screen real estate to spare, keep this window open off to the side so that you can observe how Maya keeps close track of everything that is going on in the background.
2. From a perspective view, RMB over the torso geometry and choose Isoparm from the Object Components Marking menu. Select the front polar isoparm by clicking it with the LMB. Shift-click the back polar isoparm so that both the front and back isoparms are highlighted in yellow. Choose Modeling, Edit Surfaces, Detach Surfaces, Settings, and make sure that Keep Original is Off. Then, click the Detach button to cut the sphere in two.

3. Delete the right side of the sphere. A strange anomaly occurs when you use both sides of a detached sphere for reflection sculpting in Artisan: To get a predictable result, you must mirror one of the sides. To do so, select the remaining left side and choose Edit, Duplicate, Settings. In the window, set the scale parameter for the X axis to –1. Click the Duplicate button, and a mirror object should appear. Rename it R_side, and select the right and left torso objects to prepare the model for Artisan's Sculpt Surfaces tool.

"The Sculpt Surfaces aspect to Artisan allows you to push, pull, or smooth surfaces via tablet-driven input. As with pressure-sensitive applications in 2D paint packages, a Stamp Shape is selected which determines the shape of the area of the surface that will be selected and edited in a single stroke. The Radius and Magnitude of effect is controlled by pressure, while the artist can move vertices in the directions of their normals, the global axes, or along the U and V directions of the surface. This interface for editing vertices is powerful, maintaining near real-time interactivity while working on complex models."

— A. Alvarez

4. Now you are ready to use the Reflection option in Artisan. Go to Modeling, Edit Surfaces, Sculpt Surfaces Tool, Settings. In the Stroke tab, activate the Reflection Editing checkbox, and then choose the Multiple Surfaces option. Be wary of the north and south poles of the sphere when using this tool because it can give strange and unpredictable results in these regions due to the dense proximity of isoparmeters and strange "pinch" parameterization that occurs in the spherical primitive.

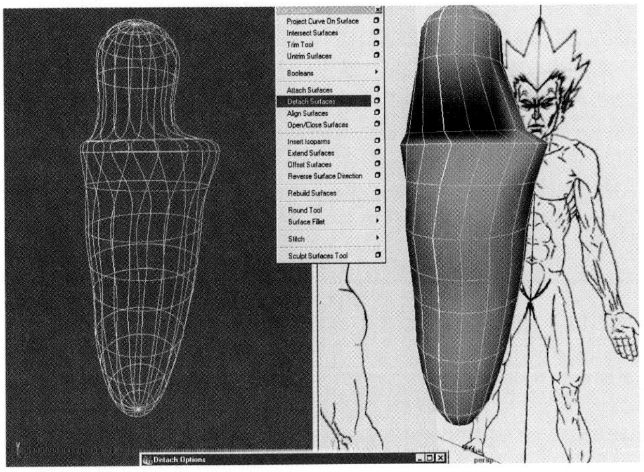

Figure 5.8
Detaching the surfaces.

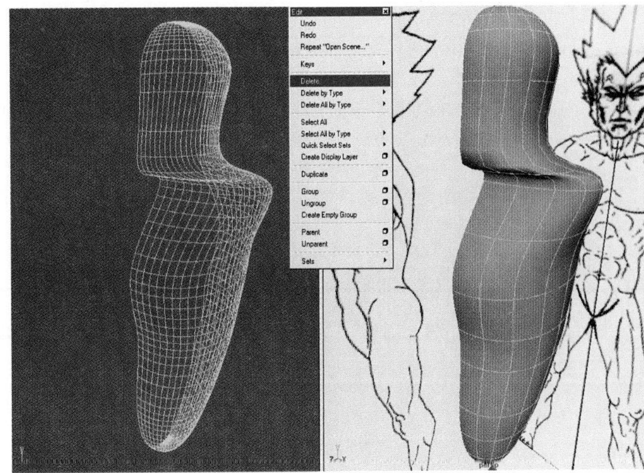

Figure 5.9
Deleting the right half.

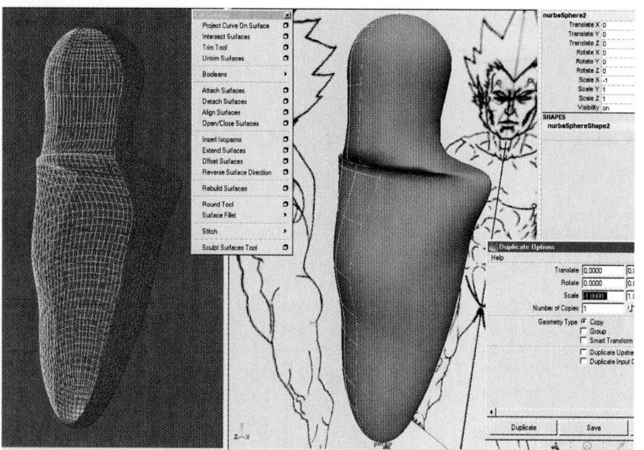

Figure 5.10
Mirroring the left side.

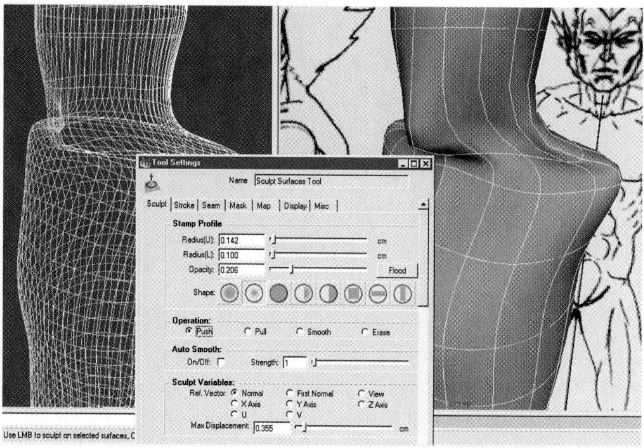

Figure 5.11
Setup for reflection modeling in Artisan.

5.1.2 Adding Detail to the Head

Creating the eyes and mouth may require some practice to perfect. Also, use care when adding more mesh resolution horizontally by adding vertical isoparms; the details span the length of the character. It is safer to add horizontal isoparms because their influence is more localized and does not tend to get in the way of the rest of the model as much.

> *"With the basic shape finished, we can begin adding details, trying to first concentrate on horizontal curvature. If a vertical isoparm is inserted while working on the head, the new isoparm flows all the way down to the bottom of the pelvis. Thus, by beginning work horizontally, we can work our way gradually down the body. When we finally need to insert vertical isoparms—say, to tuck the edges of the nostrils—it is a good idea to look around the surface to see if the new isoparm could help shape another region as well. This method keeps balance and fluidity of form while working on the model. The final model of the eye was put in place prior to doing any work in the eye region to act as a reference for the lids."*
>
> \- A. Alvarez

After editing as much of the face as possible with the existing local isoparm, it is time to add some essential yet well thought-out vertical isoparm to your model. When inserting these vertical isoparms, be frugal; place fewer at first and more when necessary. Whenever you insert isoparms and you intend to model symmetrically, you must follow certain basic guidelines. Remember to insert isoparms on only one of the halves of your model, and then mirror that half after inserting the new details.

1. In the perspective viewport, click the left side of the torso model. Press the 1 key to set the display smoothness to Low, and the 4 key to set the shading mode to Wireframe. Next, hold the RMB over the left side of the model and

choose Isoparm. Click any existing horizontal isoparm and drag the mouse up or down, stopping where you want to insert a new isoparm. A ghostlike isoparm should follow your mouse until you release the RMB. This new "yellow" curve is only a marker of where an isoparm will appear when you insert one, and is not permanent until you actually insert the isoparm. By holding the Shift key, you can click a curve you highlighted to unselect it or Shift-drag an isoparm so that you have multiple markers for new isoparms.

2. Next, go to Modeling, Edit Surfaces, Insert Isoparameter and convert all the yellow highlighted potentials into usable, editable isoparmeters. Now it is possible to add more details to the face and torso with Artisan. Re-mirror your half torso model every time you add isoparms to one half to keep the seams together. The simple mirror duplication MEL script that was introduced earlier comes in handy for this task. Adapt the script to your needs by recording any regularly repeated steps in order. Always test a new script before you finish creating it to verify that the script works properly.

3. Another equally important method involves using a negative scaled instanced duplicate. *Instancing* creates a duplicate that has a live reference to the original that forces it to update with its original. To create an instance, select the half torso model and choose Edit, Duplicate. If you need to make the instance a non-referenced object, delete its history. Remember that because the instance automatically references its original, you should use the multi-surface reflection painting in Artisan; not doing so will produce incorrect results. The first Artisan technique tends to keep the seams tangentially stable, but with the second technique it is easier to do mirrored CV level edits without using Artisan.

Note

Keep in mind that too many isoparms positioned too closely together will often result in kinks or unwanted ripples. Try to space the horizontal isoparms as evenly as possible. Adding a Blinn shader to the model helps to expose surface shape irregularities.

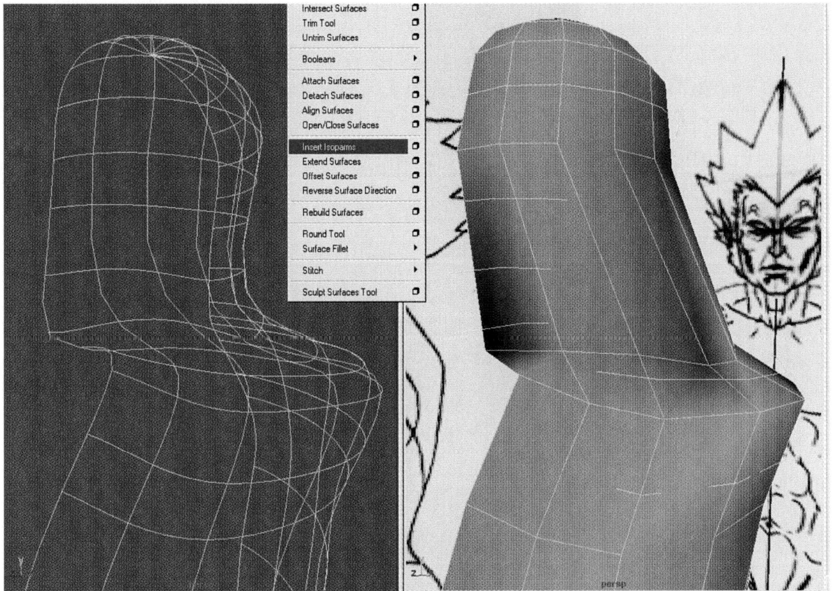

Figure 5.12
Setting the torso to low-res wireframe.

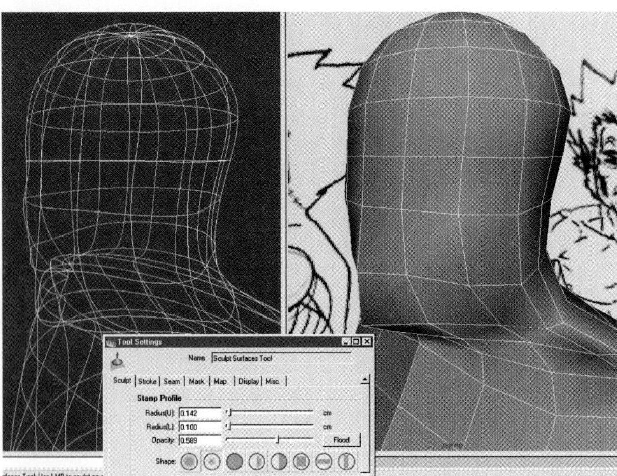

Figure 5.13
Inserting isoparms allows details to be added.

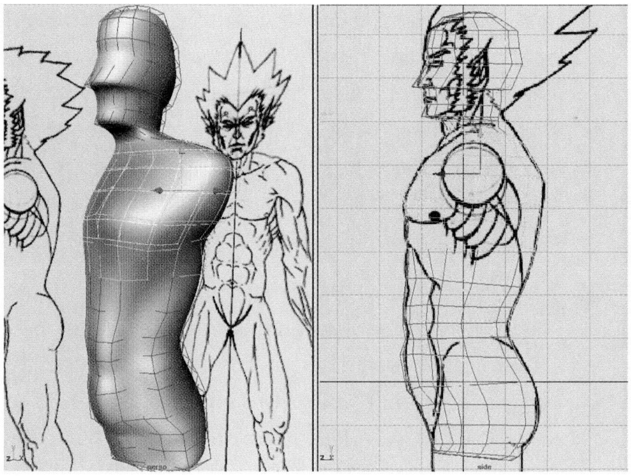

Figure 5.14
Source image templates.

5.1.2.1 The Mouth and Eyes

"Regions that pinch, such as the corners of the eyes and mouth, can be tricky due to the close proximity of isoparms that need to be inserted. Wrinkles can easily pop up, so always check your hulls to make sure they don't cross each other. Hulls are the lines that do not lie on the surface, drawn between the surface's vertices. They can be displayed using Display/NurbsComponents/Hulls. However, the wrinkle problem is easily fixed with Artisan, using Sculpt Surfaces/Smooth. This mode spreads apart isoparms, literally smoothing out the selected area. The mouth was handled by pushing a 'cave' in through the lips, creating relatively accurate lip and cheek thickness. The gums, teeth, and tongue (are) then sculpted as completely separate geometry using Revolve for the teeth and Loft for the gums and tongue."

- A. Alvarez

Creating the perfect mouth and eye sockets requires skill and patience. This is a good time to save your file and do some modeling R&D to get a feel for how all the delicate pinching and folding will work. It won't always look perfect the first time, so give yourself room to create some strange mutants along the road to perfection. Create the eye crater first:

1. When creating the mouth and eye sockets, you will want to insert the optimal isoparms that will enable you to pull large craters out of your model and still retain a smooth surface surrounding the craters. To do this, you need a few "control" isoparms near the vicinity of the outer edge of the eye and mouth craters. Hold down the RMB over the surface and then select Isoparm. Click an existing horizontal isoparm, drag it to where you want to leave a new one, and release the mouse button.
2. To add the other isoparms, hold the Shift key and click-drag a new isomarker (which shows up as a temporary yellow selected isoparm). Place at

least two iso-markers on each horizontal side and at least two on the top and two on the bottom vertically, to start out your eyebrows. Then, when your selection is complete, use Modeling, Edit Surfaces, Insert Isoparm to finalize the details. To punch in the geometry that will form the base of the crater, carefully select only the vertices you want to edit, checking in Wireframe to make sure that you highlighted only what you want. Then, translate the selected vertices in the negative Z axis. If the shape looks a little stretched, you can massage it to redistribute the CVs. Use the Pick Walking method of navigating through the mesh one CV at a time, using the arrow keys to get the most precise detail.

3. To create an eyeball model, create a sphere and turn it on its side by rotating the X axis 90 degrees so that its poles point down toward the Z axis. Then add a little eye texture by making a new Blinn shading group in the Multilister window, and map its color channel with a ramp. Set the ramp type to U Ramp, and set the colors to black, purple, and white. Localize the ramp's gradient bars near the top and bring the black down a little from the top. Now, apply this to the sphere by middle-mouse-dragging the shading group from the Multilister on to the object in the viewport.

4. Next, delete the sphere's construction history. Translate and scale the eye sphere so that it matches the reference art. Then add a Group node to the sphere, with its center set to the origin. Duplicate the eye group and then scale it –1 across the X axis to give the torso two reference eyeballs. Freeze the eyeball's transformation by choosing Modify, Freeze Transformation. You can template these if you need just their wireframes as reference. Now, as you create the eyelid regions, you have a proper reference in place.

Note

Artisan can be a little unpredictable when working in pinched areas and cavity regions. Therefore, it's safest to manually edit these regions. Remember also that the region behind the eye does not have to look perfect because the eye in the socket will obscure it.

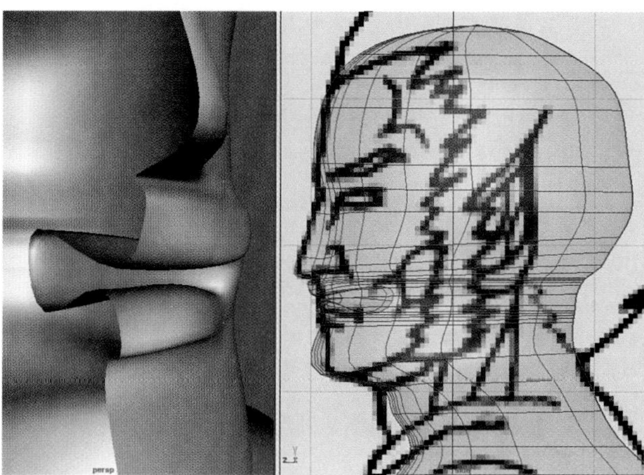

Figure 5.15
Forming the mouth cavity.

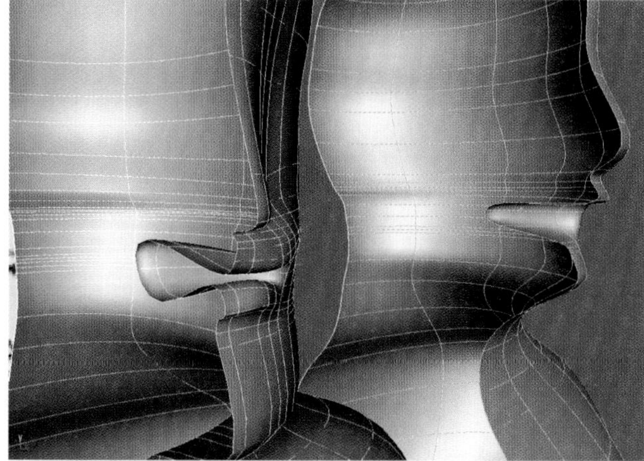

Figure 5.16
Sculpting the lips.

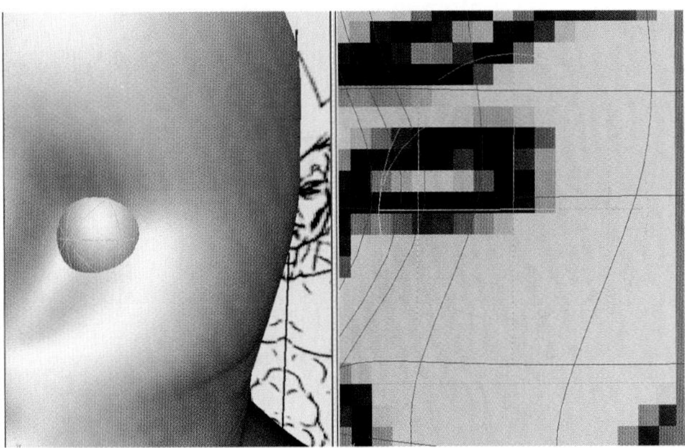

Figure 5.17
Placing a sphere for the eye.

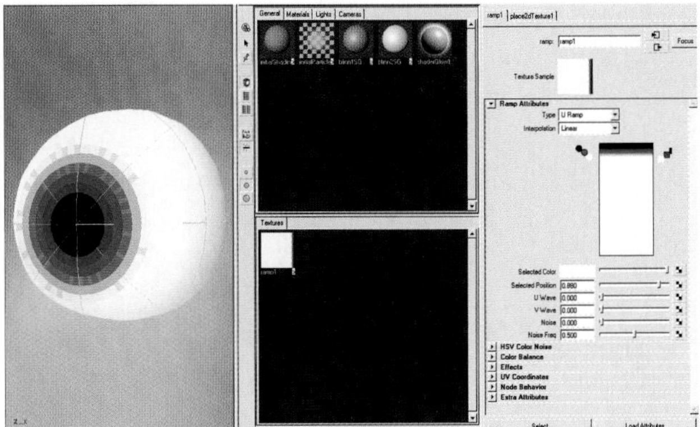

Figure 5.18
For pupil reference, add a ramp to the sphere.

5.1.2.2 The Nose

The nose can be a little tricky, requiring you to spend some time evaluating how you want to use the existing parameterization of the middle of the face without adding many custom isoparmeters.

1. The main section of the nose is a conical shape that protrudes slightly. Working close to the head region in the side view, pull out the main portion of the nose region by Shift-selecting individual CVs. If you pull out too many and the nose/nostril region looks too wide, use Artisan's Sculpt Surface with low displacement and opacity values to mold it back into place.

2. The nostrils require creative use of the surrounding isoparm details. After flaring the lower edges of the nostril area outward, be sure that there are enough isoparm CV details to pull a small nasal cavity upward.

3. Alternate between single- and multiple-object editing while adding the finishing touches so that you get a good idea of how your nose model is progressing. Although Artisan's Sculpt Surfaces tool is very useful, especially when you're using it to sculpt contours and bumps, it can be a little frustrating when you're using it to add details directly at the seams. The software's seam correction utility is helpful; however, it is processing a lot of data, so interactivity can be less responsive.

Tip

When creating any of these close-up facial details, try working on half of the model and then, using the techniques described earlier, add the mirror duplicate, gently smoothing and refining the details you added with intricate CV editing.

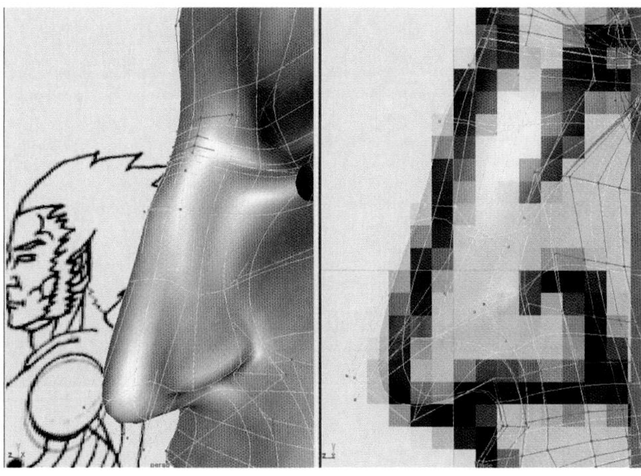

Figure 5.20
Sculpting the lips.

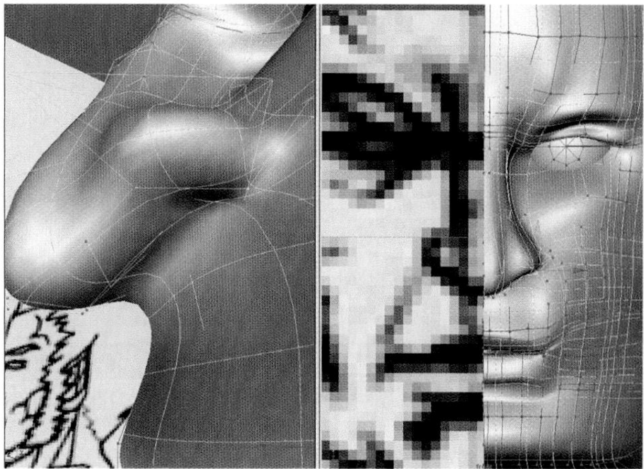

Figure 5.19
Forming the nose.

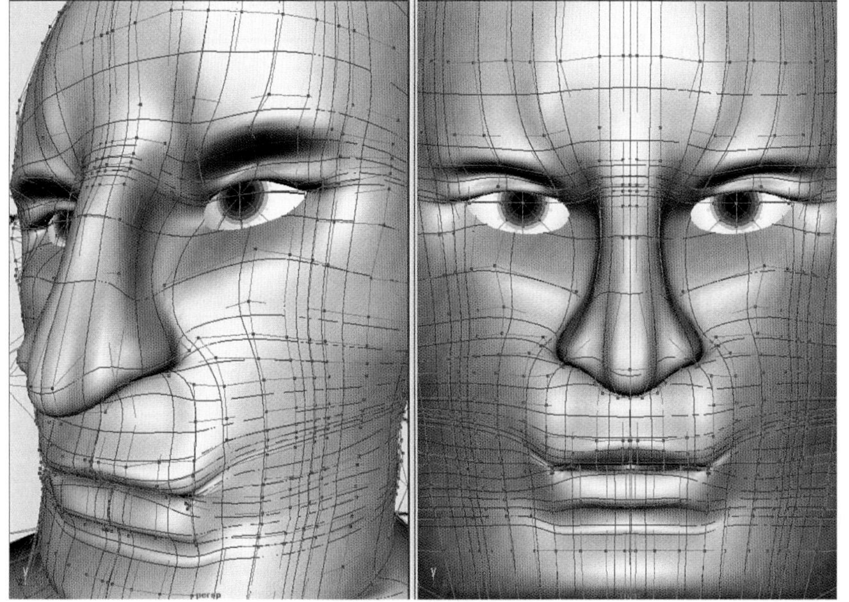

Figure 5.21
Placing a sphere for the eye.

5.1.2.3 Facial Bones and Muscles

Now it is time to model facial bone and muscle structure. Try to use as much of the existing eye and nose isoparm details as possible when sculpting the cheekbones, eyebrows, forehead, and jowls. Add details in the following order:

- Cheekbones and jowls
- Eyebrows and forehead
- Skin creases around the nose

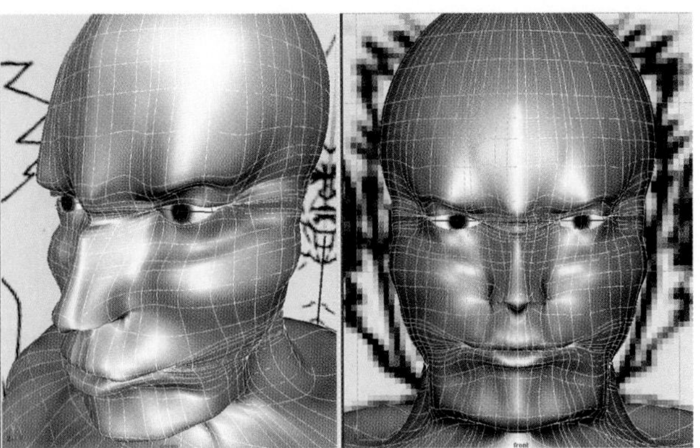

Figure 5.22
Cheekbones and jowls.

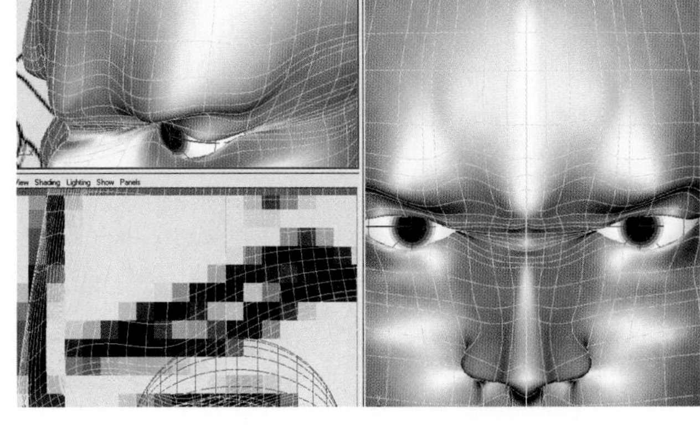

Figure 5.23
Eyebrows and forehead.

Figure 5.24
Skin creases around the nose. (Sometimes a little underlighting exposes details.)

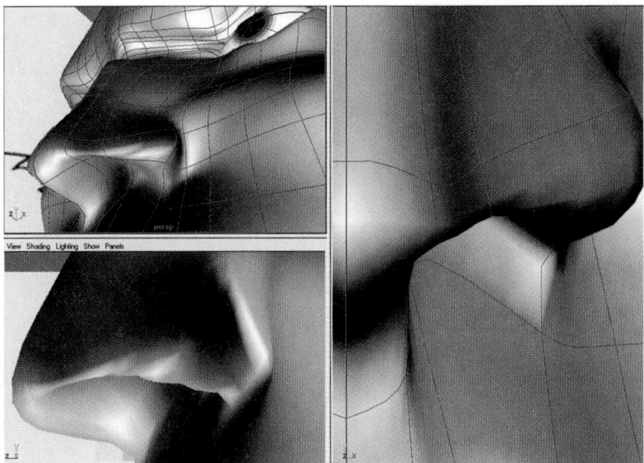

5.1.2.4 The Mouth

The mouth region is fairly complex, but if you follow this step-by-step procedure, creating it is an achievable goal.

1. Insert isoparms around the lips, adding double the number of existing isoparms located at the vertical edge of the mouth detail. These should be added above and below the lip area as control points to allow a tight change of direction as the cavity is burrowed out. It is often easier in the long run to place vertical and horizontal isoparms fairly tightly together and then spread the individual areas apart, as opposed to pulling the details together after inserting them far apart from each other. This helps prevent wrinkles from popping up.

2. Select the CVs that you will use to form the back of the mouth and to simulate the beginning of the esophagus and throat. Translate them backward, but be sure to leave room for the teeth, gums, tongue, and the roof of the mouth.

3. Smooth the transition area by spreading out the isoparm's CVs toward the top and bottom of the mouth area, inserting horizontal isoparms as needed.

4. Add isoparms to finish off the upper and lower lip regions.

5. Smooth any wrinkles that may have been caused during the sculpting and the addition of details. Use the Sculpt Surfaces smoothing option to lightly brush over the wrinkles and gently spread them apart.

5.1.2.5 Joining the Two Halves

"Now, you can attach a mirror copy to our head/torso surface, remembering to attach at both the front and back of the geometry. First, make sure to check the surface for any areas that may have accidentally been moved across the YZ plane. Simply selecting the interior edge hull and 'Grid snapping' it to the YZ plane will work. After mirroring the surface, we could add asymmetrical details to the torso, such as musculature, for example. However, with this (type of) model, I chose to do this primarily via bump maps. Next, I detached the surface at an isoparm located near the collar bone to give the model a shirt and manipulated the top hull of the detached torso geometry to give it some thickness over the skin."

- A. Alvarez

It's time to attach the two mirrored objects. This step involves first duplicating the half torso with the X axis scaled to –1, and then selecting the edge isoparms of each object by holding down the RMB, choosing Isoparm and clicking the edge. You should see a yellow line along the edge.

1. Shift-select the back side of the your torso model and then repeat the process. Go to Modeling, Edit Surfaces, Attach Surfaces. Experiment and then undo to try a few variations of the Attach Surface options.

2. Next, choose Modeling, Edit Surfaces, Open/Close Surface, Settings. Be sure that you are closing the V surface direction, and that Shape Preserve is toggled on. Be aware that the Preserve option automatically inserts a couple of extra isoparms to help preserve the shape. This can be bothersome if you do not plan for it accordingly. Attach and close the two halves of the torso after sculpting them to your satisfaction.

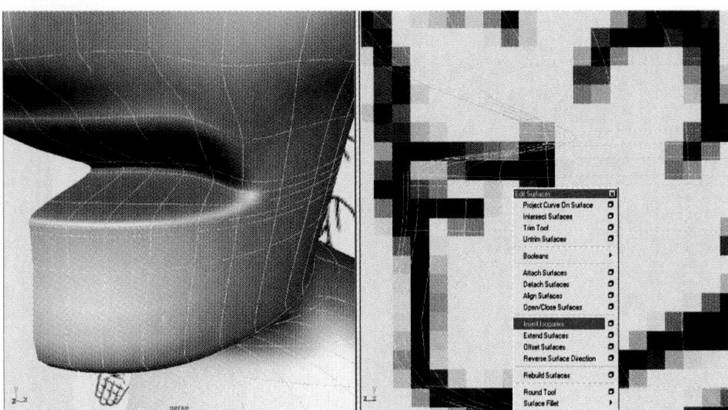

Figure 5.25
Insert isoparms around the lips.

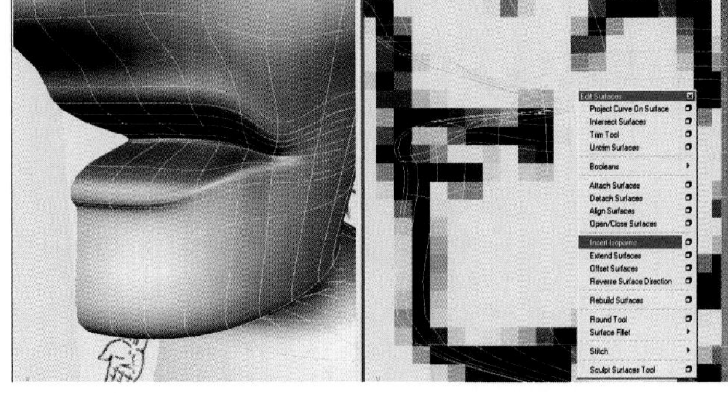

Figure 5.26
Insert details, and then move the CVs back.

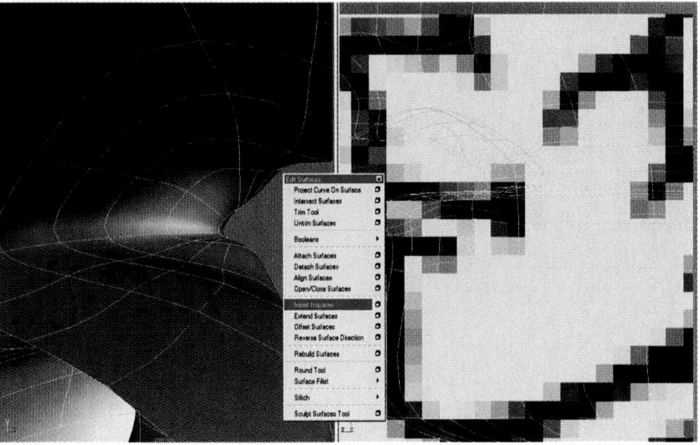

Figure 5.27
Expand the mouth to make room for the teeth.

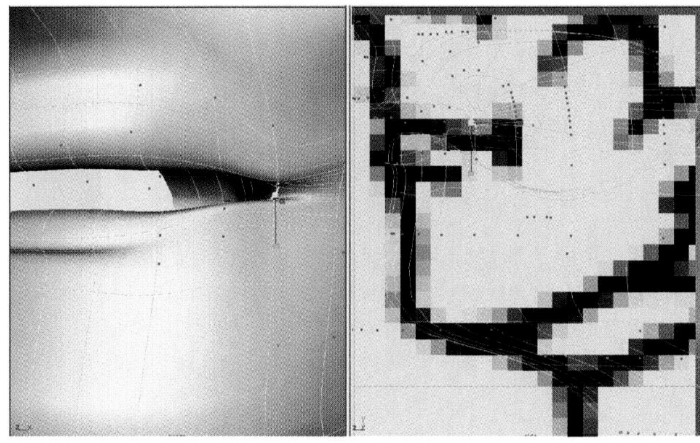

Figure 5.28
Insert details to finish off the lips.

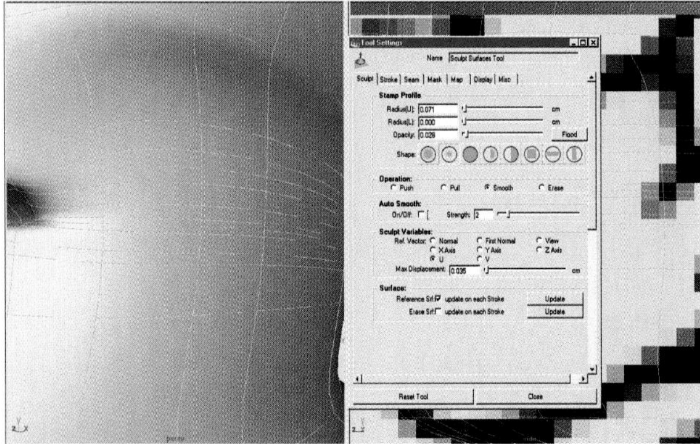

Figure 5.29
Use Artisan's smoothing brush to relax kinks.

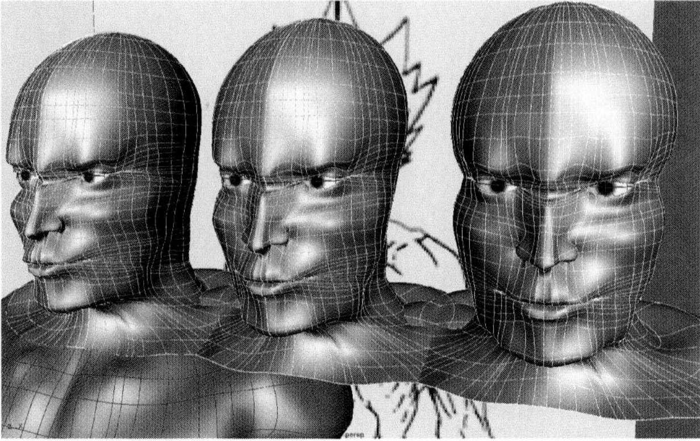

Figure 5.30
Stages to attach the head (this does not have to happen until you bind your skeleton to the geometry).

5.1.3 Modeling Legs from Primitive Cylinders

"With the head/torso geometry complete, it is time to sculpt an arm and a leg, each out of a primitive cylinder. The hand will be done separately. The same sculpting techniques as mentioned above apply, but now we need to consider the areas where the arms and legs attach to the torso, (just like) where the fingers attach(ed) to the palm. Since we are going to use the Fillet Blend tool to attach them, we can ensure good results by modeling the surfaces almost on top of each other so that they have a visually implied continuity. When this is accomplished, we can Trim, or cut away, regions of the torso within which we will create new surfaces using fillet blend. This type of geometry is solely based on the two surfaces it is blending between; thus the reason for modeling the hip areas of both the torso and leg surfaces so that they flow into each other. When creating our curves on surface in preparation for trimming, a nice new feature is the ability to make a surface 'Live' using Modify/MakeLive, which allows us to intuitively draw curves directly on the surface in the perspective window."

- A. Alvarez

Creating the leg from a primitive cylinder involves using translations and scaling to approximate the location of the leg:

1. Use a primitive cylinder set to at least 12 × 12 sections and spans.
2. Match the radius and height so that it generally looks like the reference art. Don't worry about it too much.
3. Delete history and freeze transformations to reset the leg geometry before you begin doing the hull editing.

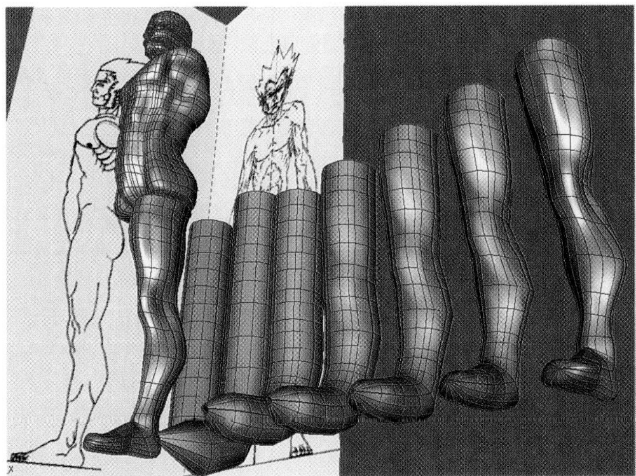

Figure 5.31
Perspective of the leg progression.

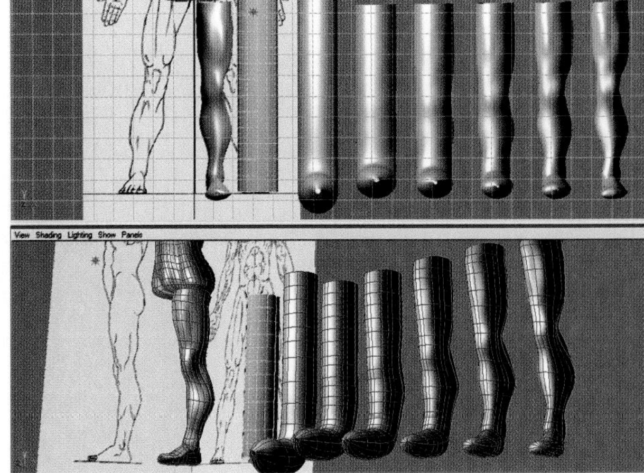

Figure 5.32
Front view of the leg progression.

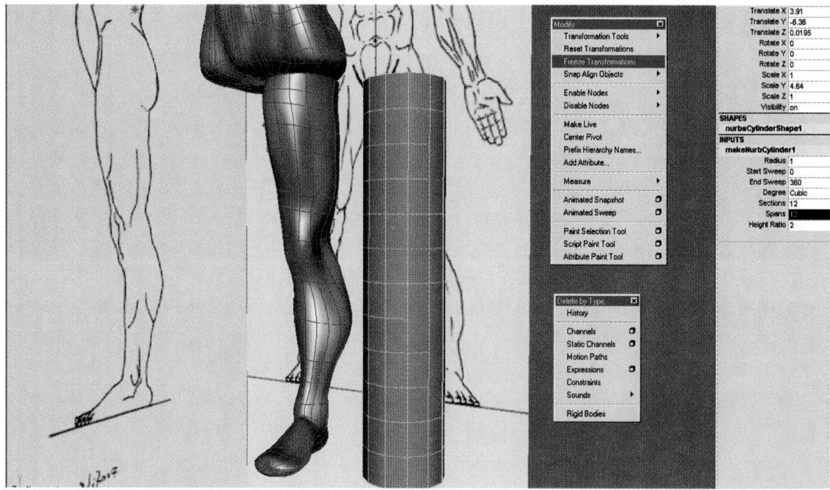

Figure 5.33
Create a cylinder and align it to the image.

5.1.3.1 Modeling the Leg and Foot

Now it is time to build a one-piece leg and foot out of the primitive. You will be using the Hull Editing method, the Insert Isoparm techniques, and the Sculpt Surfaces tool to form the initial shapes needed to sculpt the leg. Then, you will add horizontal resolution so that it approximates your image plane.

1. Begin by spreading the hulls to fit the side view.
2. Cover the reference image of the leg from the hip/pelvis region, all the way to the toe of the foot with the cylinder's geometry, one hull at a time.
3. Use the Translate tool to scale and rotate each hull to massage the geometry into place.
4. Add extra detail near the ankle and heel region by bunching up several hulls. Doing this also prepares this region to be stretched into the foot.

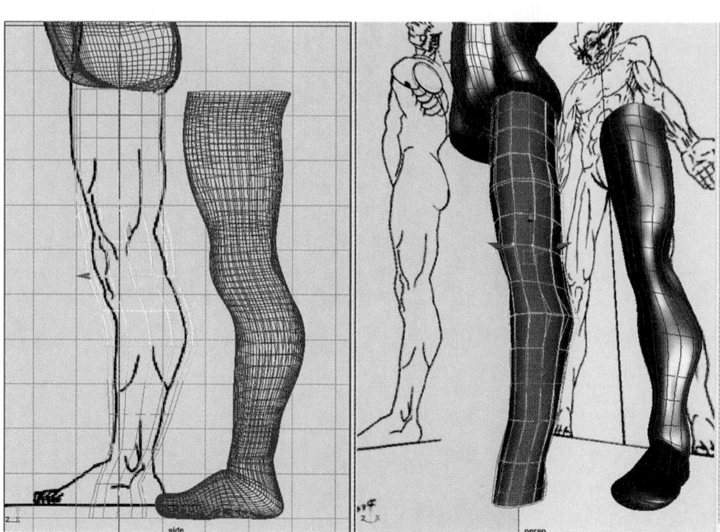

Figure 5.34
Continue hull editing to match your reference image.

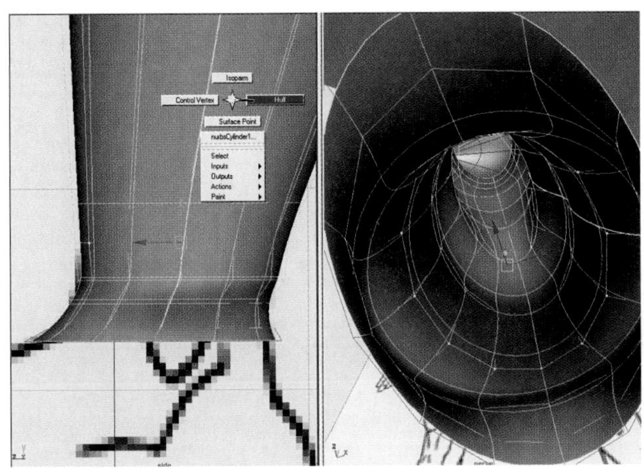

Figure 5.35
Add extra isoparm details near the ankle.

5.1.3.2 Adding Details

In this series of steps, you will add resolution to the geometry so that it matches the image plane. Keep in mind not to go overboard on mesh detail because you can add a lot of detail in the texture-mapping stage.

1. Add vertical details when placing each new isoparm. Take time to edit each new isoparm before adding the next curve; the shape of subsequent curves that are added will be closer to what you need.
2. Add horizontal details more sparsely because they influence more of the geometry.
3. Massage the added resolution into place using your favorite detailing methods. Be sure to place enough isoparms so that you can pull the foot out of the ankle area.

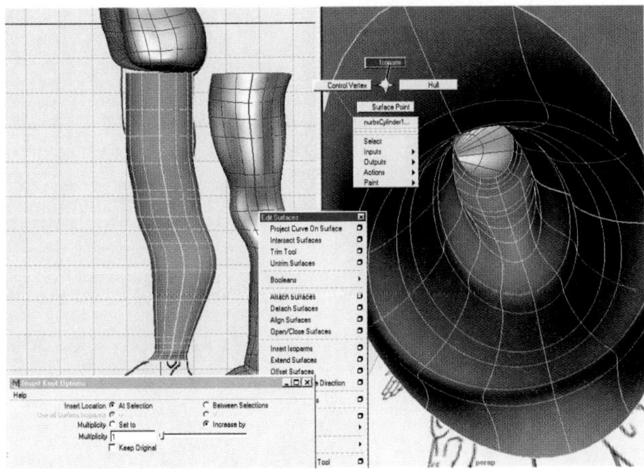

Figure 5.36
Insert horizontal isoparms.

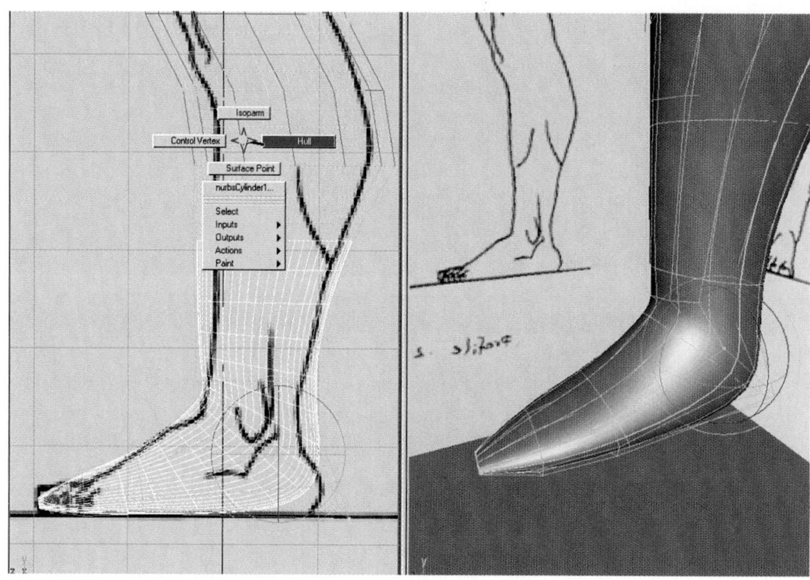

Figure 5.37
Rotate, scale, and translate the hulls to form an ankle and a heel.

5.1.3.3 Adding Muscles

Use Artisan's Sculpt Surfaces tool to add muscular details. Remember to keep the opacity and the Max Displacement set low so as to gradually build up details. When adding geometry to the back of the knee, use a pair of horizontal isoparms to help detail the skin clefts between the rear thigh and calf regions.

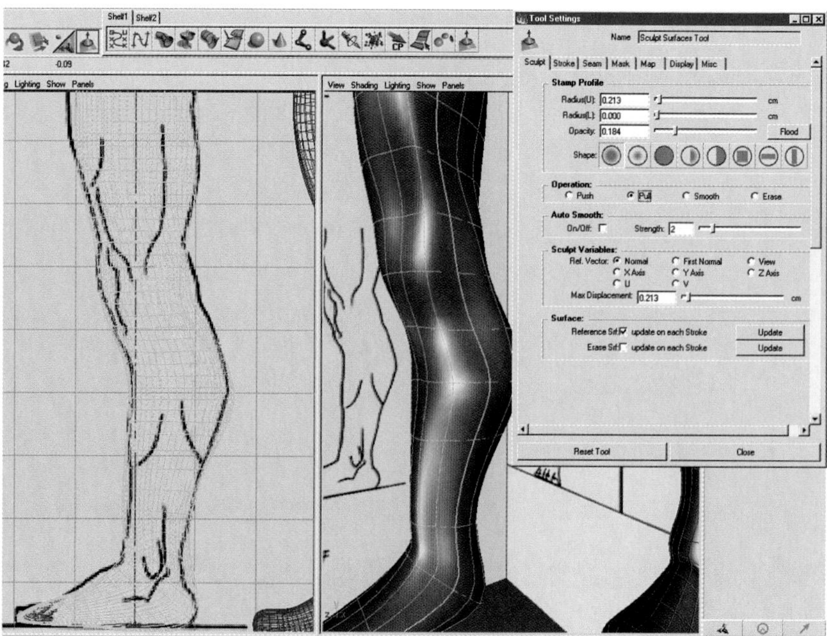

Figure 5.38
Use Artisan to mold the new geometry.

5.1.3.4 Prepping the Leg for Connection to the Torso

In this series of steps, we will prep the leg for connection to the torso. Add extra detail to the torso as necessary to help facilitate the region in which the leg transitions to the torso.

1. Align the geometry so that there is a smooth-flowing transition between models. The more similar they are, the more seamless the fillet blend will be.
2. Match the front and back groin region.
3. Make sure the buttocks are matched up by grouping the left leg. In the Group settings, be sure the group pivot is defaulted to the origin. Then duplicate the left leg group and scale to –1 X. Observe the symmetry of the two legs and how they approach the torso.
4. Model the area, making sure there is a smooth flow between the leg areas.
5. Add extra isoparms to the leg as necessary to enable the front and back areas of the torso and leg musculature to be properly contoured. If you add any isoparms, always re-duplicate any other mirrored parts to keep everything consistent.

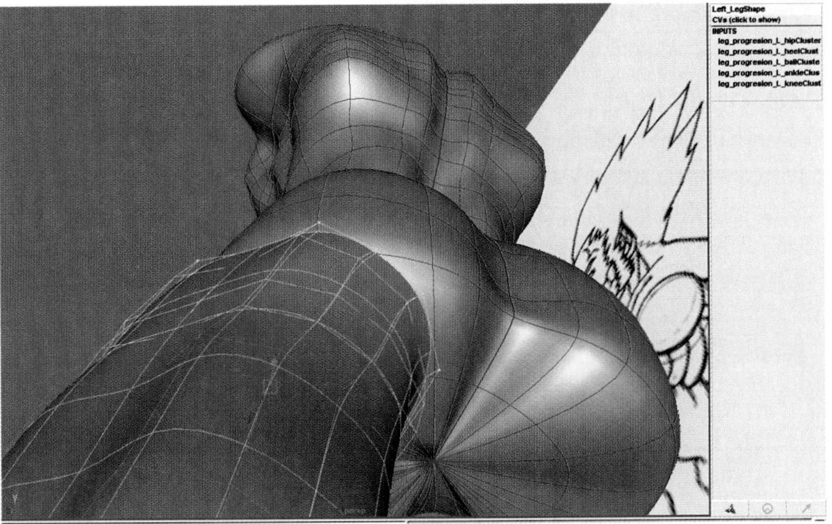

Figure 5.39
Edit the leg so that the geometry flows together.

5.1.4 Creating the Arms and Adding the Hands

We will create the arms from primitive cylinders following much the same process we used to create the leg. The heel and elbow have certain common traits—for instance, the bone protrusions have similar visual qualities.

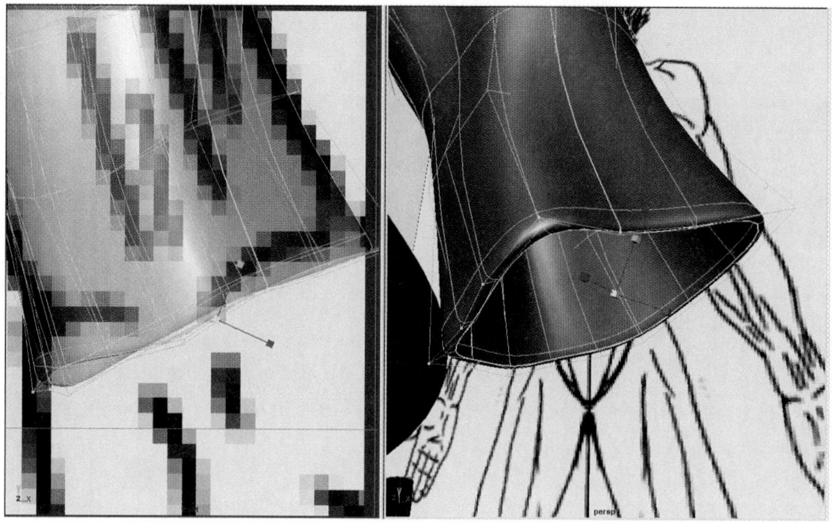

Figure 5.40
The arm's progression from primitive to anatomy.

5.1.4.1 Adding Muscles

You can add musculature detailing to your heart's content. Keep in mind that over-exaggerated muscles are a bit of a fad in the gaming markets, but a more tasteful and realistic, well-defined and muscled humanoid can still be appealing. For more reference on the subject, observe your own physiology and consult anatomy reference books for artists.

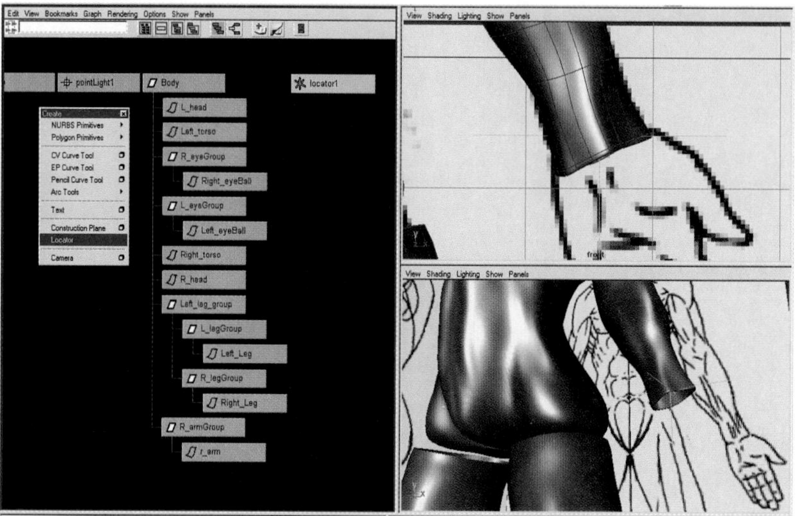

Figure 5.41
Hull editing is your mantra.

5.1.4.2 Tucking the Cuff

Make a cuff at the wrist to simulate a skin-tight, leotard-style outfit and to give you an easier way to integrate the hand. If need be, add a couple of isoparms. The Global Stitch tool, if tweaked enough, can also be used to integrate the hand.

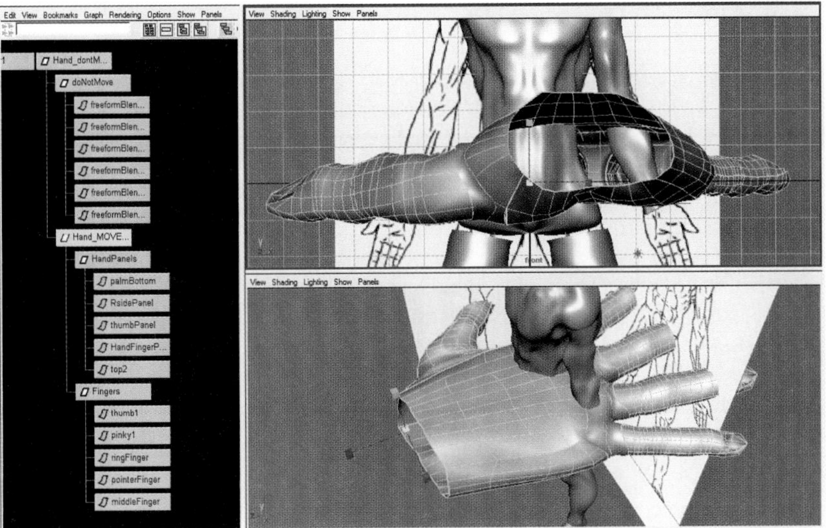

Figure 5.42
Tucking the cuff is one way to add the hand.

5.1.4.3 Preparing the File

It is time to prepare the file for integration with another file. Go into the Hypergraph and organize the geometry into a simple hierarchy so that even people who are unfamiliar with the file can understand your database. In this way, you will establish good workflow habits that will help you to organize your production pipeline.

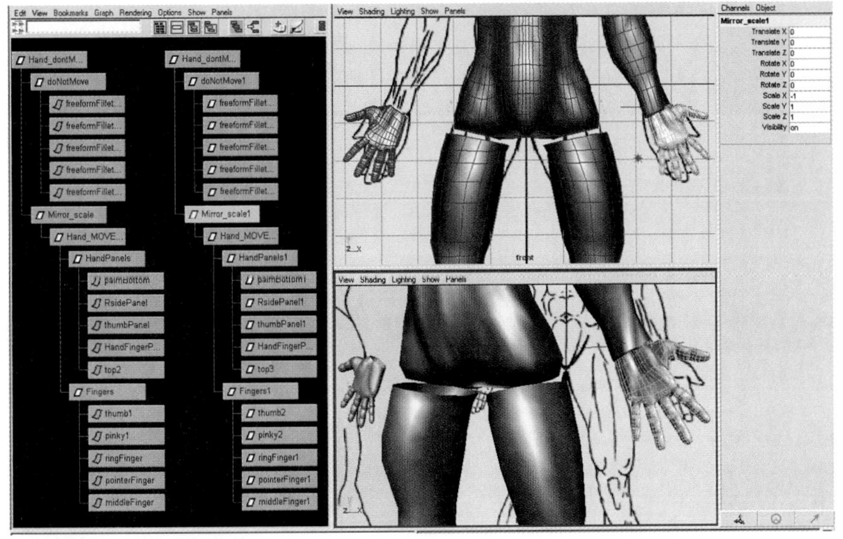

Figure 5.43
Preparing the scene's database is easy if you start before things get too complex.

5.1.4.4 Adjusting the Hand Model

You will have to adjust your hand model so that it fits the humanoid when you import it into your file. Import the hand file by going to File, Import. Locate and open the hand model that you completed, or use the ready-made file provided on the CD-ROM.You can find this file, called hand_done.mb, in your CHAPTER4/PROJECT/HAND DIRECTORY.

Note

Create your complete hand model in a separate file. You may want to export the arm to give yourself a reference for scale and placement. Refer to the hand tutorial in the previous chapter, and the character setup tutorial in the character set-up chapter. When your hand is skinned, has set-
driven keys, and is possibly even texture-mapped, import it and generally follow the same sort of grouping and scaling procedures to retain the construction history. If all you have is a hand model, delete its construction history using the earlier methods and group it for easy movement as a simple modeling reference.

1. Import the hand model. While examining the Hypergraph, become familiar with your database by selecting various objects and viewing where they are located in relation to each other hierarchically. Assess all your changes and additions. You performed most of the grouping while you were organizing the hand file in the Chapter 4 tutorials. They were prepared according to a procedure designed to simplify the duplication process while still retaining construction history. Using the Insert and Translate tools, reset the oversized hand's Hand_MOVEME group pivot to the base of the hand.

Note

When translating and duplicating highly complex models with dense, interwoven construction history, such as the hand model, you must understand the nature of construction history and how processes that retain active history, such as fillet blend, work. If you do not plan it out with a defined strategy, your model will often simply explode, suffering from predictable anomalies such as double transforms and history feedback loops.

2. Group and duplicate blend objects using the Upstream Graph option. Do not attempt to scale these groups or the blends; they derive their scale from the source geometry used to create them.

Note

Grouping and duplicating skinned objects is acceptable as long as you duplicate the skeleton and control objects at the same time, activating the Duplicate Upstream Graph option. Because the skeleton to which they are skinned determines their scale, you cannot perform any scaling on the geometry group.

3. Pick the group Hand_MOVEME that contains the hand geometry with no blends. Rotate and scale Hand_MOVEME to match the biped's arm and the background image. The hand that is provided is a right hand, and the arm geometry is on the left; therefore, align the hand to the arm for a natural, anatomically complete humanoid and then flip the hand.
4. Group Hand_MOVEME with the pivot around the origin. This prepares the geometry for mirror-scaling to create the other hand. Name this group Mirror_Scale.
5. When you duplicate the main hand group with the Upstream Graph preference, scale the hand's Mirror_Scale group –1 on the X axis, but leave the fillet blend group alone. This live construction history geometry derives its scale attributes from the source geometry from which it is created; thus fillet blend location is generated based on the location of the palm and fingers.

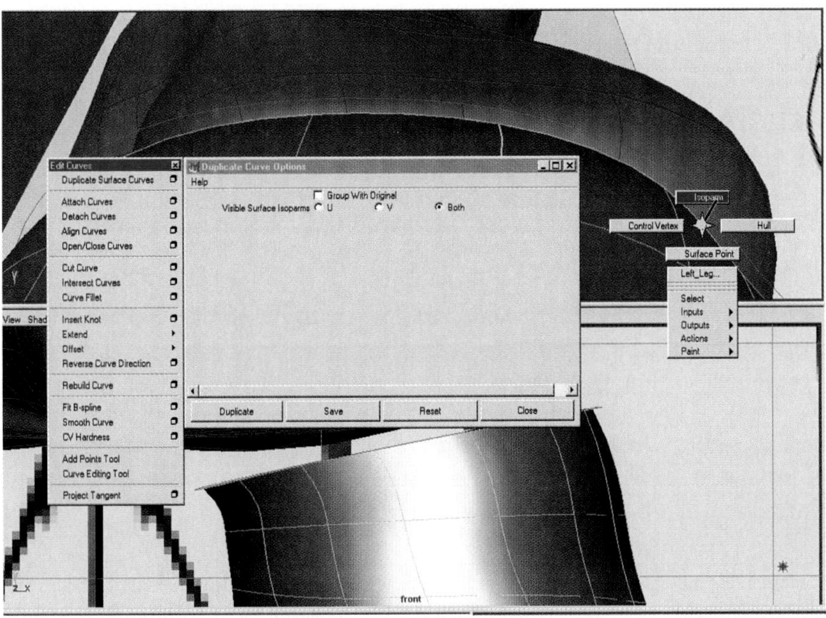

Figure 5.44
Move only the designated translations group to maintain proper construction history.

5.1.5 Using the Fillet Blend and Freeform Fillet Tools to Join the Limbs

To attach a blend to the arm, you will need to create a trim curve. Create a curve on surface at the shoulder using one of the techniques introduced in Chapter 4, "Modeling an Organinc Humanoid Hand with NURBS." One choice for curve on surface includes drawing directly on the surface utilizing the Modify, Make Live feature.

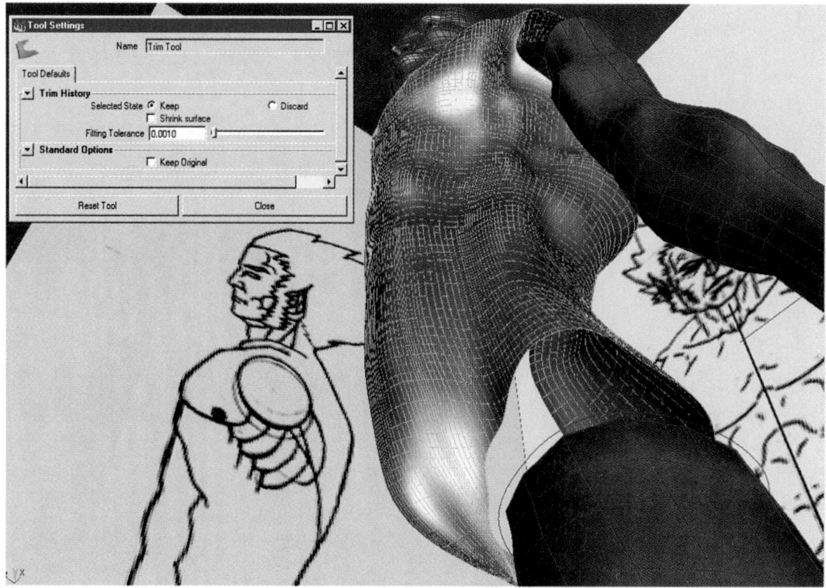

Figure 5.45
The Make Live feature provides a convenient way to create curves on surface.

5.1.5.1 Blending the Legs

1. To blend the legs, create a curve on surface at the leg nubby (the lump of flesh at the base of the torso that has been prepared to blend with the leg geometry in the previous torso-modeling steps). Hold the RMB over the leg and choose Isoparm.
2. Click the top edge isoparm, and then go to Modeling, Edit Curves, Duplicate Surface Curves. Adjust the curve so that it is slightly larger than the leg.
3. Select the torso and Shift-select the curve. Orbit the perspective view to look down the length of the leg toward the torso. Make sure you have this perspective viewport active prior to the next step. Go to Edit Surfaces, Project Curve. This command projects the curve onto the torso.

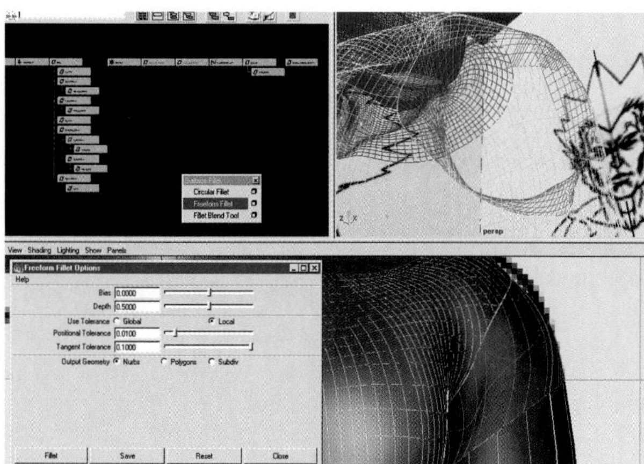

Figure 5.46
Duplicate the edge isoparm.

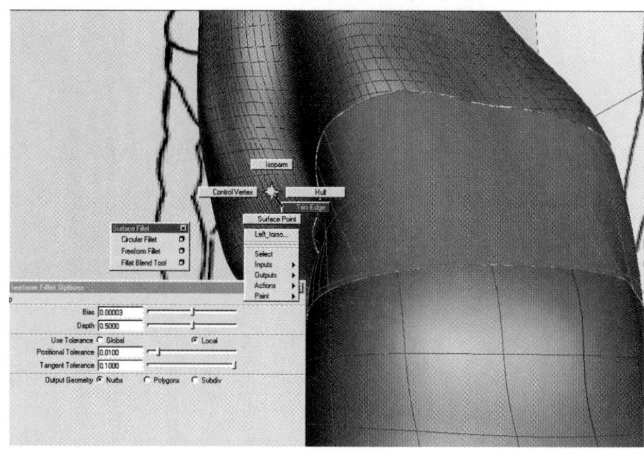

Figure 5.47
Project the curve onto the torso.

4. Trim the torso, making the holes for the arms and the legs. Then, if need be, trim the upper arm fillet blend.

5. Trim or detach surfaces to prepare the upper leg near the rump. Then, to seamlessly close the space between the parts, use either Fillet Blend or Freeform Fillet.

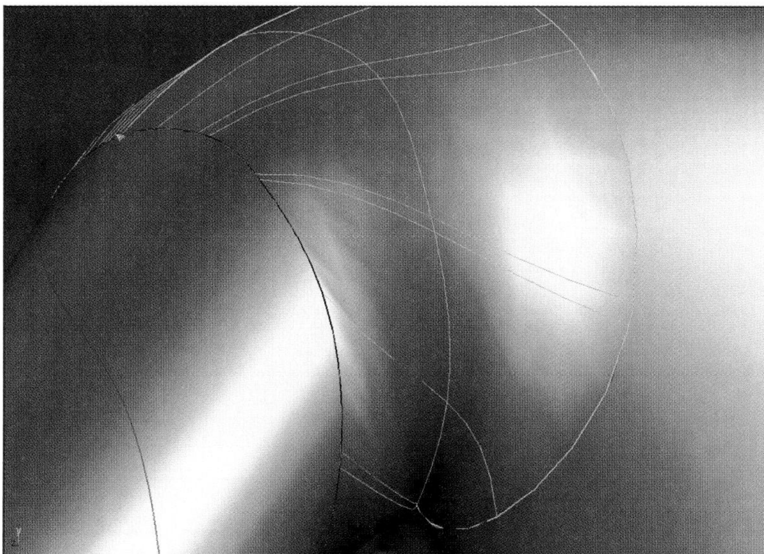

Figure 5.48
Trim the torso object.

5.1.5.2 Maintenance

1. Use the Manipulator tool and select the blend's inputs in the channel box to fine-tune the look of the blend. Click and drag the little blue dots located at each edge of the fillet blend. By moving them around the curve, you can help the blend choose the best spot to place the isoparmeters.

2. Group your objects so they will be easy to use later in other projects.

3. Edit the torso and limb geometry if necessary to perfect the blend's shape and remove any unwanted seams and holes at the blend's edges.

Note

There are several issues concerning duplication of a half biped—issues such as when and at what stage in the model's creation to do it. Retaining the skinning work during mirror duplication is a way to save time on symmetrical bipeds. Use the Duplicate with Upstream Graph option (found in the Duplicate Settings window) on the root group. Then scale to –1 on the X axis the specified group labeled Scale_ME to flip the mirrored duplicate half.

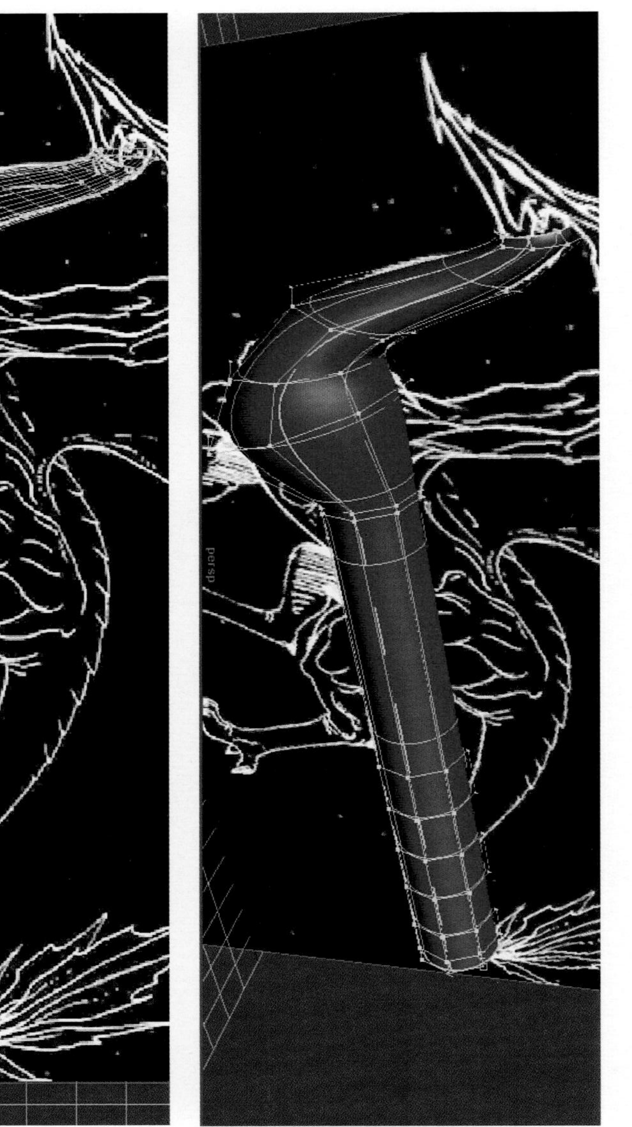
persp
side

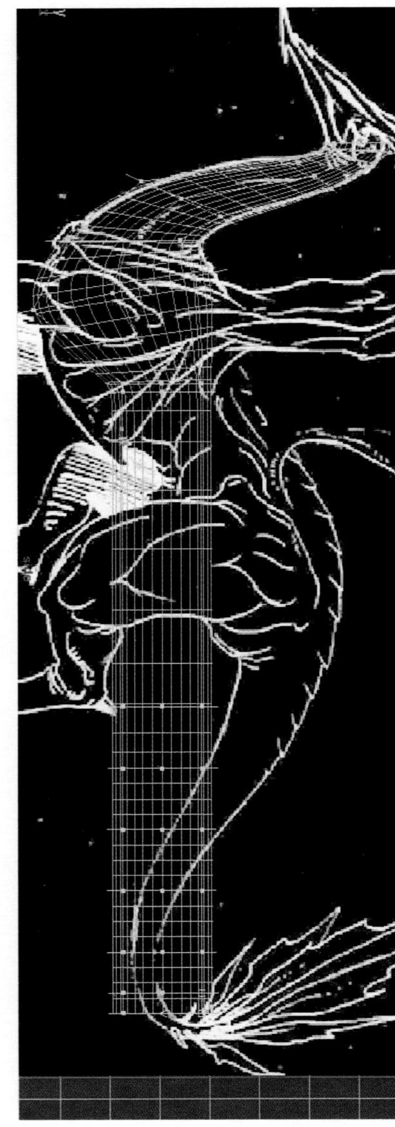

CHAPTER 6

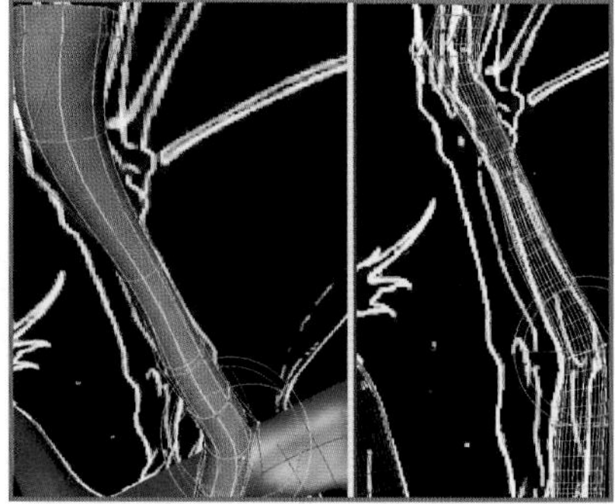

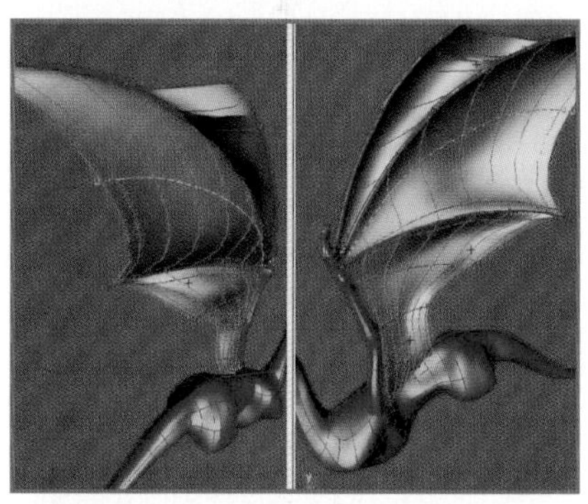

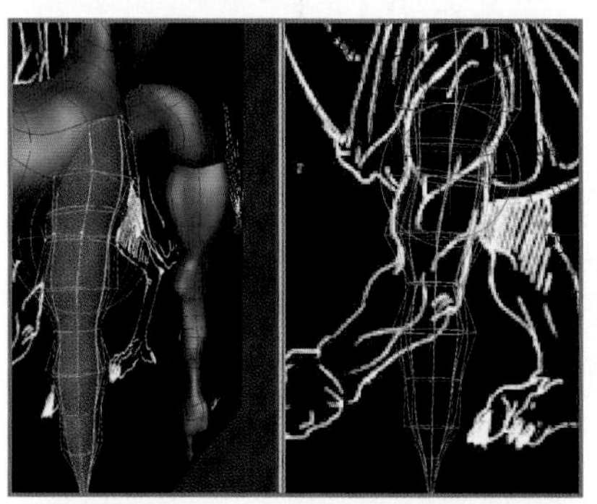

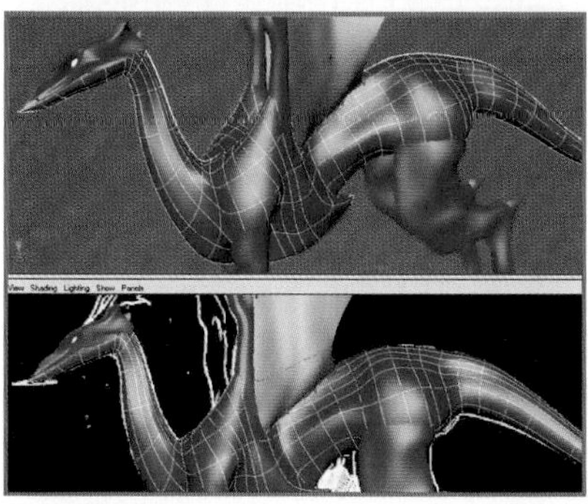

Modeling an Organic Dragon with NURBS

by Nathan Vogel

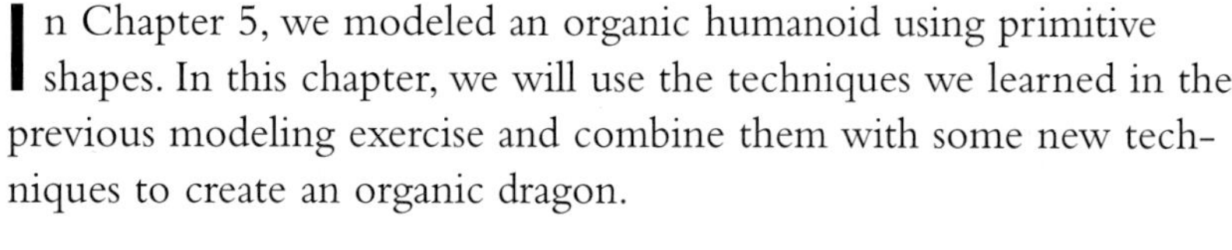

In Chapter 5, we modeled an organic humanoid using primitive shapes. In this chapter, we will use the techniques we learned in the previous modeling exercise and combine them with some new techniques to create an organic dragon.

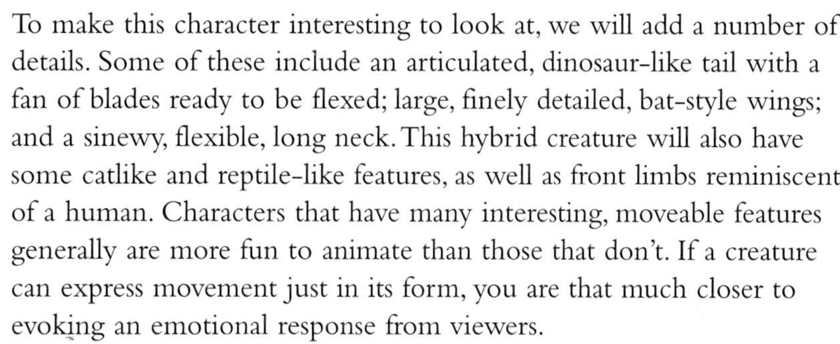

To make this character interesting to look at, we will add a number of details. Some of these include an articulated, dinosaur-like tail with a fan of blades ready to be flexed; large, finely detailed, bat-style wings; and a sinewy, flexible, long neck. This hybrid creature will also have some catlike and reptile-like features, as well as front limbs reminiscent of a human. Characters that have many interesting, moveable features generally are more fun to animate than those that don't. If a creature can express movement just in its form, you are that much closer to evoking an emotional response from viewers.

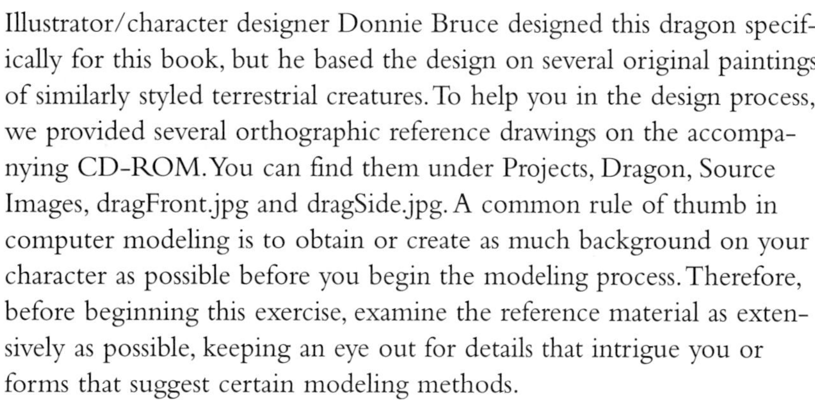

Illustrator/character designer Donnie Bruce designed this dragon specifically for this book, but he based the design on several original paintings of similarly styled terrestrial creatures. To help you in the design process, we provided several orthographic reference drawings on the accompanying CD-ROM. You can find them under Projects, Dragon, Source Images, dragFront.jpg and dragSide.jpg. A common rule of thumb in computer modeling is to obtain or create as much background on your character as possible before you begin the modeling process. Therefore, before beginning this exercise, examine the reference material as extensively as possible, keeping an eye out for details that intrigue you or forms that suggest certain modeling methods.

We will model our dragon character according to the following steps:

- Flesh out the body.
- Construct the wings.
- Model the front legs and digits.
- Form the head.
- Sculpt the tail spikes.
- Shape the rear legs, toes, and spikes.
- Make the model seamless and detailed.

Maya offers a multitude of modeling and 3D sculpting techniques, and you are urged to explore other methods to model the dragon after trying the technique we will outline here. After modeling this character, you should feel fairly comfortable modeling characters based on drawings or pictures. This tutorial builds on the all the modeling tools and methods introduced in the previous two modeling chapters. As such, many steps will be summarized and referenced in a logical manner.

6.1 Fleshing Out the Body

Using a primitive cylinder, we will form the entire outline for the main body. We will begin by stretching the hulls of the primitive cylinder from the neck to the tip of the tail. Later, we will introduce techniques for forming the head. As in the humanoid project in Chapter 5, this character's body will be split in half to facilitate symmetrical modeling. After inserting sufficient isoparms to allow the surface to be dense enough to sculpt, we will use Artisan to add muscle and bone bulges.

6.1.1 Forming the Shape of the Body

Placing the image plane is a straightforward process that accessible through the Multilister under the Camera tab.

1. Double-click the side view to open the Attribute Editor and open the sublisting called Environment. Click the Image Plane button; when prompted, browse for the dragon's side view image, labeled dragSide.jpg, and load the image.

2. Next, create a primitive cylinder that will be the foundation of the dragon's body. Using the Manipulator tool, accessed by pressing the T key, click the NURBS cylinder inputs, translate the top manipulator, and snap it to the grid on the Z axis. By doing this, you are lining up the cylinder's hulls and the isoparameterization to prepare it for shaping to the image plane.

3. Edit the makeNurbsCylinder inputs so that it reads 8 sections and 12 spans.

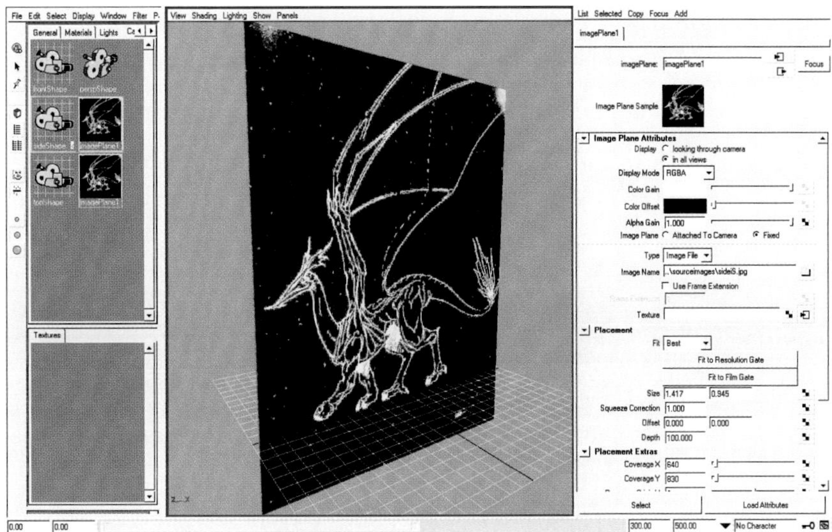

Figure 6.1
Placing the source images for the dragon templates.

4. After setting up the cylinder's parameters, delete the construction history to keep the model light. Do this by going to Edit, Delete by Type, Delete History.

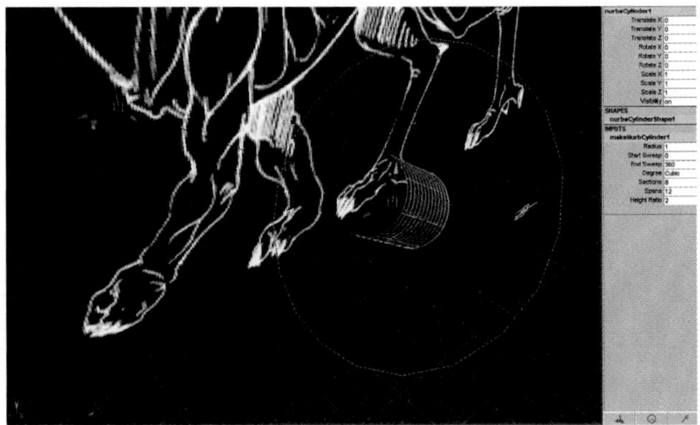

Figure 6.2
The Manipulator tool allows you to edit certain inputs that are not available in the Channels box.

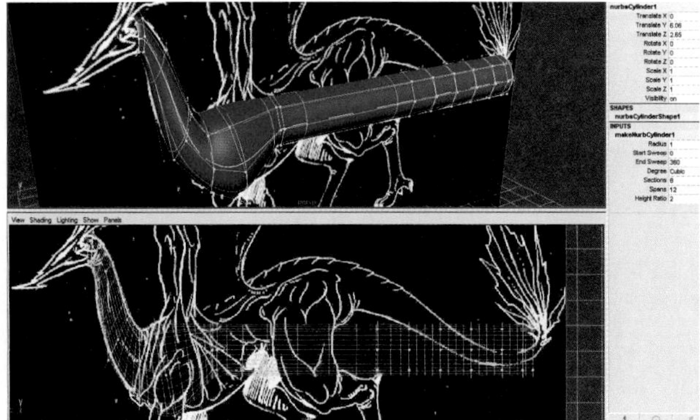

Figure 6.3
Your goal is to distribute the cylinder's hulls across the image plane.

5. You must distribute the geometry across the image plane to approximate the shape of the dragon's body. You can use basic component-level hull editing to match the image plane reference. To do so, edit the hulls from the side view by holding the RMB over the model and choosing the Hulls option.

6. Click the end hull and use the Translation tool to drag it to the edge of the dragon's tail.

7. Click the hull at the other end of the cylinder and drag it to the edge of the dragon's neck.

8. The two edge hulls have a companion hull that sits right next to them and must be translated to maintain their original proximity with regard to tangency.
9. Distribute the remaining hulls evenly across the rest of the dragon's body.
10. Match the tail region to the reference image plane, keeping the width of the tail constant to avoid strange and unwanted bulges resulting from a misshapen hull.

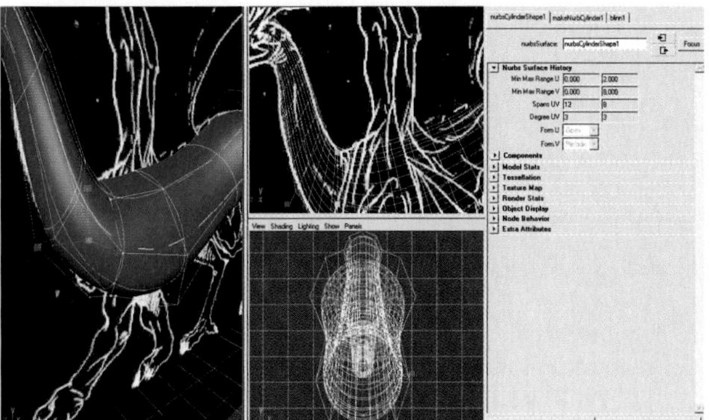

Figure 6.4
A close-up image showing early detail on the neck.

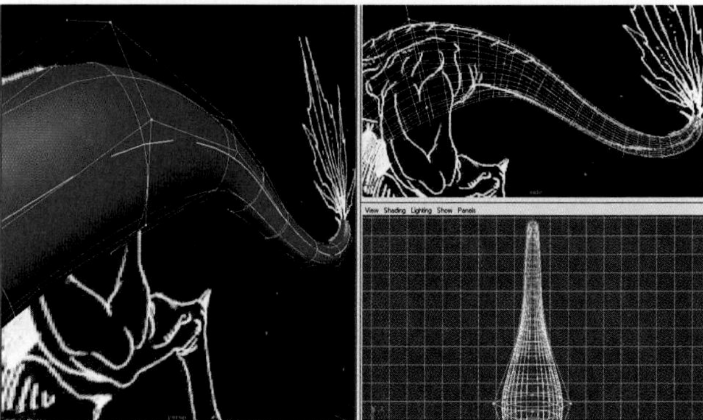

Figure 6.5
Try to create a smooth transition in scale from the torso to the tail regions.

6.1.2 Forming the Neck

When modeling the neck, you must make sure that the skin and muscle at the base of the neck have adequate geometry. You will need this geometry for ensuring articulated movements after you apply deformation skeletons to the model.

We purposely did not create the original primitive cylinder with enough isoparms to flesh out the outline. When molding geometry, it is often better to use as few details as possible when forming the general shape. Then, when the form is partially constructed and you add more isoparm details to it, you can add them only where you need them.

1. Set the display smoothness to 1 and the shading mode to Wireframe, by pressing the number 4 on the keyboard. This reduced-detail view helps isolate the existing parameterization and aids in the selection of the correct isoparms to start from.
2. Starting with the neck region, insert isoparms vertically to add resolution to the dragon's body geometry. Hold down the RMB over the beast and choose Isoparm from the Marking menu. Then choose Modeling, Edit Surfaces, Insert Isoparms.
3. Repeat this process to match the image plane until you are satisfied with the details. Try not to add any unnecessary resolution.

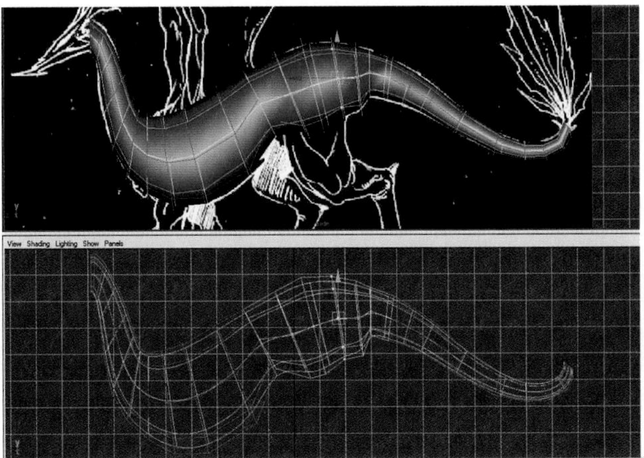

Figure 6.6
Adding resolution with the Insert Isoparm tool.

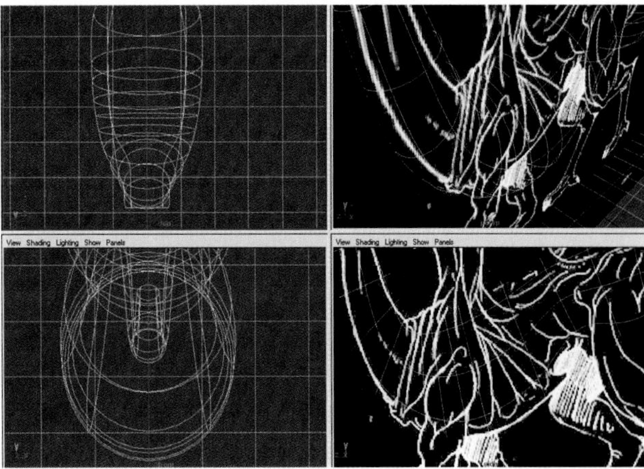

Figure 6.7
Vertical resolution adds good local control without impacting the overall mesh density too much.

Figure 6.8
Don't go overboard adding resolution, and be careful not to add isoparms right next to each other or you will create visible creasing.

6.1.3 Forming the Tail

Now we will add and distribute enough isoparms to allow the tail to bend back to its original form. In general, 3D geometry prefers to bend into place more than it prefers to bend out of place.

1. Insert more isoparms to add details to the tail region.
2. Pinch to a close the last isoparm's hull as well as the companion hull that sits next to it by scaling them to zero. Be sure to insert enough detail around the tip to allow it to close without creasing or deforming the rest of the tail area.
3. Sculpt a nubby tip that you can use as the base for the tail spikes.

Note

The best way to model from an orthographic drawing is to prepare the shape so that it is stretched out to its extents. But because the paper was too short to fit the entire model the way the artist designed it, the orthographic side view of the dragon has a curved tail instead.

At this time, we won't add the horizontal isoparms that run the length of the body, waiting until after we bisect the model in half in a later step. This keeps the sides from having uneven parameterization and helps preserve the initial symmetry that has been designed in from the beginning.

Note

Keep in mind the musculature required for a strong tail such as this. The underlying bones in the tail are short and flexible and are closer to the surface at the base of the tail, centering as they approach the tip. This kind of bone structure produces a crest or ridge on the top of the tail, especially near the base, where it attaches to the rear legs.

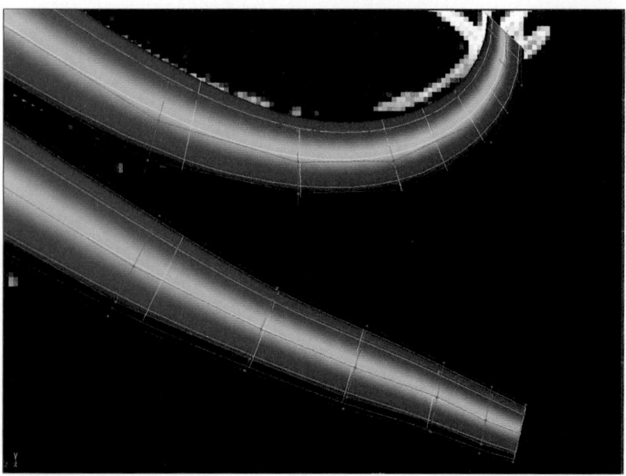

Figure 6.9
Give the tip of the tail enough resolution to allow for later animation.

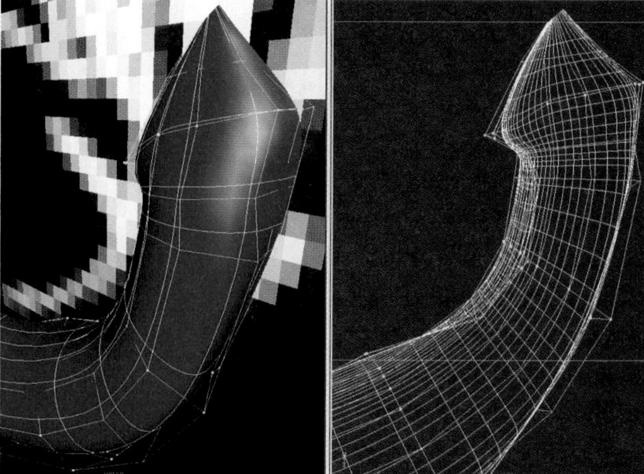

Figure 6.10
Give the tip a little nub as a foundation for the tail spikes.

6.1.4 Forming the Torso

The hulls of the torso must be scaled in the X as well as the Y axis. To do this, you will need to add some extra isoparms to attach the wing joints and to create the nubby wing extensions.

1. Fill out the central torso area and begin adding detail by inserting more isoparms. Make sure to add details to the chest area a bit at a time. The ridge on the chest that many reptiles and bird-like creatures share can be seen embedded in the design.
2. The wing nubs need drumstick-like musculature to flap wings such as these. Because this is a fictitious creature, we made up the anatomy and structure from scratch and blended them with zoological and mythical references. Use the Sculpt Surfaces brush tool in conjunction with detailed CV editing. Be sure and use the smoothing brush to prevent any kinks resulting from the CV editing.

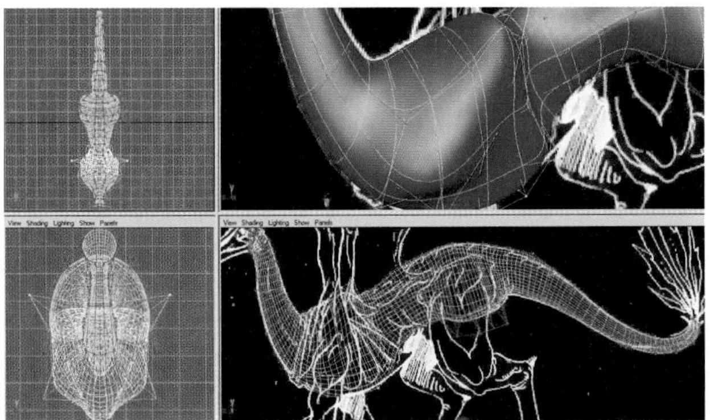

Figure 6.11
Add the wing nubs and fill out the central torso area.

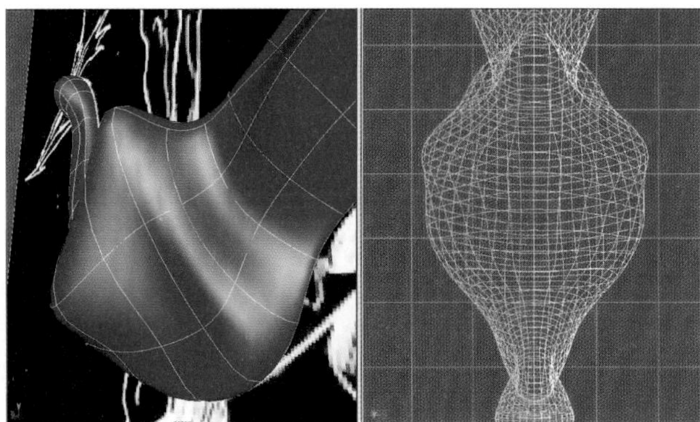

Figure 6.12
Bulge out the drumstick-like wing nubs.

6.1.5 Mirror-Painting the Muscle Details

When you're adding isoparms horizontally to your model and you want your end result to be generally symmetrical, you must keep the following in mind: If you are adding an isoparm on each side of your model and you don't use a script for precision, the two sides may be a little off. This small aberration can ripple through your model to create all sorts of strange anomalies and annoying quirks that may prevent you from doing a trim correctly.

Note

Symmetrical modeling is not new. In fact, sculptors have worked this way for centuries. Not only does symmetrical modeling save time, it also helps you to avoid making your model look too uniform, as you can easily add unique variations to the model after you fuse the two halves.

However, versions 1.0 and 1.5 of Maya are weak in this respect. Maya 1.x does not enable you to adequately re-seam both the front and back sides of a bisected shape. Instead, when attaching surfaces, you must use the Open/Close Surface command, which adds an extra isoparm to each side of the edge of the surfaces to be closed. Apparently, the software does this to preserve the model's original form. However, it also causes huge problems when you want to repeatedly attach and detach the two sides for making symmetrically editable blend-shape targets (using such tools as the Artisan reflection brush); each time you attach the sides, you add more isoparms. You can get around this by always leaving your pairs of symmetrical models unattached until the final version of your model. Save regularly and always have a recent version left untouched as a separate backup on a different computer or external media as a precaution.

With this in mind, rather than trying to add the details needed for the wing socket twice, do it once on each side of the beast. It is practical to use Artisan's mirror modeling tool, the Reflection Paint mode. Also, crossing the X axis with your edits will inevitably produce unusable results. With that in mind, perform local area tweaks that don't cause any of the CVs to cross the zero point on the X axis. You can always fix any X-axis "rebel" CVs later by enabling grid snapping by holding down the X key and dragging the point toward 0 on the X-axis working grid. Although these various fix-it techniques can help, it is much better to avoid working messy from the beginning, as the fixes can possibly introduce other subtle problems down the line.

In Alias | Wavefront's PowerAnimator and Maya 2.0, you can successfully split a symmetrical object in half. Delete the second half, and edit the first half with an Instanced Mirror Duplicate or Normal Mirrored Duplicate (covered in Chapter 5's "Creating a Torso and Head from a Sphere" section), in Artisan's Multiple Surface Reflection Paint mode.

Here we will discuss how to split the body into two pieces and use Artisan to mirror-paint the muscle details on the remaining half of the model as well as on its new mirrored duplicate.

1. Set the selected body geometry to Rough Smoothness Display by pressing the 1 key on the keyboard, and set the Shading mode to Wireframe by pressing 4. Select the top and bottom polar isoparms that run along the back ridge and the center of the belly by holding the RMB over the body. Select one and Shift+select the other. With the yellow iso-markers still highlighted, go to Modeling, Edit Surfaces, Detach Surfaces.
2. An Artisan bug anomaly does not allow two halves of the same object to be reflection-painted predictably, so delete the right half. Remember to delete the history and freeze transformations to reset the model.
3. Duplicate the left half and then scale it to –1 on the X axis to mirror it over the origin.
4. Select both sides of the body geometry. Go to Modeling, Edit Surfaces, Sculpt Surfaces Tool, Settings. Go to the Stroke tab and be sure that the settings for Reflection Painting and Multiple Surfaces are on.
5. Edit and tweak Artisan's opacity, brush size/shape, and maximum displacement preferences while sculpting the geometry.
6. To add more isoparm details, delete the R_side and return to step 3.

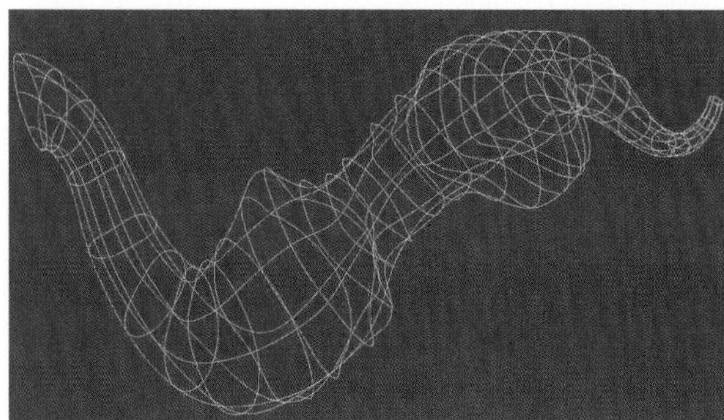

Figure 6.13
Selecting the polar isoparms to prepare the model for bisection.

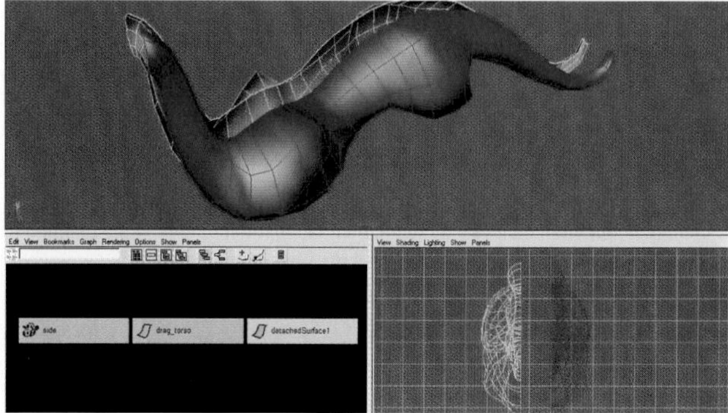

Figure 6.14
Deleting the right half.

6.2 Constructing the Wings

In this section we will use a new technique to construct the wings' membrane. We will create an armature of several sets of NURBS curves and then perform a loft across them all to create the membrane that stretches between the bony, finger-like wing tips. The wing arm and wing fingers are made of standard primitive cylinders edited into place.

6.2.1 Creating the Meat and Bones

Next, we will use a primitive cylinder to flesh out the basic form of the wing arm.

1. Create a primitive cylinder as the starting place for the new wing arm. Placement of a new primitive cylinder in the perspective view will help facilitate correct alignment in XYZ space.
2. Be sure to place the cylinder directly above the front wing lump to ensure that the isoparms easily integrate with the main body.
3. Set the makeNurbsCylinder inputs to 8 × 8 and adjust the radius to generally match the reference picture of the wing's arm area loaded in the image plane.
4. Remove the construction history by deleting the history. Afterward choose Modify, Freeze Transformations to reset the cylinder's attributes to 0 in its new origin.

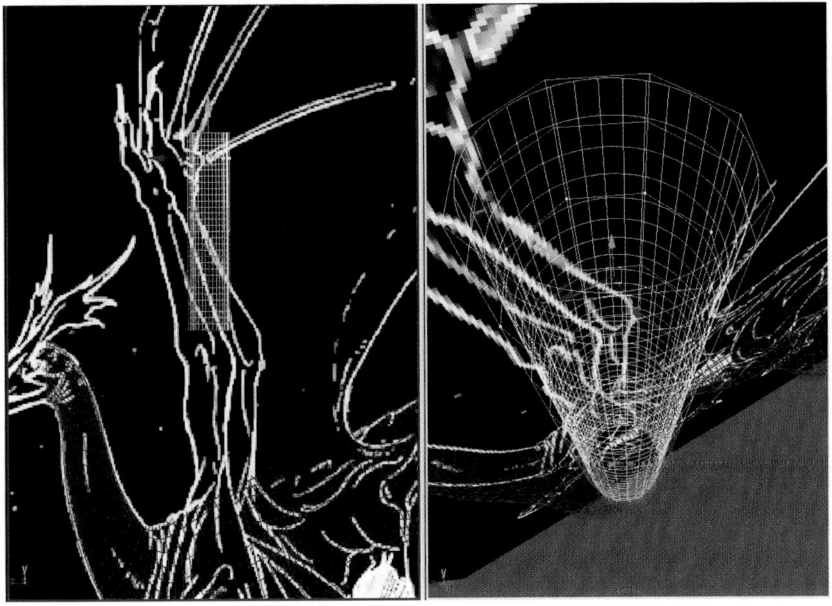

Figure 6.15
Start with a primitive for the wing-arm, and then freeze transformations to reset the zero point.

6.2.1.1 Distributing the Hulls and Adding Resolution

Here we will distribute the hulls according to the reference image plane and add resolution. Our goal is to have a mostly one-piece wing-arm model that includes the base, the quasi-wing thumb, and the three nub-like foundations for the wing fingers.

1. Hold the RMB over the new arm model and highlight the Hulls option from the Marking menu.
2. Select the ring-like vertical hull at the bottom edge of the new cylinder by clicking the pink ring hovering around the top isoparm.
3. Translate it to the wrist region. Scale, rotate, and transform the hull and all that follow to your best depiction of a wing arm, keeping in mind musculature and bone structure.

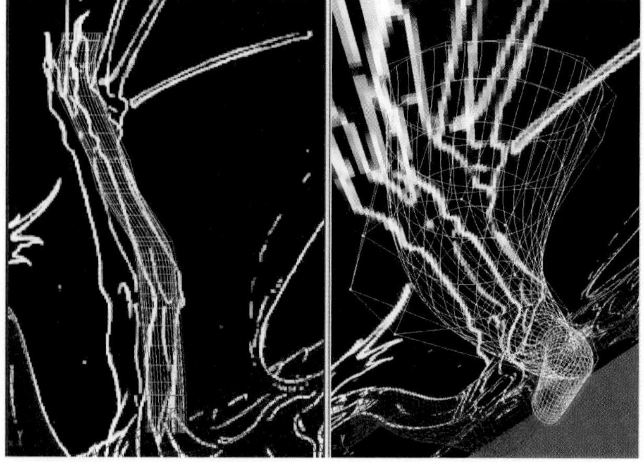

Figure 6.16
Practice your hull mantra, scale to match.

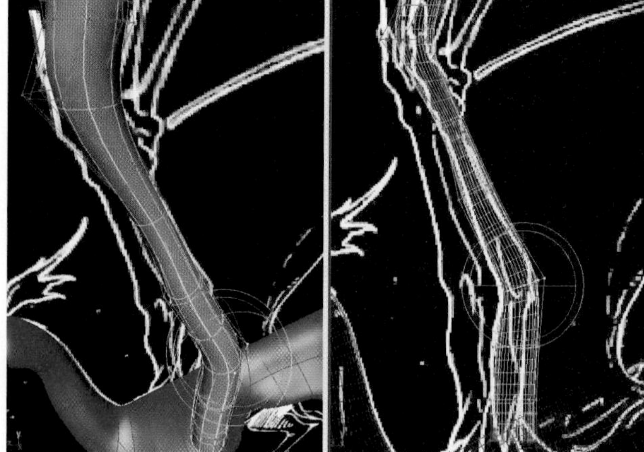

Figure 6.17
Use Rotate, Scale, and Translate to edit.

4. Repeat with each hull until the object generally resembles the wing arm. Use the arrow keys to "pick-walk" through the hulls one at a time for easy editing. Make sure that the top isoparm smoothly approaches the main body's wing nub-attaching region.
5. When you have spread out all the hulls, it is time to insert more isoparms to increase the resolution. Set the Shading style to Wireframe and the selected arm model to Rough Smoothness by pressing the number 4 and 1 keys on your keyboard, respectively. RMB on the wing and choose Isoparm.
6. Hold down the RMB and activate Isoparm mode. Click-drag and then Shift+click-drag more ring-like isoparms across the length of the wing arm. When you're satisfied, choose Modeling, Edit Surfaces, Insert Isoparms.

7. Repeat the hull-editing technique until the wrist extends to become a wing paw with a thumb-like extension already attached. Make sure the tip of the thumb/claw is smoothly pinched closed. Pull out the geometry to form a sort of knuckle nub for each of the wing fingers to use as a foundation.

8. Finish editing the muscles and contours by using Artisan to tweak the geometry of the arm.

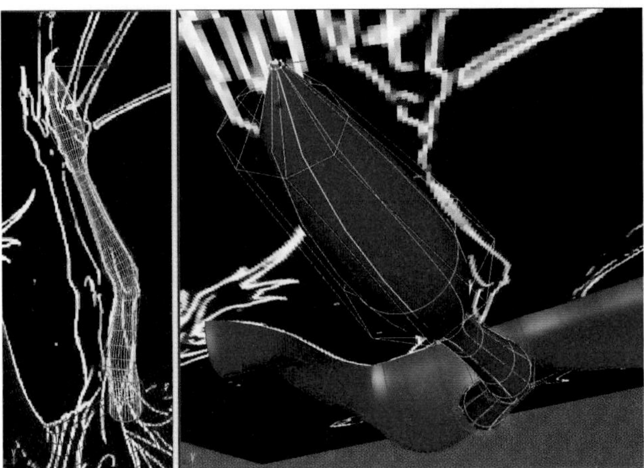

Figure 6.18
Create a pod at the tip of the wing arm, and then carefully pinch it closed.

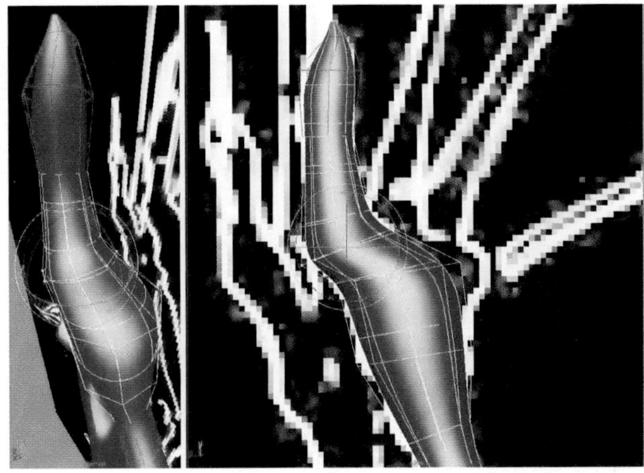

Figure 6.19
Massage the hulls to match, creating a thumb.

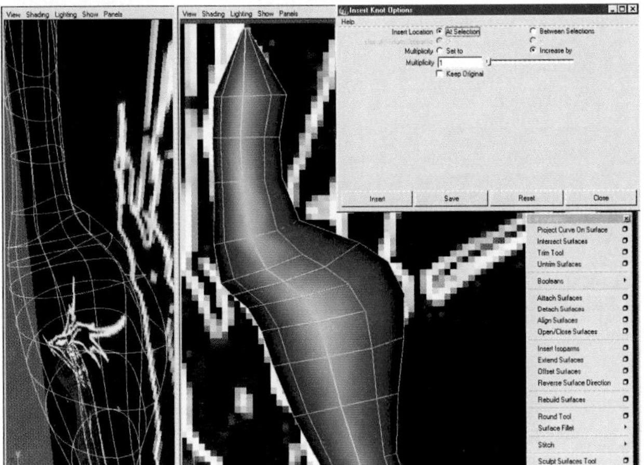

Figure 6.20
Insert isoparms to create a wing palm.

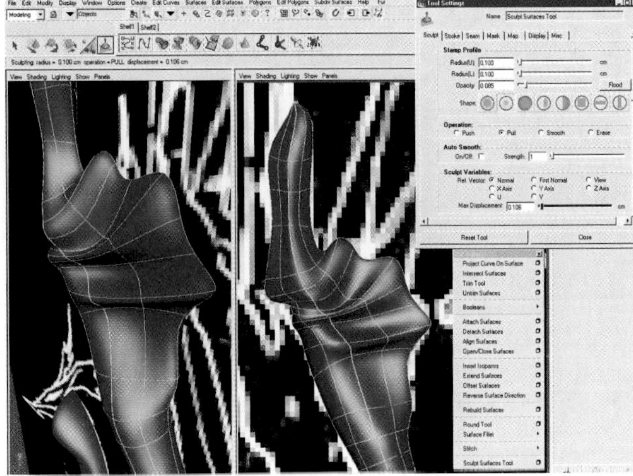

Figure 6.21
Adding resolution is necessary to allow detail work with Artisan.

6.2.1.2 Creating the Wing Fingers

Now we will create the wing fingers by sculpting a primitive cylinder and then duplicate it to make the successive fingers. Slightly offset the scale to simulate the fingers getting smaller near the tip.

1. Create the wing fingers from hull-edited primitive cylinders set to 8 × 12 spans and sections.
2. Replicate them to finish placing the wings' other digits.
3. Parent the digits of the base wing finger together for easy replication.
4. Replicate and scale the rest of the fingers to match the image plane using object duplication and manipulation techniques.
5. To help you position the fingers, set the pivot point to the base of the finger using the Translate tool and the Insert key.
6. Duplicate, scale, rotate, and translate the finished finger two more times and distribute them to match the reference art.

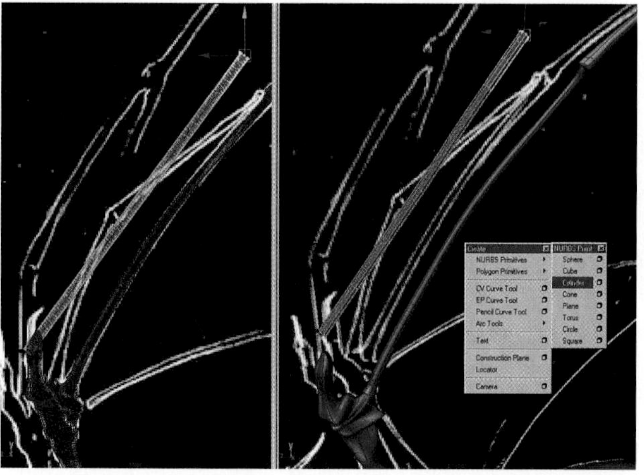

Figure 6.22
Starting with a cylinder primitive, edit the existing hulls to match the drawing.

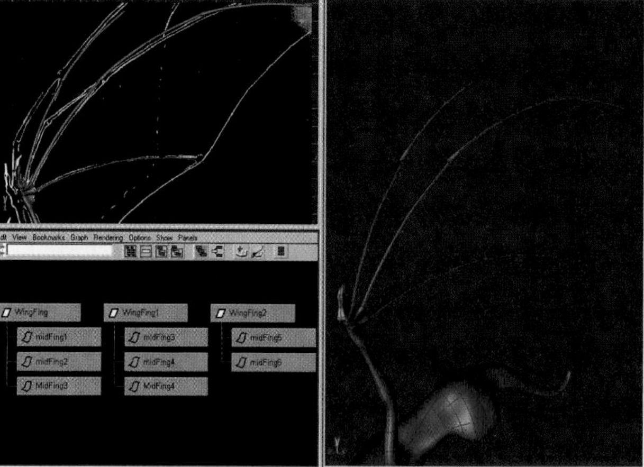

Figure 6.23
Reset the pivot point and duplicate the fingers.

6.2.2 Creating the Leathery Wing Membrane

For this step, we will create a primitive NURBS circle across the XZ axis and then scale its X axis down so that it becomes a very thin ellipse. We will then edit the points so that they resemble a crescent shape, such as what the cross-section of a wing might look like.

Before duplicating and distributing the edited circles, we will display and offset the built-in selection handles. Later, when geometry is covering the circles, you will be able to more easily select and edit them.

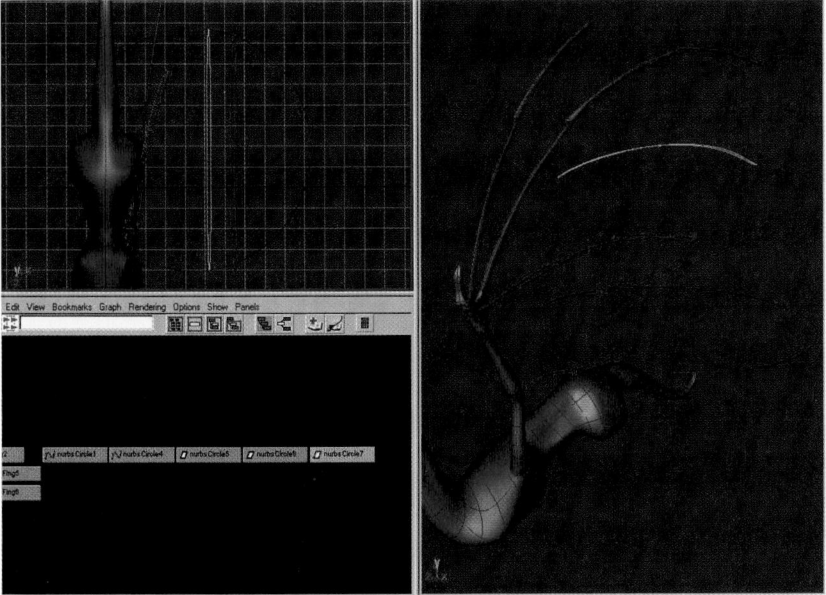

Figure 6.24
Start creating the wings by creating a circle and deforming it so that it becomes the foundation for the lofted wing membrane.

1. Select the new crescent-shaped NURBS curve and go to Display, Object Components, Selection Handles. Enter Component mode by pressing F8. Then, from the Pick Masks All On/Off pop-up menu next to the Component Mode button, choose All Off. Select the plus-shaped (+) selection handle pick mask.
2. Draw a selection marquee around the curve's new selection handle and, with the Translation tool, offset the handle in the positive X direction a few units. Press F8 to exit Component mode.
3. Duplicate, CV edit, translate, scale, and rotate the loft target curves until you have 10 of them to the represent the wing shape. Place three curves in exact proximity to the bottom finger—one above, one below, and one in the middle. Place three more curves below the bottom finger, toward the body. Place another curve halfway between the base finger and the middle finger, and one more directly by the middle wing finger. Finish by placing two more curves near the top (shorter) finger to form the upper mini-wing area.

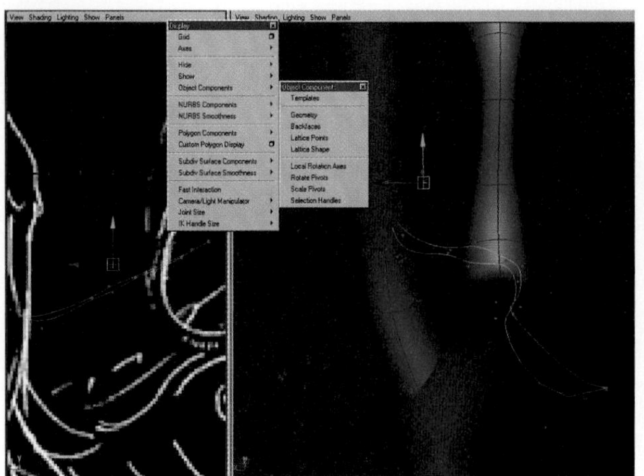

Figure 6.25
Edit the pairs of CVs simultaneously.

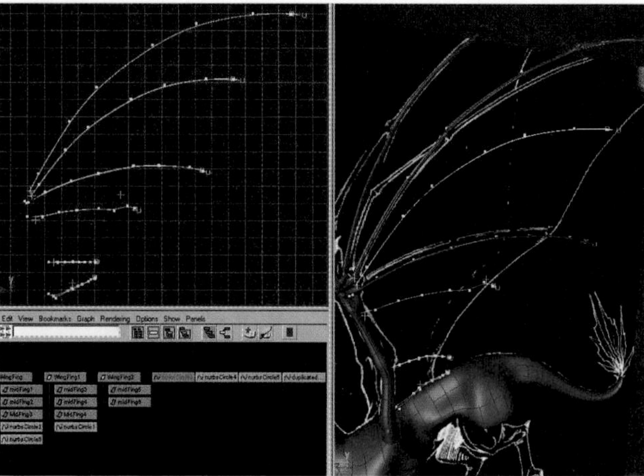

Figure 6.26
Distribute duplicates of the curve after displaying and offsetting a selection handle for easy selection later.

6.2.2.1 Creating a Curve on Surface

The next step is creating a curve on surface for the base of the wing to seamlessly blend to. We will do this by selecting the bottom wing curve and then maneuvering a camera to look directly at it with the dragon body/torso directly behind it.

1. First, duplicate the curve before projecting it. You must do this because you must delete the actual curve that you will project after to clean up the construction history list. If you don't, every time you edit the curve that you projected, the curve on surface will also change in the same way, causing unwanted changes in the wings' surface.
2. Select the duplicated curve and name it for easy identification. Then Shift+select the torso and choose Modeling, Edit Surfaces, Project Curve on Surface. Remember that the angle of your camera relative to the selected objects for projection determines how the angle of curve projection will be calculated.
3. If the projected curve does not look pleasing or useful, you may need to re-project it from a slightly different angle, or edit the duplicated curve floating above the torso itself for point-level editing. When you are satisfied that the curve will work as the base attachment for the wing to the torso, delete the duplicated projection curve.
4. Simplifying the shape will make a smoother loft. Rebuild the curve on surface, making sure that the settings are set to Keep Original On. Experiment and undo until you have about 16 knots. Then display a selection handle for it as well.

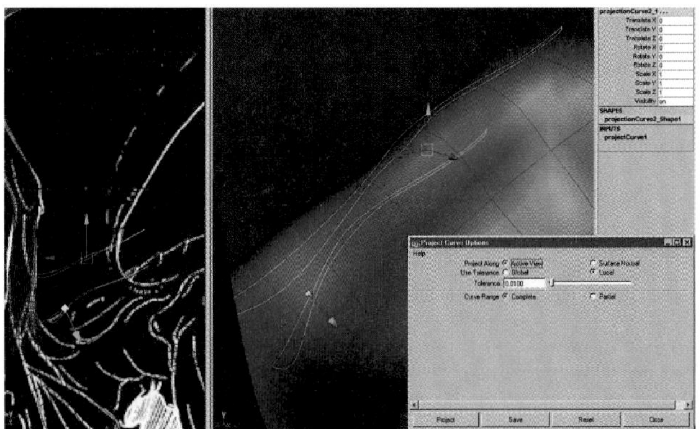

Figure 6.27
Project a curve on surface for the base of the wing membrane.

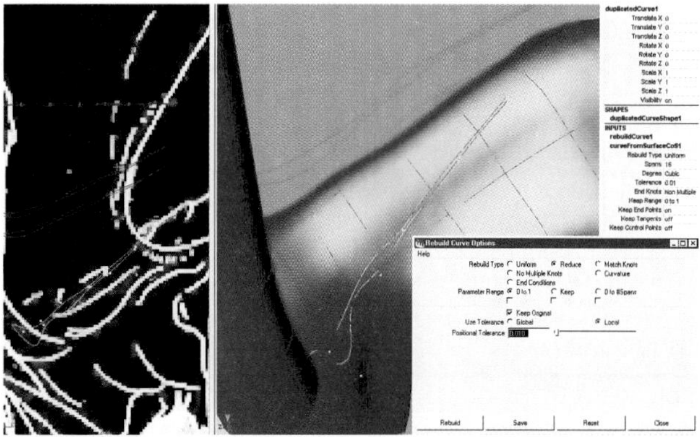

Figure 6.28
Rebuild the curve on surface to enable a smoother loft. Locate and name the curves in the Multilister for easy identification.

Figure 6.29
Display and offset a selection handle for the rebuilt curve.

6.2.2.2 Testing the Lofts

It is time to try some initial test lofts. The order in which you select the curves is important, so keep that in mind. In general, if you stay consistent, such as front to back, top to bottom, left to right, and so on, the result should look the same.

1. One by one, from top to bottom, select and then Shift+select all the wing curves, including the new curve on surface but not the top two for the upper mini-wing.

2. You can select the wing curves alternately from the HyperGraph (assuming you named the curves logically). Or, if you set the Show Preferences setting in an alternate viewport's menus so that it does not display surfaces, you can view your progress but still make easy selections.

3. If the order was wrong, or if your curves are not aligned when viewed from an angle, the membrane geometry could be messed up and, thus, not visually appealing. If this is the case, press Z and try again after identifying your error. If it basically looks right but just needs a little tweaking, go on to the next section.

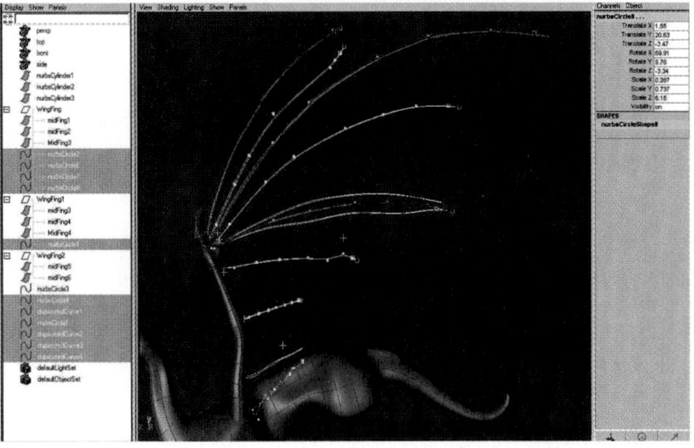

Figure 6.30
Shift+select each curve in order from bottom to top, starting with the rebuilt curve on surface at the base and ending with the topmost isoparm.

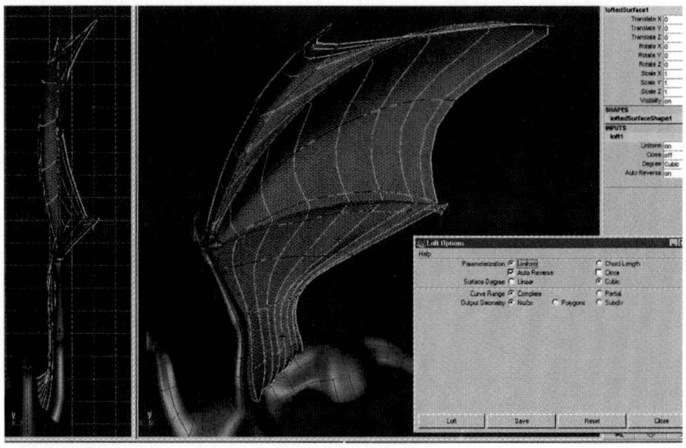

Figure 6.31
Perform a series of test lofts, undoing the loft, making minor edits to the foundation curves, and then re-lofting.

6.2.2.3 Correcting the Lofts

You can correct lofts in several ways. Here we will discuss how to edit the position of the component curves and their CVs. (You can also edit the surfaces of the new membrane geometry, as discussed in the next section.)

1. Minor curve rotations, translations, and scaling can result in significant changes overall. Be very careful when rotating, translating, and scaling curves, and keep your finger near the Z key for undos.

2. Try editing each curve's component CV by highlighting the curve, holding down the RMB, and choosing CVs. This makes the curves' components visible and editable. Try moving the CVs one at a time. After selecting a single CV, you can "pick-walk" to the next one by pressing the arrow keys. This saves time when doing minute detail editing.

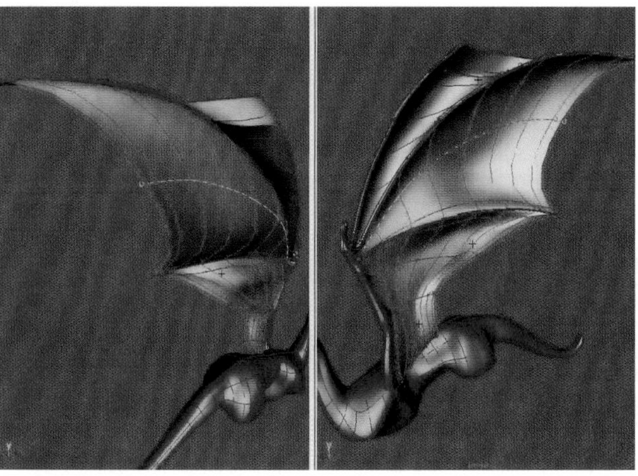

Figure 6.32
Carefully edit the membrane source curves.

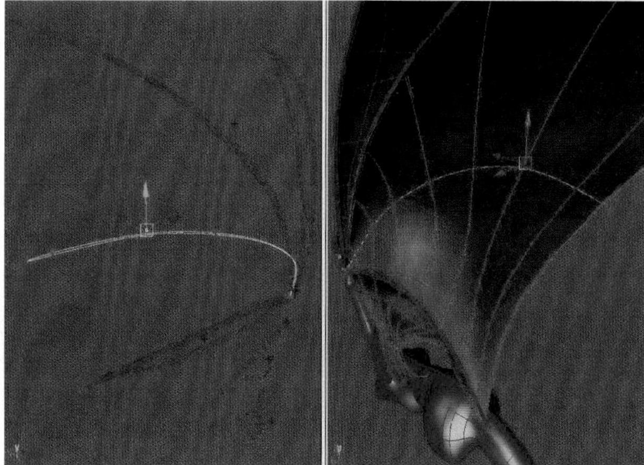

Figure 6.33
In one of the two views' mode, set Surface Display to Off to allow easy access to selection and editing of the wing curves.

6.2.2.4 Smoothing the Wing Membrane

Smoothing the wing membrane geometry, if necessary, is the last step in the wing-modeling process. Use your favorite method of editing. Keep in mind that the wing is made of two sheets of geometry that are very close to each other, making it difficult to use certain less-delicate tools.

When smoothing the wing membrane, you can use one of the following methods:

- You can use various lattice types to retain the general volume of the wing membrane.
- You can also use point-by-point editing, which often works quite well for minor detail work.

After you are finished smoothing, you can "Save As" and experiment a little by embellishing the geometry with skin wrinkles and other close-up details. Otherwise, you can save that kind of work for the texture-mapping pass, during which you can add a bump and/or displacement maps to achieve the same result, but with lighter geometry that is easier to set up.

Note

Artisan has a preference for detecting multiple surfaces that are nearly intersecting, and for editing them in a similar fashion for a consistent appearance. This prevents the two wing membrane pieces from pulling away from each other. When using Artisan, you must remember to make sure that one side of the wing does not get too thick.

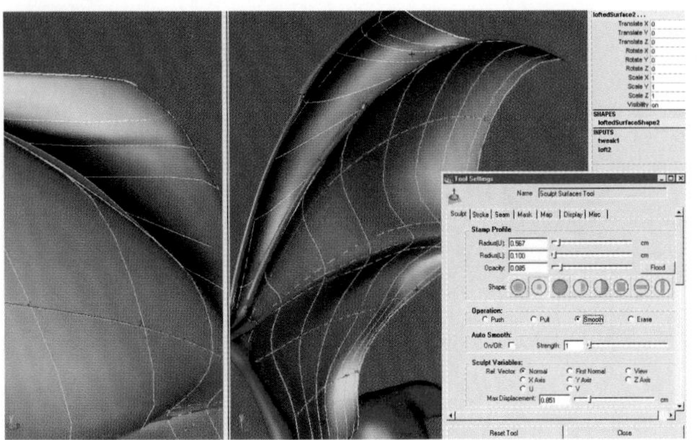

Figure 6.34
Smooth the wing membrane with Artisan's smoothing brush.

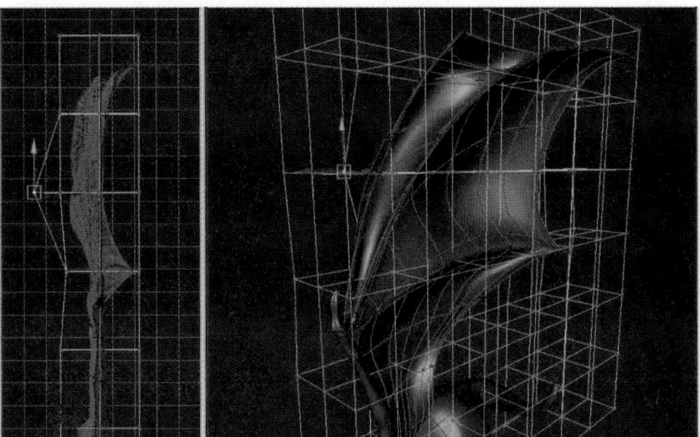

Figure 6.35
Try using a lattice deformer to retain volume and to keep the geometry's seams together.

6.3 Modeling the Front Legs and Digits

The front and back limbs are constructed very similarly in the initial stages. Keep in mind when modeling that the front limbs resemble a human's arms, and that the rear legs resemble the legs of a horse or a big cat. For stylistic reasons, we have decided to model our dragon with a three-fingered front paw and a two-toed rear foot.

6.3.1 Sculpting the Front Legs

We will create a primitive cylinder as the starting point for the new arm-like front leg. Placing a new primitive cylinder in the perspective view will help facilitate correct alignment in XYZ space.

1. Place the cylinder directly beneath the front leg lump to ensure that the isoparms will easily integrate with the main body.
2. Set the makeNurbsCylinder inputs to 8 × 8 and adjust the radius to generally match the reference picture loaded in the image plane. The exact attributes don't matter because the hulls will be dramatically modified later.
3. Remove the construction history by deleting the history. Afterward, choose Modify, Freeze Transformations to reset the cylinder's attributes to zero in its new origin.

6.3.1.1 Distributing the Hulls

The next project is to distribute the hulls according to the reference image plane and then to add resolution. The end goal is to have a mostly one-piece leg model that includes the upper thigh to the tip of the inner toe.

1. Hold the RMB over the new leg model and then highlight the Hulls option from the Marking menu.
2. Select the ring-like vertical hull at the bottom edge of the new cylinder by clicking the pink ring hovering around the bottom isoparm.
3. Translate it to the ankle region. Scale, rotate, and transform the hull and all that follow to your best depiction of a leg, keeping in mind musculature and bone structure.
4. Repeat with each hull until the object generally resembles the leg. Use the arrow keys to "pick-walk" through the hulls one at a time for easy editing. Make sure that the top isoparm smoothly approaches the main body's leg attach region.
5. When you have spread out all the hulls, it is time to insert more isoparms to increase the resolution. Set the shading style to Wireframe and the selected leg model to Rough Smoothness by pressing the 4 and 1 keys on your keyboard, respectively. RMB on the leg and choose Isoparm.
6. Click-drag and then Shift+click-drag more ring-like isoparms across the length of the leg. When you're satisfied, choose Modeling, Edit Surfaces, Insert Isoparms.
7. Repeat the hull-editing technique until the wrist extends to become a palm/front paw with an inside finger/claw already attached. Make sure the finger/claw is smoothly pinched closed.

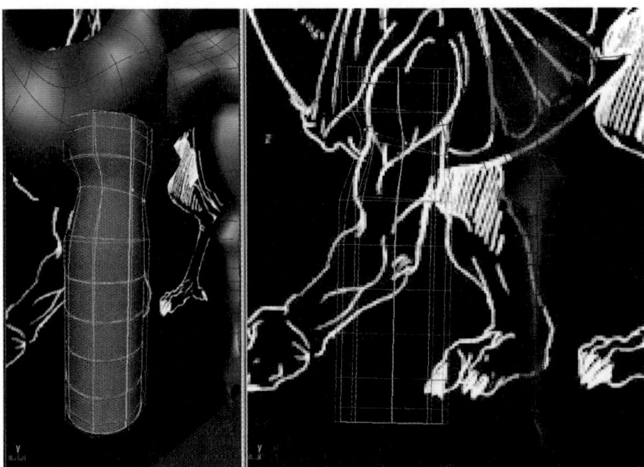

Figure 6.36
Match the hulls to the muscles.

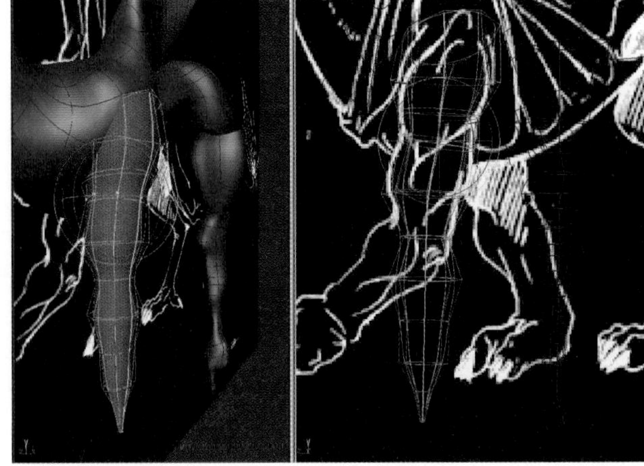

Figure 6.37
Pinch the tip closed.

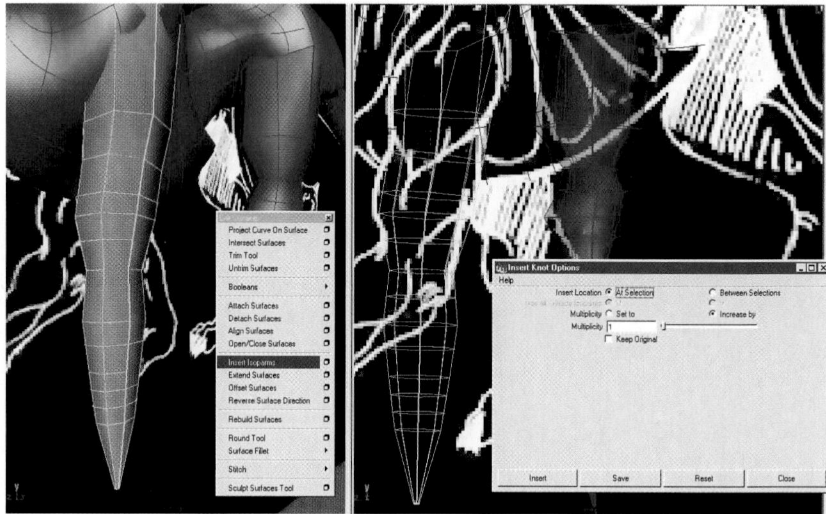

Figure 6.38
Insert more isoparms to up-res the model.

6.3.2 Sculpting the Arm

Make sure that a solid palm/paw is formed as you sculpt the arm, and that the knuckle area is sufficiently flared so that you can attach a thumb and pinky. Use CV and hull editing for precise geometry manipulation.

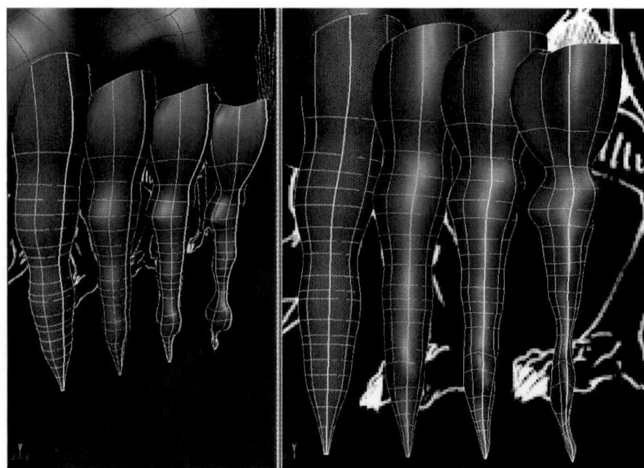

Figure 6.39
A perspective and side view of the arm progression.

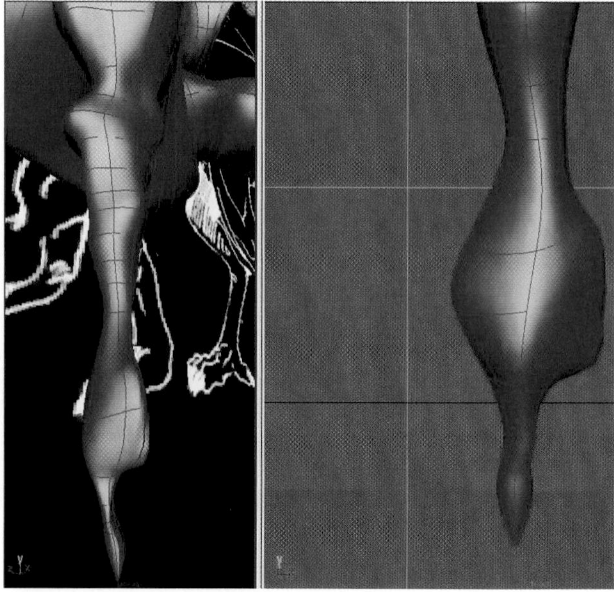

Figure 6.40
Forming the finger and palm from the base of the arm.

6.3.2.1 Adding Detail to the Arm

To add detail to the arm, you must sculpt muscles and add detail to the torso to aid in musculature integration. You must also adjust the torso so that the arm and shoulder appear to function together naturally. To do this, use the Sculpt Surfaces tool. If you must insert additional isoparms for added detail, do so, keeping in mind that light geometry is preferable.

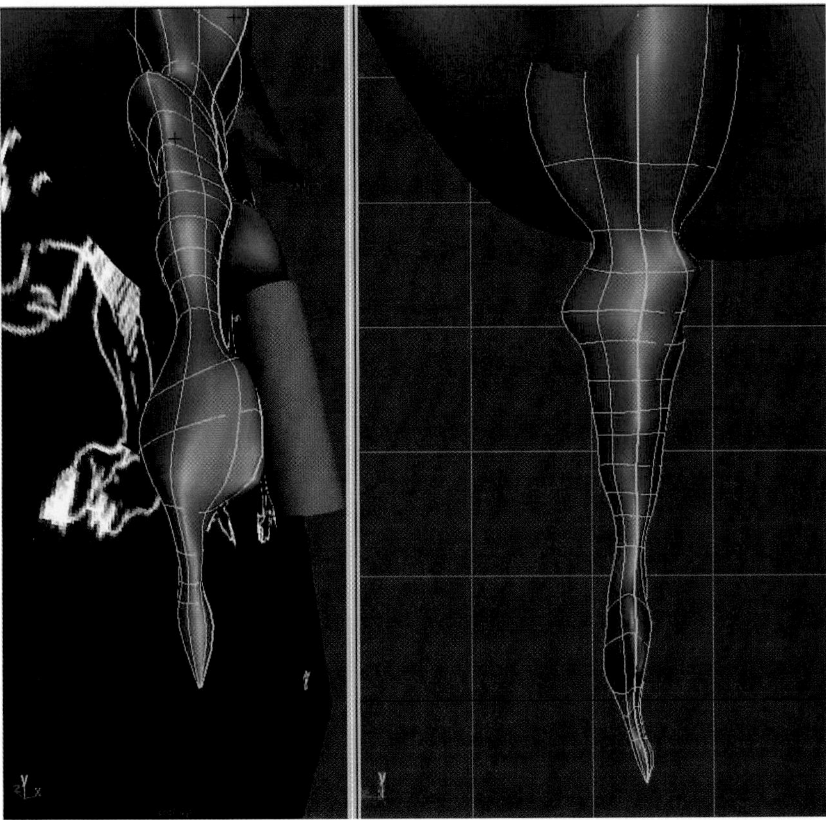

Figure 6.41
Adjust the shoulder and arm so that the arm flows smoothly as it transitions to the torso.

6.3.3 Creating a Pinky and Thumb

The next step is to create a pinky and thumb for the arm. We will create the finger objects by duplicating the front leg and then detaching the middle finger, which we will use as the source for new digits.

1. To create the outside fingers, duplicate the leg model and then scale it to –1 on the X axis. Move it to an area that is easy to access and select an isoparm at the base of the knuckle. Choose Modeling, Edit Surfaces, Detach Surfaces, Settings. Make sure that the Keep Original setting is Off.

2. Detach and then delete the extra arm geometry. Move the finger back to the palm region of the original arm. Repeat the process without X scaling, or duplicate and alter the pinky to create the thumb. Either way, try using the work that you have already completed as a source for work that you still need to do.

3. Using hull-editing techniques, merge the finger near the arm. Later you will decide whether to use fillet blends to create geometric integration, or to use the transparency masking blend method. You will base your decision on how you plan to view the creature in the camera shots. Super close-ups will require fillet blends, and normal distance shots will enable a lighter modeling method without trims or blends.

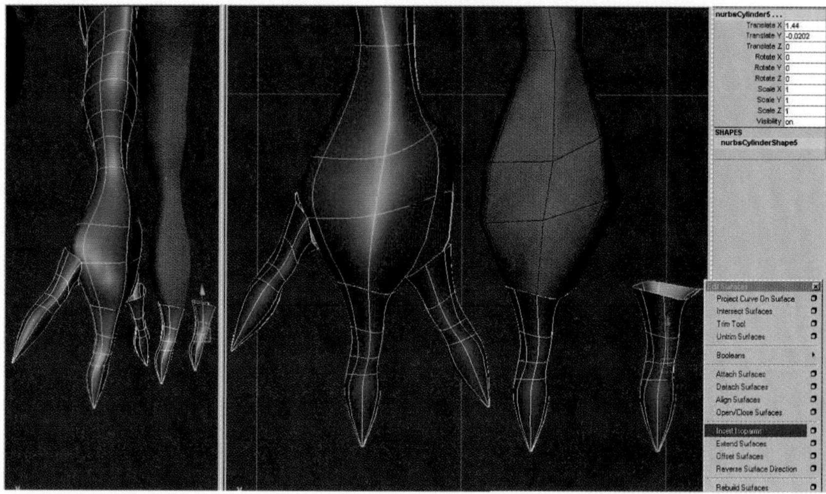

Figure 6.42
Select a knuckle isoparm on a duplicate arm, detach the surface, and then delete the arm fragment to generate fingers and thumbs.

6.4 Forming the Head

Now we will create the bottom of the skull from the end of the neck-torso model. First we will delete the R_side, add the new isoparms, and then hull-edit to match the image plane. Then we will mirror the result so that we can prepare our model to be sculpted into shape.

1. Insert isoparms by holding down the RMB over the model and choosing Isoparm from the Object Component Marking menu that appears. Mark several new isoparms that will provide enough resolution so that you can easily pull out the details required.

2. Edit the hulls to match the drawing. Use the Translate tool and scale only in the Y axis to preserve the half shape's X axis CVs. The fine-detail sculpting will occur when the mirror side is in place.

3. Mirror the L_side by using either the MEL script or the Duplicate function set at –1 scale on the X axis. Make sure that the new object is properly named. Now you are ready for some more Artisan sculpting with the reflection brush.

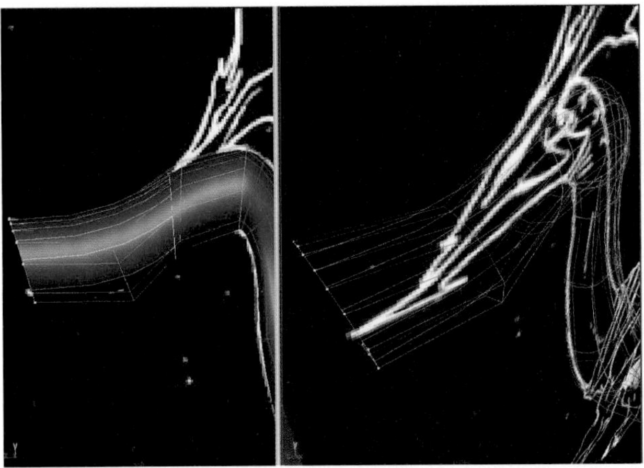

Figure 6.43
Extend the neck to become the head, hull-editing the new isoparms that have been inserted.

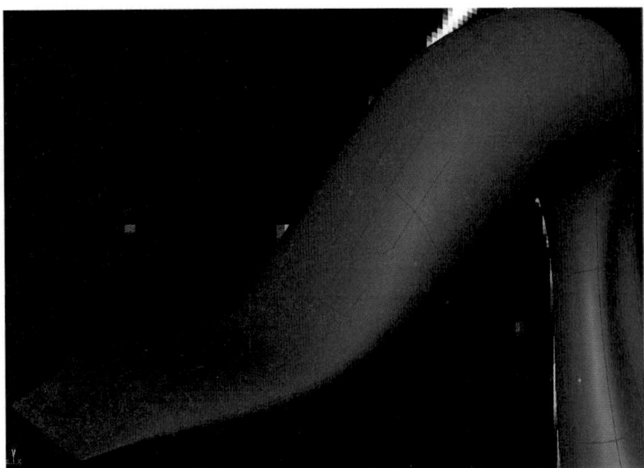

Figure 6.44
Flatten the nose to prepare the foundation for the head.

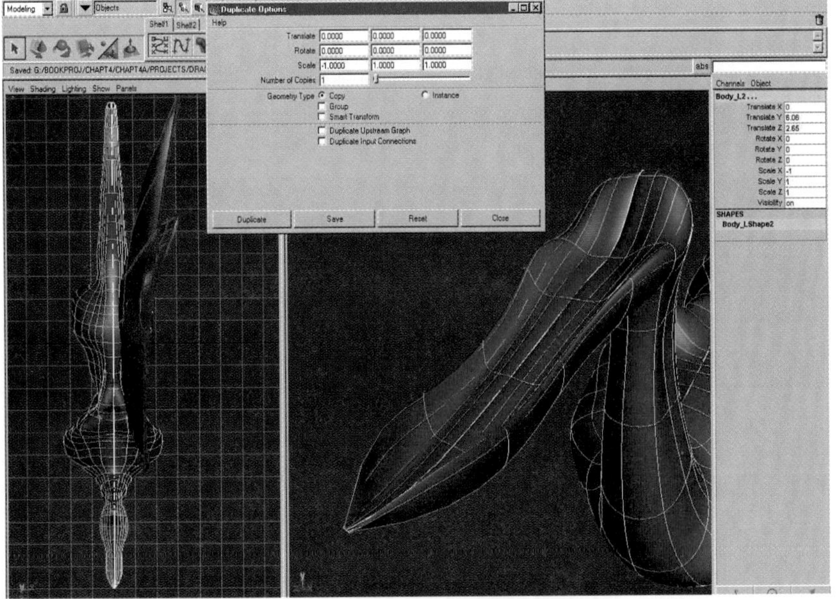

Figure 6.45
Mirror the torso to check your progress.

6.4.1 Editing the Head Model

To edit the bottom of the head area, use a combination of Artisan's Sculpt Surfaces tool and a lattice, only to the region that needs editing.

1. Make sure both halves are selected. Activate Artisan's Sculpt Surfaces tool, and proceed to edit the contours of the lower jaw region. (You can instead add a deformation lattice to selected CVs for both halves. This enables easy, simultaneous editing of both sides.)

2. In the side view, go to Component Editing mode by pressing F8. Make sure the component pick masks are set to CVs Only. Then select the head region's CVs.

3. Go to a perspective view to verify that both sides have selected CVs. If they do, go to Animation, Deformation, Lattice, Settings. Make sure that Group with Base and Auto-Center On are activated. Now, when you add the lattice, it should be centered on the selected CVs. If the lattice needs to be moved around, be sure and select the lattice's group node by hitting the up arrow with the lattice selected. Then use the Translate, Rotation, or Scale tool to maneuver the lattice to the desired location. When editing the inputs of the number of lattice points you will need, keep in mind the level of detail that you require for the edits and add them before doing any lattice point tweaks. Then scale, rotate, and transform the lattice points until you are satisfied. When you are sure that you don't need the lattice anymore, select both sides and choose Edit, Delete by Type, Delete History.

Note

You can also manually select a hull on one side and then Shift+select a hull on the other side. This method takes longer, but it is the most precise. When both hulls are selected, the Scale tool works correctly, allowing uniform scaling around the proper center of the object.

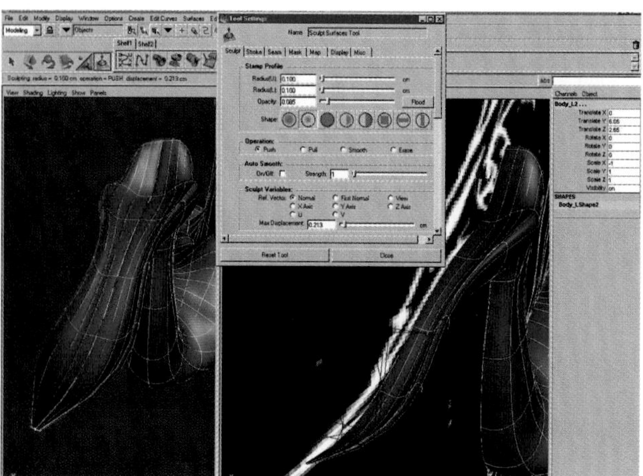

Figure 6.46
Use Artisan to sculpt the lower head area, making a cavity for teeth and gums.

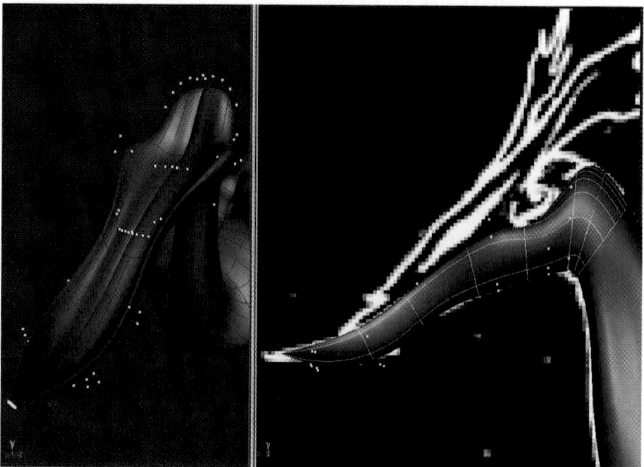

Figure 6.47
Try direct CV editing to move chunks of geometry around. Be precise and symmetrical about which CVs you pick at this time.

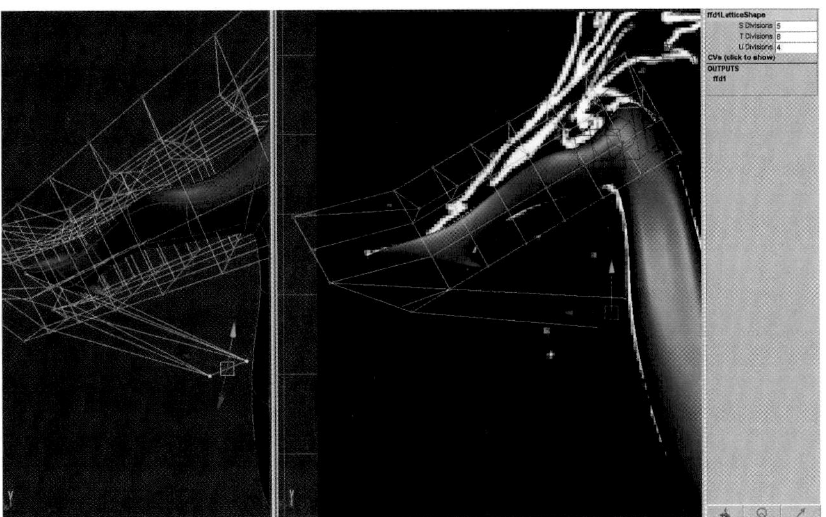

Figure 6.48
Place a lattice around the highlighted CVs in the head region. Translate, rotate, and scale symmetrical groups of lattice points to achieve useful results.

6.4.1.1 The Chin, Lower Jaw, and Skull

1. Zoom in to the chin area to verify that the tip of the lower jaw is pinched seamlessly. If it isn't, scale the hull or chin CVs until it does. Be sure to zoom out to check for any surface errors caused by overlapping geometry.
2. Collapse the whole lower jaw region up to the base of the neck. Imagine a circle. Now imagine taking the CVs from the top half and aligning them with the bottom half, but slightly above so that there are no intersecting vertices. That way, the lower jaw is significantly concave to enable placement of the teeth, gums, and tongue. Do this until the lower head region is sufficiently carved away, and then proceed to the next section.

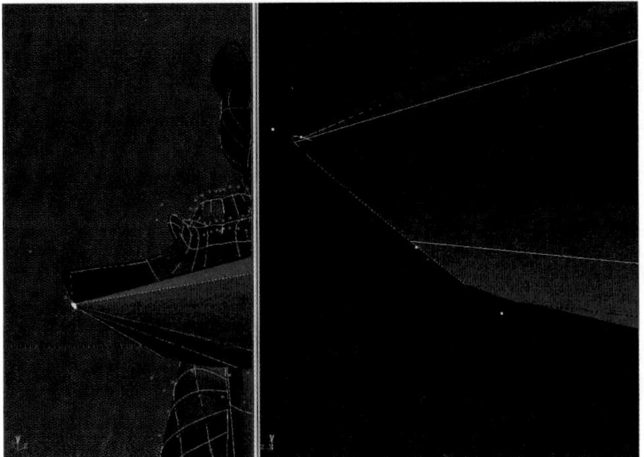

Figure 6.49
Pinch the chin seamlessly by zooming in very close. Make sure that you do not scale in the negative units.

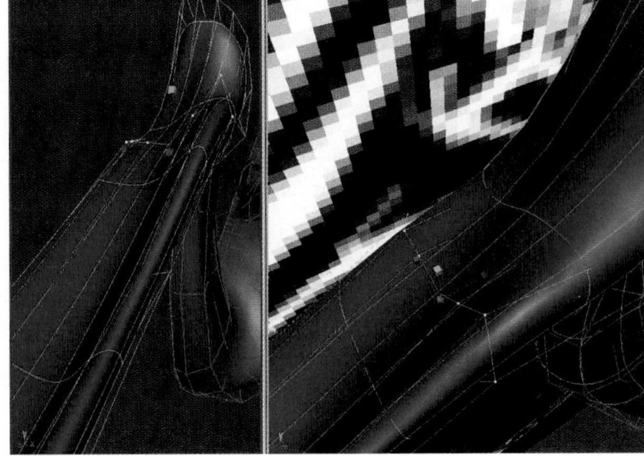

Figure 6.50
A close-up view of the inner jaw region on the lower head.

3. Create a primitive sphere for the top of the skull. Using the Translation, Rotation, and Scale tools, form the primitive sphere into the general shape of the head.

Note

Do not use component-level editing until after you have initially shaped the sphere and decided on the mesh resolution. If you component-edit before you change the level of detail of the primitive sphere, when you alter the sphere's input mesh resolution you will get strange results. This is because the sphere's components have a list of parameters that have labels. When new components are added to a sphere because of an increase or decrease in the number of spans and sections or in the start/end sweep degrees, the topology is altered but the remaining geometry is numbered as it was. New geometry is added or removed from the base of the object when the parameters are altered.

If it is too late and you want to add resolution, simply use the Rebuild Surfaces tool, found under Modeling, Edit Surface, Rebuild Surfaces, Settings. With Rebuild Surfaces, you can add or subtract resolution. The tool is a bit dangerous, however, because it will often give you results that don't look exactly the way you want. You can use the tool fairly interactively by leaving the Tool Settings window, alternately trying different settings and pressing Z to undo the results before closing the dialog box. This way, you can experiment until you reach a setting that you can feel will achieve your goals.

Assuming that you follow this example, the Rebuild Surfaces tool will not be necessary at this time. Always keep in mind that the Insert Isoparm tool is a much more precise way to add resolution to a model.

4. After shaping the sphere to approximate the top of the skull, go to Edit, Delete by Type, Delete History to clear the model's construction history. Then choose Edit, Reset Transformation.

5. Insert several isoparms horizontally to prepare the sphere to be edited. Hold the RMB down over the skull object, choose Isoparm, and then mark several isoparms between the nose and the neck, left to right, by Shift+clicking on an existing isoparm and dragging out additional yellow isoparm markers. Choose Modeling, Edit Surfaces, Insert Isoparms.

6. Use the hull-editing technique to adjust the skull's isoparms with the standard transformation tools until they closely approximate the image plane. If the skull needs more resolution, reference the steps involved in the early torso modeling stages, in which you bisected the geometry, deleting one half with the Detach Surfaces tool and adding resolution with the Insert Isoparameter tool, afterward mirroring the duplicated result.

7. Before adding any vertical isoparms (the ones that go the length of the skull), to remain symmetrical you must first cut the sphere in half, much as you did with the body earlier. Select the top and bottom isoparms as if inserting new ones, and then choose Edit Surfaces, Detach Surfaces. Delete the right half, select the other half, duplicate it, and –1 X axis scale the new object to mirror its original half.

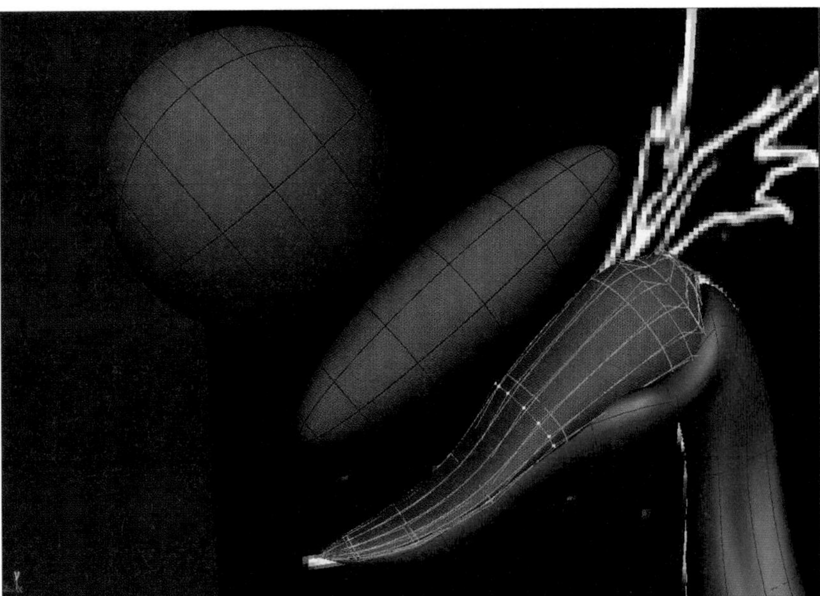

Figure 6.51
The top of the head begins as a primitive sphere.

6.4.1.2 Shaping the Skull

Now we will shape the top of the skull, using Artisan's reflection tool and other detail-adding techniques.

1. Select both sides of the skull geometry. Go to Modeling, Edit Surfaces, Sculpt Surfaces Tool, Settings. Click the Stroke tab to ensure that the settings for Reflection Painting and Multiple Surfaces are both on.
2. Proceed to push and pull details until the top of the head looks acceptable. Remember to add nostrils, eye sockets (a reference eyeball sphere is good to put in place), face spikes, and head spikes.
3. For some extra details, try applying a lattice to the upper skull models and the CVs in the lower jaw region of the main body geometry. Then, with a combination of lattice point and CV point pulling, edit the form of the entire head. After completing the work with the Lattice tool, select all the geometry that the lattice was placed on and choose Edit, Delete by Type, Delete History to remove the lattice, yet retain the deformation effect desired.
4. Finally, to add special details to the head, such as nubs that will be used to attach the horns, straight-up CV editing is best. Remember to delete half of the head and do the point editing on one side, and then mirror the result to save you time and to look consistent.

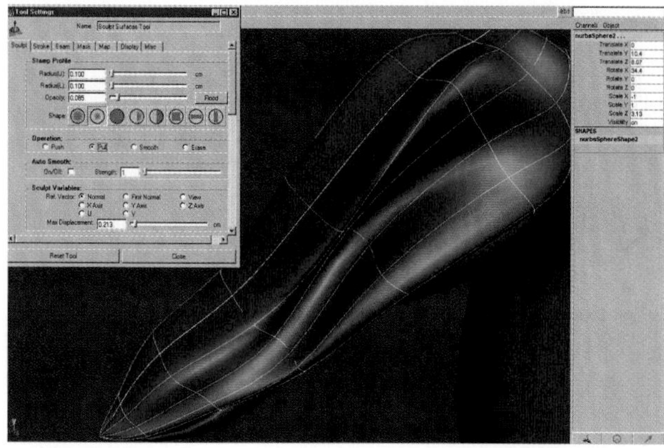

Figure 6.52
Bisect and then mirror-duplicate the head to allow reflective editing in Artisan.

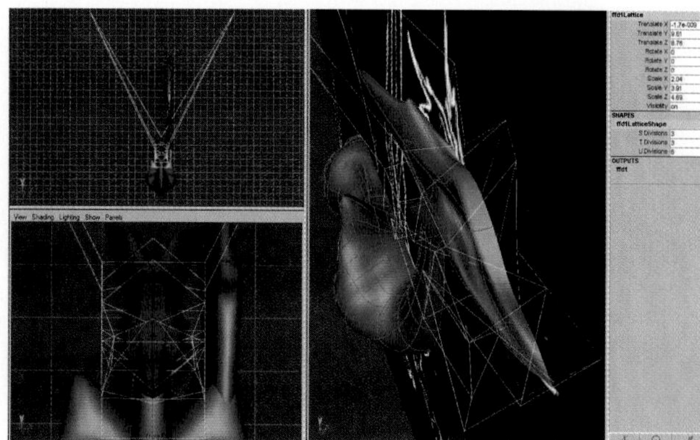

Figure 6.53
Use a lattice on the head for interesting effects. Try a lattice around both of the CVs at the bottom of the head, with the top of the head in one lattice.

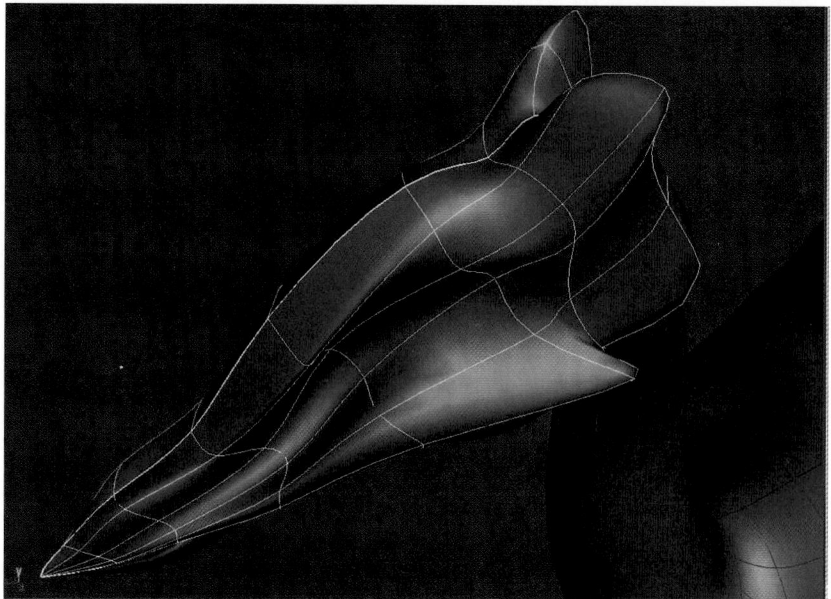

Figure 6.54
At this stage the model still needs a lot of work!

6.5 Sculpting the Tail Spikes

Beginning with a NURBS primitive cone, sculpt the first tail spike by doing the following:

1. Set the sections and spans to 8 × 4.
2. Translate on the Z and Y axes to place the cone over the image plane reference art.
3. Adjust the radius to the approximate width of the widest spike.

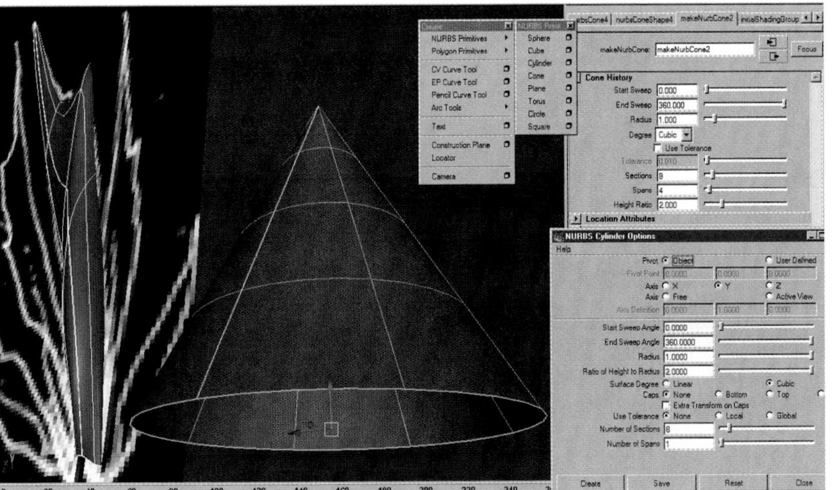

Figure 6.55
Create a new primitive cone. After creating the object, edit its inputs to start it off with enough geometry.

4. Scale the X axis down so that the tail spike is much thinner, and then delete the history and freeze the transformations.
5. Pull points in the Component mode so that they match the image plane. Think about rotating isoparms on their sides to stretch out some of the vertical spikes.

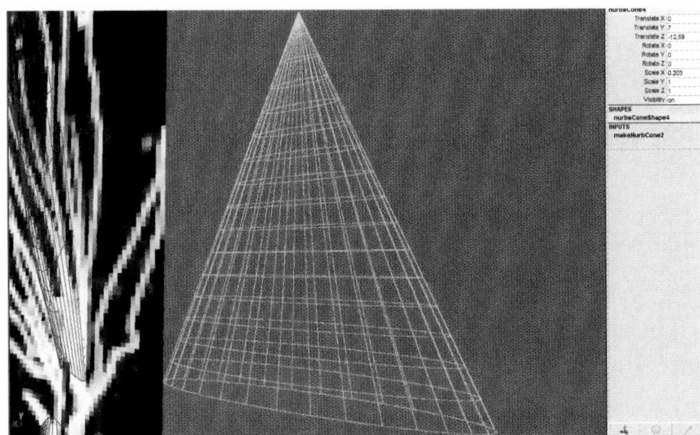

Figure 6.56
A little X axis scaling goes a long way.

Figure 6.57
Translate, rotate, and scale the hull points to match the drawing.

6.5.1 Forming the Remaining Tail Spikes

Now we will use the same technique to create the rest of the tail spikes.

1. Offset the pivot point and then duplicate and edit the spikes to distribute and personalize the other tail spikes.

2. Make sure the Translation tool is on and then select the cone. Press the Insert key to place the pivot at the base of the cone spike. Make sure that you move the pivot only in the Y and Z axes by using the green- and blue-colored arrow icons on the Translation Manipulator.
3. Duplicate, rotate, and then CV-edit to distribute and personalize the next spike so that it matches the reference art, or manipulate it to your own preference.
4. Repeat step 2 three more times to complete the tail fan.

Figure 6.58
Voila! A rendition of the tail spikes. A cool set-driven key could make them open and close.

6.6 Shaping the Rear Legs, Toes, and Spikes

Now we will create a primitive cylinder as the starting place for the new leg.

1. Place your new primitive cylinder in the perspective view to help facilitate correct alignment in XYZ space. Be sure to place the cylinder directly beneath the rear leg lump to ensure that the isoparms will easily integrate with the main body.
2. Set the makeNurbsCylinder inputs to 8×8 and adjust the radius so that it generally matches the reference picture loaded in the image plane. The exact settings are not important; the shape will be completely reformed hull by hull.
3. Remove the construction history by deleting the history. Afterward, choose Modify, Freeze Transformations to reset the cylinder's attributes to zero in its new origin.

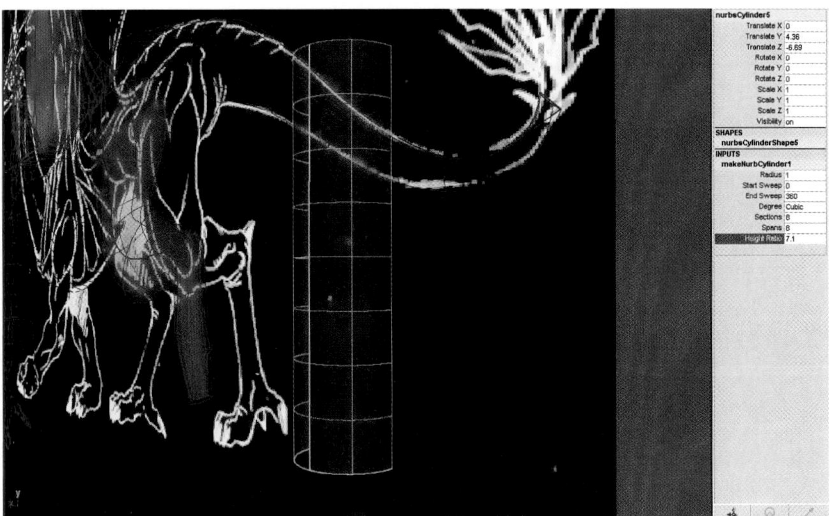

Figure 6.59
A new primitive is again born into your world. Align it with the drawing, and then give it the freeze.

6.6.1 Distributing the Hulls

The next step is to distribute the hulls according to the reference image plane, and then to add resolution. The end goal will be to have a mostly one-piece leg model that includes the upper thigh to the tip of the inner toe.

1. Hold down the RMB over the new leg model and then highlight the Hulls option from the Marking menu.
2. Select the ring-like vertical hull at the bottom edge of the new cylinder by clicking the pink ring hovering around the top isoparm.
3. Translate it to the ankle region. Scale, rotate, and transform the hull and all that follow to your best depiction of a leg, keeping in mind musculature and bone structure.
4. Repeat with each hull until the object generally resembles a leg. Use the arrow keys to "pick-walk" through the hulls one at a time for easy editing. Make sure that the top isoparm smoothly approaches the main body's leg attach region.
5. When you have spread out all the hulls, it is time to insert more isoparms to increase the resolution. Set the shading style to Wireframe and the selected leg model to Rough Smoothness by pressing 4 and 1, respectively. RMB on the leg and choose Isoparm.
6. Click+drag and then Shift+click-drag more ring-like isoparms across the length of the leg. When satisfied, choose Modeling, Edit Surfaces, Insert Isoparms.
7. Repeat the hull editing technique until the ankle extends to become a foot with an inside toe already attached. Make sure the toe is smoothly pinched closed.

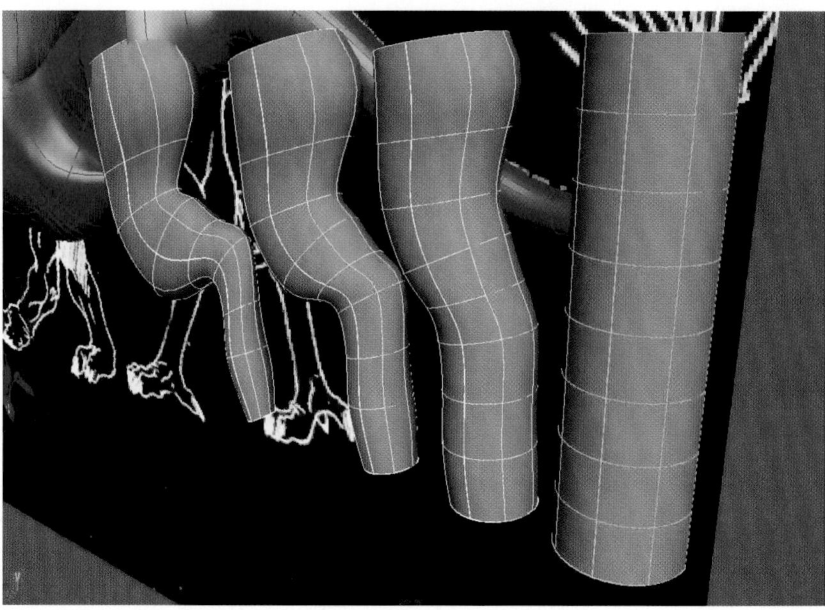

Figure 6.60
Progression of the leg through its evolution.

6.6.2 Adding Detail to the Leg

Next we will create the embellishments that will add some detail to the leg. First we will create the heel and ankle spike from primitives. Then we will create the outside toe object by duplicating the leg and then extracting the toe component.

1. Use a primitive cone for the heel and ankle spikes. Use the Scale, Rotate, and Transform tools to approximate the general shape of the horny protrusions. Component-edit the hulls to give each spike some unique character.
2. To create the outside toe, duplicate the leg model and then scale it to –1 on the X axis. Move it to an area that is easy to get at and select an isoparm at the base of the toe. Choose Modeling, Edit Surfaces, Detach Surfaces, Settings. Make sure that Keep Original is off. Detach and then delete the extra leg geometry. Move the toe back to the foot region of the original leg.
3. Using hull-editing techniques, merge the toe near the leg. Later you will decide whether to use fillet blends to create geometric integration, or to use the transparency masking blend method. You will base this decision on how you plan to view the creature in the camera shots. Super close-ups require fillet blends, and normal distance shots allow a lighter modeling method without trims or blends.

Figure 6.61
Heel and ankle ornaments begin as primitives.

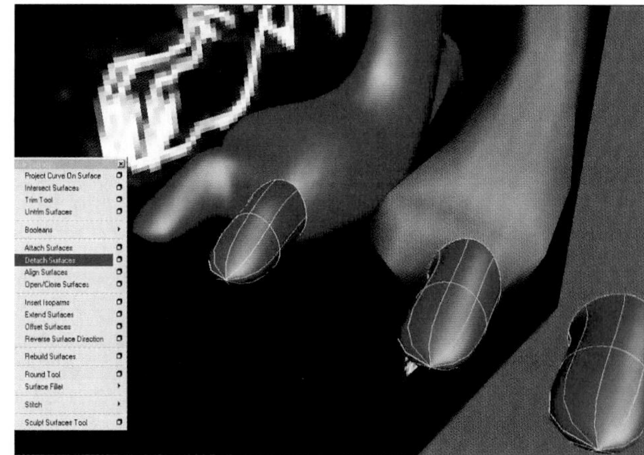

Figure 6.62
Create the toe by duplicating the leg and using Detach Surfaces.

6.7 Adding Details to the Torso

To actualize a smooth and flowing body design, you must sculpt muscles and add detail to the main body torso to aid in musculature integration.

1. Use Artisan's Sculpt Surfaces tool to fluidly add muscle details to the leg. Remember to keep Opacity and Max Displacement set low to avoid any extreme changes that may result in unwanted lumps.
2. If you want to add more detail, add some extra horizontal isoparms (the ones extending from toe to thigh). Return to sculpting with Artisan's brush set.

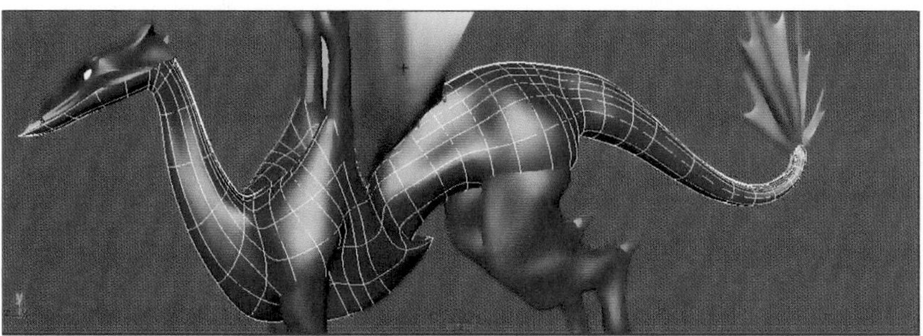

Figure 6.63
Adding body details with Artisan.

Figure 6.64
Verify that you are pulling points symmetrically on both sides.

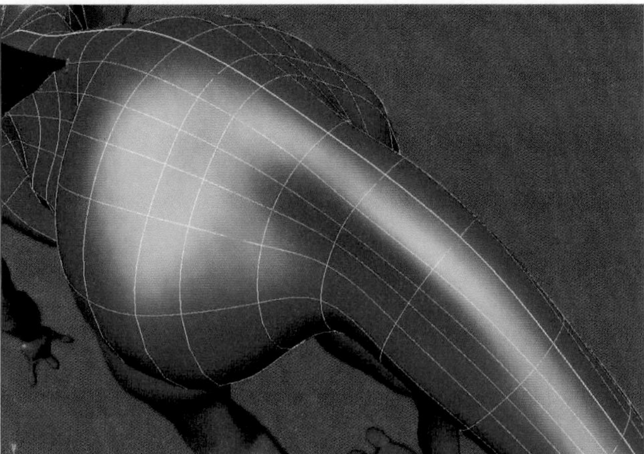

Figure 6.65
Remember to do a detail pass to add realism and accentuate details.

3. Select the main body torso models and begin to add some extra detail to the rear leg/rump region using Artisan's Sculpt Surfaces tool. Remember to activate the Reflection and Multisurface Stroke options so that you can paint on both halves at the same time.
4. Verify that the geometry flows together, visually going back and forth between editing the main body and the leg with Artisan's brushes.

6.8 Making the Model Seamless and Detailed

Making the model seamless and detailed requires integrating various portions of the model, using the Fillet Blend or Transparency Mask tool.

6.8.1 Toes, Fingers, and Legs

To integrate the toes and fingers to the leg and arms using the Fillet Blend or Transparency Mask tool, do the following:

1. Use the techniques introduced in Chapter 4,"Modeling an Organic Humanoid Hand with NURBS," which cover the construction of the hand, to create a curve on surface on the paw or foot to be used as the foundation of the fillet blend.
2. After creating the curve on surface, trim it and create a fillet blend that spans the gap to blend the geometries together.

Tip

Rather than using fillet blends, you can also build a gradient ramp or grayscale image to make a transparency map that causes the geometry to artfully go transparent where the blend is to occur. However, this technique is tricky, and the model must be finessed for this effect—in other words, you must make sure that the surfaces have just enough overlap to achieve the effect.

To integrate the leg to the body using the Fillet Blend or Transparency Mask tool, refer to Chapter 4, which covers hand modeling, for all sorts of detailed modeling techniques for making a curve on surface act as the blend foundation. Then trim and blend the pieces together.

Tip

Transparency maps can save processing time and offer lighter geometry, but they are a tricky way to work. Give yourself adequate time for experimentation.

6.8.2 Horns and Spikes

You can add horns and spikes to make your dragon look more interesting. The Animated Sweep tool allows for interesting spiral and twisted horn variations.

1. First, keyframe a closed curve so that it moves in a way that resembles the shape of the horns you want to fabricate. Give yourself at least 100 frames on the timeline to work with.
2. Animate the closed curve so that it scales to zero. This will close the horn at one end.
3. Choose Modify, Animated Sweep, Settings and experiment with the step count by changing the By Time feature. This is the parameter that determines how many times a cross section is sampled down the animation path. The Start and End times must be set to the point on the time line where you made the start and end keyframe for the curve.

At this point you are finished with the first stage of creating a model. From here on out, use your judgment as an artist to add all the unique details you see fit. In Chapter 7, "Texture Mapping Organic NURBS Characters," we discuss how to create a high-resolution multi-layered texture map that is pre-fit for your experimentation.

Texture Mapping Organic NURBS Characters

CHAPTER 7

Texture Mapping Organic NURBS Characters

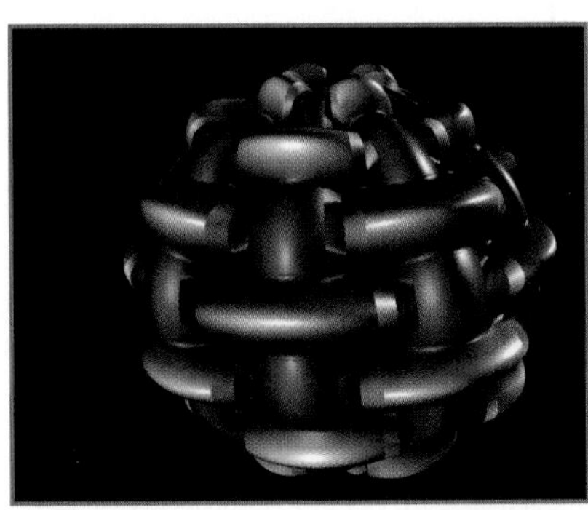

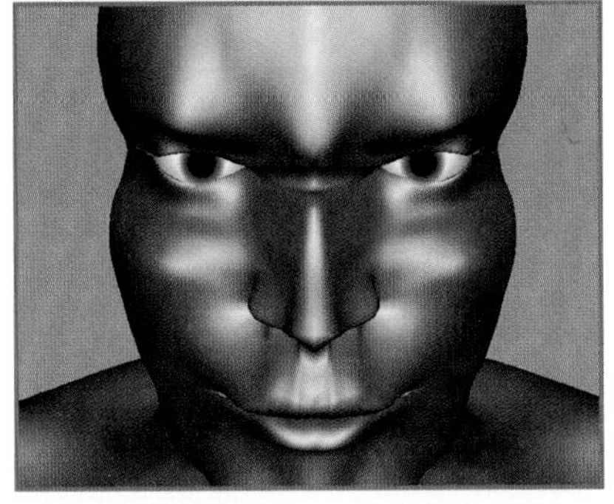

by Nathan Vogel

Forming the geometry of your model is only the first step in designing a computer-generated 3D character. Adding the color and texture that give a character detail is an important and necessary process. By Alias | Wavefront's own admission, texture mapping in Maya is often slow, tedious, and time consuming. Several solutions are available that can help speed up the initial stages of adding specifically assigned, precisely placed, user-controlled texture maps.

This chapter leads you through a series of methodologies intended to help you gain experience and confidence so that you can create your own custom texture-mapping techniques. The chapter also incorporates techniques gleaned from several professional texture-mapping resources and presents these techniques as optional and useful complimentary methodologies.

7.1 Introduction to Texture Mapping

To come up with the best creative solutions for any project, you must approach the process of texture mapping organic characters with an open mind. Organic characters present their own unique challenges and dilemmas when it comes to texture mapping. However, most conventional organic characters have several key texture-mapping steps in common that you can use to solve a large part of your general texture-mapping needs. By familiarizing yourself with these techniques you will find useful ways to overcome the challenge of texture placement and the dilemma of properly layering your textures.

You can create and map textures to your organic character in many ways. Projection mapping from orthographic views is a common technique that can result in a predictable and consistent mapping result. After completing a basic planar projection to provide a base for further layering, you will have a firm idea of what it takes to establish an initial texture-mapping formula. From that point, there are various routes that you can take to further enhance the rendered look of your character. This chapter discusses techniques covering integration of the resulting textures with a layered shader node. Methodologies for using the new reference texture mode are also addressed, along with the proper use of texturing utilities such as the switch node.

7.1.1 The Difference Between Shading Networks and Texture Maps

The first step in working with texture mapping in Maya is to understand its metaphor for organizing surface appearances. Maya uses the metaphor of a Shading Network consisting of many levels of organization that connect all the various aspects concerning texture mapping.

The first level of the network is the Shading group, which acts as a holder for the three inputs, called *ports*, that hold a connection to various aspects of a surface's description.

The first of the Shading group's three ports is the Surface Material port. The second is the Volume Material port, which is generally used to shade volumes of transparent area for certain software particle rendering modes and vaporous fog regions. The third is the Displacement Mapping port, a slightly tricky but very rewarding parameter that determines how the surface of your object is actually altered in three dimensions. The Displacement Map attribute is akin to the bump map Material attribute, but it actually displaces your geometry, or moves it around, to create real bumps whereas bump maps don't move the geometry around at all.

The next level of the network is that which contains surface materials. *Surface materials* describes the shading parameters that determine the way light interacts with the surface of an object. In Maya 2.0, there are nine types of surface materials. Five of them share a handful of common parameters that allow you to interact with the shader's appearance. For example, the Blinn surface material contains a host of common material attributes. To almost all the material attributes you can apply a texture map that you can use to influence the final result of the surface material's appearance.

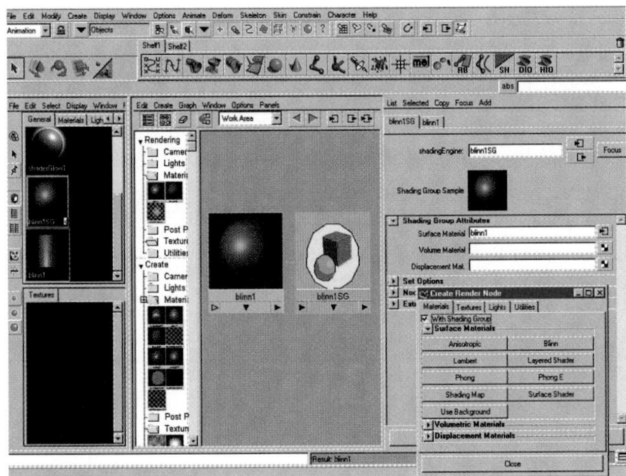

Figure 7.1
The Shading Group's three input ports.

Figure 7.2
The displacement map on a NURBS compared to a polygonal model.

> **Note**
>
> For polygonal objects, the displacement attribute can affect only the number of vertices the object possesses. NURBS objects use the tesselation criteria object attribute found in the attribute editor when selecting a NURBS model to determine the level of detail that the displacement can attain. Thus NURBS objects are much more compatible for displacement mapping than light polygonal surfaces (i.e., non-dense meshes). Be aware, however, that high tesselation values cause longer rendering time.

The following is a brief description of common material attributes in the Blinn surface material.

The Color parameter controls what color the object will be. This parameter is commonly mapped with some sort of texture to determine the color of the material's surface. This attribute is directly influenced by the diffuse parameter, which affects the color much like a scale of color intensity. Color needs a light for it to be seen.

The Transparency parameter controls how much light passes through the surface material, allowing the area behind the object to be seen.

Note

Maya interprets black and white values opposite to the way leading paint packages and most other 3D applications interpret them. In Maya, white is considered transparent and black is considered opaque, thus reversing the whole color and grayscale spectrum.

Ambient Color is the attribute that determines what color and value the darkest dark can be. A light source tends to bounce all over an area, lighting up even the shadows to a degree. Thus, pure black is something that should be applied carefully for any daytime shots.

Note

When compositing 3D characters or objects into real-world sets, it is important that you synchronize your Ambient Color to match existing ambient values from the footage you are using. Using the color sampler from Maya 2.5, Maya Composer, or Adobe Photoshop is a good way to find the color and darkness of the shadowed part of objects in the area of the live footage into which you are placing the character.

Incandescence is a special-effect parameter that enables you to simulate that the object somehow is a light source itself. The Incandescence parameter does not cause the scene to be illuminated. If you set incandescence to bright red, the object appears bright red, even if lights are shining on the object. For example, a fantastic beast's red glowing eyes being seen even though it's in a dark shadow demonstrates a good use of the incandescence parameter. At half value, the scene's lights would have half influence over the brightness of the materials appearance if it had a 100% incandescence value.

Bump Mapping is a special attribute that needs a texture map of some sort for its effect to be seen. When an image or procedural texture is mapped to the bump channel, this parameter and its subsequent attributes enable you to control where you perturb or "bump" the surface of your model on a pixel level. This enables you to simulate the appearance of raised and low areas by placing a shaded area and a highlight where the bump-mapped regions have variance. A 75-degree light source and an opposite 75-degree camera angle tend to provide the best results for showing off the effect.

Note

When viewed from close up or at extreme 90-degree angles, bump maps can appear flat. This is a weakness of the bump mapping metaphor in general. Round objects with many surface normals tend to hold up better to bump mapping than objects with faceted flat surfaces. The lack of any bumps along an object's profiled edges can be remedied somewhat by using a displacement map (as noted earlier).

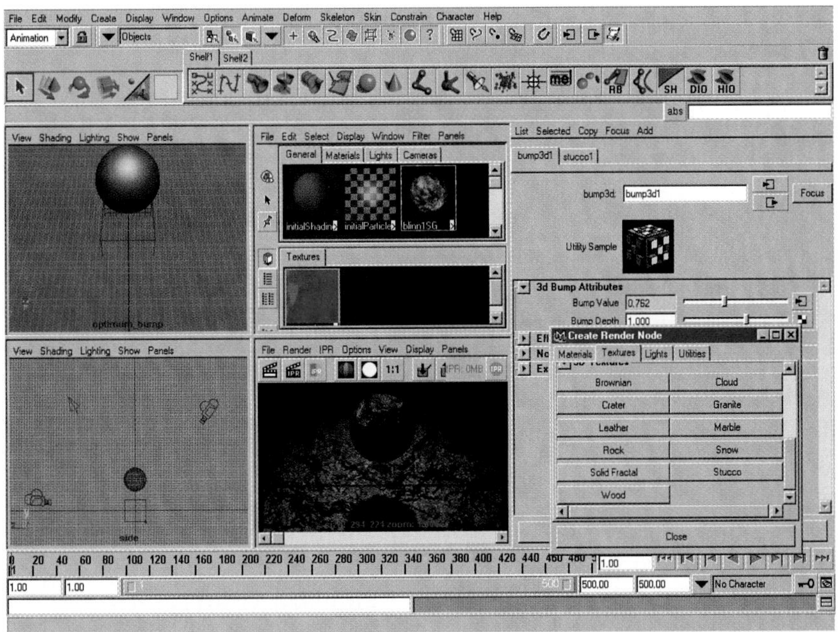

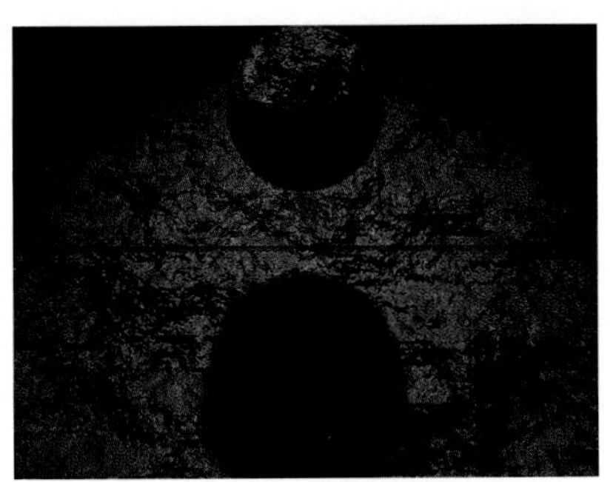

Figure 7.3
Avoid flattening a bump map by positioning the cameras and lights at non-90-degree angles to the surface.

The Diffuse attribute influences how bright your surface is in light. This parameter does not affect the ambient or specular values directly, but it does influence the color attribute.

The Translucence attribute controls how much a surface transmits light. Translucence is a slightly tricky parameter to control, but it can be useful for certain effects that need a see-through, but not totally clear, look, such as skin, hair, and fur. Ambient lights do not affect the Translucency value.

The Specular attribute controls how a bright, shiny spotlight will look on various surface types. Specular shading is determined differently by the four kinds of surface materials that use this parameter. Phong, Phong E, and Blinn shaders have a fairly common way of influencing the look, although their controls are each slightly different. The differences are observed when viewing their controls in the Attributes Editor below the standard controls in the specular sub-section. The

Anistropic shader is different in that it creates rectangular or linear-looking specular artifacts rather than the standard circular artifacts common to the other surface types, which is very handy for creating the long, stringy, specular light reflected on many kinds of hair. Each shader has its own stylistic look that is best appreciated by experimenting with each one under various geometry and lighting situations.

The Specular Color component determines what color the highlight will be. Metal reflects its own color, whereas plastic has a white highlight if it is lit by pure white light. Human skin can be rough or shiny, depending on where it is on the body. Because of the high visibility of the Specular attribute, mapping this parameter with a texture map can produce good results, simulating a variance in the surface's shiny and rough areas. When combined with a bump map, realistic effects can be achieved.

The Reflectivity attribute influences how much a surface will reflect an environment map or the surrounding objects if raytracing is activated for the object. Map this attribute with a texture to simulate reflective and non-reflective areas of the same object. For example, a smooth marble's surface is reflective except where the natural veins are exposed leaving a non-reflective rough area. The rough area would be black and the shiny area white to simulate this with a reflection map.

The Reflected Color is used to map environment maps and textures to simulate a reflected scene without the rendering hit needed for raytracing, or just to give the object a solid reflected color. As an object moves, the Reflected Color image stays, still giving the reflective object the appearance that it is moving past something stationary. This parameter is directly influenced by the Reflectivity attribute. If the Reflectivity attribute is set to zero, no reflection color or map is seen; if it is set to a high number, that reflection color is seen all over the surface.

The Special Effects section of the shaders attributes editor has a parameter called Glow that calculates a fuzzy haze that the renderer post-processes and then composites over the object's surface. Unfortunately, the Maya glow effect is a single layer of effects and is non-shader specific. Thus, an animated glow on one shader will affect the non-animated glow of another shader. The only way to get around this is to render all objects that have animated or specific and unique effects separately from the animated glows in the scene.

By applying texture maps to each of the above attributes, you can control precisely where and how each attribute will look, and you can produce powerful distinct looks or subtle effects.

Note

To match a reflective character or object to a real-world set, use elements from the background plate to create a panoramic image to reflect. If you are on location, photograph or videotape imagery from all six cardinal directions—N,S,E,W, up, and down— to be used as source imagery for a realistic reflection effect.

7.2 Projection Maps Versus UV Maps

7.2.1 Projection Maps

Over the years, many techniques have been created specifically to address the needs of texture mapping three-dimensional geometry. Many of the traditional "ABCs," the so-called classical mapping metaphors, were invented many years ago when the first 3D technology was being pioneered. Users of 3D applications know of these classical mapping metaphors as *projection maps*.

Various primitive shapes are used as projection forms. An object receives portions of the texture map from the angle that the projection form is facing relative to itself. This is very useful for a simple object that is similar to its texture primitives, such as mapping a linear tube with a cylinder projection or a ball with a spherical projection. But when dealing with complex geometry, the traditional metaphors need something more intuitive. The most common projection shapes are as follows:

- Planar maps, such as a slide projector
- Cylindrical maps, which wrap around the object, such as a sticker on a pole
- Spherical maps, which distort the top and bottom of the texture map where the sphere pinches at the poles
- Cubical maps, such as six duplicate planar maps shaped like a cube projecting inward

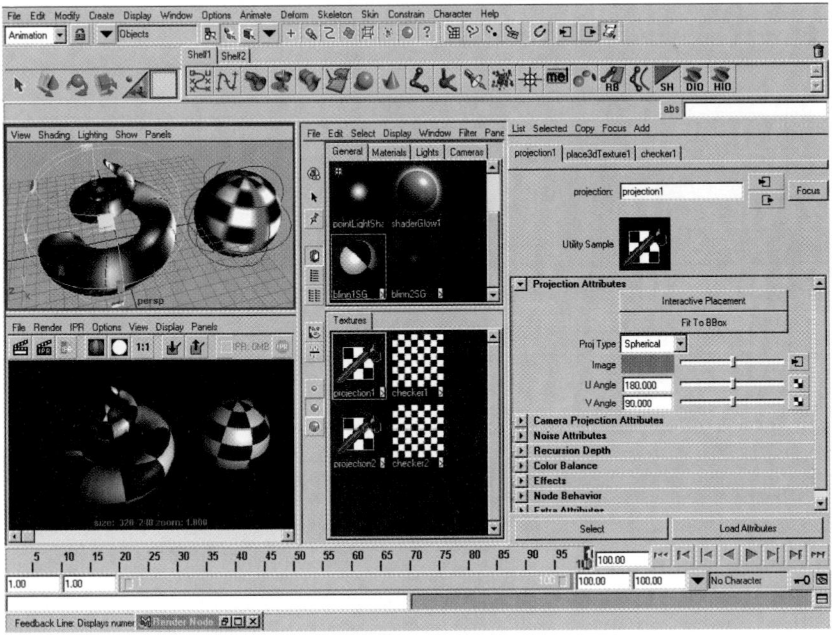

Figure 7.4
Texture primitives around a simple and a complex object with and without stretching.

7.2.2 UV Maps

UV mapping, or *intrinsic mapping*, is another way to deal with a model that has more complex forms. Imagine a nautilus shell with a spiral shape. You could texture such an object using standard projection mapping, but this would require that you match the image to the surface parameterization to avoid stretching and texture distortions. Suppose you want a simple tile's striped pattern to follow the direction of the geometry itself, down the shell's twisting tube-like form. This intrinsic, "intuitive" direction is commonly referred to as UV mapping, in which U and V coordinates are plotted on a 2D graph across a 3D form, taking into account the surface area only. Imagine unwrapping a sphere into a plane and assigning a grid to the surface, with the bottom-left corner being the zero point of both the width and the height. The bottom right corner is marked 1 for the width, and the top left corner is marked 1 for the height. Thus any point on the surface has two decimal numbers between 0 and 1 plotting out the position anywhere on the unwrapped surface.

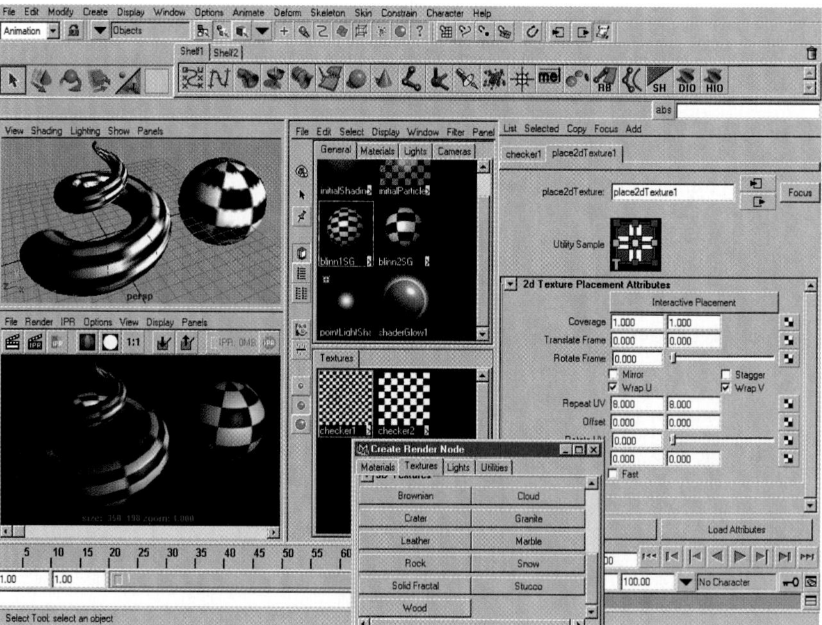

Figure 7.5
A 3D deformed surface with a pattern that shows UV parameters as a texture.

7.2.3 A Hybrid Method

The modern 3D artist is not satisfied with these mapping methods alone. Many artists have adopted a hybrid system that mixes and matches the above techniques with other more niche methodologies that are used to cover the other's weakness and promote the strengths of each as well. Often, an artist creates a texture map by projecting a series of images onto a surface and then converting the imagery into a single "pre-distorted" image. The artist then refines it within a paint package and then imports it back to the 3D environment for final layering and tweaking.

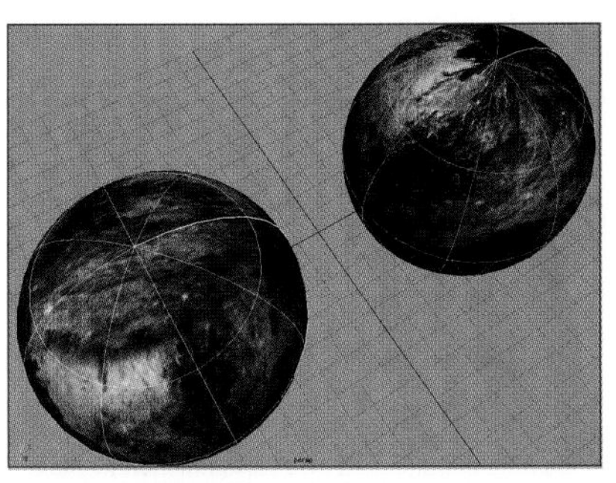

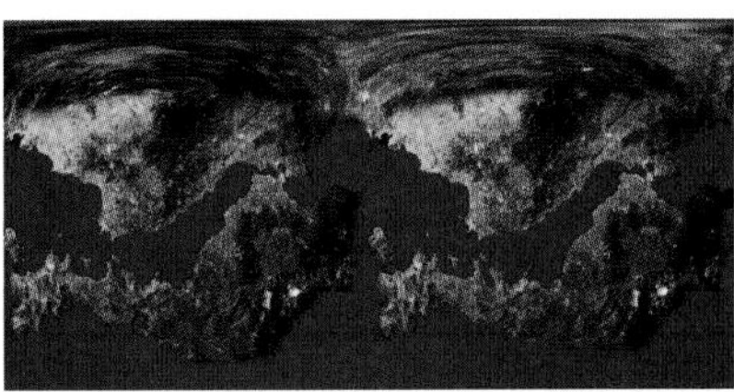

Figures 7.6
A pre-distorted texture, next to a texture that has not been pre-distorted, both on a NURBS model. Note the pinching on the left sphere with a texture that was not pre-distorted.

To help facilitate this task, the industry has produced several well-integrated 3D paint packages as well as a host of plug-in and stand-alone texture mapping utility programs. Each tool has its advantages and disadvantages. But overall, with or without the help of external programs, you can use them to create many fine results. All you need is patience, creativity, perceptive attention to detail, and experience gained by practicing a mix-and-match metaphor when creating your textures.

7.3 Texture Mapping Organic Characters with Projection

Projection mapping can be a useful and necessary step when mapping organic characters. It does have its weaknesses and limitations, however. Imagine a slide projector aimed at a dancer standing still. The image is projected across the dancer's body. If the dancer moves even a little, the projected image will be altered. This is the case with projection mapping.

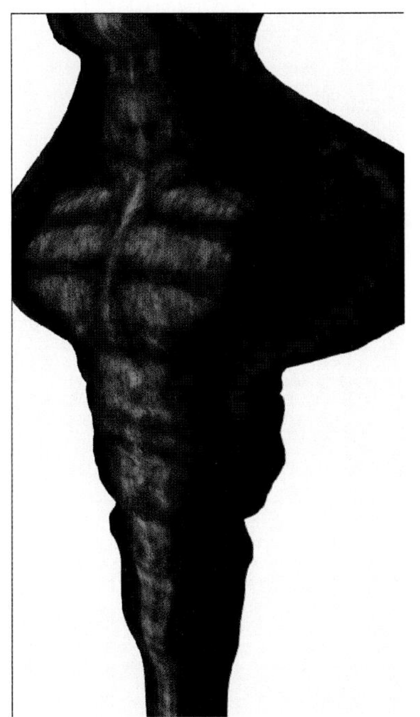

Figure 7.7
A projection map, before and after misalignment on a snake monster.

To deal with this fact, a workflow has been developed with several alternatives offering a customizable solution. You must decide early on which method you will use so that the rest of your workflow can be optimized to make this process go smoothly. The two common ways to proceed using Maya's built-in tools are either converting the projection to a solid map or creating and using a texture reference object.

7.3.1 Converting a Projection Using Convert Solid Texture

Using the standard projection mapping tools, such as planar, cylindrical, and spherical, apply texture maps to an optimized pose of your character. Choose your pose so that it allows an unobstructed view of your character and the areas that you are working on. Create this pose with a duplicate of your document so that you do not inadvertently interfere with any work that you may have to do. After you have created the texturing properly by test rendering, you can easily reapply your character's Shading groups to your original file by exporting the shading groups to disk from the multilister's File menu.

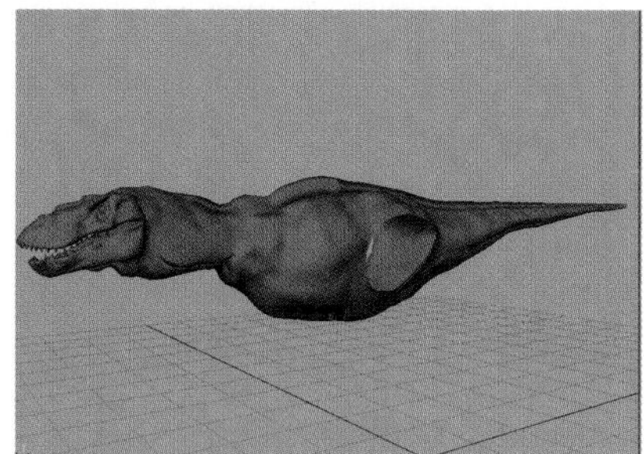

Figure 7.8
A posed character (left) and a non-posed character.

If you have not already created a skeleton for your character, read Chapter 10, "Animating Characters and Keyframing," to learn how to bind the skin to your character to allow for easy deformation to an optimized pose. Otherwise, use other deformation tools, such as the lattice, to assist in posing your model.

To pose the dinosaur, I stretched its body to create a horizontal straight line from its nose to its tail. Doing this enables an easy line of site to allow an unobstructed view of the surface that you wish to texture map.

After posing your character, you must convert the projection node that you connected to the Shading group that you have created for your model. For organic characters that are made of many different parts, such as fillet-blended regions and stitched elements, you must convert the solid texture for each piece of geometry to a solid texture separately setting the pixel sizes according to the dimensions of your object.

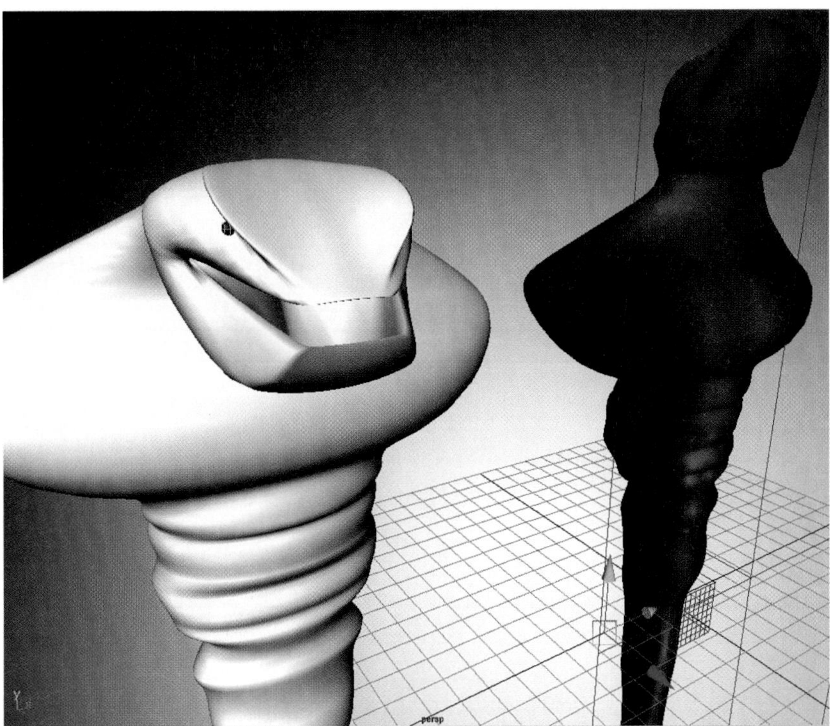

Figure 7.9
When posing this snake monster, I bent the head so that it looks straight up. This ensures clear access to the neck and chest areas.

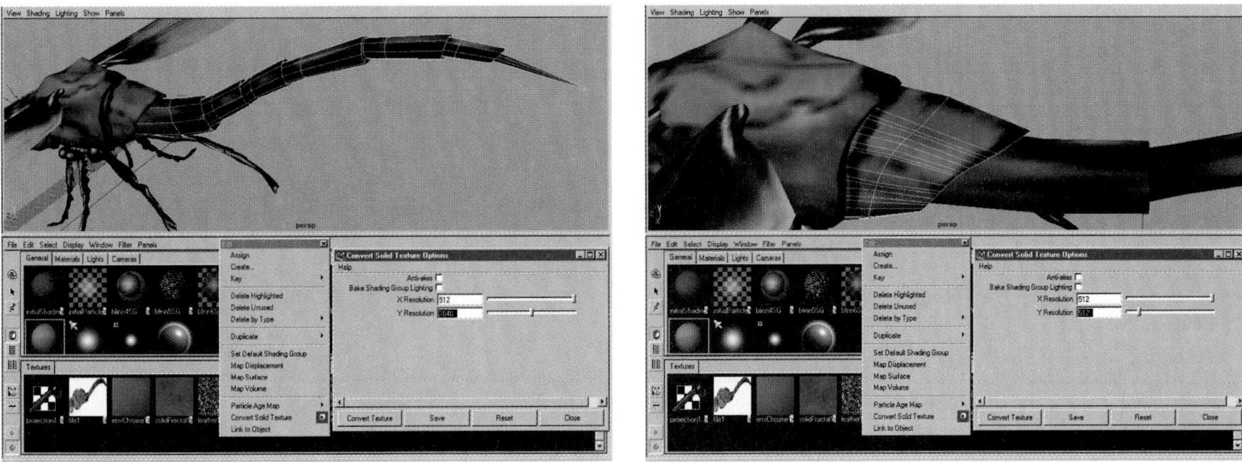

Figure 7.10
These are the Convert Solid Texture Pixel Ratio settings for a long object vs. the settings for a short object.

To properly organize this procedure, it is a very good idea to name all the objects and shaders with short but easy-to-recognize naming conventions, as the converted texture uses both the shader and the object names for generating the name for the new texture. Because you will inevitably have to re-map the files to an alternate location on the drive at some time, it is a good idea to familiarize yourself with every texture that you convert.

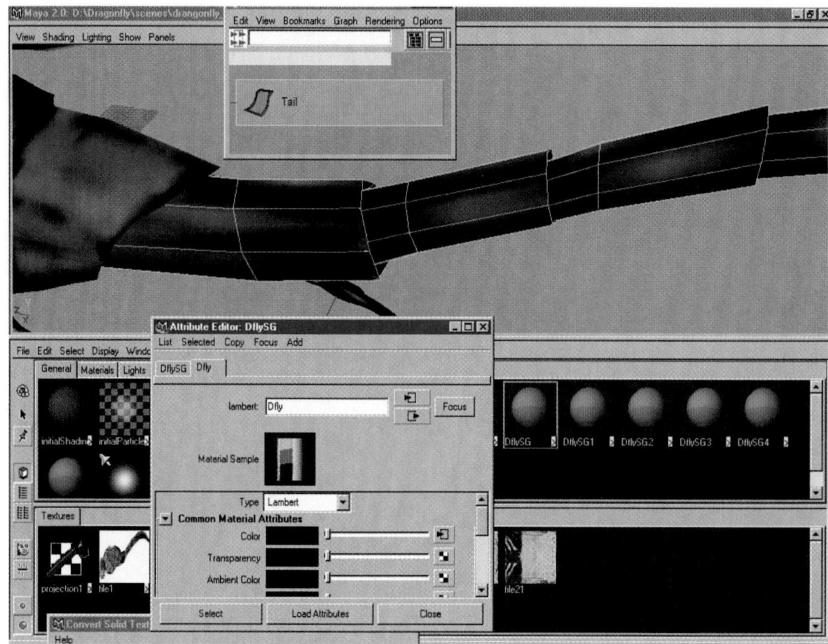

Figure 7.11
An example of some short, easy names seen in the multilister that you can use for shaders and objects.

It is a good idea to check the appearance of each element after the conversion to decide if it looks clean. If it is blurry, you will probably have to reconvert the texture at a higher resolution. For this reason, and to save disk space on unnecessarily big textures, go through the process one object at a time and convert only to the resolution that you need.

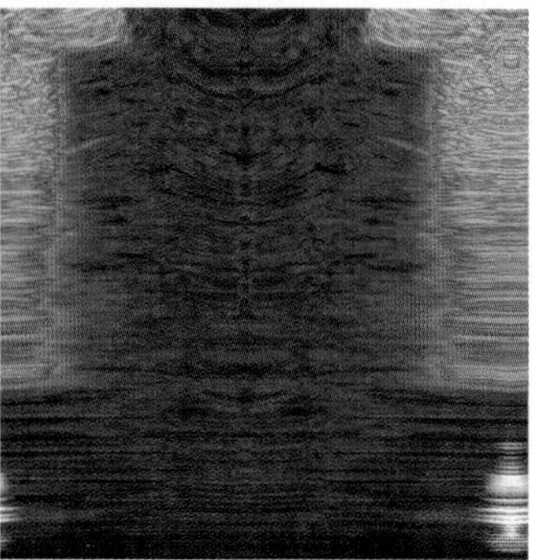

Figure 7.12
A blurry texture (left) and a sharp texture (right).

Converting solids is a good way to work for many reasons. First, you can directly edit the final, predistorted image in a paint package. This enables you to fine tune and tweak the look.

Another benefit of a converted solid texture is that it will display correctly in hardware rendering, instead of giving you a blurry view, as with projection mapping. This is important when you need to determine what your character will look like without having to do a test render.

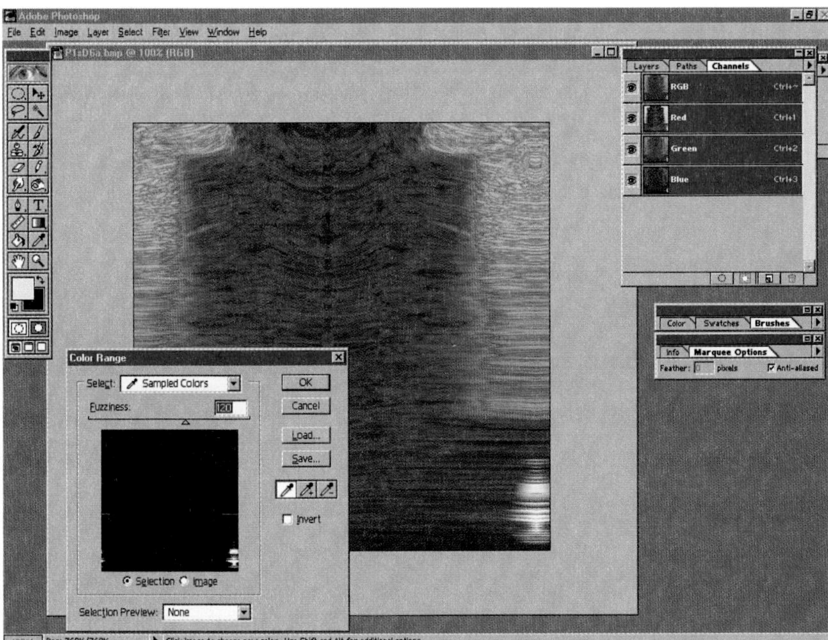

Figure 7.13
Editing the converted image in Adobe Photoshop.

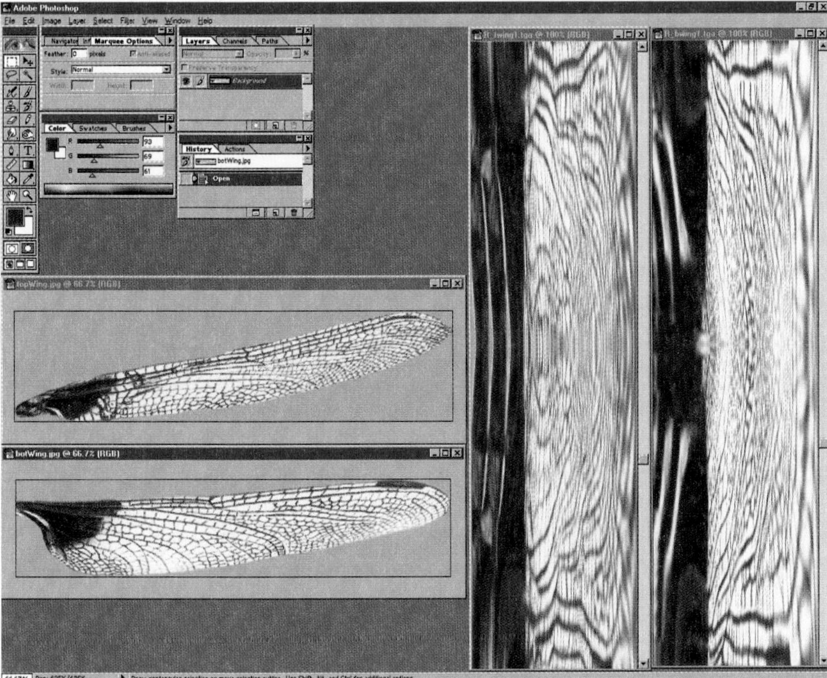

Figure 7.14
A projected texture vs. a solid texture.

A character is more portable (not needing to retain a texture reference duplicate) when you have converted the solid textures. All you need is a single copy of the character and the texture images in the source images folder and you are ready to go. This way, you don't have to have texture reference doubles of your character imported or referenced to the final master file that you are working with.

As long as your conversion resolution is not too high (it should not be over 1024×1024 unless it is absolutely necessary) causing oversized file sizes (over 4MB) converted textures tend to render a little faster than projections.

Depending on the topology of your geometry (perhaps an object with both dense and sparse isoparameter distribution in the same model, such as the unusual topology often associated with faces) when Maya converts a solid texture, the areas of dense isoparms will get more samples, making the textures in the areas of dense geometry appear to be sharper and the areas of light geometry to be more blurry.

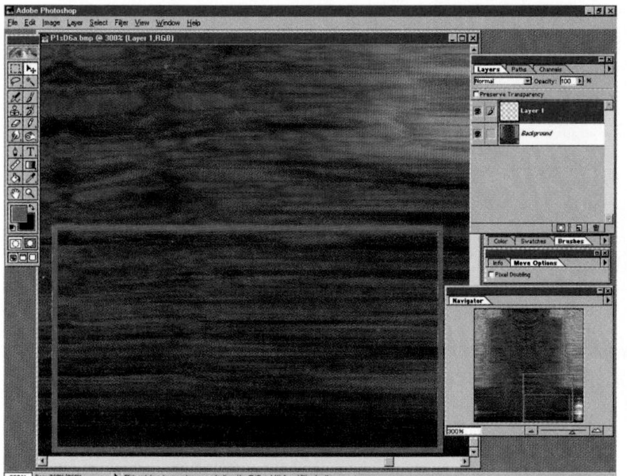

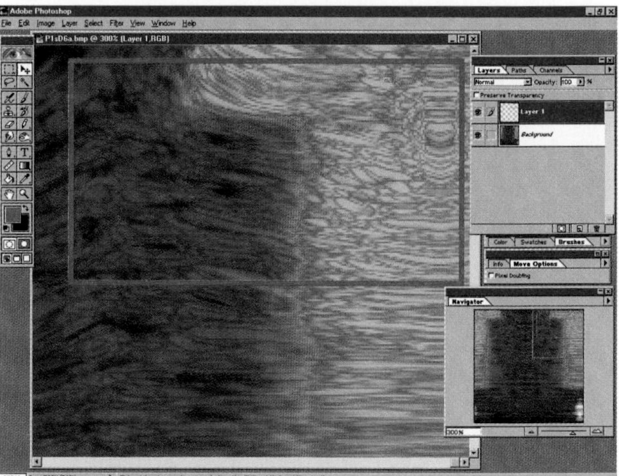

Figure 7.15
Detail of low and high sampled areas of a converted solid texture.

It is important that you apply the Shading groups directly to the geometry and not to a group to which the geometry is parented. The conversion process needs an object to be selected, along with its corresponding shader, to be successfully converted.

Of course, the extra RAM needed for rendering and the disk space issues mentioned earlier are important considerations when converting many texture layers.

One of the more important things to consider when converting to solid textures is mapping channels and when to do so. If you add a separate texture for some of your various texture channels, the convert solid process will be applied and will convert a picture separately for each attribute. If you have the same image mapped to several attributes, Maya will still convert it once for each time it is used. This can be a waste of time and disk space if you are unaware of this aspect of the process.

If you are aware of this aspect, it can also be useful if you need to do custom tweaks to every channel, which is often the case. If you want to minimize your conversion time, apply the texture to only the color channel (and to the transparency if you are building a layer that will later require a mask). This enables you to create each attribute map in Photoshop based on the original converted image. This saves conversion time and allows maximum user tweaking.

The downside is that if you have many individual objects, you will have to maintain consistency across all the images. For this reason, doing your tweaks using a batch processing package such as Adobe Photoshop 5.0 or greater, Adobe After Effects, or even Equilibrium's Debabelizer can save you from having to memorize too many obscure "tweaky" steps. If you are not limited by disk space, conversion times, or multiple converted attribute maps, Maya will simply convert each attribute separately even if it is a duplicate. The exercises in the later section "Texture-Mapping Tutorials" cover this subject in a detailed, step-by-step manner.

Note

By default, Maya creates a new Shading group for each of the character's objects. If you were to tweak any parameters afterward, you would have to manually set each of the groups to match your edits.

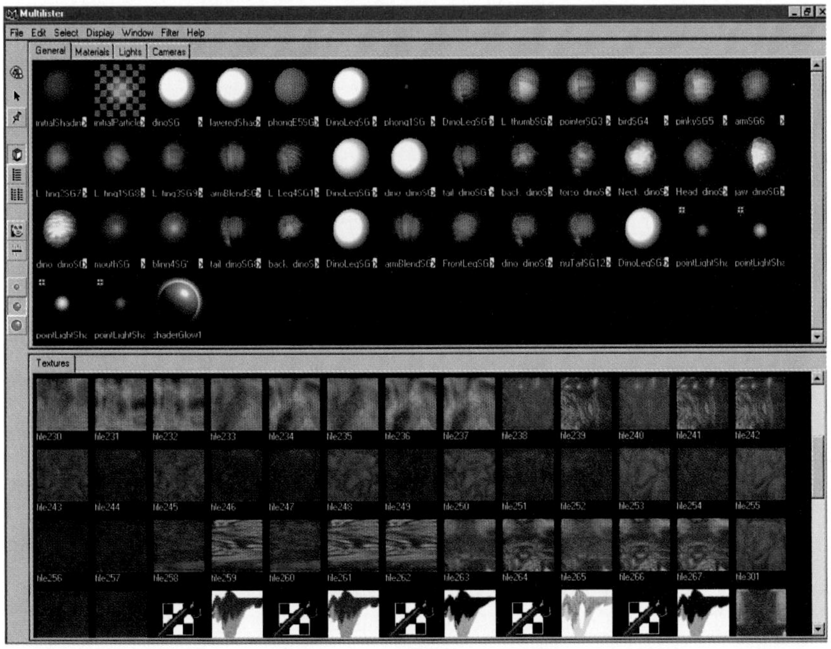

Figure 7.16
The multilister is seen displaying a scene with too many separate Shading Groups to be easily manageable.

Using a Switch node will enable you to use one shading group for all the pieces and then switch just the texture for each of the pieces. See "Additional Texture Tips" later in this chapter for information on how to integrate the switch node into your shading network.

7.4 Creating and Using a Texture Reference Object

If you intend to use a texture reference object, you must follow a certain workflow from of the moment you set up your character to obtain the intended results. Remember, this process will duplicate your character's geometry; thus, your file's size will increase accordingly. Read the following section fully to help you determine if your scene would benefit from the use of a texture reference object.

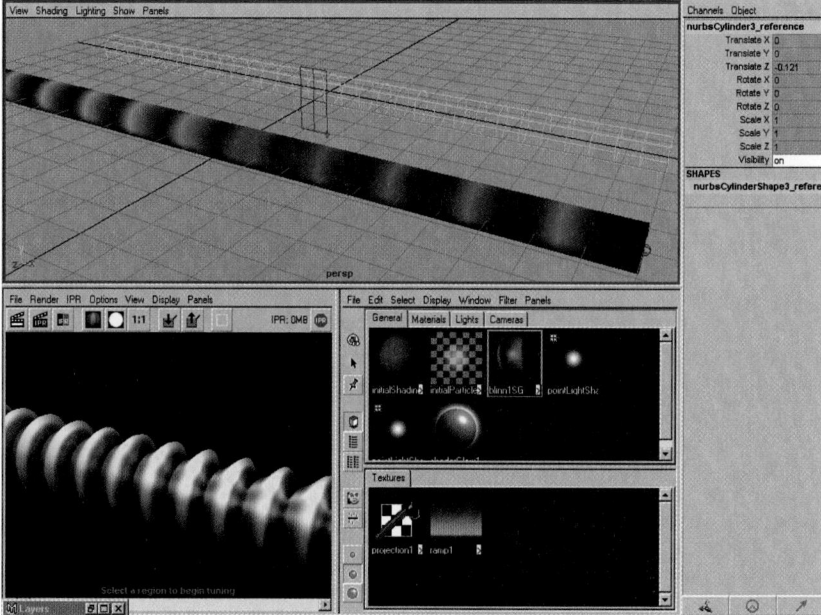

Figure 7.17
A model and its texture reference object.

Method I: If you are creating a texture reference object before you have applied any deformers on the skeletal system and before you have applied any textures, follow these steps:

1. Select your character geometry (not the group) and then go to Rendering, Shading, Create Texture Reference Object. You will see a template duplicate appear nearby. This is the texture reference object.

Note

When you create a texture reference, it has all its transform attributes automatically locked. Therefore, if you want to move it around easily, the Channel box must already be showing. If it isn't, you have to switch your attributes display to the channel box. To do so, hold down your space bar and, when the hotbox shows up, go to the far right and click and hold the mouse button. Then choose Attributes, Channel Box.

2. Click your master object and choose Rendering, Shading, Select Texture Reference Object. Next, go to the Channel box, highlight the grayed-out transformation attributes, hold down the RMB over them, and choose Unlock. Now you are free to move the texture reference object to the optimum location for your texture-mapping phase.

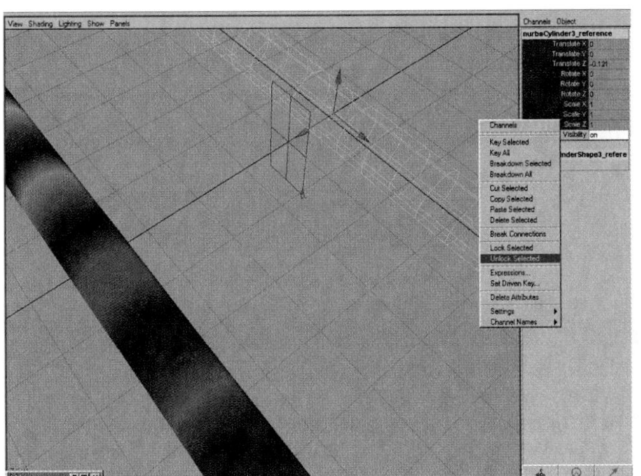

Figure 7.18
Unlocking the transforms of the new texture reference object.

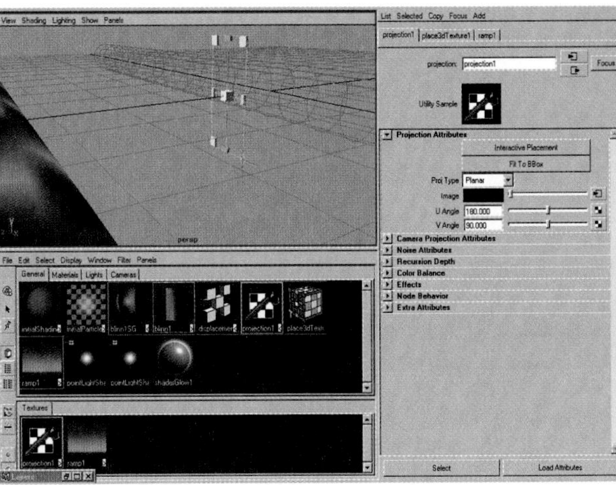

Figure 7.19
Placing your projection icons around your texture reference object.

3. At this stage, proceed to do all your texture mapping relative to the templated texture reference object. This means that you apply texture mapping parameters on the new referenced geometry. From this point on you can use all the texture mapping tutorials included, but keep these steps in mind when conversion to solids is discussed.

When the texture-mapping phase is completed to your satisfaction, you can go on to do a test bind of your geometry to verify that the texture reference object is working properly. Unlike with the Convert Solid Texture procedure, you can continue to tweak your texture mapping pretty much any way you need to later.

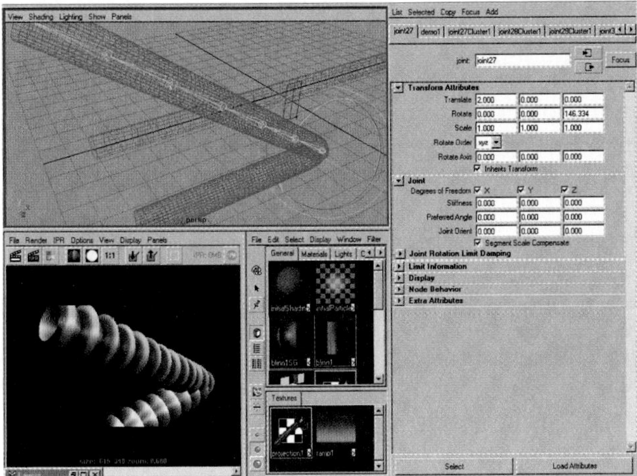

Figure 7.20
Test bind your character, and then perform some sort of deformation to observe your success.

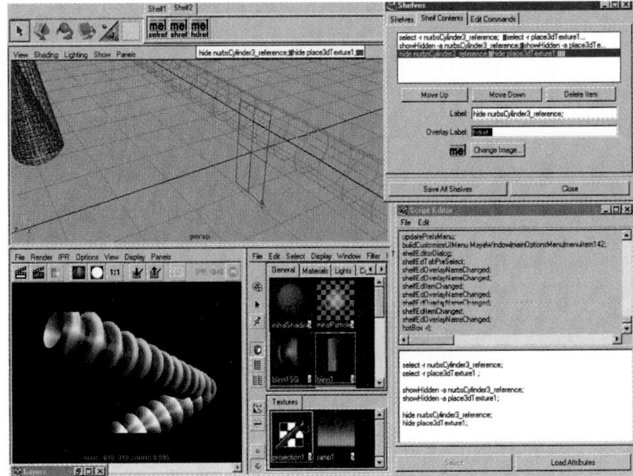

Figure 7.21
Use shelf MEL buttons to hide and show the texture reference object and its corresponding texture paraphernalia.

It is a good idea to hide the texture reference geometry and the projection nodes so that they don't distract you or get accidentally moved. Make a couple of shelf buttons by copying the actions of you hiding and displaying your selections and then selecting Script and dragging it to your shelf with the middle mouse button.

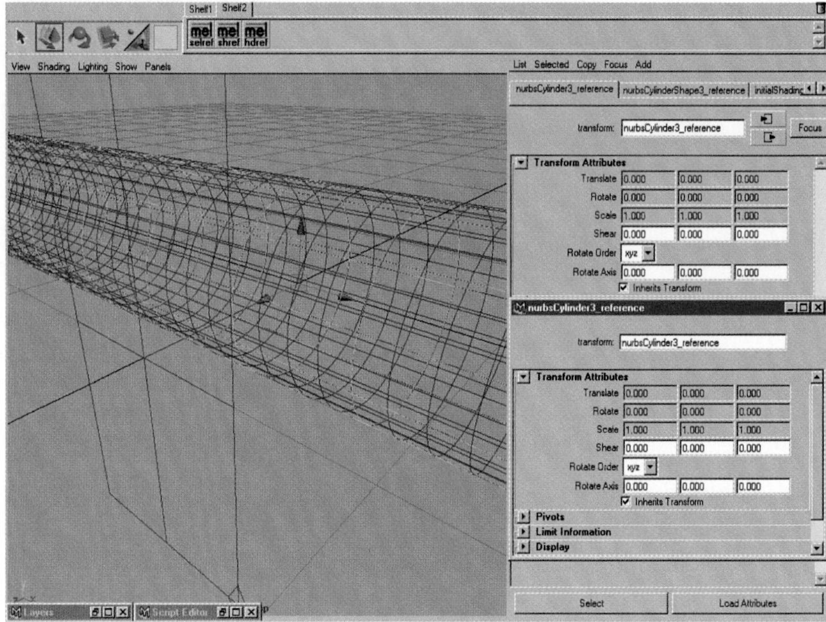

Figure 7.22
If you have already texture-mapped your object, you must synchronize the location of the reference object. Otherwise, all your texture maps will be oriented incorrectly. To do this, simply save a keyframe of all your animation control objects at frame -10 on the timeline to prepare your work for integration. Then you can save subsequent positions for your posed texture object.

Method II: If you are creating a texture reference object before you have applied the skeletal system or any other deformers, but after you have applied your textures, follow these steps:

1. Select your character geometry (not the group) and then go to Rendering, Shading, Create Texture Reference Object. You will see a template duplicate appear nearby. This is the texture reference object.
2. When the templated reference object appears, you will have to move it so that all the projections and solid textures match the original. Follow step 2 in Method I to unlock the texture reference object's transformation attributes.
3. To complete the process, follow the instructions outlined in step 3 in Method I.

Method III: If you are creating a texture reference object after you have applied deformers or a skeleton to your character, follow these steps:

1. Deform your character into the pose that is most optimized for projecting texture maps (see the section "Converting a Projection Using Convert Solid Texture," earlier in this chapter). Use your skeleton to make this pose as symmetrical and exposed as possible, optimizing the projectable area for maximum effect.
2. Select the character's geometry and choose Rendering, Shading, Create Texture Reference Object. Following the procedure outlined in step 2 in Method I, unlock the texture reference object and move it to a desirable location.
3. To complete the process, follow the instructions outlined in step 3 in Method I. Parent any Place3dTexture nodes used in the texturing to the transform node of the master object. Check out this network in the hypergraph by clicking the object and clicking the Up and Down stream connections buttons in the toolbar. Observe how everything is inter-connected, showing how the objects reference one another.

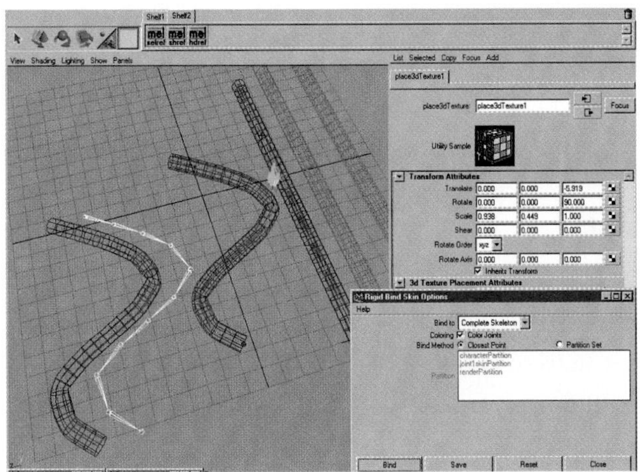

Figure 7.23
Here is shown the texture-mapping pose created to give easy access to tricky texture map areas such as the torso.

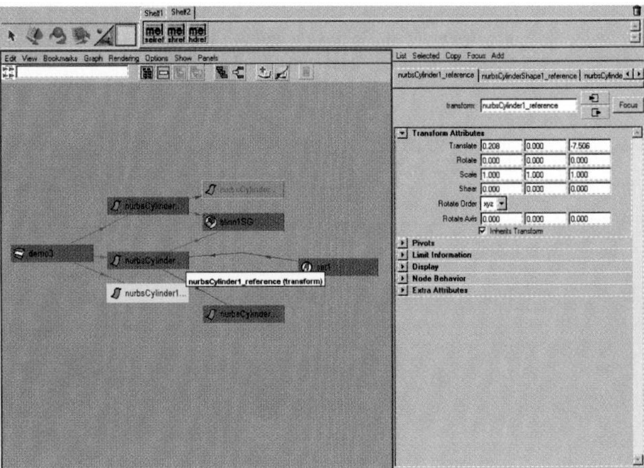

Figure 7.24
The Hypergraph view of a texture reference object.

If used properly, texture reference objects can solve many problems involving texture mapping using projections and solid shaders. You can animate any texture parameters you want, while at the same time save disk space and rendering RAM.

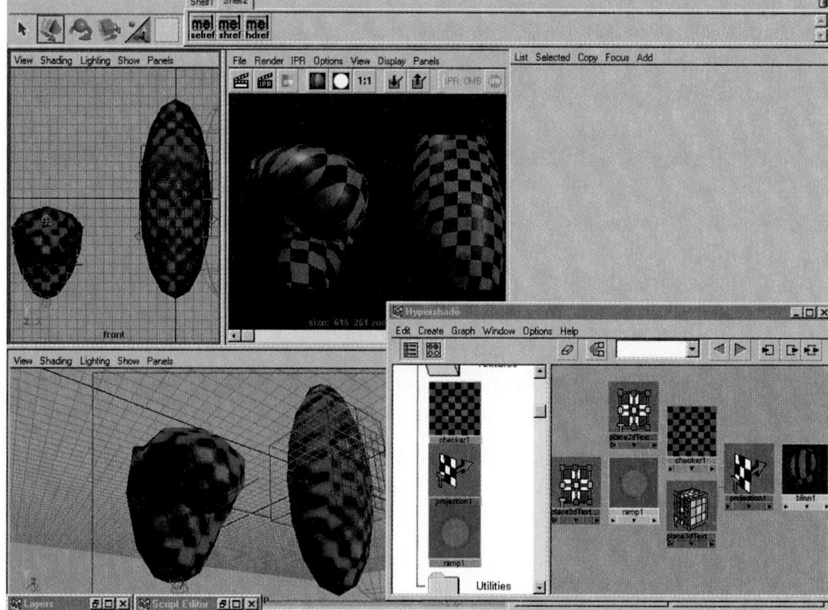

Figure 7.25
The sphere on the left is the original object, and the one on the right is the texture reference object. They should look identical, but the texture is skewed incorrectly on the sphere on the right.

7.5 Creating a Projection Texture in Photoshop

Here is a brief demonstration of how to use a texture reference object after binding a simple character to a skeleton. This subject is covered in depth later in this chapter, but there are some important steps to establish before you even get to Photoshop to create the texture map.

7.5.1 Screen Resolution Versus Texture Resolution

The size of your texture map is directly related but not equal to both the largest size that you will see any specific part of your character and the final output medium; otherwise, you will have wasted assets such as rendering time and disk space.

Following is an exercise that might help you come up with your own custom texture mapping solution for your particular needs.

A classic analogy is the case of the "zoom up on your character's eye from a distance" shot.

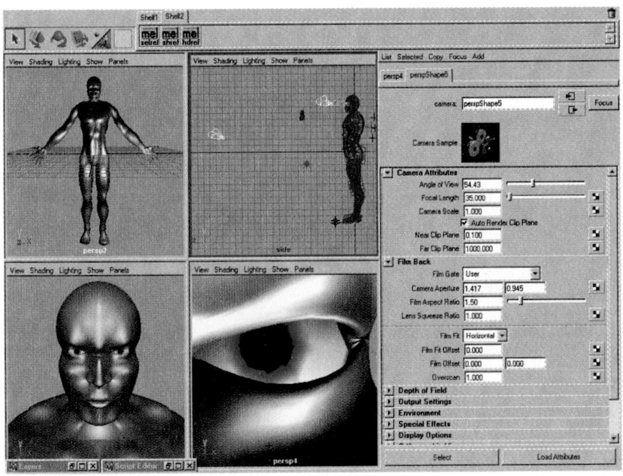

Figure 7.26
Small thumbnails of an iris zoom shot.

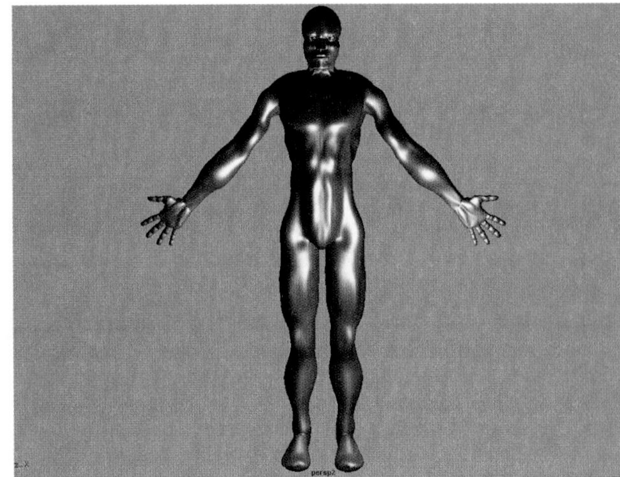

Figure 7.27
This image shows the subject from a long distance.

From a distance you will see the whole character. Thus, the whole front side must be textured.

As the zoom nears the head, the facial textures will be seen in full screen.

As the zoom centers on the character's eye, the retina and iris will be the final subject as the pupil engulfs the camera.

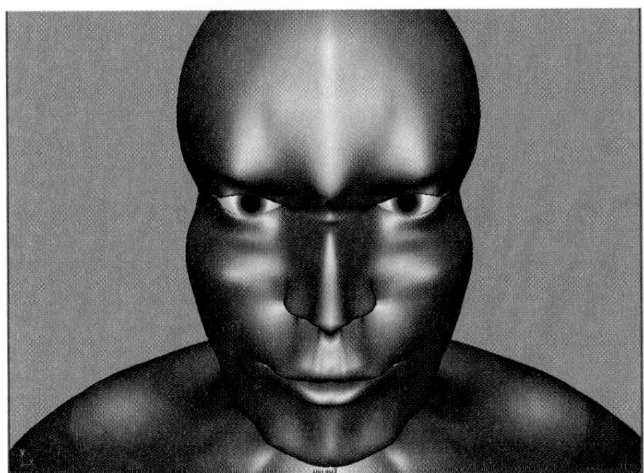

Figure 7.28
The face close up.

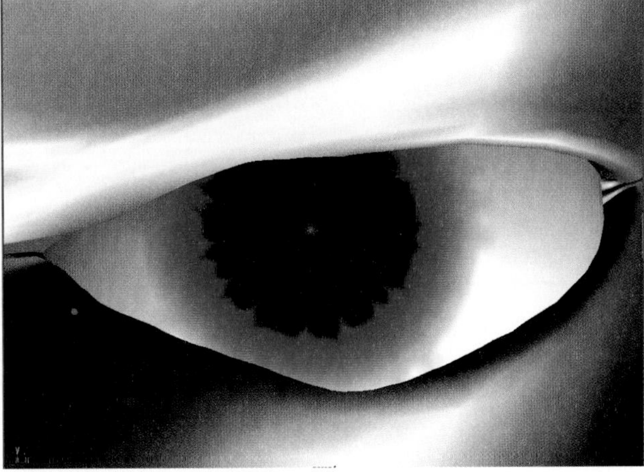

Figure 7.29
An extreme close-up of the pupil.

You must complete at least three high-resolution texturing steps for the image to look right. First, you must plan the process by taking some common-sense measurements of your resolution requirements. Let's assume that our output will be 720×486 pixel NTSC video. Let's also assume for quality reasons that we need for our texture at least 1.5 times the screen resolution in pixels so that the renderer has a little something extra to work with.

The full body will need a high-resolution texture simply because it has to cover so much area. At its closest point, the upper torso is seen with the shoulders cropped out. That means that if measured in screen pixels, there are at least 720 pixels across being viewed. Add in at least half that much (360 pixels) on either side for the shoulders, doubling the width to 1440 pixels. Factor in the 1.5 multiplier and you get 2160 pixels. Add another 20 pixels on either side for bleed over and you are at an even 2200 pixels in width.

An average humanoid is at least three times its width in height, giving us about 6600 pixels tall. This could be brought down to 2200 if you choose to just texture the torso separately. Because the zoom to the face bypasses much of the body detail, the lower half of the body could be a separate texture—as low as half the resolution of the upper torso, or 1100 pixels wide by 2200 pixels tall.

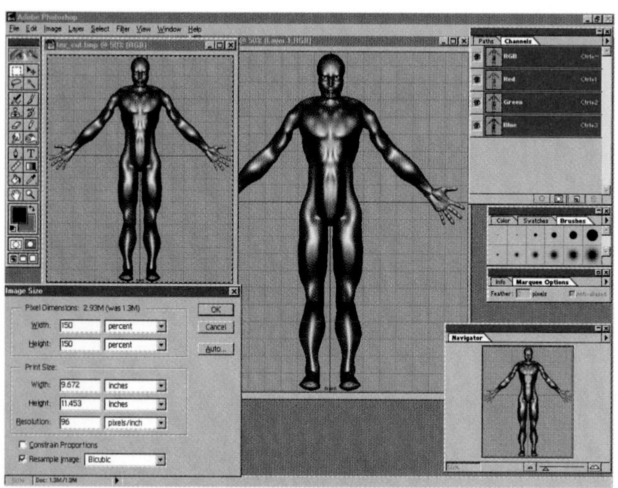

Figure 7.30
Texture size vs. screen resolution size.

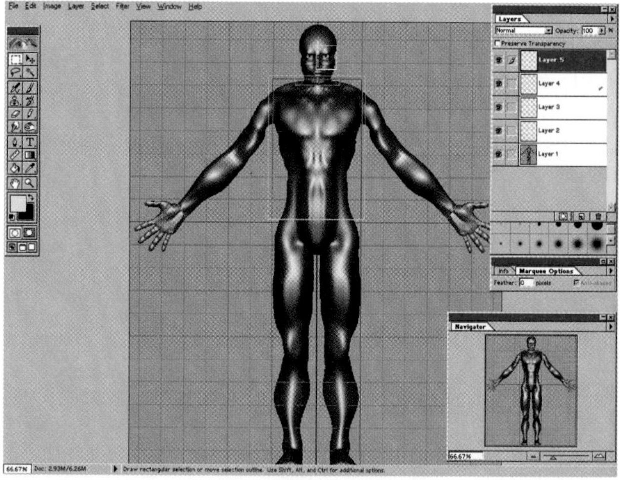

Figure 7.31
Outlining your texture regions.

As the face becomes full screen we focus on the area near the eye. This covers an area vertically from the eyebrow ridge to the base of the nose and from the ear to the nasal crest horizontally. That would be double 720 times 1.5 to be 2160 pixels (round up to 2200 pixels) by approximately 3300 pixels vertically to include the mouth and chin.

The eye could probably get away with a texture as small as 1k square, or 1024×1024 pixels. The eye may also need a fairly high-resolution reflection map as well if you want a detailed environment to be seen in the wet, reflective surface of the eye.

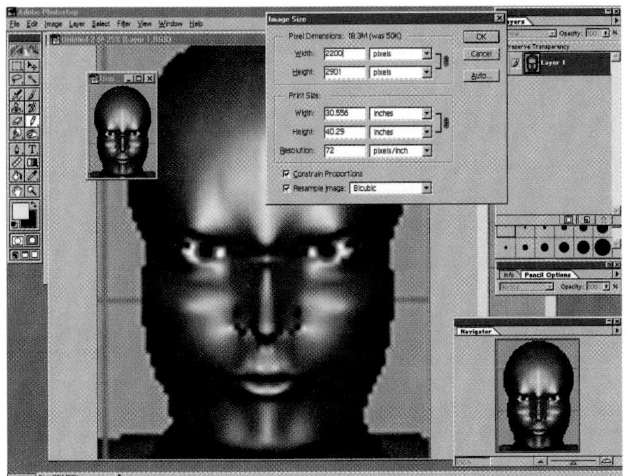

Figure 7.32
Detail requirements for the face.

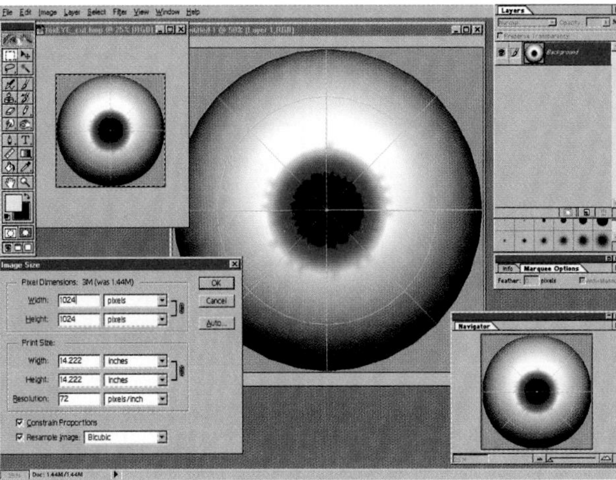

Figure 7.33
Eyeball detail requirements.

Remember, all these measurements are for the front side of the face. You must still be concerned with the back side and any other fixes you might have to do on the side regions to cover up stretching textures. Also, remember that keeping your image resolution in the powers of two help the rendering software to create quality imagery by simplifying its job of averaging pixels.

7.5.2 Mapping Scenarios and Texture Regions

Now that your resolution concerns have been addressed, it is time to plan the specific methodology needed for your situation. Each part of your character must be broken down into a clear set of needs with tasks assigned to solve them. Consider the head, for example. The later section "Texture-Mapping Tutorials" offers one solution for mapping some of the facial details. Other issues, such as the eyes, mouth parts, and hair, all need separate attention as well.

You may be able to use a frontal planar projection for doing the face rather than the 3D head texture scanner technique by using the projection mapping exercises included in the "Texture-Mapping Tutorials" section. Then the issue would be making proper transparency maps to cut off any stretching textures. For complex textures requiring many layers of various projections, each with a transparency, you will need to consider using the layered texture with a Stencil node. For the most complete results with the most versatile of applications, try to think cubically when considering if you have covered all the possibilities. If you form a cluster of projection nodes around a subject, you will be more likely to cover all the sides of your character. The trick is making the proper masks for each layer to blend together just right. Also think about a base layer that is the foundation for further texture mapping. The tutorials later in this chapter focus on creating solid base layers as foundations for further detailing.

7.5.3 Identifying the Character's Sweet Spot

When framing and posing your character's best possible viewing angle ("sweet spot"), keep in mind how the movie masters made costume and make-up designs come to life on the silver screen.

Consider Darth Vader's costume in the movie *Star Wars*, for example. In the movie we see a big, dark, menacing cyborg warrior who crushes people with his thoughts. But when you look at the character up close for more than a few moments, you can instantly see much of the suit that either was not shown at all or was briefed upon, leaving your imagination to fill in the details.

The point is that your first impression of a character, along with subsequent proper framing and presentation, can go a long way toward bringing a character to life. You don't always have to make every bit of the character's texture map absolutely perfect for every possible shot. You should, however, do your best at dressing up the character for the "sweet spot" shots that show off your care and work that you have done for the best possible wow factor.

7.5.4 Displacement Maps Versus Bump Maps

You should determine whether you will use displacement maps or bump maps early in your planning stage. If you will be going close to bump-mapped detail and lingering with the camera, the fake contour afforded by the bump map will be obvious and look like the work of a novice. If the extra tuning and rendering time required by adding a displacement map can be allocated, by all means go for it.

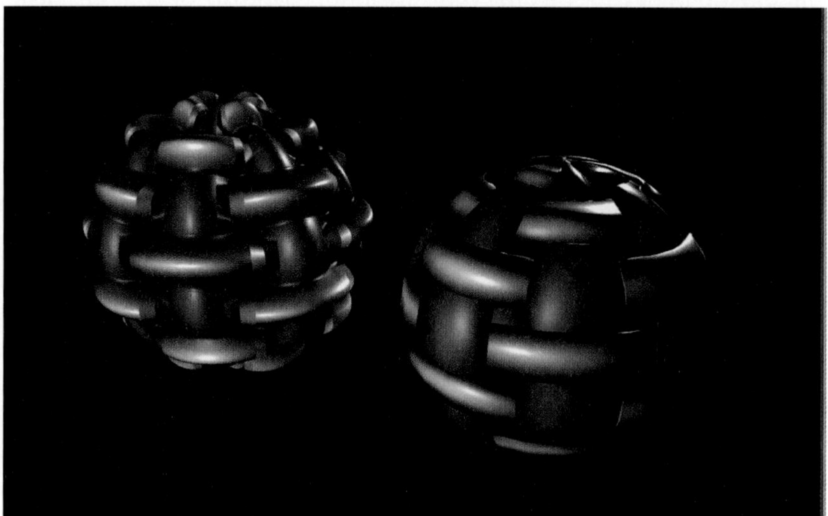

Figure 7.34
Bump-mapped geometry vs. displacement-mapped geometry.

Some people have been known to use displacement mapping as a way to model much of their set using pictures and low-resolution NURBS planes. They create detailed set interiors with some well-planned displacement maps that were fully thought out and integrated to save time and get extra quality from having details that would have taken far too long to model.

In summary, it is important to follow a plan of action to help answer all the questions that you need to determine early on. This will help you to organize and simplify the sometimes tedious job of using Photoshop to create detailed, easy-to-place texture maps that are reliable and consistent.

7.5.5 Layered Shaders Versus Layered Textures

The Layered Shader and the Layered Texture are very similar shading groups; each has its time and place. Both begin with a Layered Shader shading group. The Layered Shader mixes several Material nodes together, blending the various components to create the result.

The Layered Texture uses several texture nodes together as the source for a single Material node. That way, it doesn't mix several Material types, just the texture information. Just like the Layered Shader, the Layered Texture uses a base texture and then builds each level up as another layer on top.

Note

Whenever you use layers of shaders or textures, you must be sure that the default color in the Color Balance section of both the File node and the Projection node is set to solid black. Otherwise, the image will add a gray film over any layer beneath it.

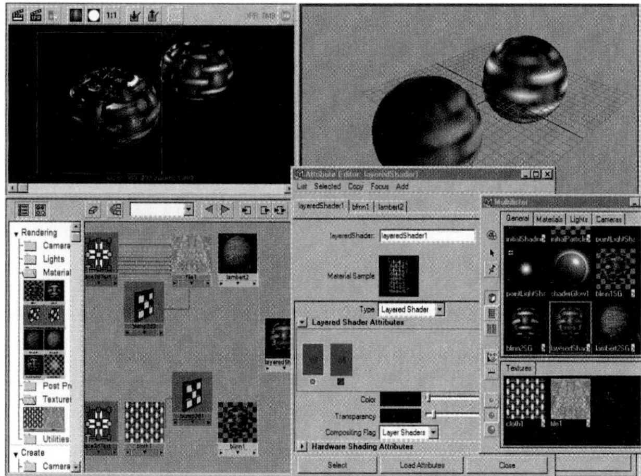

Figure 7.35
The Layered Shader method.

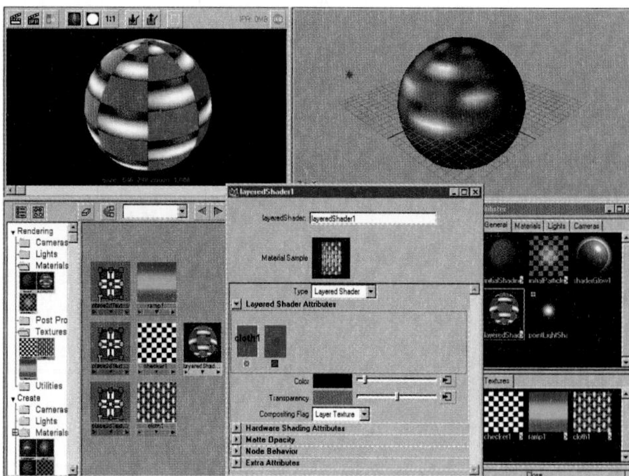

Figure 7.36
The Layered Texture method.

Both layered shading types use transparency as a way to transmit color information to the layered shading Network. Maya provides various options for compositing textures on top of one another. The two most commonly used ways are mapping the transparency attribute and creating and using a Stencil texture node.

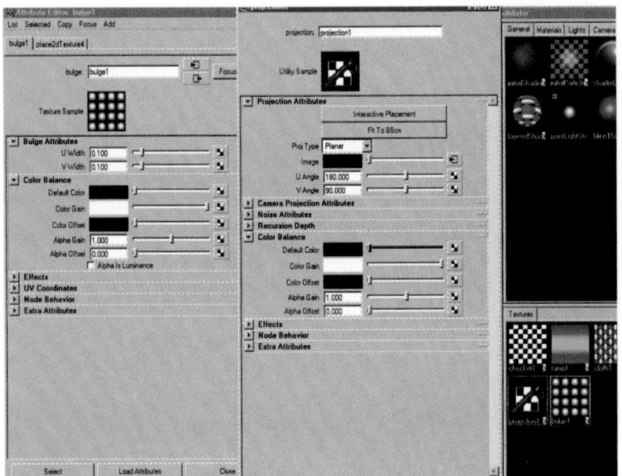

Figure 7.37
It is essential to change the default color to black in both the File node and the Projection node.

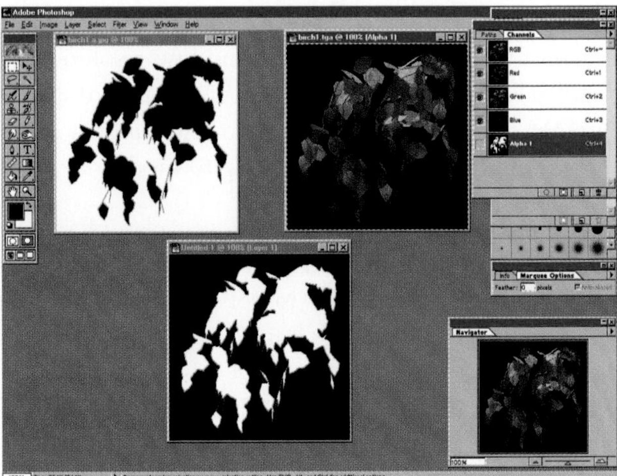

Figure 7.38
The reverse mask is for the transparency channel, in which the usual industry accepted black transparent style mask is used for the stencil.

Transparency is easiest because it is an attribute that is built into all the standard shading groups. Stencils, on the other hand, have more options and thus can be more useful when you integrate them into your production pipeline. Stencils end up being a very important part of the process for integrated layered textures. More detailed information regarding some of the nuances in controlling stencils is provided in the section "Using Stencil Nodes with Projected Texture Maps as Components of Layered Shaders," later in this chapter.

7.6 Advanced Texturing Methodologies

The following tips are some insights that introduce concepts for further experimentation. Please feel free to mix and match the ideas presented to help you design your own custom solution to your own texture mapping projects.

7.6.1 Mapping the Various Texture Channels for Realism

You can assign a texture map to each of the attributes in a surface material. This concept was briefly addressed during the descriptions of the surface attributes specifications at the beginning of the chapter. You can map each of the attributes so that you get user-defined control of every pixel of information rendered to the screen. With a grayscale or color image mapped into any specific attribute, the slider bar becomes a grid of pixels controlling the values on a pixel-by-pixel level.

7.6.2 Applying Realistic Details for Natural Effect

To simulate reality, adding dirt and grime is an essential ingredient during the texture-mapping phase. Many techniques are available, but some are more accessible than others.

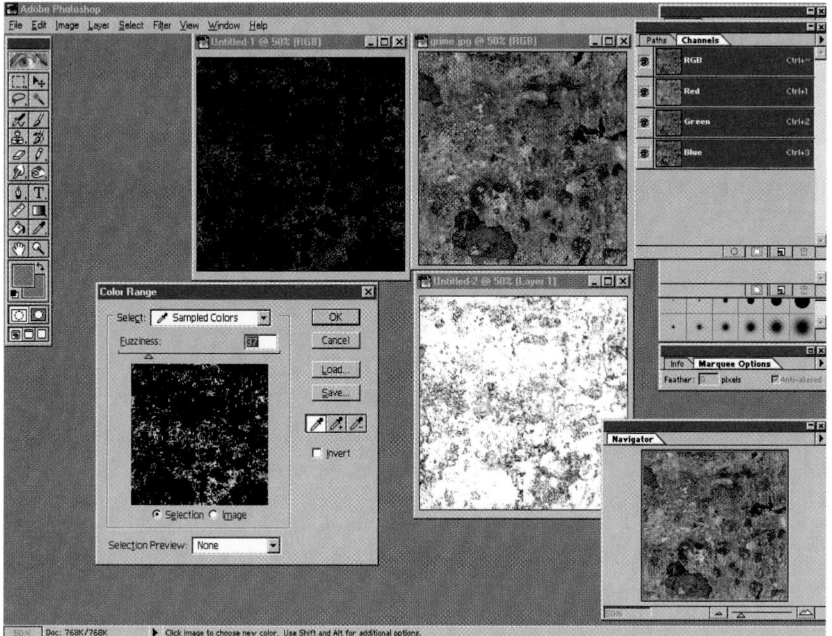

Figure 7.39
A light dash of grime can be used as a diffuse map, a color map, a specular map, a bump map, and a separate transparency map to make it all blend together.

Use a layer of very transparent filth imagery as part of the top of your layered shader networks. Use these filth maps as part of a subtle specular, diffuse, and bump map that subtly alters the clean look of CG to a more pleasing, artistic, used look.

7.6.3 Alex Lindsay's Texture Tips on Aging an Object

Alex Lindsay, a highly regarded expert in the field of realistic texture mapping, has come up with a well-thought-out methodology for determining the specific details that you might find on any 3D model of a character. He asks you to simply make up a rough history of an object to help tell a story, with a texture map giving insight into the object's past.

Ask yourself a few questions about the surface that you are trying to simulate, and use your answers as the basis of a process for aging the subject to appear as real as possible.

1. How old is this object?
2. Where did this object come from?
3. How did this object get here?

4. What kinds of weather has the object experienced?
5. What parts of the object receive wear and tear?
6. Has the object ever been fixed or upgraded?
7. Why was the object created?

Use these texture questions as an example interview of an object. Let's choose a belt around a character's waist.

1. The belt is 150 years old.
2. A leather worker created the belt.
3. The belt was owned by a pirate but was won in a card game by a soldier who later died wearing it in a battle. A child found it and then used it as an adult (your character).
4. The belt saw a lot of ocean air for many years and was occasionally soaked in seawater every time the pirate got wet. Then the soldier had to march in the hot sun, which dried it up and left cracked edges. The child adorned it with additional markings and a few pieces of colored string.
5. The buckle holes get particular abuse. The buckle itself is tarnished where hands don't touch it but otherwise has shiny regions polished by use.
6. The belt almost tore once but was stitched together with thick thread in an area.
7. The object was originally custom designed for the pirate.

Now that you know the history of the otherwise generic "belt," you can make a list of details based on each answer, giving you a wealth of possible things to add to your object to enhance its visual appeal.

1. Old belt, well-worn, cracked, peeling, with a tarnished silver skull-and-crossbones buckle.
2. Non–machined, slightly uneven placement of buckle holes. Bump-mapped leather press patterns embossed in the surface but with soft, rounded edges from use.
3. Pirate markings, such as plundered ships being counted and skull-and-crossbones pendant tied to the leather part. The soldier didn't add any extra details, but he did bleed all over it, leaving a slight reddish blotchy stain over some of the leather. Some blood also coagulated and dried in the crevices of the buckle. The child added personal effects and childish writing.
4. The weather question suggests ways of adding surface details that will enhance the realism.

5. Adding areas that are used usually makes that area stand out somehow from the rest of the surrounding details. A diffuse map could lighten the used area, while also giving the less-used surfaces a slightly darker look. Also, the used area might have a smoother bump map and a corresponding higher specular value in that area.
6. Repairs will give part of the subject a new quality, making that area really stand out but also reminding the viewer how aged the other areas are.
7. Determining the original intention for the object reinforces the design aesthetic. The pirate wanted the belt a certain way, and he paid extra money to be sure it was unique and thus a higher status symbol, and so on.

Now you have a sense of history that far outweighs the original call for a "belt." Now just thinking of the belt evokes a sense of history and character. Extrapolate this technique if you have time to give everything you make a greater sense of fitting into the "big picture."

7.7 Specialty Texture-Mapping Products

The following are some complimentry products that can aid in the creation and editing of texture maps.

7.7.1 Studio Paint

Studio Paint is Alias | Wavefront's own 3D paint package. To date, it offers the best support for Maya when compared to existing 3D paint packages. Studio Paint is easy to use. Using an IRIX SGI computer, export your model as an Alias Wire file and import it into Studio Paint. You may need go to Windows, General Editors, Plug-In Manager and activate the Alias Export option. While you're there, turn on the Obj and DXF export options: You will need them to prepare geometry for other 3D texturing programs that you may prefer to use that are not mentioned.

When in Studio Paint, you can use all the powerful tool sets to create your texture. Then, when you are ready to bring it all back to Maya, you re-export an Alias Wire with textures and Maya reads the file. Studio Paint actually turns the NURBS model into a UV texture-verticed duplicate polygon model that matches your NURBS character perfectly for texturing purposes. Maya reads these UV textures and is ready for you to reassign them to your original NURBS character. In the back of the *Learning Maya* manual from Alias | Wavefront you will find a brief demo for working with Studio Paint and Maya. In addition, a MEL script is demonstrated that refreshes the texture from your hard drive when activated. This MEL script could just as easily be used to refresh changes to your texture map from other complimentary paint and 3D paint packages.

Studio Paint does not provide the down-home familiarity that Adobe Photoshop offers most digital artists, but it does work and it has a proven track record. If Studio Paint were to use more of Maya's shortcut and hotkey conventions, this would definitely increase its appeal as an integrated part of the texture-mapping pipeline. For now, Studio Paint remains a tool abandoned to the dwindling SGI platform. Because of its obscure nature, it remains practically ignored and forgotten by the plug-in developers who can't justify supporting a tool that does not have Photoshop plug-in compatibility, or connect with the Photoshop 3.0 layers compatibility standardization adopted by much of the paint package industry.

7.7.2 Amazon Paint

Amazon Paint, from Interactive Effects, is one of the original 3D paint forefathers, and thus it still has much of the antiquated interface and general functionality that was common to early 3D era high-end software. It works, but it isn't pretty. Amazon supports Alias|Wavefront's texture-mapped Obj file format for connection to Maya. It is not quite as easy to integrate with Maya as Studio Paint. It does offer some unique tools, however, the best of which is the brush feature that allows one to paint on the normals of the object.

The brush feature is similar to the Maya Artisan brush interface. The Paint on Normals option is surprisingly lacking from the Alias|Wavefront product Studio Paint, in which the brush simply projects from the screen angle, forcing the artist to constantly turn the object a little at a time when working.

Amazon has a solid reputation, but it is also known to be frustrating as a general-purpose paint package. Nonetheless, if you have access to it, learning it could benefit your work.

7.7.3 Mesh Paint

Developed by Positron Publishing, Mesh Paint has the integration factor with other industry tools. It is a plug-in component for Avid SoftImage as well as for Discreet's 3D Studio Max. Positron has a stand-alone version for both the Macintosh and the PC, and it has a range of supported industry formats.

Sadly, Mesh Paint is no Photoshop, but for what it is, it works pretty well. Strong cross-platform integration and the internal nature of the plug-in versions make Mesh Paint a viable contender.

7.7.4 Detailer

MetaCreations' Detailer is cross-platform to the Mac and Windows environments. It does not have any fancy connections to the Maya software package, but it does have a MetaTools-style painterly interface, complete with many essential tools directly borrowed from its older brother, Painter.

Detailer even supports the Photoshop 3.0 layers specification for easy connection to your favorite tools.

7.7.5 SurfaceSuite Pro

This package is not a 3D paint package, so in a way it is the odd one in this collection of 3D texturing products. It does, however, offer a special and unique perspective to the 3D texture-mapping methodology. Sven Technologies' SurfaceSuite essentially creates texture-warping coordinates that you place on a picture you have imported. Then you match points on the model to a corresponding location on the image—for instance, the eyes to the eyes. The texture map warps to fit between your established anchor points. After layering a composition of imagery, you export a final pre-distorted version of the texture composition to a single flattened and warped image. When this image is applied to the NURBS model from which you based the texture, it matches the original UV texture vertices.

Sven has announced a special Maya plug-in called LiveLink that enables concurrent work to be accomplished in Maya, SurfaceSuite, and Photoshop. As you finish each stage of work, you press a common update key that will force all the images to be synchronize, in each of the corresponding applications.

In addition, versions of the LiveLink plug-in are (will be) available for SoftImage and MAX to keep the SurfaceSuite texture database's maximum compatibility with Maya between multiple platforms.

If this tool sounds useful to you it could theoretically be used to speed up your workflow considerably. It could conceivably be an integrated part of your texture-mapping pipeline.

7.8 Texture-Mapping Tutorials

The exercises presented in this section present various types of organic character-mapping techniques. Each one provides various solutions and workarounds to common texture-mapping scenarios. All the digital assets needed for these tutorials are provided on the CD-ROM in each stage of completion from beginning to end. You can either use these documents or build your own from scratch.

The following is a summary detailing some highlights about the various characters used in the texture map tutorials.

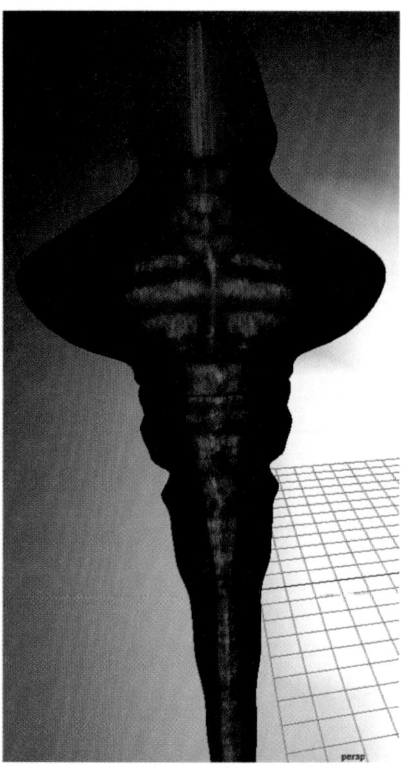

Figure 7.40
Snake Monster detail.

- **Snake Monster.** This character is a single mesh made from a NURBS cylinder primitive. The original geometry must be prepared. Its head needs to be deformed to point upward, thus preparing its geometry for projection mapping by removing any areas that

Figure 7.41
Dragonfly detail.

Figure 7.42
Dinosaur detail.

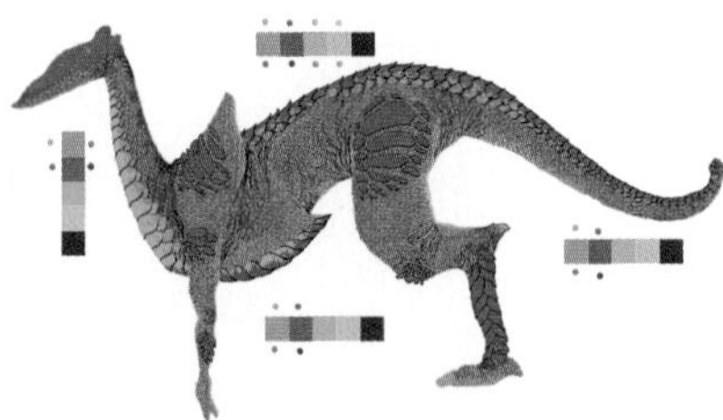

Figure 7.43
Dragon detail.

Figure 7.44
Organic head detail.

would cause projection overlap. *Overlap* occurs when the projection node is covering an area of geometry that bends over itself, making an area receive the projected texture pixels more than once. To avoid overlapping pixels, the character was pre-posed into place. Note: The final model does not need to be distorted this way; the final solid UV coordinates that are generated from this procedure will map correctly, no matter what pose the character is in.

- **Dragonfly.** This character is built from several NURBS objects because insects are segmented creatures. The legs and antenna will not be the focus of the tutorial. The dragonfly's main body and wings will most benefit from the planar projection mapping technique, so these body parts will be the focus of the exercise. The character was built in such a manner that it did not need any special tweaking to be texture-mapped.
- **Dinosaur.** This complex NURBS model was purchased from Viewpoint Digital as is. In some areas the model uses trim curves and fillet blends to attach the arms. The legs, however, appear to have been built by matching the isoparms of several stitched objects to create nearly seamless hind quarters. This model will require various extra steps to achieve a final useable result. Many of the steps needed should be familiar by now, with the added complexity of having many small objects that are combined, thus requiring more time and attention to detail.
- **Dragon.** The Dragon from Chapter 6, "Modeling an Organic Dragon with NURBS," will be the subject of this self-paced exercise. This one is for you to adapt the lessons to which you have been introduced so that you can go through the process yourself and apply the base foundation layer as an exercise. This character requires special preparation for the wings, but the body is very much akin to the dinosaur because they both have trimmed and blended body parts attached.
- **Organic Head.** This head texture-mapping demo has a unique workflow and offers a complete and self-contained step-by-step process. It is presented in the original form the guest author Tadao M. intended it. Tadao modeled a head for the character himself. Any roughly cylindrical head model will work for this demo. Any hair or extra details, such as long curved horns or antenna, should probably be textured separately for optimum effect.

The first four characters (the snake monster, dragonfly, dinosaur, and dragon) share similar physical attributes and, thus, utilize the same initial steps to create a texture map. Here is a rough summary of these essential initial steps to begin creating texture maps using the projection method.

This workflow is intended to enhance Maya's built-in texture-mapping tools by using precise orthographic projections as templates for texture generation.

After creating a high-resolution texture in your favorite paint program (in my case, I am using Photoshop), bring back the image into Maya and apply it as a projection texture. Convert that texture map into the more flexible and useful UV map, which can be used in a wider variety of final render situations than the projection method.

After creating this pre-distorted UV map, give the image a companion mask file, or transparency map, in the paint program. Each separate side of the character to which you want to apply a unique texture should be put through the same steps. You can use the prepared UV-mapped images together with a layered shader, creating a seamless combination of the source imagery.

After concluding your parameter editing, convert the layered shader to a single UV texture map to further optimize the final rendering.

Upon completing these steps, you will have a texture-mapped organic model that is ready for character animation. You will also be familiar with the workflow used to create the texture. Although these techniques were demonstrated using an organic NURBS model, they can be adapted to many other texture-mapping situations. Keep an open mind and be aware of any steps that might be altered to work better in your particular texture-mapping situation.

7.8.1 Snake Monster

Exercise 7.1: Snake Monster

1. The first step is to prepare exact orthographic views of the snake monster model. I like to set up my views in the panels as two side by side, as this makes it easier to align the views for longer models such as this snake monster, "The Virus Lord." This is an important time to decide how much detail your model needs. You determine this by evaluating several criteria:

   Stage 1.1: Will your model be seen from all sides? This determines how many orthographic views you will need—one for each side to be viewed (the side views might be interchangeable depending on what the shape of your model is and whether it has symmetrical sides).

   Stage 1.2: Will your model be viewed in close-up detail as well as in long shots?

   If only one of those is the case, determine what resolution you will need and where you need it—for example, is the close-up view of the creature's face or of the tip of its tail? This could be a good reason to make two textures, one high resolution for close-ups and the other low resolution for distant shots.

Exercise 7.1: continued

Stage 1.3: Will your model need to be influenced by a deformer and/or bones?

This tutorial assumes this last question to be true; otherwise, the section dealing with converting to UV maps is irrelevant. The main benefit of a UV map over a projection map is that the UV can be deformed and the projection can't. Think of projection maps as slide projectors. Imagine this slide projector pointing at you. If you move around, the slide's image will remain stationary and different parts of the imagery would cover you depending on where you moved. Now imagine the slide's image being painted on your body like a tattoo. When you bend your arm, the tattoo bends with your skin. This is what happens when the projection is turned into a UV-mapped image. Thus, if your model is not going to be deformed, you can simply parent the projection map Texture3dPlacement node to the model itself, saving the lengthy process of converting to UV.

Note

When preparing orthographic views of your model, try to line up the grid patterns before outputting the views. This will help you to precisely paint your texture by giving you easy-to-reference placement markers.

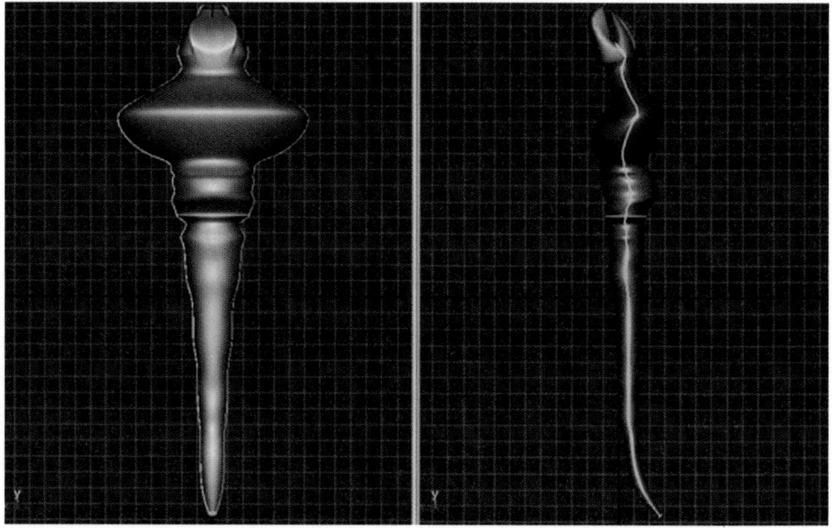

Figure 7.45
The first step in the creation of a texture is the orthographic reference imagery captured from a screen grab of the model.

2. Use Playblast to export frames to Photoshop. First, open your render globals, set the range to begin and end on 1, and set the output image format type to Targa (my favorite because of its high compatibility and solid cross-platform recognition). Next, open your Playblast settings dialog box.

Stage 2.1: Set output format to Fcheck. This option outputs to single frames rather than to AVI or SGI movie-type formats. This is needed so that you can bring the image into your paint program as a template.

Stage 2.2: Set the display size pop-up menu to From Window. This option specifies that the custom views that you have set up with the grid backgrounds match each other. The other settings would work, but they will not guarantee that the grids match.

Stage 2.3: Set the scale to 1.00. This simply sets the output to a 1:1 ratio. What you see on the screen is what will be output. Optionally you can get a specific screen size by either using the resolution settings in the render globals or the input values. These do, however, make matching the various views slightly more difficult.

Having set the parameters correctly, Playblast once from each viewport, leaving the screen in two views, side by side.

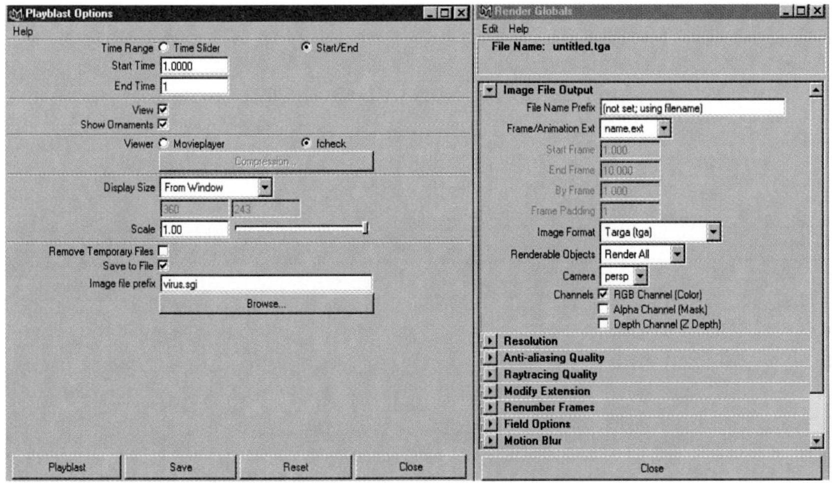

Figure 7.46
Generating the screen grabs to use as templates for Photoshop.

3. In Photoshop, import the images for preparation. Focus first on the front view. This is an important time to make the final decision as to the resolution in which you are going to paint your texture. This resolution setting will affect your final resolution constraints when rendering. Usually, I prefer that an image be too large rather than too small. Remember, you can always lower the resolution later, but after you have painted a texture it is not a good idea to scale it larger because this would further blur the image and thus be actually lowering the resolution. This resolution also helps determine what size custom brushes you will need to create for highly detailed texture work.

   With your new custom brush library that you make at the proper resolution for this image size, proceed to paint the texture map:

Exercise 7.1: continued

Stage 3.1: Layer-mask half of the image if the model is symmetrical. That way, later you can flip the two halves together, saving you time on the initial stage. Don't worry: After mirroring the images, you can still go back and add unique detail to each half because everyone knows that in most cases, two halves are not 100% alike. Also, be aware of the bounding box around the edges of your model. A bounding box is the smallest possible rectangle that your organic model can fit into. This rectangle is what Maya will use when receiving the texture map that you will soon be projecting.

Note

Crop your texture to the model's bounding box to help facilitate precise projection.

Stage 3.2: Now paint in layers the different aspects of your model's texture using the image of the model as a template guideline to work from. It is a good idea to allow a fair amount of bleed color and texture to spill over the sides. This will help pad the convergence of the other side view's texture maps when we merge them together.

Stage 3.3: When the images have been completed, merge the layers and output them to SGI or Targa format (depending on the platform you most commonly use).

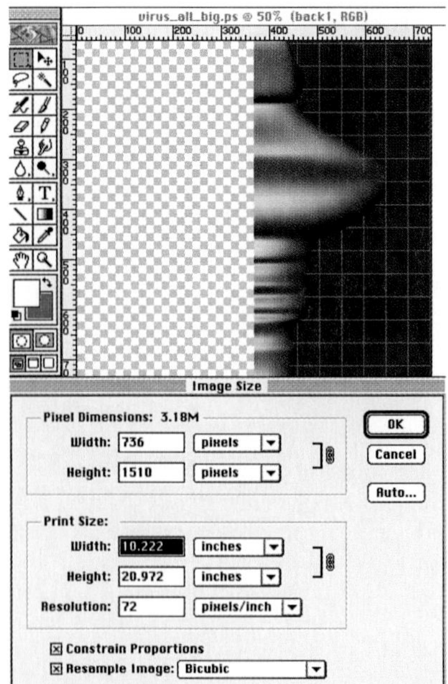

Figure 7.47
Working in Photoshop with the orthographic templates.

4. When the painting stage is complete and the texture map has been saved into the sourceimages directory, you are ready to create a new shading group and map an image file node to the color parameter.

   Stage 4.1: Open the Multilister by going to Windows, Rendering Editors, Multilister. In the Multilister Edit menu, choose Create. This opens the Create Render Node, a palette full of options to create various combinations of new rendering utilities and shaders. Click the Materials tab and then click the Blinn and Lambert buttons (you will need both later).

   Stage 4.2: Click the Textures tab. Choose the little round button at the top of the window marked As Projection, and then click the File button. By choosing As Projection first, the new file node automatically is attached to a node called projection1 that deals with projection style texture placement. To make the projection node connect to the color node of the Blinn shader, you must middle-mouse-drag the projection node from the Multilister to the color box in the Open Attribute Editor window displaying the Blinn shader and its various mapping channels.

5. To apply the newly mapped shader to the geometry, select the model in the viewport. Then, in the Multilister, select the shading group and, from the edit menu, choose Apply. Make sure that a viewport is active by pressing the Ctrl key while the mouse is over

the viewport of your choice. Then press the 5 and 6 keys in succession. The 5 key activates shaded mode and the 6 key activates textured view.

6. Test render the character or finished body part each time you get to a stage that you feel confident about. A final render pass at the final resolution at which you will be outputting will help you determine early on the quality of the texture resolution.

7.8.2 Dragonfly

Exercise 7.2: Dragonfly

1. The Dragonfly uses an orthographic side view as the basis of the body texture. The legs and antenna have a simple solid shiny black shader—either a Blinn or Phong E, with no special mapping needed, except maybe a reflection color map.

   Stage 1.1: First, artist Donnie Bruce hid the eyes, wings, legs, and antenna in Maya.

   After the side and top view were exported to Photoshop, Bruce painted over the model's image, creating the basis for the texture. He saved a copy in Photoshop.

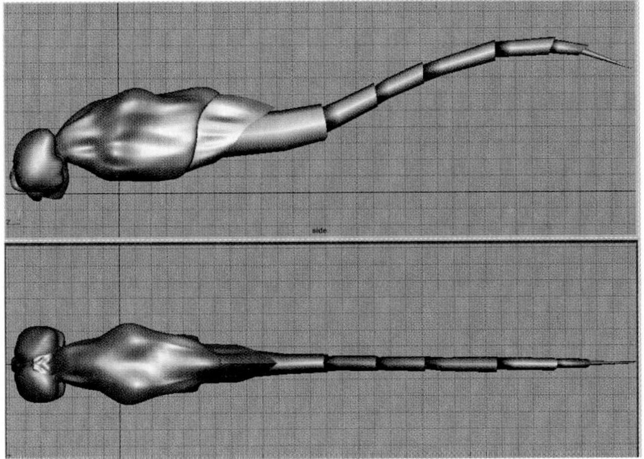

Figure 7.48
Orthographic views of prepared Dragonfly character.

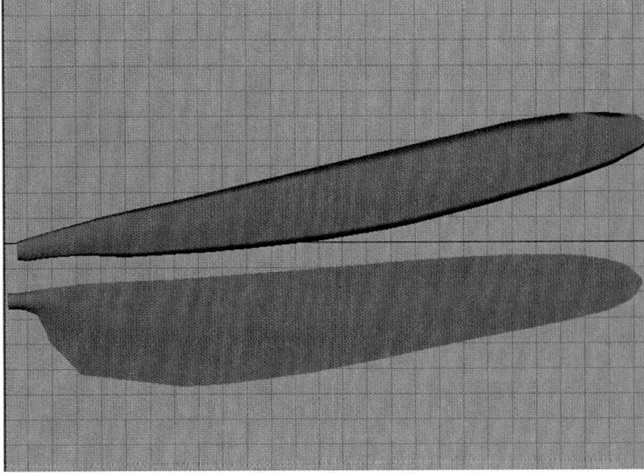

Figure 7.49
Orthographic views of the wings.

   After selecting the head, main body, tail base, and tail in Maya, he applied a new Blinn shader and loaded a standard planar projection map into the color attribute. The fly_side.jpg image was used as the projection source.

Exercise 7.2: continued

Stage 1.2: After orienting the planar projection's place3Dtexture node to correctly cover the character, he conducted test renders with a simple lighting setup to evaluate the projection alignment. Necessary adjustments were completed and the source image was edited in Photoshop to cover some areas that had some white bleed-over.

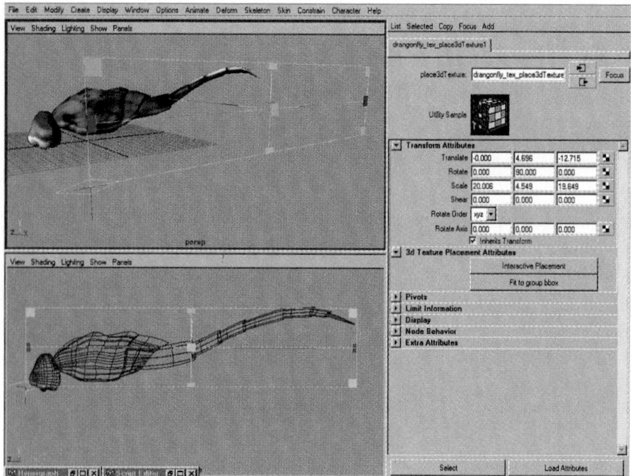

Figure 7.50
A new Blinn Shader was added, along with a corresponding planar projection with file node loaded into its color attribute. The planar node was oriented with the fit to Group Bounding Box button in the Attributes Editor.

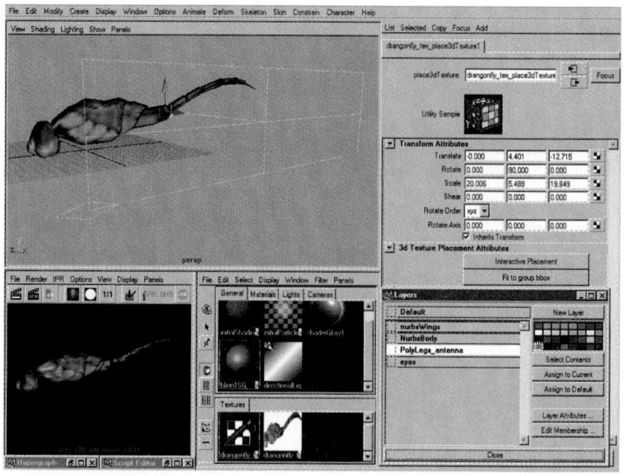

Figure 7.51
A light and test render were added to evaluate projection. The projection node was re-adjusted and more bleed was added in Photoshop to regions that didn't quite work.

2. A Convert Solid Texture was used on each object one at a time. Symmetrical half objects only had one half converted because the other half used the same texture as the first and the mapping was identical. The convert-to-solid settings were optimized for each object so that the pixel sizes would reflect the dimensions of the object.

Note

The pixel sizes seem to be reversed, with the X value controlling the vertical pixel scale and the Y value controlling the horizontal value. So the tail resolution was set to a ratio such as 512×1024, whereas the tail base was only 512×512.

Step 2.1: Convert Solid Texture makes a texture that wraps around the whole object, so the artist took into account the whole shape of the object when setting the pixel values. To tell if the process worked, he test-rendered at the final output resolution. If the texture didn't look like the original projection when test-rendered, he reconverted it until it looked right.

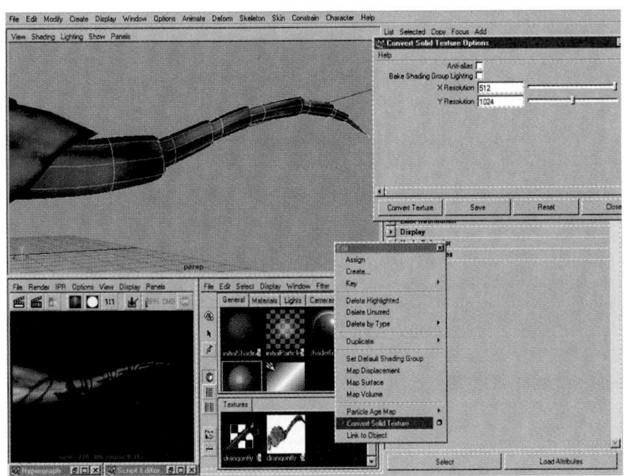

Figure 7.52
Two different Convert Solid Texture settings for the tail and the tail base.

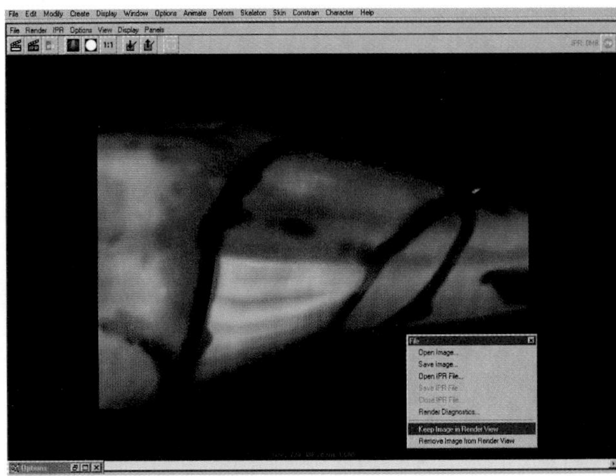

Figure 7.53
The artist test-rendered the newly converted texture map for clarity. He reconverted it to eliminate blur.

Note

Anti-aliasing smoothes out the final texture by rendering twice the resolution and averaging the pixels. If you need sharp details, anti-aliasing should be turned off.

Step 2.2: After each image was created, they were all converted to Targa format and edited in Photoshop. The conversion was accomplished using Fcheck on the PC and either the command line utility imgcvt or Maya Composer on the IRIX platform. Each image was cleaned of any of the unwanted background color and then the red channel of each of the images was extracted to a new document. With the Actions palette set to Record mode, the images were value-corrected and processed to emphasize details to create a nice bump map.

Some situations call for the use of a separate texture image for each attribute to specify precise values. Other times you can get away with using the bump for a specular and sometimes even a reflectivity map. Sometimes the color can also be used as a diffuse map as well to give a more worn look. The artist brought these images back into Maya and mapped them into the specific attributes. He then experimented with various combinations with the IPR window active to give rapid final rendered results.

Exercise 7.2: continued

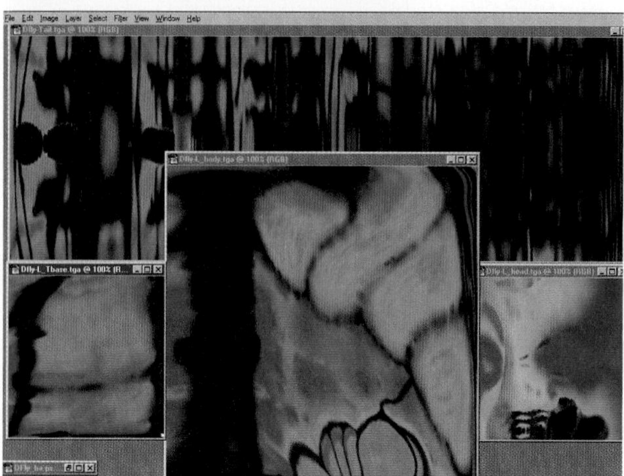

Figure 7.54
The artist ensured that the color correction was consistent with a batch-processing paint package.

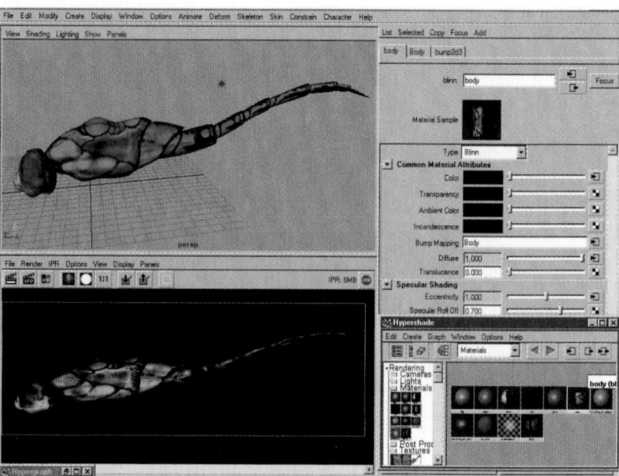

Figure 7.55
Tweaking your shader with the IPR window is a joy that every Maya user will appreciate.

When a look was established, the artist used a switch node. A switch node allows the use of one shading group for all the objects, switching only the individual attribute-mapped images specific to the objects. See the explanation of the switch node in the "Additional Texture Tips" section, later in this chapter.

3. The artist then hid the Dragonfly body, leaving only the wings on the right side visible. Each of the wings will use its own planar projection node, with a separate shading group for each of the front and back wings.

   Stage 3.1: The first wing was completed and then its Shading Network was duplicated. The newly updated shader was then applied to the back wing, and the duplicate Shading Group was adapted by replacing the old wing image with the original for the new projection nodes. They were then renamed and positioned to correspond with the back wing.

   Stage 3.2: First, one side was textured and then the two were grouped and duplicated with a –1 X axis scale value. The wings are made of single pieces of geometry made from lofting profiles, so the texture projection literally goes right through the model and an inverse duplicate is seen on the other side.

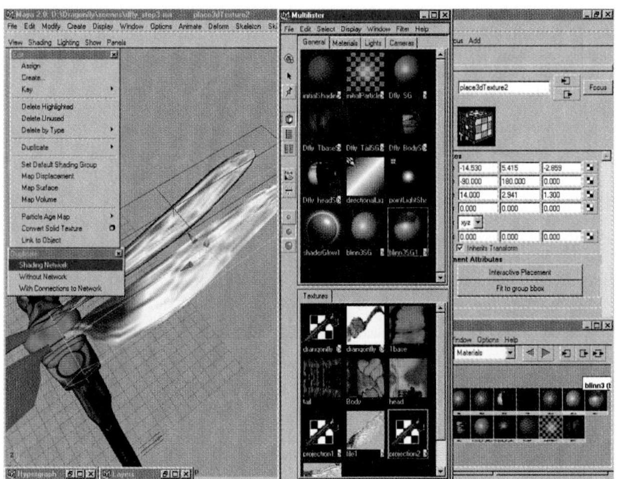

Figure 7.56
The whole shading network was duplicated for the front wing and adapted for the back wing.

Figure 7.57
Projection maps pass through the front and back sides of the model so that each side has the image of the wing.

Stage 3.3: Identifying the pass-through effect is a common artifact of projection mapping and determining whether it can be useful or a hindrance. This wing benefits from the effect. If you achieve a negative artifact, scale the Place3Dtexture icon so that it stops before reaching the other side.

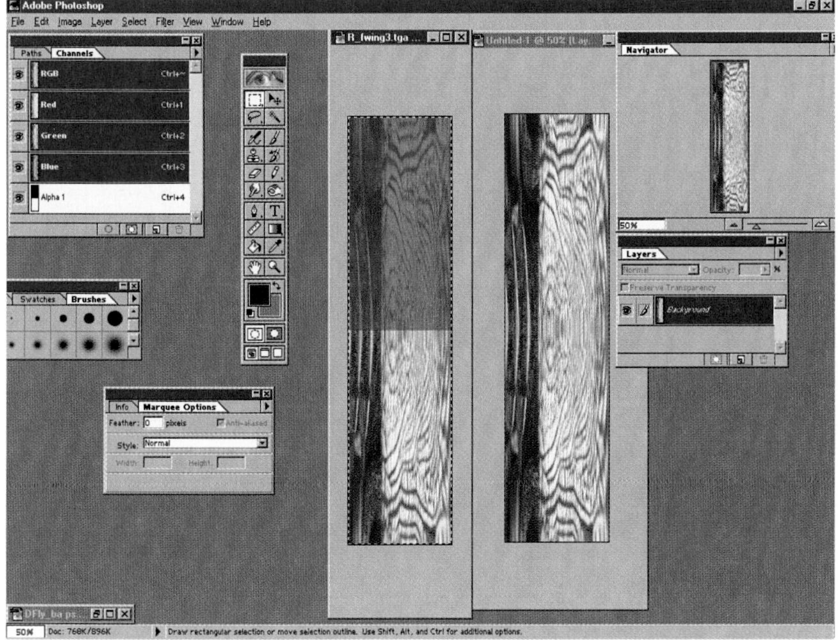

Figure 7.58
Scaling the Projection icon to determine projection range.

Exercise 7.2: continued

Stage 3.4: Afterward, the image was converted using Convert Solid Texture and edited in Photoshop to create a bump map that lifts the veins out, giving the wings a more dimensional appearance. The bump map was also processed somewhat to give certain areas a slightly stippled effect. In Maya, a Blinn shader was used to start things off. The transparency attribute can also use the image for the bump in this case. The veins were less transparent than the cell area. You can experiment with the attributes to get the look that you desire.

Note

If you prepare your character properly, the wings don't actually need to be deformed, depending on the types of animation that your bug is going to do. If you decide not to bind the bones, you can parent the wings to the bones to avoid using Bind Skin. Then you can simply parent the place3Dtexture node to each wing, and converting solid texture is not necessary. If the wings do have a bind skin, you could still parent the place3Dtexture node to the bones driving their rotation.

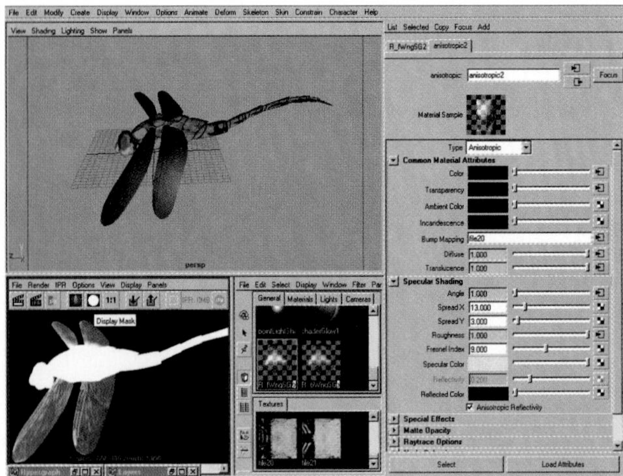

Figure 7.59
Transparent effect achieved from using the bump image to map the transparency attribute.

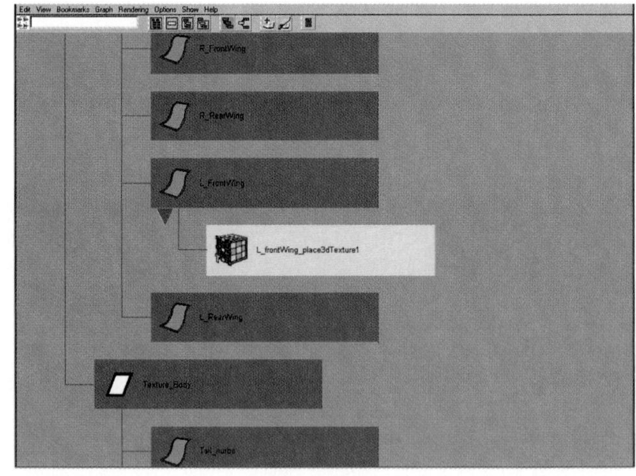

Figure 7.60
Parenting (not binding) the placement node along with the wing geometry to the bone to retain the projection alignment vs. converting solid texture and binding the wings to the skeleton as optional workflow choices.

7.8.3 Dinosaur

Exercise 7.3: Dinosaur

The following is an outline of how this model was textured for a project. The texture map is provided for educational purposes only courtesy of Dimension 7 and Minds Eye Media.

1. The dinosaur model provided for educational use only from Viewpoint Digital came pre segmented and trimmed though you will have to name the body parts your self. The arms had active trim surfaces, but the blended shapes from the arms no longer had any history attaching them to the trims. The body was made from many individual pieces. In hindsight, the texture reference object path may have been a better route to go if I were to do this again, but the version of Maya that was used at the time to create the texture did not support the option, though it could easily still be done.

   Stage 1.1: I hid the legs and arms after the big beast was posed into an optimized texture position. I wanted as straight a line as possible from nose to tail. Hiding the limbs gave access to the underbelly and gave a smaller texturing area for more optimized image sizes.

   Stage 1.2: I generated side, top, and front orthographic views and exported them to Photoshop. The legs were exported as separated images from the same viewports. The legs were then put back on as layers in Photoshop.

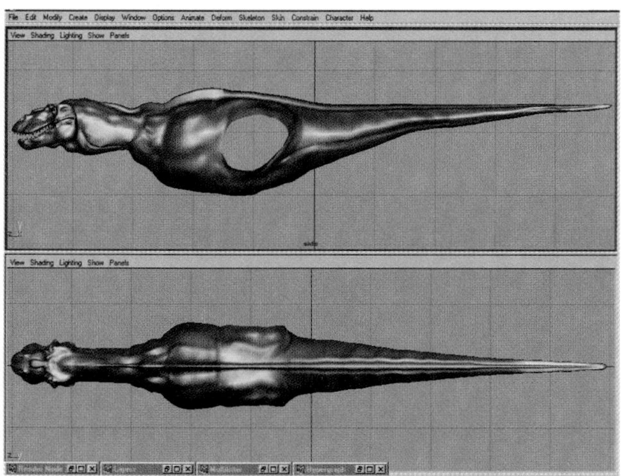

Figure 7.61
Preparing separate orthographic shots for the body and the legs in their new texture pose.

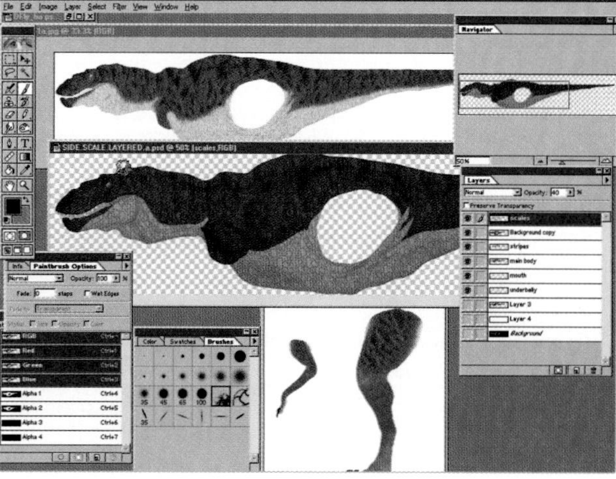

Figure 7.62
The legs were reintegrated with the body in Photoshop and a texture map was painted in Photoshop.

   Stage 1.3: After careful consideration for the shots in which the dinosaur was going to appear, I decided to use only the side view. For a more detailed texture for the belly and the back, top and bottom textures can be generated. Time is money; unless you are practicing these techniques on your own time, you will find that there is limited time for absolute perfection in the production pipeline. The only thing that matters for film and video is what the camera sees. For real-time 3D games, however, you may need to make a 360-degree map that will work on all sides, though it probably will not need to be as high in resolution or detail. Each situation calls for a unique solution. For

Exercise 7.3: continued

the purposes of this exercise, we will assume that the side view orthographic projection is a good start for a base layer of foundation texture.

2. I created the images using the standard techniques outline so far. The red channel was a good source for the bump map.

   A dark scaly layer was superimposed over the bump source to punch out the scaly details. Because of the multitude of pieces in this texture job, the bump map was loaded into the shader previous to conversion of solid texture. This meant that the Convert Solid Texture tool automatically converted both the images for the attribute maps one at a time.

   Each part of the beast was optimized for the X/Y pixel sizes for the Convert Solid Texture resolution. For these multiple object conversions, it was important to name both the geometry and the shading groups with short, concise, and easy-to-understand naming conventions. When the conversion occurs, the process uses both elements to generate the name of the image. For some odd reason, Maya writes the image to the main level of the current project directory. The images are also only written as 24-bit Maya .iff files. This means that you have to use Fcheck, or you can use the imgcvt command line tool on the IRIX platform, to convert each of the Convert Solid Texture images to a useable format such as Targa for image processing and other editing.

Note

In Maya 1.x, the Convert Solid Texture worked only on the color attribute. That meant that you had to load the other attribute maps into the color channel and turn off all but a specially tuned ambient light just to convert them (now there is a check box for the Light Baking option in the Convert Solid Texture box Settings box).

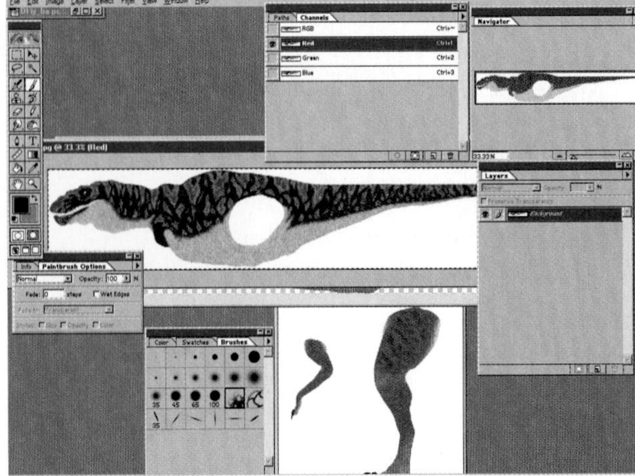

Figure 7.63
Extracting the red channel as a method for starting off the bump map.

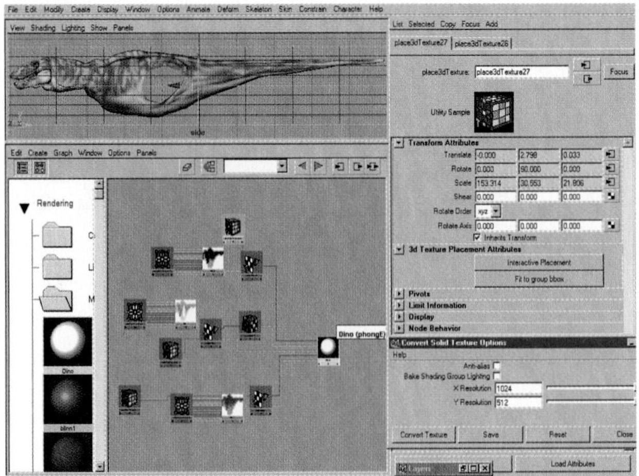

Figure 7.64
Depending on the complexity of the model, you must decide whether to convert all your corresponding attribute maps automatically or create them from the converted source manually by image processing in Photoshop.

3. The biggest issue that arose for this stage of texture mapping was the need to align the texture convergence for the limbs with the body. Various steps were used to solve this issue. The best way that was eventually used was to reduce the amount of sharp-edged patterns that overlap the area in Photoshop using the clone tool. Most of the editing was done on the limbs' texture, but some needed to be reduced on the main body image as well. The teeth, claws, inner mouth, and eyes were not textured with this pass and a series of simple solid colored shaders were used until a later time to texture them was available.

7.8.4 Dragon

Exercise 7.4: Dragon

The Dragon texture was created by Donnie Bruce. The wings were textured separately. The model was built out of two symmetrical halves, completely identical in every way. So the texturing process is done on one half of the geometry and then reused for the other half of the model.

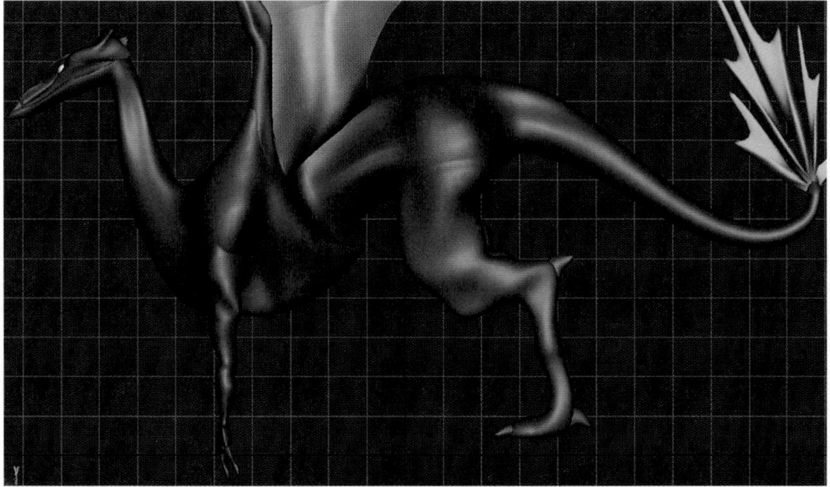

Figure 7.65
Orthographic source imagery.

1. The major difference between this texture map job and the previous three is the fact that this model has been symmetrically cut in half during the modeling stage. This simplifies the texturing process. For preparation, the model compares to a mix between the Dinosaur and the Dragonfly.

2. The source image is included, as is the dragon geometry, for you to practice with. Try to integrate the steps presented, innovating any new strategies that you come up with. You may even want to go back into the Photoshop document and examine the layers, doing any edits that you see fit to create the right effect.

Exercise 7.4: continued

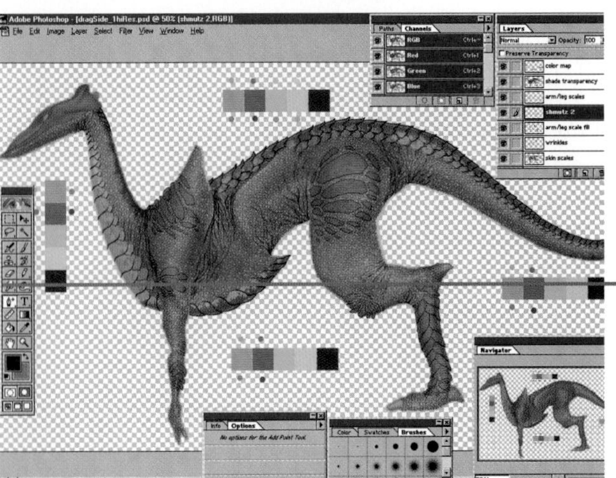

Figure 7.66
Photoshop painted with detailed masks and layers.

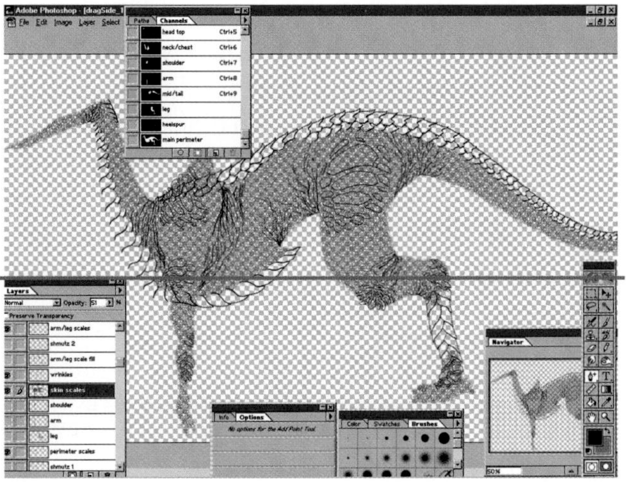

Figure 7.67
More details of the masks and layers.

7.8.5 Human Head: Tadao's Groovy Texture Scanner

by Tadao M.

One of the most difficult tasks in character animation productions is making faces. Faces are very important because they influence the impressions of the characters. Doubtlessly, making a realistic, highly detailed face model requires a very advanced skill level, and even if your face model is great, bad texture can ruin it. Making good-looking textures is also a very difficult task. Because a face has such complex structure, it won't look believable if you just use tileable textures. Especially in the case of a woman wearing make-up, you have to put lipstick or eye shadow on certain areas. It is extremely hard to figure out where those areas are in a simple square image.

To solve this problem, many methods have been invented. For example, one of the most popular ones is using a grid map. However, you have to switch back and forth between the 3D program and paint program many times. Also, it is very hard to be precise. You can also use 3D paint programs. However, most of them support only polygonal models. So if your model is made up of NURBS, you'd have to convert it to polygons. I would caution doing this because you have to be very precise in matching textures between NURBS and polygons.

One 3D paint program from Alias | Wavefront, the maker of Maya, called Studio Paint, can work with Maya seamlessly, and it can paint directly onto NURBS surfaces. Yet, it's expensive and is available only on the SGI platform, and you need a pretty fast computer to run it comfortably. Besides, you'd have to spend some time to learn the new software application.

One day, when I was working on a project as the head of the texture department for the SFAAC Manga project, I got an assignment to make textures for faces of characters, and I tried to find the best way to do it. Back then, I didn't know how to use Studio Paint, so using it was quite time consuming, and we couldn't afford to waste time. What I did at first was take a screen shot of the face from the front view, bring it into Photoshop, and paint a face texture by using it as a template. It seemed as if it worked pretty well, until I test-rendered the side of the face. It was getting a stretched effect, which can easily happen with planer projection maps. Especially with the bump map, small dots had become ugly scratches. I had to stop what I was doing and think about possible solutions.

That's when I came up with this "Virtual 3D Scanner" technique. The concept is actually fairly simple. I got the stretch effects because I projected the map from only one direction. If I could project the map from all directions, meaning 360 degrees all around the face (which is cylindrical projection), I could solve the problem. The key was how to take screen shots from all the directions (how to make a template image). Then, I remembered the "3D scanners" that are used at virtual reality laboratories in Hollywood such as Cyberscan. The machine has a laser head that goes around a person and makes a flattened image of the face. All I had to do was simulate the same process in 3D programs. I spent a half day writing down the process on a piece of paper, and then completed the texture map. Through this approach, I eliminated most of the problems. Here is how to do it.

Exercise 7.5: Preparing the Face Model

1. Launch Maya, and open your face model. Hide all surfaces that you don't want to see in the unwrapped image. If you would like to see where separate surfaces (hair, eyebrows, eyelids, and so on) are going to be, leave them visible. I recommend doing this because it makes the painting process easier.

2. Select all the visible surfaces. Select Edit, Duplicate, Display, Hide, Hide Unselected to hide the original surfaces. Deselect all, and select the face surface. Click Edit, Group. In the Option window, set Group Pivot to Center. Click Group and then Reset. Close the Option window. Deselect and select all the visible surfaces except the face surface. With the Shift key held down, select the face surface and press the Up arrow key to select its group node. Select Edit, Parent. Click the face surface and press the Up arrow key to select only the group node. Move the whole group along the X axis with grid snap on (hold down the X key) from the front view to make sure that the pivot point of the group is at X=0 in the world coordinate.

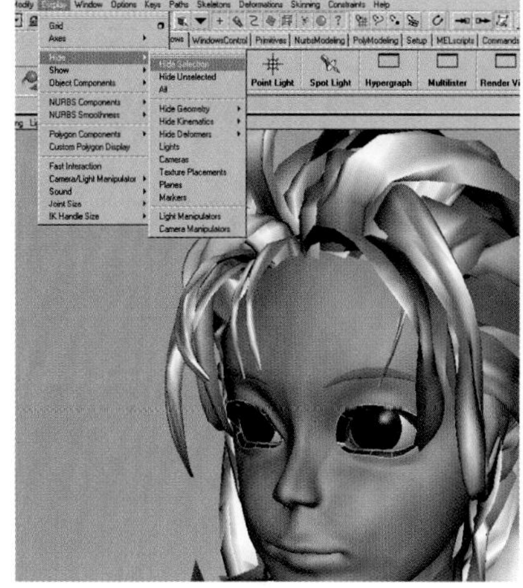

Figure 7.68
Hide surfaces that don't need in the unwrapped image.

Exercise 7.5: continued

3. Next, go to pane menu Shading, Smooth Shade All, Lighting, Use Default Lighting (or press 5). If you would like to see isoparms in the unwrapped image, select Shading, Shade Options, Wireframe on Shaded.

Exercise 7.6: Creating Projection Texture

To measure the height of the unwrapped image, you must create a projection texture.

1. Bring up the Multilister, and create a new Shading Group. (I like Blinn shaders for human skin). Name the newly created shading group **faceSG**. Assign it to the face surface.
2. Bring up the Attributes Editor for the faceSG. Click the Map button for the Color attribute.
3. In the Create Render Node window under the Textures tab, click File with the New Texture Placement check box set to On As Projection.

> **Note**
>
> Do *not* assign any other surfaces (for example, hair, eyebrows, and so on) because the texture is for the skin only.

4. Name the newly created projection node **projection_faceColor**. Set the following attributes: Proj Type: cylindrical; U Angle: 360. Click the Fit to BBox button.

> **Note**
>
> In the case of a face with multiple patches, the 3D placement icon won't fit to the surface as you want. You would have to use the Interactive Placement function to do it manually. There are two important points:
>
> The location of the pivot point of the 3D placement icon and the face group node must be exactly the same.
>
> The height of the 3D placement icon must be tall enough to cover the whole face surface.

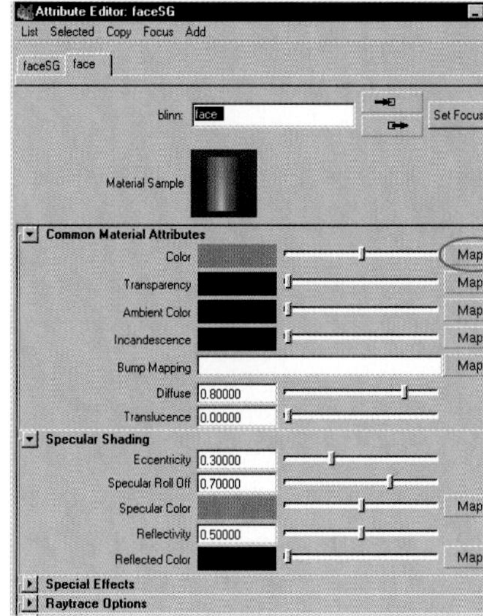

Figure 7.69
Click the Map button in the Attributes Editor for the faceSG.

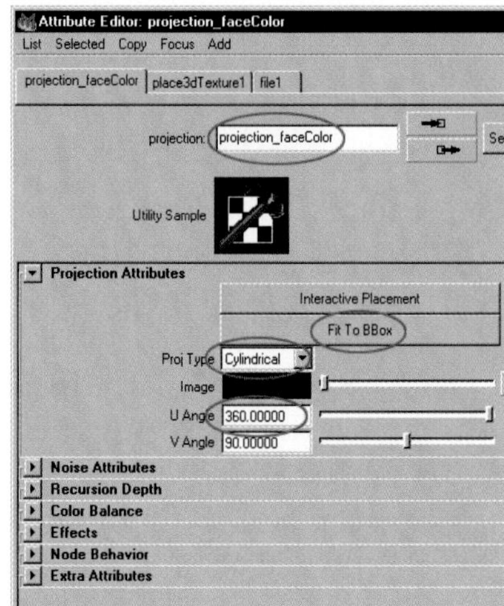

Figure 7.70
Detail of the projection node's Attributes Editor.

Exercise 7.7: Creating unwrapCamera

Create a camera that goes around the head to capture screen shots from all directions.

1. Click Panels, Orthographic, New, Front. Name the newly created camera **unwrapCamera**.
2. Select the unwrapCamera. Click Edit, Group. Name the group node **RotateCenter**.
3. Select the 3D placement icon in the modeling views. Note the values of translate X,Y, and Z. Give the same values to the RotateCenter node.

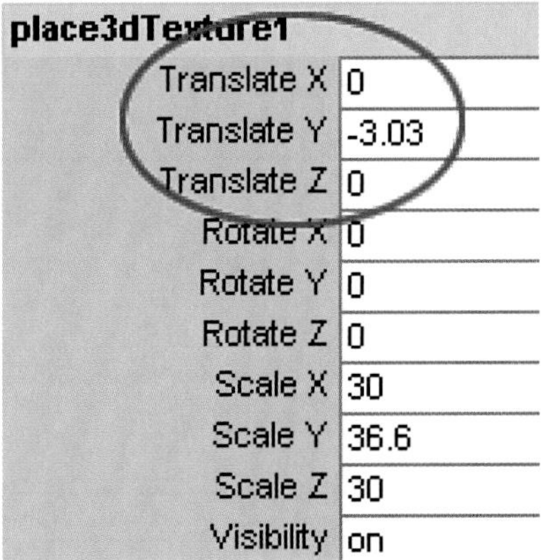

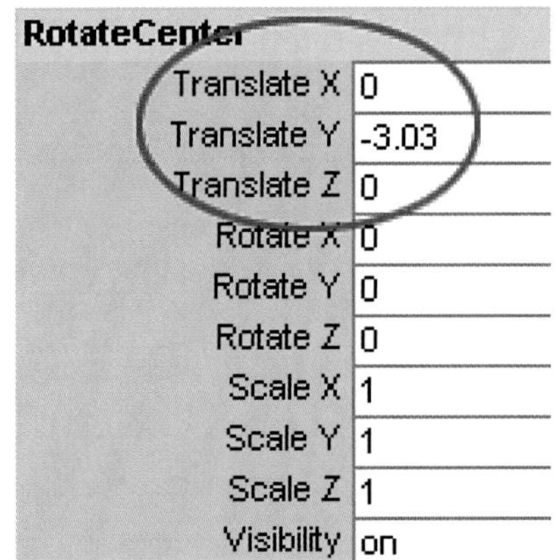

Figure 7.71
Synchronize your Translate Y axis.

4. Move the unwrapCamera along with the Z axis toward the face so that it sits right in front of the face.
5. Click Panels, Layouts, 2 Side by Side. Set the left window to the unwrapCamera, and the right window to the persp. Move the window divider to the left to make the left window (the unwrapCamera) very thin.
6. In the left window, dolly in or out until the top and the bottom of the 3D placement icon almost touch the edge of the window.

Exercise 7.7: continued

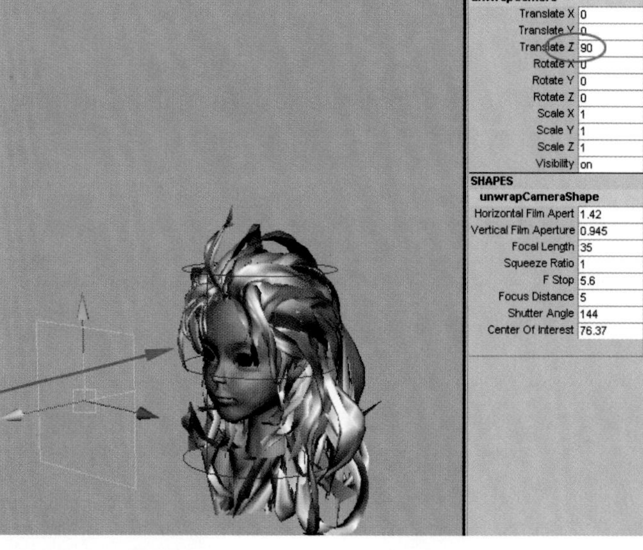

Figure 7.72
Positioning the unwrapCamera correctly.

Note

Make sure the translate X, Y, and Z values of the unwrapCamera are still 0.z

Figure 7.73
Move the window divider in Maya.

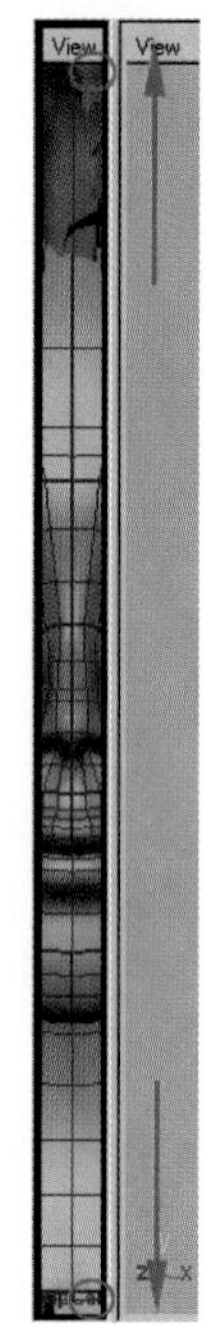

Figure 7.74
Dolly out in the front view to frame the 3D placement icon.

7. Bring up the Render Globals window. Set Image Format to JPEG. Close the Render Globals window.
8. Open the General Preferences window, and click the Animations tab. Set the following settings:

   Time Slider: from 1 to 720

   Range Slider: from 1 to 720

   Playback Speed: Free

   Playback By: 3.0
9. Go to frame 1, and select the RotateCenter node. Set the Rotate Y to 0. Set key.
10. Go to frame 720. Set the Rotate Y to 359.5. Set key. Deselect all.

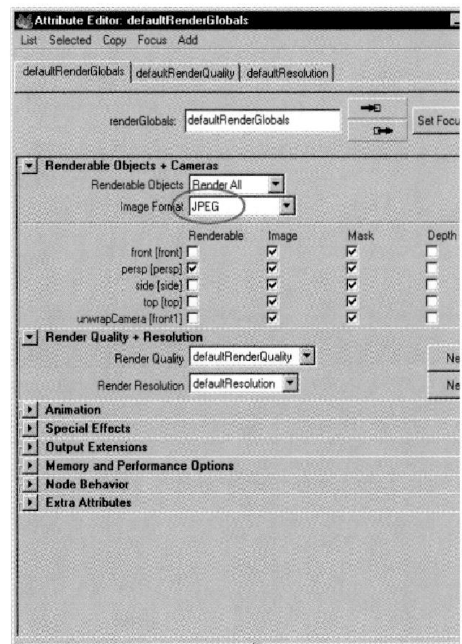

Figure 7.75
Set the Render Globals to output as JPEG file format.

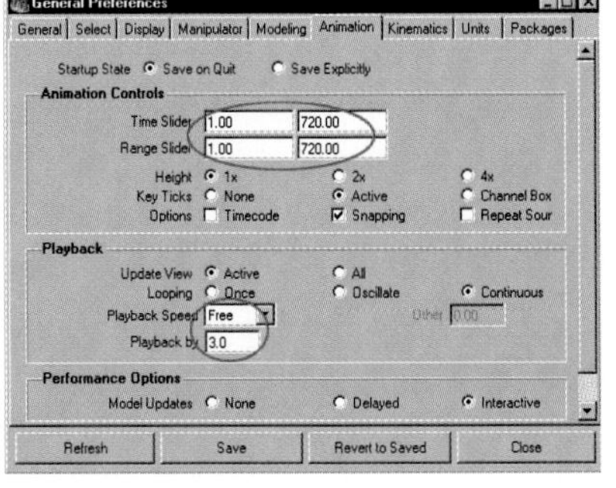

Figure 7.76
Set the playback range and Animation Playback options.

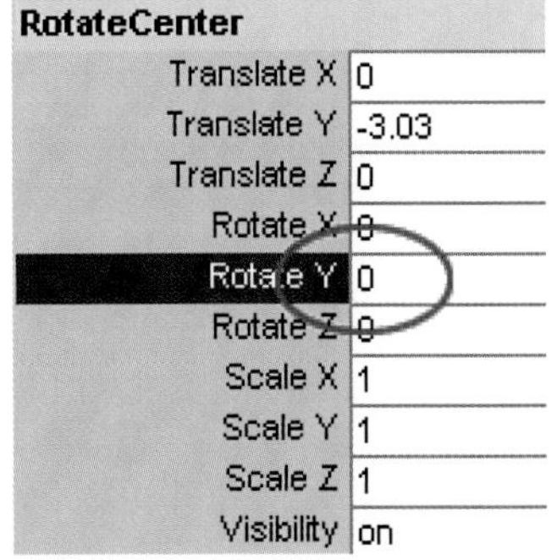

Figure 7.77
Set key on the Y axis of the RotateCenter node.

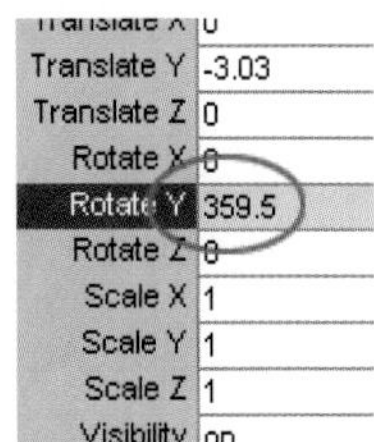

Figure 7.78
Now at the last frame set the Rotate Y to 359.5.

Exercise 7.8: Setting the Playblast Settings

1. Move the mouse pointer above the left window (the unwrapCamera), and press the Ctrl key (or click the MMB) to make it active.
2. Now go to the menu item Windows, Playblast. Set the following settings:

   View: Off

   Show Ornaments: Off

   Output Format: fcheck

   Display Size: From Window

   Scale: 1.0

   Save to File: On
3. Click the Browse button and specify the save location and the file name.
4. Click Playblast to start Playblast.
5. When Playblast is done, save the scene and close Maya.

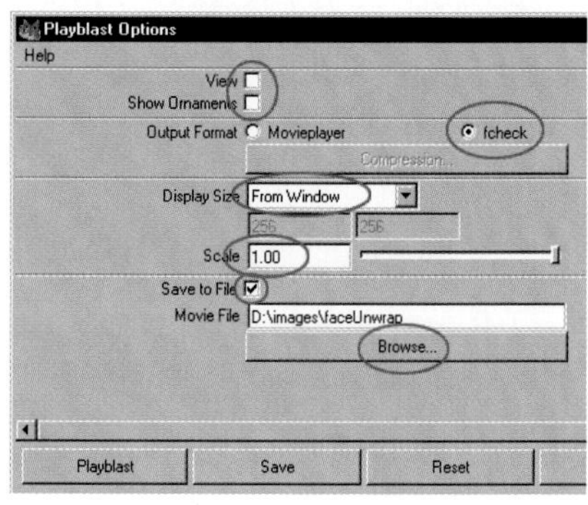

Figure 7.79
Now verify your playblast options.

Exercise 7.9: Creating an Action in Photoshop

1. Launch Photoshop, and open the first Playblast image (for example, face.0001.jpg).
2. Click Select, All. Read the image height in the Info palette.
3. Create a new image with the following settings:

   Name: unwrapFace.psd

   Width: 720

   Height: Same as the Playblast image

   Resolution: 72 pixels/inch

   Mode: RGB Color
4. Click File, Save As to save the newly created image in Photoshop format. Leave the image open.
5. With the first Playblast image still open, select New Action from the sub-menu in the Actions palette. Name it **unwrap** and click Record.
6. Select Image, Canvas Size. Set the following settings:

   Width: 3

   Height: Leave as is

   Anchor: The center

   Click OK.
7. Select Select, All.
8. Select Edit, Copy.
9. Select File, Save.

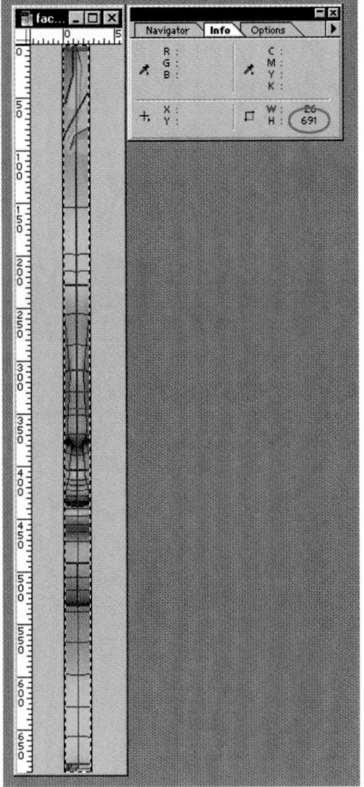

Figure 7.80
Read the image height in the Photoshop Info palette.

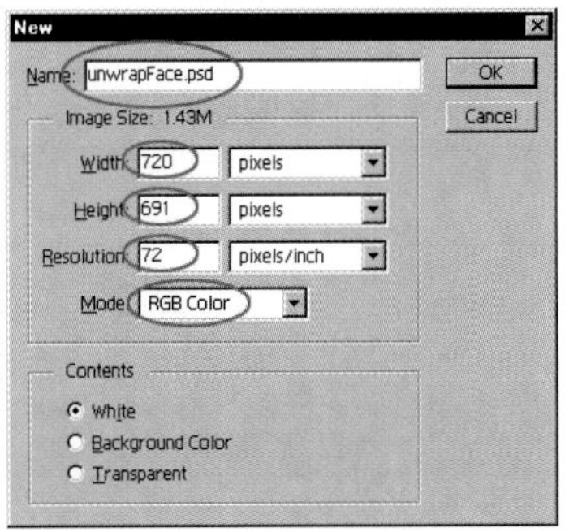

Figure 7.81
Create a new image in Photoshop.

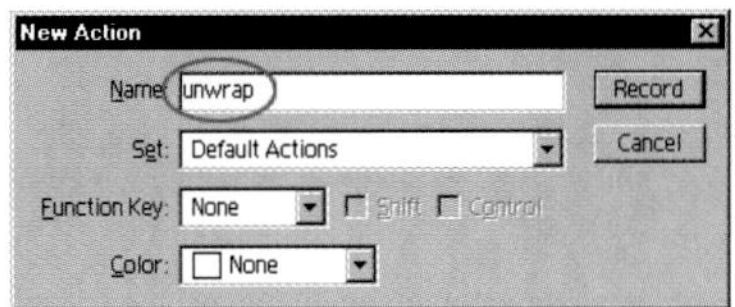

Figure 7.82
Create a new Action in Photoshop.

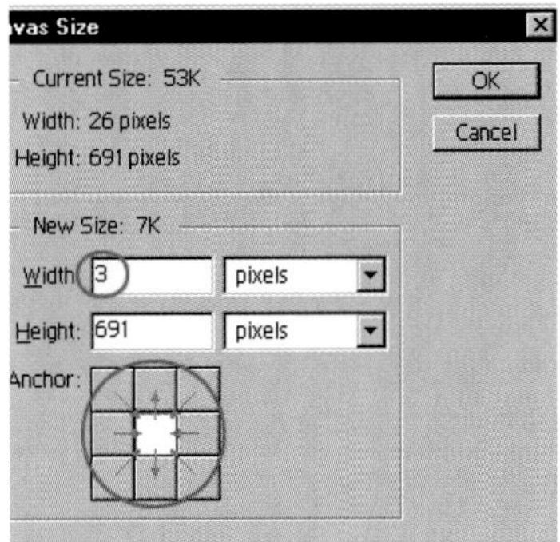

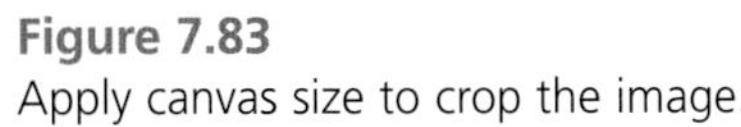

Figure 7.83
Apply canvas size to crop the image.

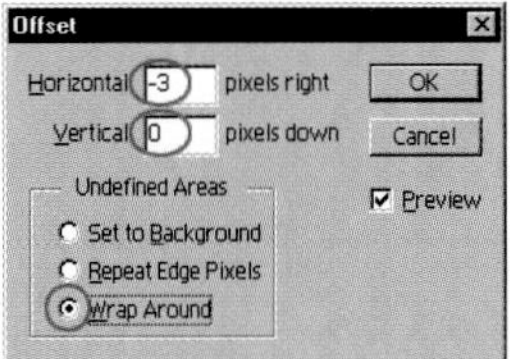

Figure 7.84
Using the Offset filter.

10. Select File, Close. This will make the unwrapFace.psd window active.
11. Select Edit, Paste.
12. Select Layer, Flatten Image.
13. Click Filter, Other, Offset. Set the following settings:

    Horizontal: –3

    Vertical: 0

    Undefined Areas: Wrap Around

14. Click OK.
15. Click File, Save.
16. Click the Stop Playing/Recording button in the Actions palette to stop recording.

Exercise 7.10: Performing Batch Operation

1. Minimize Photoshop. Open the folder that contains all the Playblast images. Create a new folder, and name it **`pb_images`**. Move all the Playblast images except the first one into the pb_images folder.

Note

In Photoshop 4.0, you can find the same command (Batch) under the sub-menu in the Actions palette.

2. Go back to Photoshop. Click File, Automate, Batch. Set the following settings:

   Action: Unwrap

   Source: Folder

3. Click Choose and select the pb_images folder. Set the following setting:

   Destination: None

4. Wait until Photoshop finishes the batch process.

Note

It really depends on how fast your computer is, but the batch process usually takes 10–15 minutes.

5. Using the green lines of the 3D placement icon, crop the unwrapped image to get rid of unnecessary space at the top and the bottom.
6. If the image is too tall or too wide, use the Image Size command to adjust the proportion.
7. Click File, Save. Now paint your skin texture.
8. Create new layers over the unwrapped image, and start painting textures.

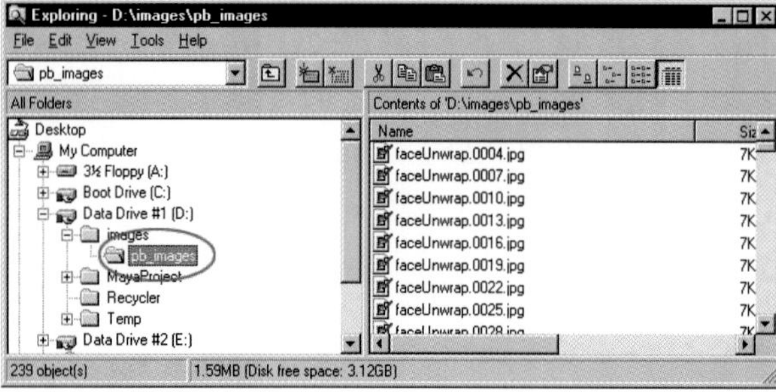

Figure 7.85
Create a folder for your images.

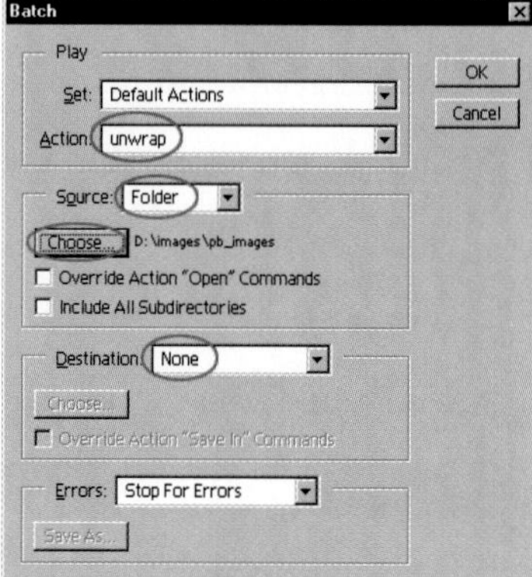

Figure 7.86
Fire off the batch process in Photoshop.

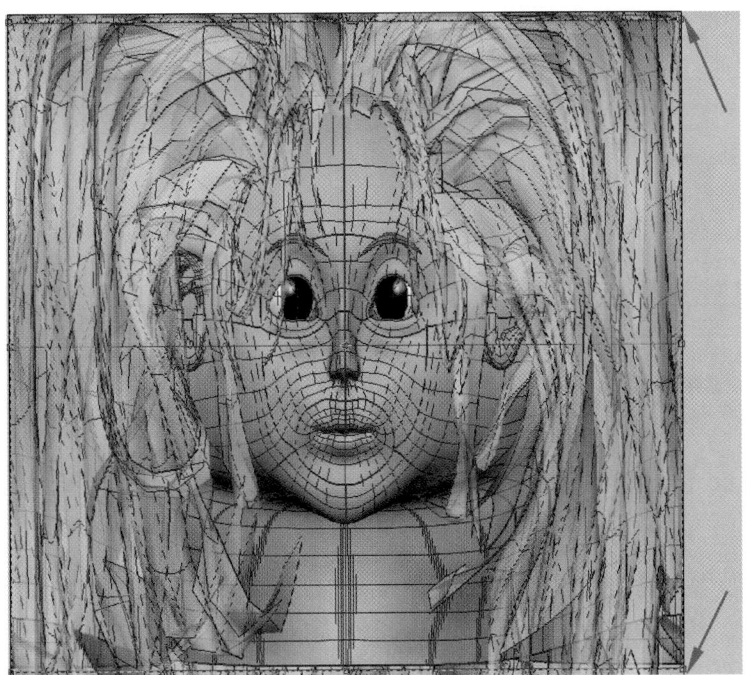

Figure 7.87
Crop the unwrapped image to get rid of unnecessary details.

9. Duplicate the Background layer (the unwrapped image), move it to the top, and make it half transparent to use it as a template.
10. When it's done, make the template image invisible. Click File, Save a Copy to export a flattened image. Choose an image format that Maya can read (for example, TIFF or Targa).
11. Repeat the previous steps for painting the skin to create bump maps and specular maps.

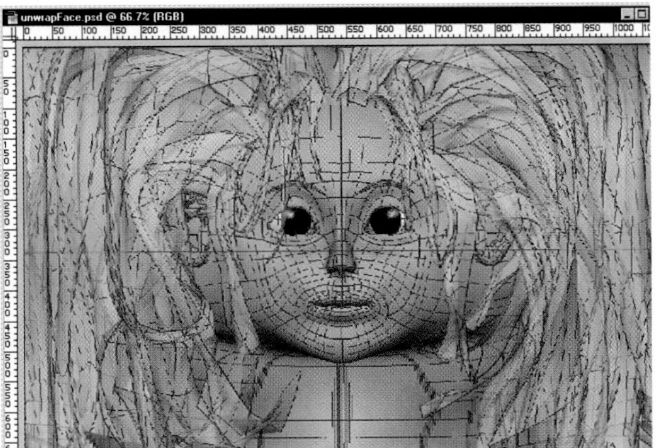

Figure 7.88
Image size the result.

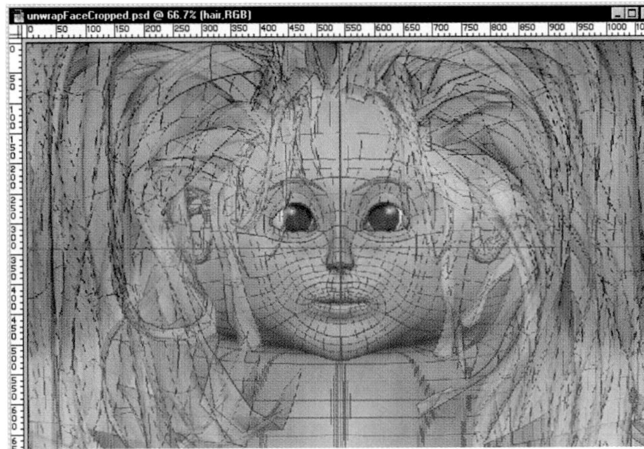

Figure 7.89
Create new layers over the unwrapped image, and start painting textures.

Exercise 7.11: Applying Texture Maps

1. Launch Maya, and open the scene file you used to create the unwrapped image.
2. Bring up the Multilister. Bring up the Attributes Editor for the file texture node that is connected to the color attribute of the faceSG. Click the Browse button and read the exported flattened image. Repeat the same steps for the bump maps and the specular maps.
3. Run a test render. If it needs adjustment, repeat the same steps for fine-tunings. Test render from many angles to check if everything is okay. When it's done, save the files.
4. If you leave the textures as projection maps, you will have a problem when you give your character facial expressions. Projection maps won't follow surface deformations unless you convert them to UV maps (or use a texture reference object if you are using Maya 2.0 or greater) by using the Convert Solid Texture command.
5. Assign the shading groups created by the Convert Solid Textures operations to the original face surfaces. Delete the surfaces that were duplicated for the unwrapping process.

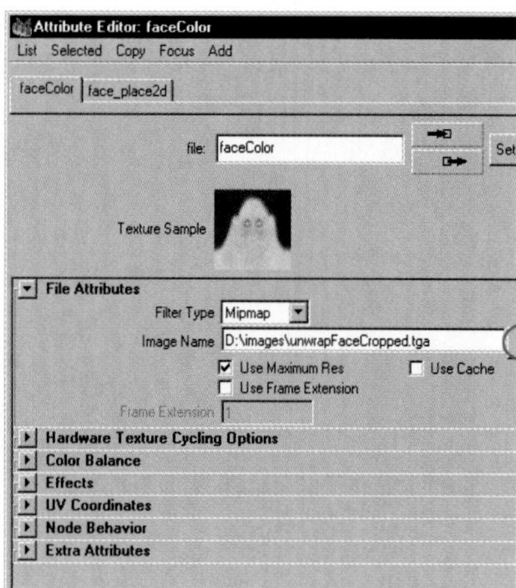

Figure 7.90
Load in the resulting texture in Maya.

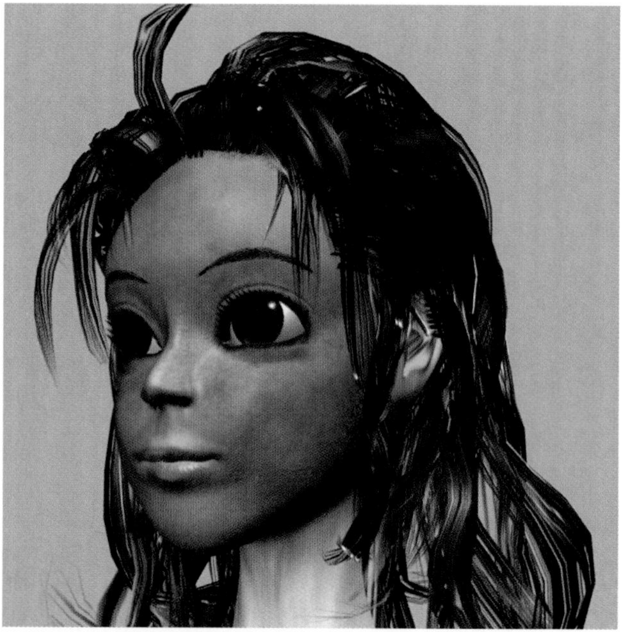

Figure 7.91
The texture rendered on the object.

7.8.5.1 Summary of Tadao's Texture Scanner

The greatest benefit of this technique is that you can use Photoshop to paint textures. With a familiar interface and tools, you can concentrate only on the creative work. I also like the fact that it can unwrap NURBS surfaces without converting to polygons. You might be thinking that this technique takes a long time, but after using it a couple of times you'll be able to go through the whole process pretty quickly.

Even though I think this technique is nearly perfect, there are a few problems. Probably the biggest problem is the pole effect at the top of the head. You can't avoid it because of the nature of cylindrical projection. If your character is bald, you'd have to combine this technique with another projection map to solve the problem.

7.9 Additional Texture Tips

The following are two important utility nodes and short demos of how they're used to solve various texture-mapping dilemmas.

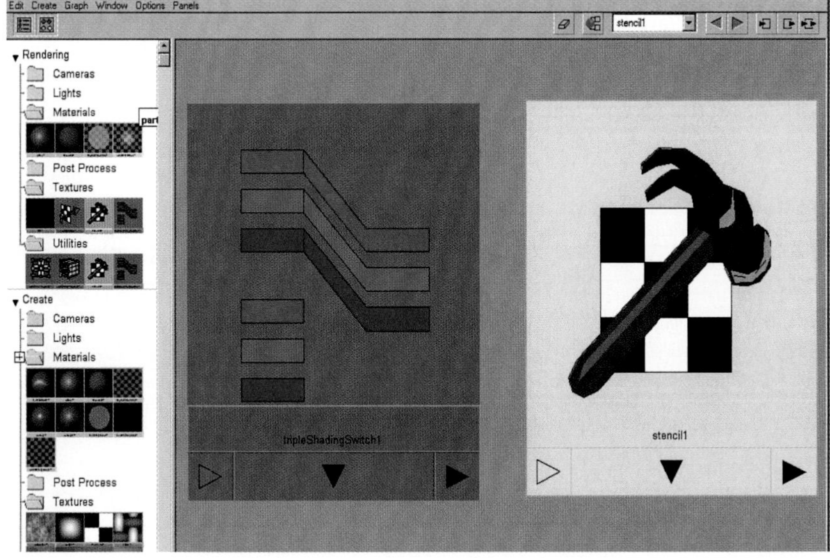

Figure 7.92
The switch and stencil shading utility nodes.

7.9.1 How to Use a Switch Node

The switch node is a useful utility that enables you to specify a shading group for many objects and then choose a single object to have a unique characteristic within the shaded members. Oftentimes when you are converting solid textures, for many objects the process creates many duplicate shading groups with only one difference: the textures that they have loaded into the specific parameters. If a switch utility was used, you could have all the objects share a common shading group and use the switch utility to organize which object will receive which texture map.

There are three different types of switch nodes:

- **The single shading switch.** Allows the switching of single float attributes, such as bump depth.
- **The double shading switch.** Allows the switching of dual-channel data, such as Repeat UV
- **The triple shading switch.** Allows the switching of three-channel values, such as the RGB color attribute. With this switch you can have a single shader control the general surface attributes of many objects but use the switcher to enable a unique texture for each object.

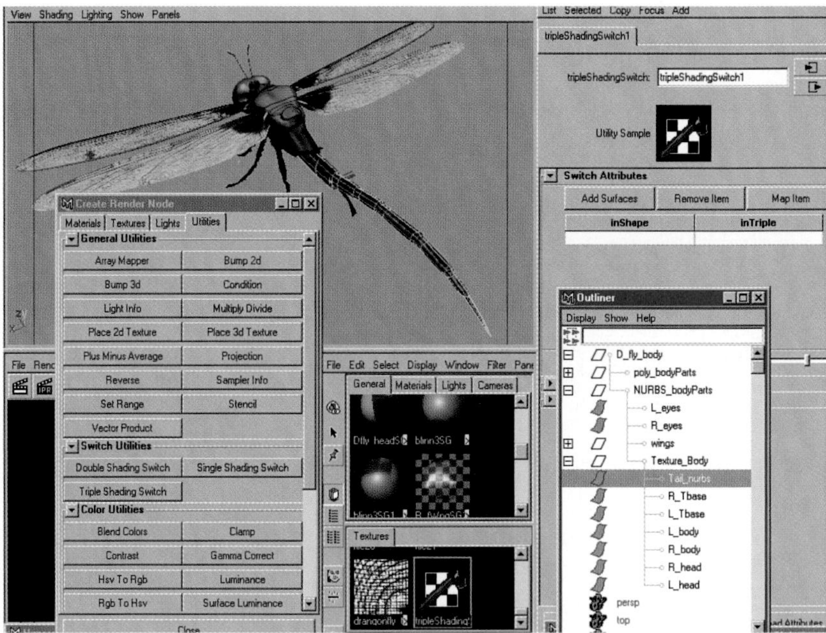

Figure 7.93
The Create Render Node Utilities tab.

Here is how to use the switch in a common workflow. The Dragonfly character has seven objects that need to share a common shading group. The catch is they each need to have different image files for their attribute maps.

1. Create a Blinn shader in the Multilister. Click the Map button next to the color attribute. When the Create Render Node window opens, go to the Utilities tab and choose Triple Shading Switch.
2. While the Create Node window is still open, create four file-in nodes in the texture section. Load the following files from the project: Dragonfly: Sourceimages: head.tga, tail.tga, Lbody.tga, and tailBase.tga.

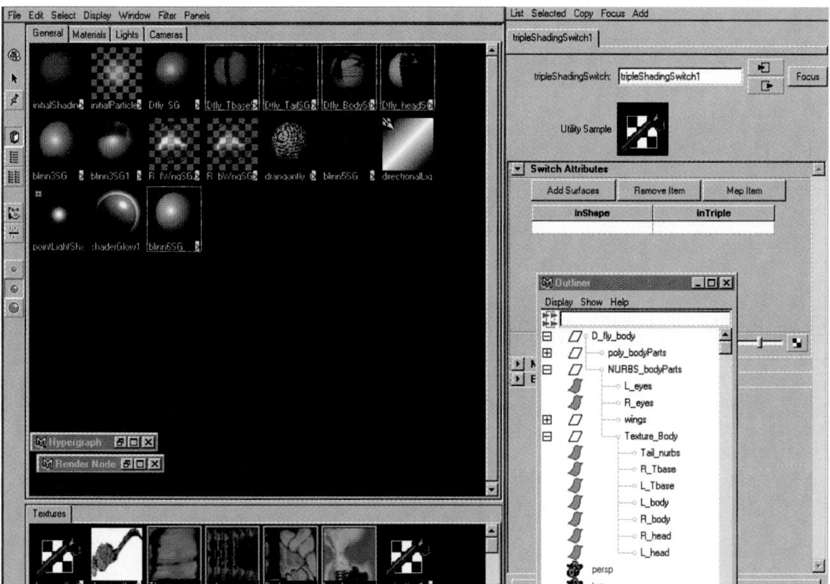

Figure 7.94
Many textures for one shader dilemma.

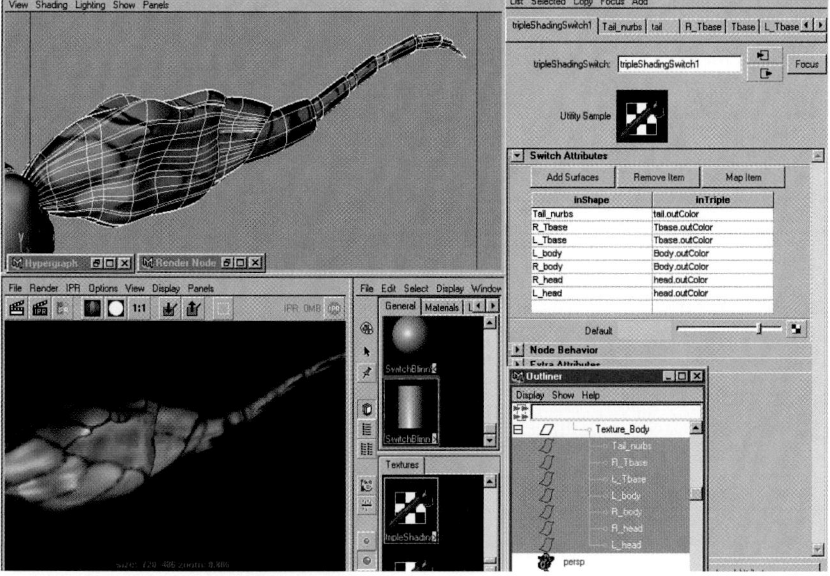

Figure 7.95
Four file-in nodes loaded with the proper source images.

3. Open the Outliner and find the names of your body parts that need to be textured. Double-click the Triple Shading Switch node in the Multilister. When the Attributes Editor opens up, middle-mouse-drag the name of one of the objects to the object's field in the switch utility. Then, from the Multilister, middle–mouse-drag the appropriate texture to the switch utility field on the right. Repeat this process for each object to which you want to add a unique texture map.

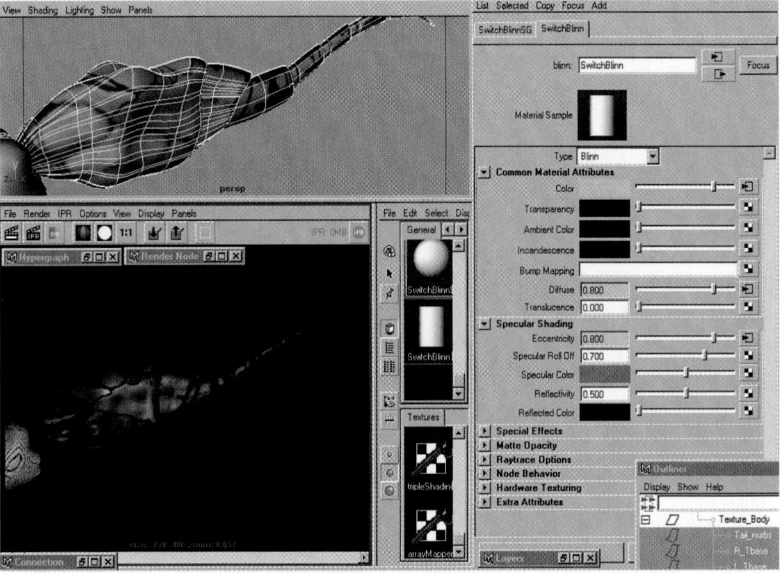

Figure 7.96
Four file-in nodes loaded with the proper source images.

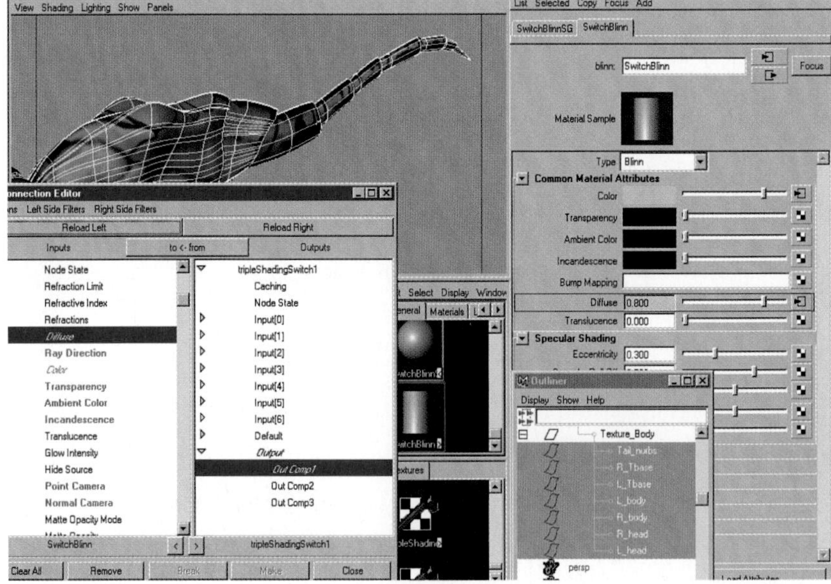

Figure 7.97
Re-use the switch if the images are compatible. Duplicate the shading network and then replace only the fields that you need.

4. If you want to re-use the Shading switch group afterward on another attribute, just drag the switch utility on to the next attribute of your choice that can receive the triple shading information.

5. If you want the attribute holding the switch utility to continue working normally on the rest of the objects using the shader, map the switch's default color attribute with the appropriate texture.

6. At times this node can save you time by organizing your shading networks. If you need to break the connection to a switch utility, just hold down the

RMB next to the switch-mapped attribute and choose Break Connection, the way you would break connections on any attribute. The attribute will return to the default color afterward.

7.9.2 Using Stencil Nodes with Projected Texture Maps as Components of Layered Shaders

A *stencil* node is a specialty node that helps the process of layering textures using alpha channels or HSV color key values to determine how to mix the imagery of the picture with any layers below or above it in a layered shader or a layered texture.

You can use the stencil masks in three ways. By default, if an image file is a 32-bit file with an alpha channel, the stencil will automatically recognize it. If your image uses a separate file for the mask, you load a grayscale picture into the mask attribute and the image is automatically aligned with the texture image. If you choose to create a color keying effect, switch on the HSV color key option and choose a color range that you want to become transparent. We will use a stencil in a typical workflow to build a layered texture.

Note

Whenever you use layers of shaders or textures, you must be sure that the default color in the color balance section of both the file node and the projection node is set to solid black. Otherwise, the image will add a gray film over any layer beneath it.

1. Create a Blinn and a layered shader, switching its compositing flag to Texture in the Attributes Editor. Now open the Create Render Node palette, click the Utilities tab, and create a new Stencil node. While the Create Render Node palette is open, create a new planar projection node just for this exercise and name it **stencil_1**.

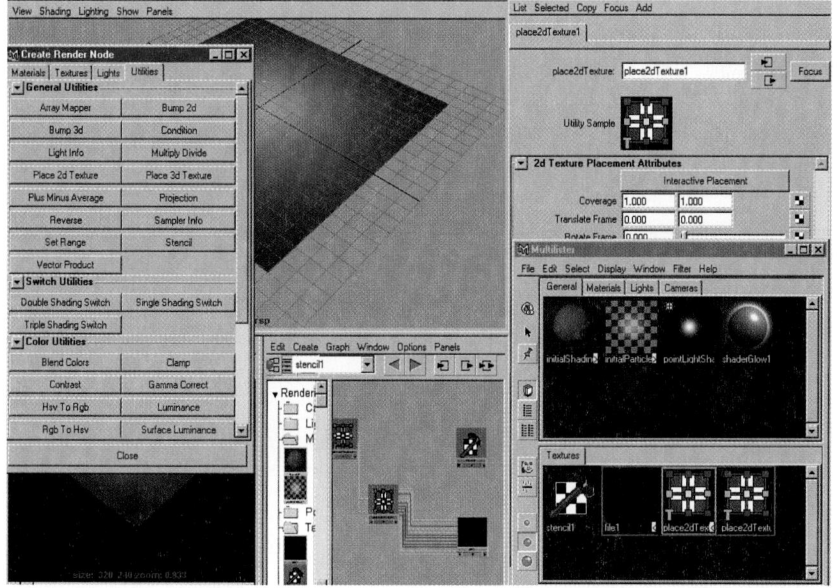

Figure 7.98
One way to create a stencil is through the utilities. This is different from making a new texture with the stencil mode on.

Tip

An alternate method of using projection nodes with the stencil node is to create a normal stencil texture and then middle-mouse-drag the stencil to the projection image map.

2. Drag the projection node from the Multilister to the stencil's color channel in the Attributes Editor. This will connect the projection map's placement tools right to the stencil for easier compositing. Now double-click the layered texture node and middle-mouse-drag the stencil node onto the texture map area in the layered node's attributes. Click the green X check box to delete the default layer.
3. Create another file texture and load in an image, such as grime.tga, to use as a base layer. Drag it in to the layered texture node. Make sure that the bottom texture is all the way to the left, and that the rightmost image is on top. Now load in an image with an alpha channel to use as the stencil's projection source image—for instance, the amethystAlpha1.tga image. In the Projection Node settings, click the Fit to Bounding Box option, and then use the place3Dtexture node to fine-tune the placement.
4. Apply the layered texture to the color channel of the Blinn shader. Then apply the shader to a NURBS plane and add a point light above it so that you can test-render it. If everything went as planned, you should have part of the amethyst picture obscuring the grime image beneath. Try adding a second stencil overlaying the first to test the technique. Try using a separate image or built-in 2D/3D texture as the mask source. Also try the HSV keying options. Use the stencil node to enhance the layered texturing process.

7.10 Using Photoshop to Create Textures for Organic NURBS Characters

In this section you'll learn to use Photoshop to create textures for organic NURBS characters, starting with orthographic profiles.

7.10.1 Preparing Orthographic Profiles and Painting Texture Maps

Here is a brief summary of the recommended steps involved in creating base-layer texture maps.

1. Import your character profiles and then scale the images to your desired output resolution.
2. Layer your profiles with selection masks, cutting the image into the appropriate selection regions to make working on the texture as precise and streamlined as possible.
3. Build yourself a custom brush library to enhance the realism of any tool that uses the brushes.

4. Use half of a symmetrical profile, such as the top and bottom or the front and back, to save time. Then add any asymmetrical details later, after the major symmetrical work has been accomplished.
5. Output the image with a separate reverse gray-scale alpha channel if you plan to use transparency mapping; otherwise, include the non-reversed alpha as part of the image if you plan to use the stencil-type texture node.
6. Test the map in Maya on your character using the projection mapping node loaded into the color attribute of Surface Material. Test-render to check how the final resolution version looks on your object.
7. Edit the profile images until the map looks good on your character, going back to step 6 until the process is complete.
8. After verifying that the color map is finished, create bump, diffuse, and specular maps, and so on, based on the color texture after initial mapping is corrected. This is a good time to start thinking about the mask channel either loaded into the transparency attribute as a reversed grayscale or as a normal alpha used as a mask for a stencil node. Transparency is only needed for the layered texture or layered shader when compositing textures together on the same object, but it is generally not as necessary for the base layer.

7.10.2 Creating Custom Brush Libraries

Creating custom brushes is an essential part of painting realistic and highly detailed texture maps. Brushes are created at whatever pixel resolution you provide as a selection. If you create a brush that is small, when used on a large resolution image it will probably be too small. So create brushes for the resolution you need by evaluating how large the target resolution's detail needs to be.

1. In a separate Photoshop document, make a pattern that you want to use as a brush. With the Selection Marquee tool selected and set to Oval shape with the feather amount set somewhere between 4 and 30 pixels, hold down the Ctrl and Shift keys and click in the center of the image you want to use for a brush. Click and drag outward. The marquee should be constrained to a circular shape and should be expanding from the center outward.
2. Go to the brush palette options and choose Define Brush. Test the brush on a new layer of your texture map file to verify that the brush looks good at the resolution in which you are working. Check the brush's spacing options by double-clicking the brush you just made in the Brushes palette. Repeat these steps multiple times to create a vast array of brushes for your convenience.

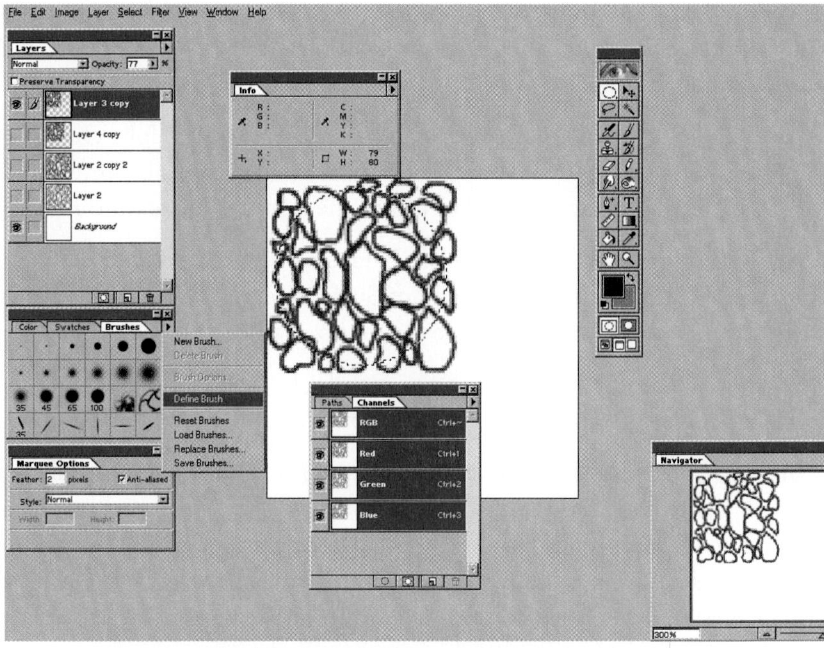

Figure 7.99
Make a feather selection around an interesting bit of gray-scale grime (color works too, but it is less predictable because it paints a colored stroke).

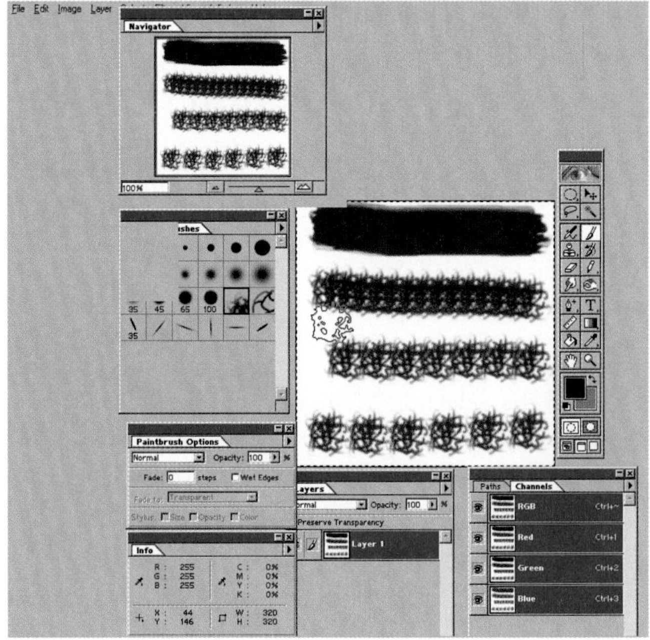

Figure 7.100
Test your new brush on a layer of your texture map to check its resolution and spacing parameters.

Using Alex Lindsay's Surface of Reality texture CD-ROM, you can make literally limitless grime and filth brushes that will help you for every brush-based action imaginable. Your brush library will help when using the Rubber Stamp Cloning tool, when painting pattern maps for scales, and for any other process involving realistic-looking results. Remember, nature is full of imperfection and schmutz, and it's our job as artists to add those living details to an otherwise sterile CG world.

7.10.3 Making Source Imagery Tile-Safe

Using the Offset filter in Photoshop, you can easily check all the edges for inconsistent tiling at once in the center of the screen. This gives you convenient access to correcting the seams with the Clone tool to add detail that will cover any visible edges that make tiling look obvious and unnatural.

1. It is very important to work with the file size of a resolution that is divisible by two. This allows the rendering software to make easier, more accurate decisions when scaling your texture for any reason. In Photoshop an even number will make it easier to apply the offset filter accurately. So check your pixel size.
2. Next, go to the Filters menu and choose Other, Offset. When the option box opens, you can interactively adjust the numbers and vary preferences before continuing. Input a number equal to exactly half of each height and width pixel value. Choose the Wrap Around option and click on the Preview check box. You will see all the edges backed up next to each other. Click OK in the Offset dialog box.
3. Now use the Rubber Stamp tool to clone image details over the edges. When the seams have been removed, use the Offset tool to put the picture back to its original appearance. Use the same Offset preferences that were used previously. You may have to do a little bit of additional fixing to the re-Offset image with the clone tool.

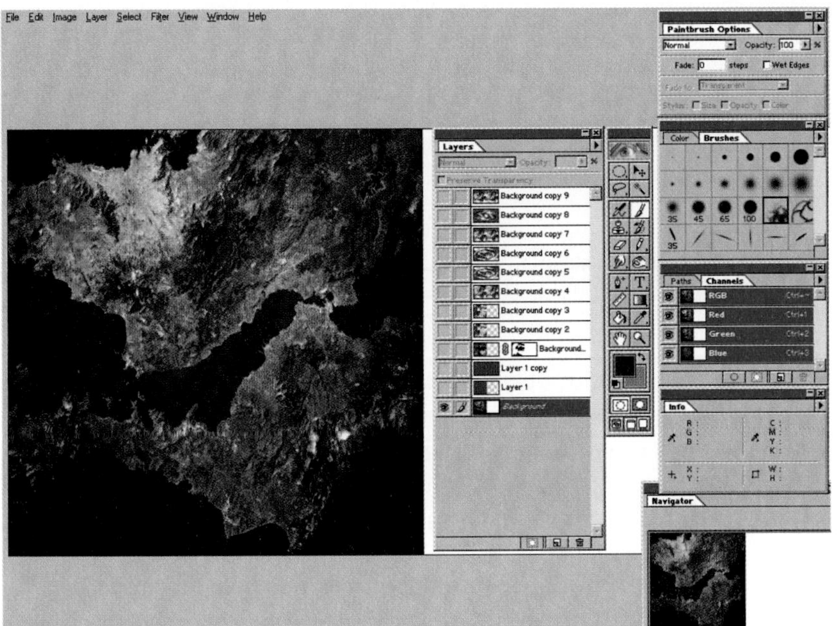

Figure 7.101
Even pixel-size imagery is a good idea when working with computer graphics.

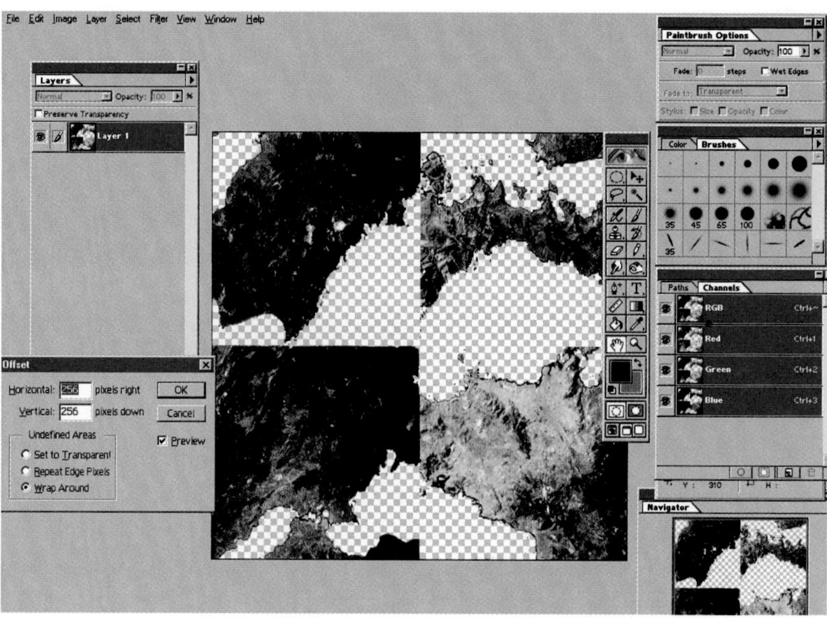

Figure 7.102
Offset half the pixel size of your image on both axes with the Wrap option On to tile your image's edges to the center.

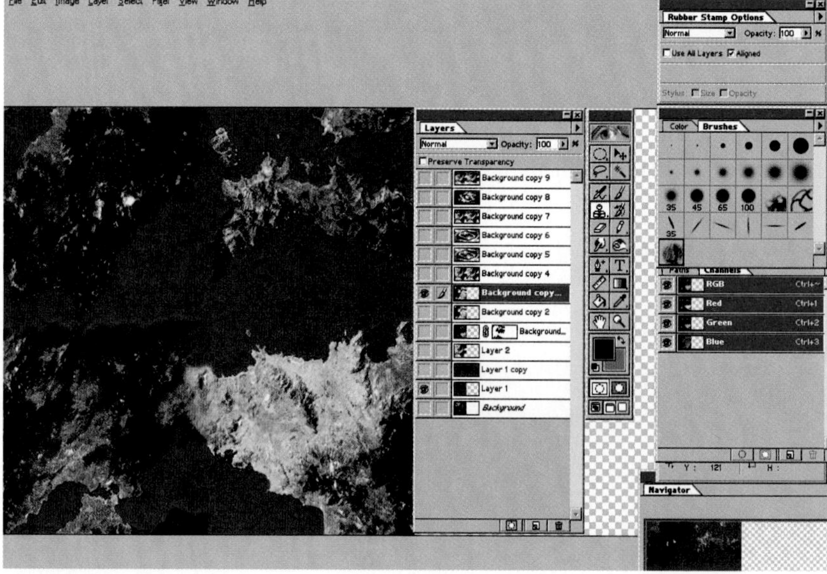

Figure 7.103
Clone the edges away by placing imagery from nearby to cover the obvious edges.

Your picture will now tile without obvious seams. Does this mean that no one will be able to tell that you are using a tiled texture? No, but it will be a lot better than if your image had obvious sharp, straight-edged seams between every tile.

7.10.4 Removing Polar Pinching

Way back in the old days, when Photoshop was still being dreamed up, one of the Knoll brothers wrote a plug-in that would pre-distort an image for the spherical pinch that occurred when images were wrapped around three-dimensional spheres. This distortion filter is called Polar Coordinates, and it has two distinct conversion modes.

1. If you do plan to texture-map a spherical object, there are some basic ground rules to remember to make things go smoothly. First, when wrapping an image around a sphere, the resolution must be twice as wide as it is tall for an even 2:1 ratio, such as 1024×512. After making this texture pinch correctly, it will no longer work if the image is tiled in Maya with the Repeat option in the Place2Dtexture for the file-in node.

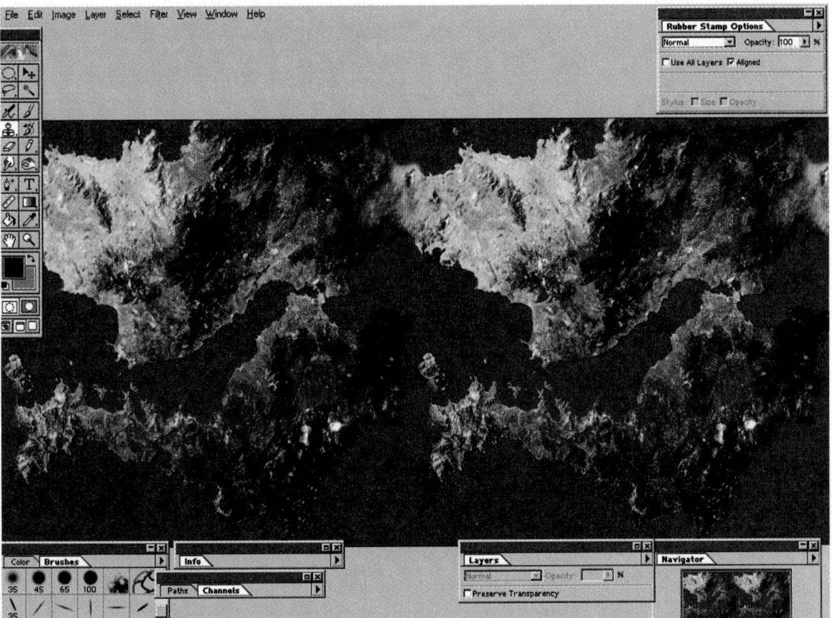

Figure 7.104
Starting with source imagery that is scaled to the proper ratio is important to give your texture map a solid foundation.

2. Next, load the picture into Photoshop and duplicate the background layer, so you can reference the original while you work. Name the new layer RectToPolar. Now go to Filters, Distort, Polar Coordinates. In the dialog box choose Rectangular to Polar conversion, and then click OK. Your image should be thoroughly distorted by this point. The left and right edge have been wrapped around to meet each other, and the resulting image inside has been distorted to appear as it would if it had been mapped to a sphere.

3. Now choose the Circular Selection Marquee and dial in some feather pixels in the Marquee Options box. The feather is set so that when you copy and paste pixels from the undistorted layer beneath this one, the edges will be smooth and natural. So don't use too high or too low a value for the feather, as the pixel size needs to be bigger in relation to the resolution. Holding down the Alt key (the Option key on the Mac) enables you to start the selection from the center instead of from a corner. Click the centermost spot of

distortion, the *pole*, and drag outward to select over the region of maximum pinching. Now, with the selection still highlighted, hide the RectToPolar layer and select the background from the layer palette. Go to the Edit menu and choose Copy. This will copy an undistorted portion to the clipboard. Now click the New Layer button and, with the oval selection still active, click Edit, Paste.

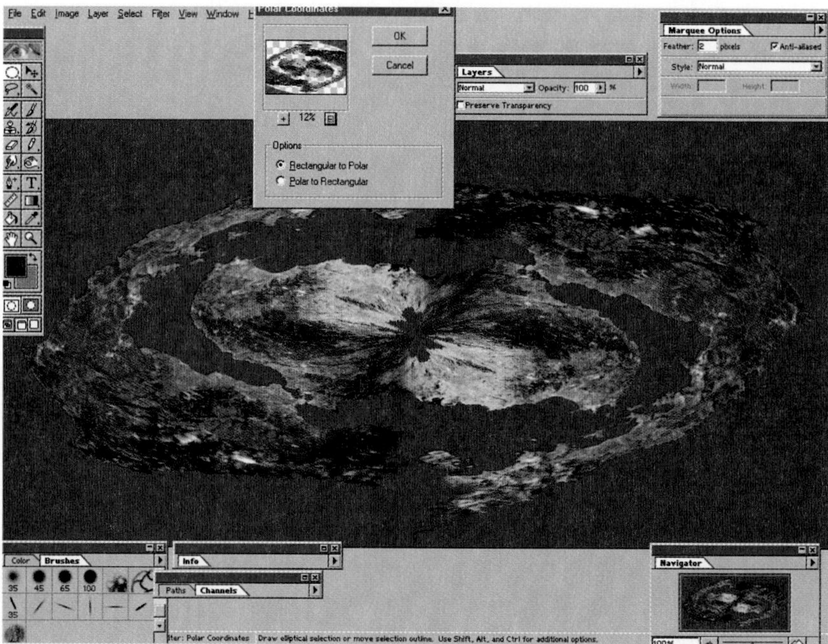

Figure 7.105
Apply the Polar Coordinates filter to your duplicate layer. You will need to have an undistorted original to reference later.

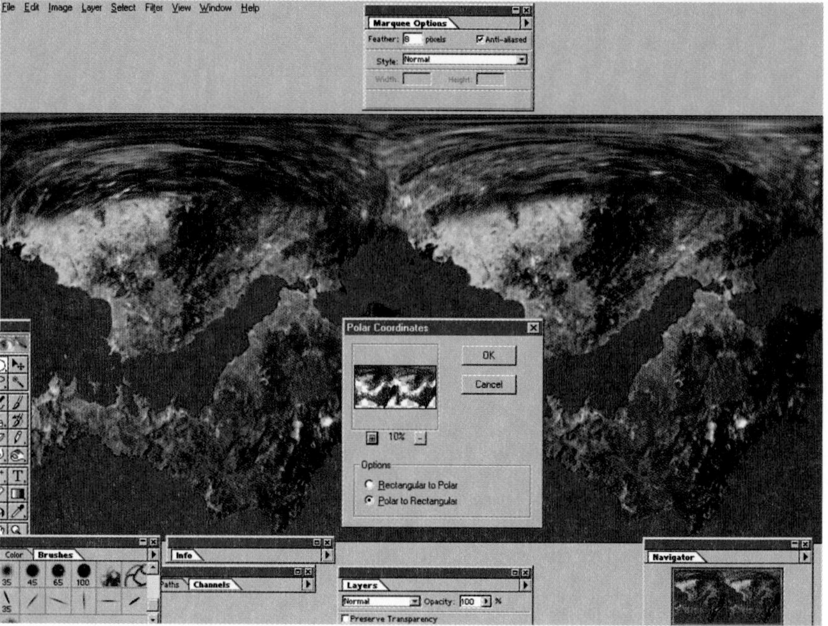

Figure 7.106
The copy of the unstretched oval region will be used to replace the stretched area in the center of the image.

4. Duplicate and then hide all but the new RectToPolar layer and the new oval layer. In the Layer Palette options choose Merge Visible. Now name this newly merged layer PolarFix and duplicate it. Make sure that there is nothing selected. With the duplicate, go to Filters, Distort, Polar Coordinates and this time choose the Polar to Rectangular option. Name this newly undistorted image **`PolarToRect`**, and then duplicate it. Use Edit, Transform, Flip Vertical to flip the duplicate. Name it **`RectToPolarFlip`**.

5. Apply the Polar Coordinates filter to the RectToPolarFlip layer with the Rectangle to Polar preference and then repeat the process of selecting an oval from the background and pasting it over the pinched region. Clone any inconsistency so that the edges blend well. Duplicate and name the appropriate layers, merging them as we did before, and then apply the Polar Coordinates again to return the image from Polar to Rectangular. A last vertical flip from the Edit, Transform menu and the image should be ready for importing to Maya.

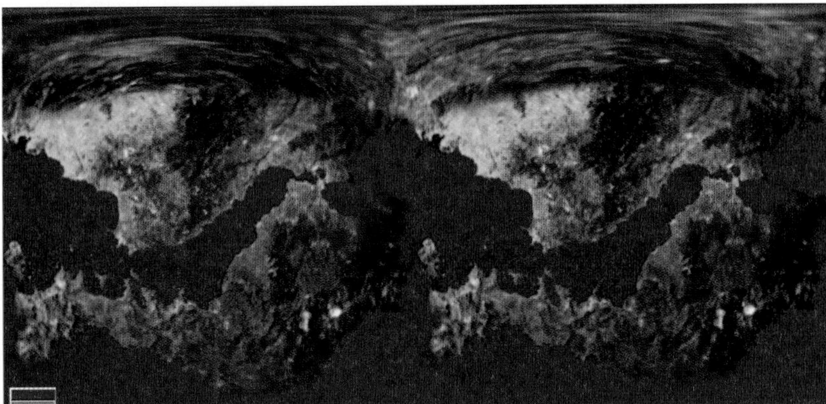

Figure 7.107
Undistorting the image will convert most of the image back to normal. A small region near the top will look stretched, but this is all right.

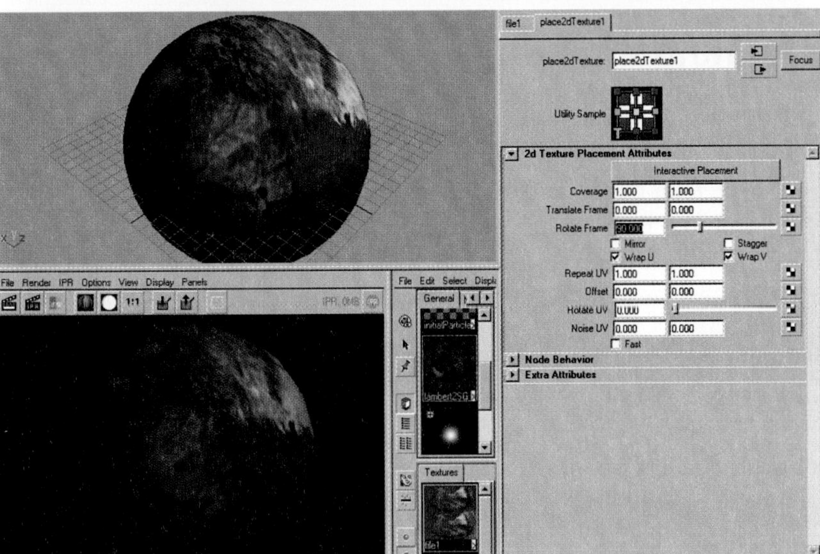

Figure 7.108
Flipping the image and duplicating the early steps will ensure that both the top and bottom poles are corrected.

In Maya you must rotate the image 90 degrees in the Place2Dtexture placement node for the image to map correctly (you could do this in Photoshop as well). Remember that this image has been specifically prepared to correct the pinch at the top and bottom of spherical-type objects. This image will not tile properly, but when it is used as directed, it should be seamless. If it is not seamless, use the included tiling technique to tile the X axis (the width) but not the Y axis (the height), as the polar coordinates deal with that aspect. This technique works well for spherical reflection maps and planet textures, as well as for correcting the pinching that occurs when texture mapping head models.

7.11 Texture Map Placement and Planning Methods

In the following sections, you'll learn to place your texture maps and create a scenario for mapping them.

7.11.1 The UV Methods

Using a special grid that is well marked on every individual grid section is a way to use the automatic UV mapping to tell where your texture map will show up on a model's surface. Then you paint or place your texture in Photoshop, aligning the details to match the grid area that you determined is mapped to the appropriate area.

This would be a frustrating way to texture a face, but it would greatly aid placement of large details on sets and backgrounds. These test grids also let you visualize the distortion regions across any model that are caused by dense/sparse regions of isoparms.

7.11.2 Planning a Texture-Mapping Scenario

Planning a texture scenario is an important part of the texture-mapping process, as it forces you to plan an overview of your strategy. Be prepared to go back to the drawing board more than once when developing this strategy, giving yourself the room to experiment and make mistakes during the early process. Remember, it takes longer than you may think to create a complete and seamless texture map scenario for your character.

1. Identify the possible trouble spots after applying the base foundation layer. Trouble spots are anywhere that you can see a visible stretch on the material's surface. These often occur where the edge of the projection map reaches the edge of the geometry and it starts duplicating the pixels to continue mapping

1A	1B	1C	1D	1E	1F	1G	1H
2A	2B	2C	2D	2E	2F	2G	2H
3A	3B	3C	3D	3E	3F	3G	3H
4A	4B	4C	4D	4E	4F	4G	4H
5A	5B	5C	5D	5E	5F	5G	5H
6A	6B	6C	6D	6E	6F	6G	6H
7A	7B	7C	7D	7E	7F	7G	7H
8A	8B	8C	8D	8E	8F	8G	8 H

Figure 7.109
Use a test grid to visualize the location of texture regions and to see distortion regions on geometry.

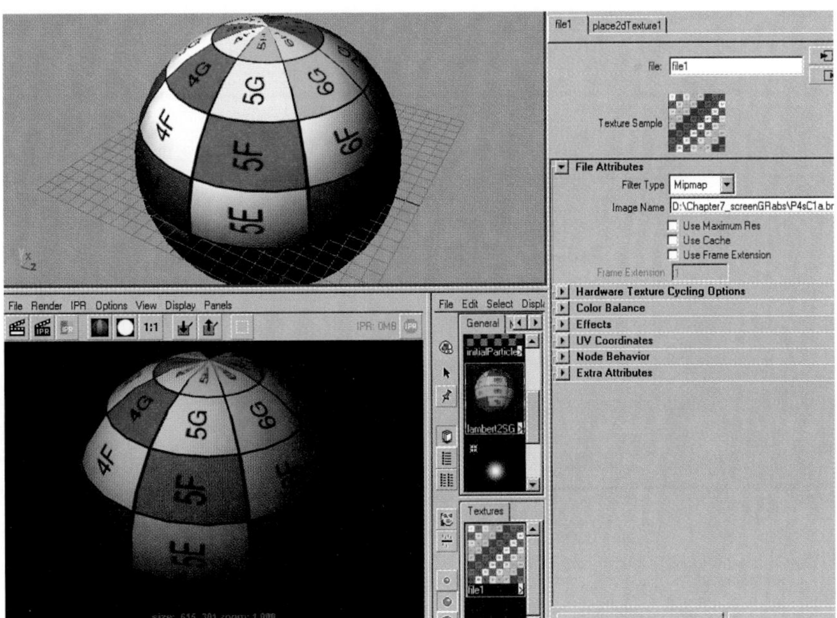

the other side of the model. This smearing can be covered or removed. If you choose to remove the smearing by matting out the stretched region with a transparency map, you will still need to put some sort of surface in the empty spot.

Tip

Whenever you use layers of shaders or textures, you must be sure that the default color in the color balance section of both the file node and the projection node is set to solid black. Otherwise, the image will add a gray film over any layer beneath it.

2. Create custom texture maps that will cover any problem spots. When making layered maps, you must turn any wrapping options off in the Place2Dtexture node or else the image will continue to tile in all directions when the edge of your texture has been reached.
3. When designing cover-up texture maps, add extra resolution to parts that will be deforming to ensure that when they are bent to their new shape, the textures will have extra pixels to help keep the stretching regions sharp.

7.12 Conclusion

These tutorials are by no means a complete solution to all your texture-mapping needs, but if you follow them thoroughly, they contain the essential building blocks to creating high-resolution texture maps that precisely map and deform with your character properly.

CHAPTER 8

Digital Lighting in Maya

by Kevin Cain

The environments you build for your animated characters are ultimately as important as the animation of the characters. Lighting, a major part of your digital environments, is an especially crucial element: It can either make or break all your hard work as an animator. Lighting has many uses. It can draw the eye, provide the right mood, elicit an emotion, or indicate the time of day or the season of year. It's crucial that your animated characters look as if they belong in your scenes, and lighting will help you do just that. Although approaches to digital lighting could easily fill a book, this chapter will provide an overview of the lighting process for interested animators who want to gain more control over their scenes.

8.1 Thinking Through Light

Maya classes lights by type: spot, directional, point, and ambient. After you understand the technical distinctions between these, you can further distinguish lights by their use and build a lighting design with this knowledge.

Before you light your digital shots, however—a word about looking at light. Production houses working on CG effects go to great lengths to make sure that the colors they see on their monitors match the colors that will be seen on the final film or video tape. Calibrated monitors, color loops, controlled labs, and the constant attention of trained "color timers" are all needed to pull this off. If you don't have all that, it pays to at least make sure that you view your lighting on the same monitor and with the same settings at all times. Getting to know your monitor's display characteristics, or profile, is a good start. Also, set your desktop background to pure black, and turn out the lights in the room when you look at your digital lighting. Any other colors on screen or light sources in the room will affect how you see your digital lighting. Finally, take your final media format into account by running test outputs. If your final output will be to video, print frames to tape frequently and examine them for differences in gamma and color. Similarly, if you are outputting to film, printing your scene to film and projecting the result is the only true test of the lighting in your scene.

8.1.1 Maya's Raw Materials for Lighting

The lighting design process starts on the conceptual level. What feelings does the director want the audience to experience? Which characters are most important in frame? What visual style is enforced? After answering these conceptual questions, the designer then works out her lighting ideas on a technical level. Creating a design involves translating lighting ideas into the different types of lights available in Maya. This section introduces these basic lighting types. After establishing how to use individual lights, we turn to their use as building blocks in a full lighting design.

8.1.1.1 Spot Lights

Spot lights emanate from the point of a cone; the size of the cone is determined by the cone angle attribute of the light. All spot lights have a *hot spot* in this angle cone, defined as the region where the light is at its most intense. With a dropoff value of zero (as shown in Figure 8.1), the entire angle cone is seen at the same intensity. Increasing the dropoff attribute progressively darkens the edges of the angle cone, decreasing the light's hot spot.

Outside of the angle cone, you can set a separate *penumbra* cone. Here, the light falls off in intensity until its level reaches zero at the edges of the beam. Together, the angle and penumbra cones make up a light's entire output cone, or *influence*.

Note

A good working size for the angle cone is 30 degrees; 8 degrees is a good starting point for the penumbra. An inverse relationship of the angle and penumbra cone sizes determines the sharpness, or coherence, of the light. When lighting designers talk about the coherence of light, or how organized a beam is, they are describing how hard or soft the light appears. A spot with zero penumbra and no dropoff has the distinctive sharp edge of a hard, or highly coherent, light. Alternatively, a spot with a 30-degree penumbra and/or a high dropoff value looks very soft and is called less coherent. Maya spots can also decay with distance. In the real world, light falls off at the square of its distance; for the equivalent in Maya, set the Decay Rate to quadratic falloff. Linear Decay Rate is another option, steadily diminishing the light output. As a default, Decay Rate is set to none.

Figure 8.1
A Maya spot light with no dropoff or penumbra.

8.1.1.2 Directional Lights

Most light beams spread outward as they are broadcast; this makes it possible to cast large shadows from small objects. Parallel light is a case when light doesn't radiate from its source. In Maya, parallel light sources are called directional lights. Instead of diverging from the source, *directional* lights broadcast their light without spread. When a directional light beam hits an object, the shadow it casts reflects the true size of the shadowing object.

Directional sources yield highly coherent, parallel light, such as the sun on a clear day. (The sun's light rays aren't really parallel, but the light source is so far from us that the angle of divergence is practically zero.) Not surprisingly, directional lights are used mainly to simulate sunlight. Directional lights have color, intensity, and direction but no obvious source in the scene. Directional lights do not decay with distance and are sometimes called *infinite* lights because their influence is not bounded by a cone.

In Maya, changing the translation of a directional light doesn't change how it plays in a scene: The vector given by the rotation of a directional light is the only value that matters. This means that a directional light can cast a shadow on an object placed "behind" it in a scene. Because translation is not relevant for directional lights, you can place all your scene's directional lights together near the origin for convenience.

Figure 8.2
A directional light in Maya, commonly used to simulate sunlight.

8.1.1.3 Point Lights

Point lights are like incandescent light bulbs; they throw off light in all directions. As the name implies, point lights radiate from a central coordinate in 3D space, illuminating any objects within their sphere-shaped influence zones. By casting light in a localized part of the scene, point lights can establish a sense of depth in the scene. Point lights are also good for streetlights, architectural accents such as sconces or torches, and otherworldly apparitions that take up a volume of space with light.

Unlike in the real world, lighting in Maya can have negative values. Negative lights subtract light from the scene, and are the best way to take spill off of an object, suck up light under a table, or plunge a part of your scene into darkness. Use several negative point lights with intensity values from –.2 to –1 so that the black holes created won't be too obvious.

8.1.1.4 Ambient Lights

Ambient lights are similar to point lights except that only a portion of their illumination emanates from the point. The remainder of the illumination comes from all directions and lights everything uniformly. In Maya, ambient sources add light to a scene from all directions, effectively raising the darkest value in your scene and squeezing the delta between your darkest and lightest values. As a default, keep the whole dynamic range in play by turning ambient lights off. Some designers find ambient lights useful as a soft fill light for a key/fill setup.

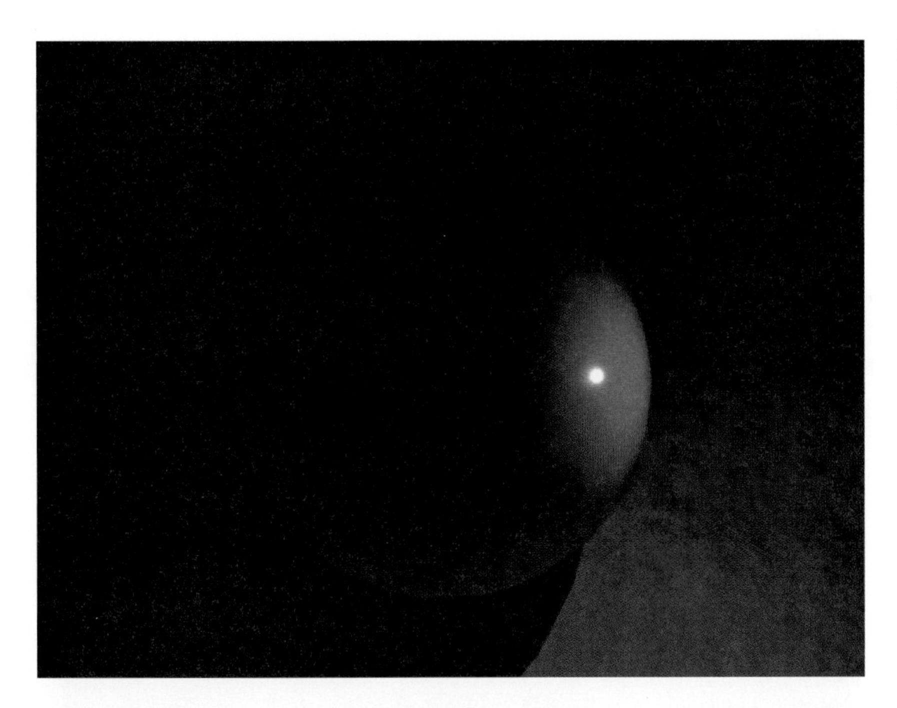

Figure 8.3
A Maya point light, used for local illumination.

Figure 8.4
Maya ambient light.

8.2 Lighting Roles

Understanding the technical properties of lighting instruments is the first step in getting ready to design lighting: The next step is knowing the roles that a light can play in a scene. Like coherence and intensity, the angle of the light is important to understand. The angle of a light is determined from a specific camera viewpoint in the Maya scene file. Front, side, and back light, the most commonly used angles, are explained in the following sections.

8.2.1 Front Light

Front light is the default, both artistically and technically. Place a directional light behind your render camera: This is a good example of pure front light—light that's directly in line with your view plane. Although front light offers good visibility, it also flattens out details (think of high school yearbook photos with their soft, frontal lighting). Very diffused, less coherent light will act like front light even if it's not: Supermarket lighting, for example, is usually broadcast from the ceiling but has the same uniform visibility as front light.

Figure 8.5
Front light is used to provide general visibility.

8.2.2 Side Light

Side light accentuates the 3D detail of a subject and, if used by itself, gives a good sculptural sense of the subject's form. As side light rakes across a surface, it reveals a lot about the shape and texture of its subject. The dramatic quality of side light has been used for hundreds of years by painters who exploited its ability to draw a 3D feeling out of a 2D surface. Compositionally, side light is used to establish image depth with planes of light. A foreground plane of light includes the objects closest to the camera; mid- and background planes include other objects farther from the camera. Each of these planes is separately lighted, creating an alternating pattern of light and dark receding from the camera. The implied 3D space between these planes of light enforces a feeling of depth in a scene.

Figure 8.6
Side light reveals the sculptural form of an object.

8.2.3 Back Light

Back light boosts definition in an image by creating a lighted edge around the subject. Also called *rim* lights, *edge* lights, or *liners*, back lights pop their subjects out from the rest of the scene, providing three-dimensional separation and a color accent. Designers mainly use back light as a way to separate foreground objects from their background; but back light is also used to give a color cast to a whole scene. Because the angle of incidence for a light is equal to the angle of reflection, back light is an effective way to push a color onto a scene and into the render camera. For instance, a back light will register a floor's color more vibrantly than front light will for the same scene. In general, back light is more about drama and less about the color relationships in a shot.

The previous three lighting roles are based on the angle of the light. The following lighting roles are described by their uses.

Figure 8.7
Back, or rim, light is used to separate an object from its background.

8.2.4 Fill Light

Fill light is a soft wash of light that seemingly comes from all directions. More easily achieved in the computer than in live-action lighting, fill or *ambient* light is used to give a color cast to the scene as a whole. Also, using fill light ensures that no part of your rendered frame will have completely black pixels, which provides greater flexibility in the compositing stage.

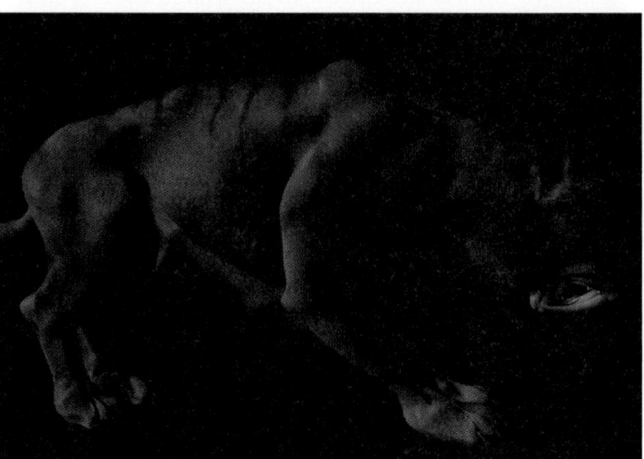

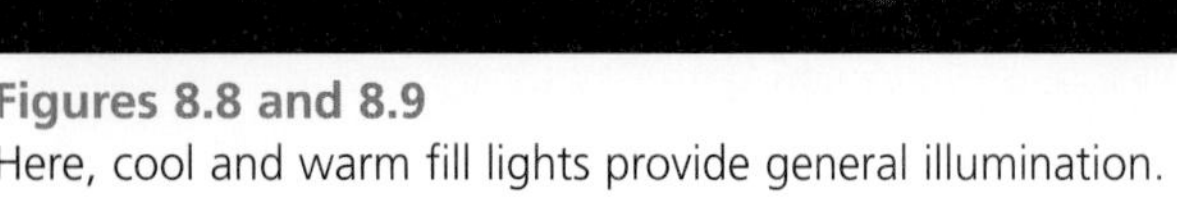

Figures 8.8 and 8.9
Here, cool and warm fill lights provide general illumination.

8.2.5 Kickers

Kickers are added as needed to boost an object's visibility or add definition to the frame. *Kickers* are a part of a family of lights that designers may refer to as *liners*, *shinbusters*, *skirt* lights, or *glow* lights. The basic position for a digital kicker is in a three-quarters-back position, almost in line with the camera. The resulting light can add an edge highlight to an object or provide an accent to bring an area up in value: Either way, the idea is to bring out part of the composition by accenting specific geometry. Kickers are also called *rovers* because they may move from shot to shot as needed. They are kept at lower intensities than the rest of the scene lighting so that their comings and goings don't upset the lighting continuity from shot to shot.

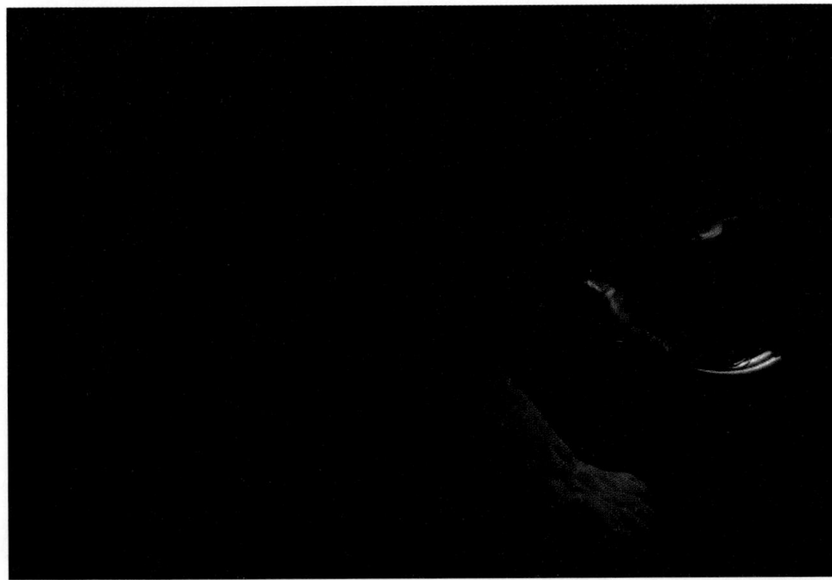

Figure 8.10
A line of light, or kicker, is used to accent parts of objects.

8.2.6 Specials

Specials are another family of lights. A *special* is any light that has a specific function: Examples include an eye light placed to accent a character's face, a pool of light that is used for a dramatic scene, or a light placed outside a window to simulate sunlight streaming into an interior.

Specials are useful for drawing the eye to specific parts of the image, or focus zones. Strong compositions with light offer clear focus zones to attract the eye—the primary focus establishes the important character/object in the frame based on the story point being made in the shot. The secondary zone highlights another important character/object and provides visual balance in the frame. Although multiple focus zones are possible, having too many quickly becomes confusing for the viewer. Test your focus zones by squinting at the image—if the primary and secondary zones are instantly recognizable, your definitions are clear.

8.2.7 Bounce Lights

Bounce lights are used to simulate the reflected light seen when a light source bounces off a surface. In the real world, when light reflects off of a surface, it ends up lighting other surfaces. Although Maya doesn't support this real-world global illumination, the effect of these reflected lights can be simulated with well-placed bounce lights. If you are lighting bright floors or walls, bounce lights should be placed to simulate the reflected light we'd expect to see in reality. When placing bounce lights, remember that the angle of incidence for a light is equal to its angle of reflection. By definition, bounce lights are subtle. Their color is based on the color of the shader from which they simulate bounce (for example, light striking a red wall needs a red bounce light).

Figure 8.11
Like kickers, specials are used to draw the eye to a specific part of the frame.

Figure 8.12
Bounce lights simulate the reflected lighting we are used to seeing in the real world.

8.3 Approaches to Lighting

The next step after understanding the roles of individual lights is to combine many lights into a full design. The following three approaches have been useful to many lighting designers and will give you a place to start. Don't think that only preordained solutions exist, though; experimentation is crucial to good lighting.

8.3.1 Motivated Lighting

Some designers start a design by simulating the lights they'd actually see if they were dealing with a real scene. Consider a woman seated in front of a fireplace at night; a standing lamp at her side illuminates a book she's reading. In this case you would have point sources emitted from the lamp, light from the fire itself that will play into the rest of the scene, and light outside to mimic moonlight. When lights are designed to account for scene light from a real, visible source, they're called *motivated* lights. That is, the lights are motivated by an object in the scene that would emit light.

Figure 8.13
In this example of motivated lighting, Maya lights replicate the angle and color of the setting sun.
Credit: Sangsoon Park.

In a motivated lighting design, *practicles* are lights that mimic the source of a light. In our evening scene we'd have a practicle shader glow on the light bulb to make it believable, even though it's not actually contributing light to the scene, and a practicle particle effect to simulate the fire. Again, bounce light has to be simulated. Because in reality the light from our floor lamp would bounce off the book and uplight the seated woman's face, we have to create a bounce light with just enough intensity to read in the scene.

8.3.2 Key and Fill Lighting

The practice of key and fill light has a large following among digital lighters. With the key/fill approach, the *key* light is used as the primary source for a scene, and the *fill* light is used to fill in the shadows left by the key light. The idea is most useful for relatively simple lighting schemes. If you're lighting an outdoor shot, the sun is your key light and a color-corrected fill light or two works as your fill. The result is a scene predominantly lit by sunlight, but with a soft accent light to soften the shadows, balance the scene, and catch detail.

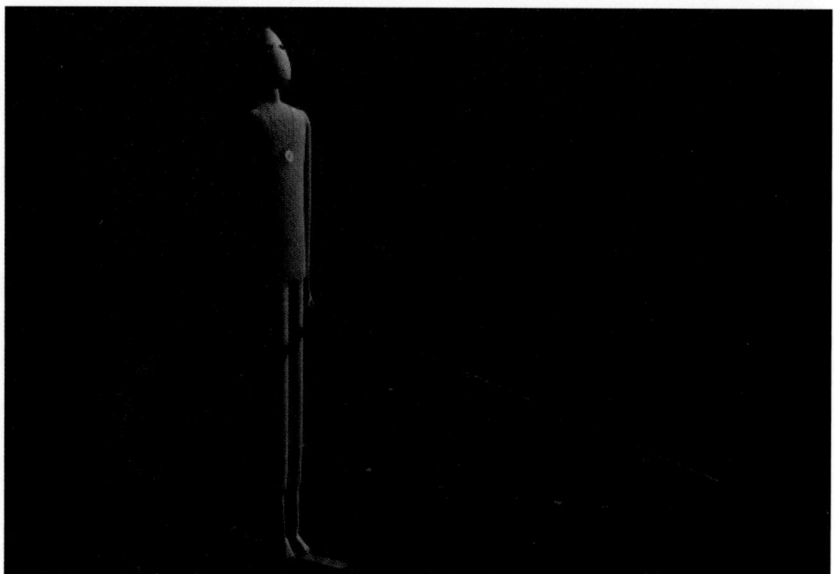

Figure 8.14
A very simple key/fill lighting setup.
Credit: Hong-Bea Shim

Note

When creating digital key and fill setups, the key light is usually approximately 45 degrees off to the side of the subject and a bit above the camera's elevation. The fill light is then rotated between 140 and 200 degrees from the key light around the central pivot in the scene. Because key/fill lighting is associated with the low- or no-color lighting plots of television, it's typical that your key and fill lights will be the same color. Warm key/cool fill complements are also common.

8.3.3 Three-Point Lighting

Three-point lighting is a more stylized approach to lighting a scene. The *three-point* approach, also called *McCandless* lighting as a tribute to its originator, generally consists of two front lights and a single back light. In the following figure, the front lights are placed 45 degrees up and over the subject. These angles give balanced lighting with some sculptural interest when compared with straightforward front light. For variation, one of the front lights is often a warm color (rose, amber) and the other is cool (pale blue/green).

The back light component of the three-point approach, also tipped at 45 degrees or more in relation to its subject's height, is usually projected without an angle in plan. The result is that the back light accents the edge of the subject. Sometimes the back light color is neither warm nor cool, but instead a pivotal color such as lavender that complements both warm and cool ranges.

Figure 8.15
A simple three-point lighting setup.
Credit: Karl LeDoux

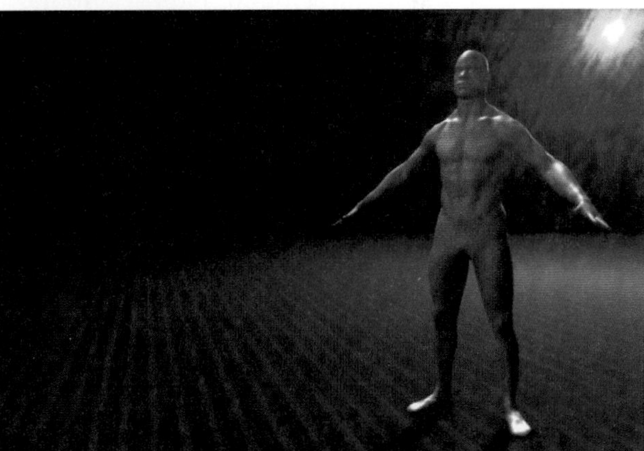

Figures 8.16 and 8.17
Two three-point lighting variations; note the use of back light in both images.
Credit: Karl LeDoux

8.4 Production Tips for Digital Lighting

There are lots of ways to create good lighting. These are guidelines only, so feel free to break these rules when you need to, and remember to keep an open eye to other lighters' work. Lots of people realize that painters and photographers understand lighting and that we in the digital world can plunder their good ideas. As T.S. Elliot said, "Steal liberally but steal from the best!"

8.4.1 Remember the Director's Visual Ideas

Usually you won't be working on your own on a film—there's a director, and you're part of a team that is trying to bring off that person's ideas. So, ask yourself: Where is the director going with your shots? It's your job to make sure that the lighting serves the mood and spirit of the film as a whole. You complement the story points that are established in the shooting script, storyboards, or story reel that everyone on the team is working from. Obeying the art direction implied by these primary documents means that the resulting film will show continuity between the story, sets, animation, and lighting. This is what you're aiming for.

8.4.2 Use Texture for Depth and Shape

On the most basic level, lighting affects the composition of a shot by telling the viewer what she can or can't see, and separating or merging foreground, middleground, and background elements. Lighting has a direct role in establishing how well a scene plays in three-dimensional space. One example is the use of planes of light to create the perception of depth in an image. Another way to create image depth is to map a bitmap or procedural texture onto a light, creating a textured light. Think of textured lighting as a tool that lets you reveal the sculptural, 3D qualities of your scene. When linked to lights, bitmap textures create such effects as the patchwork of bright and dark cast by tree leaves, or the grid-like division of space cast by the muntins of an elaborate window. Using 2D or 3D procedural textures in a light can simulate the light caustics that occur through smoke, fog, or water. Because the cone of light in these cases isn't a single open field, the areas where light fades in or out give the viewer depth cues to interpret the relative 3D depth from object to object in your scene. Humans are well made to pick up on this kind of 3D depth cue information. Textured light can help break up monolithic objects in the frame; the textured lighting field helps break up imposing forms so that they can fit better in the composition as a whole.

8.4.3 Don't Pull Focus

Unless you are working on a shot where the lighting should stand out, your lighting shouldn't be noticed by the typical first-time viewer in your audience. Saturated colors and obvious shadow lines definitely have their place, but if they

Figure 8.18
A 2D texture applied to a spot light accentuates the feeling of depth and form.

Credit: Karl LeDoux

pull focus from the story in the shot, you've called attention to yourself in the worst way possible. Strong back light washes, obvious eye/scenery accents or other little points of light, and deep colors can all call attention to the lighting for a scene in a negative way.

8.4.4 Learn from Found Light

As you start lighting a scene, stay objective about how and what you light. It takes a special effort to see a happy accident in your scene that you can incorporate into your lighting design; it's worth the trouble. Use Files, Export, Active to pull lights from a Maya scene you've finaled into one you're just starting. Seeing the unplanned light on your new scene can spark an idea.

8.4.5 Violate Shot Continuity

Shot-to-shot continuity is important in lighting—you do want your shots to feel like they're plausibly connected. However, it's common to shift rim lights and kickers on a shot-by-shot basis. It's all right for you to add an eye light in one shot and take it out in the next, or to do the same with a kicker to boost texture in one shot without tracking that change globally through the whole scene. As long as you can maintain the subjective perception of continuity, you can cheat your levels and modify lighting positions on a shot-by-shot basis. It's more important for you to support the story points than it is to be a slave to continuity.

8.4.6 Sometimes Less Really Is More

Theater lighting designs can call for hundreds of instruments, and lighting for TV or film can be equally demanding in terms of the number of lights used. In computer graphics, lights add significantly to render time, and working with hundreds of lights would crush even the fastest computer. Keep your design as conceptually simple as you can to keep down the number of lights required. If you need to have 100 separate flares of light uplighting a long corridor, render a single frame and use it to build shaders that can replace the lights themselves for future renders.

8.5 Color Perception for Digital Match Lighting

Maya was written to mimic the way light behaves in the real world. To understand how to match light digital objects against a live-action background plate, you need to understand real-world color temperature. In the following figure, the two rocks in frame right were modeled and lighted in Maya, and then composited onto a film plate background. The light angles and colors selected in Maya must match the color temperature of the live-action lights: The following notes introduce color theory to help make this match lighting easier.

In the wave model of light, color is a function of its wavelength. The wavelength is related to the *frequency*, or the number of times per second a light wave's crest passes a point of observation. Together, the wavelength multiplied by the frequency is equal to the speed of light. When our eyes see electromagnetic radiation within a certain range of wavelengths, we interpret this information as color.

Figure 8.19
The color of the digital lighting used above matches the live action sunlight captured on the film plate.

Blue waves are shorter than red wavelengths. The usual way to measure wavelengths is in *Angstrom* units, 1000 of which make 1 millimeter. (A unit that lighting designers in film use often is *nanometers*: 10 Angstroms is equal to 1 nanometer.) Considering the whole electromagnetic spectrum, it's a big range: The part of the spectrum we can see is very narrow compared to the whole range.

The electromagnetic spectrum covers a range of colors, and a range of energies. There's more energy in blue light than red because energy is directly related to the frequency of light—which is to say that wavelength goes up as the frequency goes down.

Color *temperature* is used to talk about how hot or cool a light appears to be for the live-action camera or the Maya renderer. Although a scientist would measure the wavelength for the sun at 5100 Angstroms, a photographer would instead measure the same thing a different way by talking about the sun's color temperature.

Color temperature is measured in *Kelvin* (shown below as K). Natural light sources tend toward higher temperatures. Stage and film lights can approach these cooler, higher ranges, but most regular incandescent sources have much warmer, lower temperatures. Coloring a "white" light with a red filter brings its temperature down, while coloring the same light with a blue filter raises its temperature. Following is a list of some common lights with their color temperatures in Kelvin. Use these values to match Maya lights with live-action background plates, from the light blue of 5700K sunlight to the amber of a 2400K light bulb. Color swatch books with color temperature listings are available to help convert Kelvin numbers into colors for Maya lights.

Sunlight	5700K
Florescent light	4300K
Sodium vapor street light	3600K
Slide projector lamp	3300K
1000-watt halogen lamp	3200K
Architectural halogen	3000K
60-watt incandescent bulb	2400K

8.6 Lighting, Rendering, and Compositing

Rendering is an important part of the lighting pipeline. A few suggestions follow for using the render processes available in Maya. First, lighters don't often make a lot of changes between renders because it's hard to track simultaneous changes, even for someone with experience. Instead, you're better off trying out values one light at a time. You might start with the relationship of a key light to a fill light. After locking the attributes of these first two lights, you could add rims, then bounce lights, then specials. There's no set order. Some designers start with back light every time, others with front light; but they work each discrete lighting idea out before going on.

As you progressively tune your lights, you can either use *IPR (Interactive Photorealistic Rendering)* or a compositing approach. IPR gives you the power to make real-time, incremental adjustments to your lights and textures without having to render frames repeatedly. IPR even lets you interactively render lights as you move them or change their attributes. In the IPR render window, Shift+clicking on an object will allow you to select any light playing on that object from the pop-up list. Once the light is selected, you can interactively translate and rotate the light and tune its attributes. (Also, remember that you can drag and drop textures directly from the Hypershade window.)

Another tactic that saves you from re-rendering the whole frame involves tuning lighting at the compositing stage. If you render the lighting for a scene in passes, those separate images can be composited and modified without going back to Maya to render. Any software that gives you a good, fast way to export the changes you make back into Maya will work. Examples in Composer and Photoshop are shown in the following sections, after an introduction to additive mixing.

8.6.1 Compositing with Additive Color Mixing

Additive compositing methodologies have for years been a crucial part of the production pipeline at professional studios; they've proven to be great strategies in production, especially when memory and processor time are considerations. To achieve real-time lighting changes in IPR on a modest machine, you'll have to settle for a small, lower-quality viewing window. In contrast, a compositing approach is much less memory-intensive: It lets you see a full-resolution frame and isn't slowed down by higher render globals.

Maya renders lighting additively, just as light mixes in the real world. Contrary to what you may have been taught in school, the primaries of real-world pigments such as paint are amber, magenta, and cyan, as shown in the figure on the next page. The secondaries (created by mixing the primaries) are red, green, and blue. In light, the primaries are red, green, and blue, which correspond to the secondary pigments, and the secondaries of light are amber, magenta, and cyan.

Pigments such as paint mix *subtractively*. When pigments are mixed, the new color is the intersection of the initial colors. In other words, if you mix amber and cyan you should get green; if you mix all the colors together you should get black. (Actually, because we don't have perfect pigments in the real world, mixing all three primaries yields a muddy gray.)

In light, colors mix *additively*. Mixing two lights gives the resulting light the union of their colors: When you mix all colors together, you get white light. This means that whether you render your Maya scene in lighting layers and additively composite them or render your whole scene in one pass, you'll get exactly the same result.

Maya user Carlos Pedroza has written a MEL script for this book that will let you apply additive compositing to build your lighting in Photoshop and then update your Maya lights to match. (Called `photoConv`, this script is found on the accompanying CD-ROM.)

You don't have to use Photoshop to do this kind of compositing. The advantage of using Composer, After Effects, and so on, is that you can bring in an animation sequence as easily as a still. This lets you quickly see lighting playing in a moving scene—something you can't do in Photoshop.

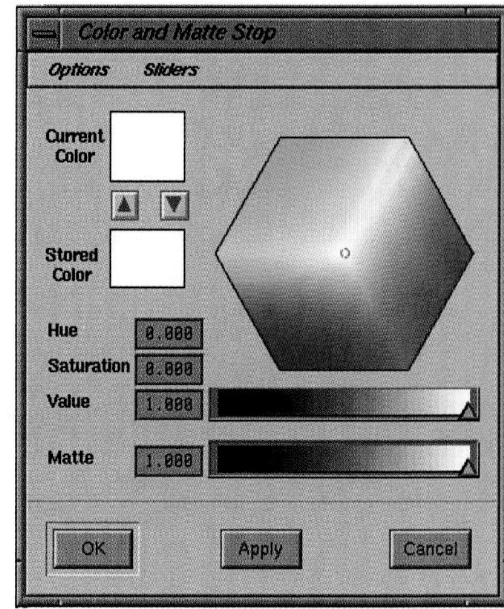

Figure 8.20
The Maya Composer color picker illustrates additive color mixing.

8.7 Fast Previewing of Maya Lighting

You can quickly preview changes to your Maya scene lighting by additively compositing multiple render passes. You can apply this process to any of the major compositing packages found today. For this exercise we present two examples: one using Alias | Wavefront's Composer and one using Adobe's Photoshop. Both aim to give you an overview of how powerful the additive compositing method can be for previewing digital lighting changes. (To try the Photoshop example, you must first install the Light Converter MEL script included on the CD-ROM.)

Exercise 8.1: Previewing Maya Lighting in Maya Composer

1. The process starts in Maya. In your scene file, Display, Hide all lights except for a single light (for example, your key, fill, or bounce light). Render this scene as a full-resolution image with just this light visible. Repeat the process with your other lights, creating a render pass for each light in your scene. If you have a system of lights that work together as a whole (such as a series of architectural accent lights), render a pass with all these lights visible rather than rendering them all individually.
2. Moving to Composer, File, import all your lighting passes into a new script. Select each imported image; add an Effect, Color Effect, Brightness event. These events will enable you to preview changes in your Maya light intensities; rename each event to provide faster access.

Exercise 8.1: continued

3. Select one of your new Brightness events (for instance, your key lighting pass) and Shift+select another Brightness event (your fill pass). Use Effect, Layer, Add to additively composite these two. Select the resulting Add event and Shift+select the next of your Brightness events. Again, use Effect, Layer, Add to composite the two. Continue the process until all your lighting pass events are incorporated.

4. Testing the top Add event in your script will yield a frame that includes all your lights. One by one, you can change intensities for individual lights and test the results without having to re-render the frame in Maya. Select, Ignore can be used to remove one or more render layers: This is useful to test what your scene looks like with fewer lights. The following figures show these two tests composited using different lights.

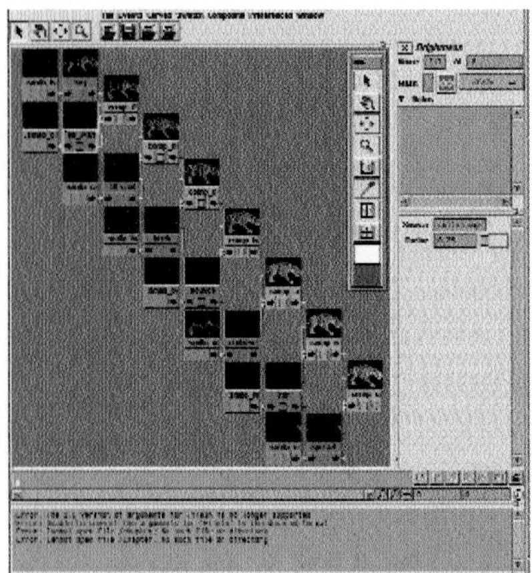

Figure 8.21
This Maya Composer FlowGraph window contains multiple lighting passes composited with Add events.

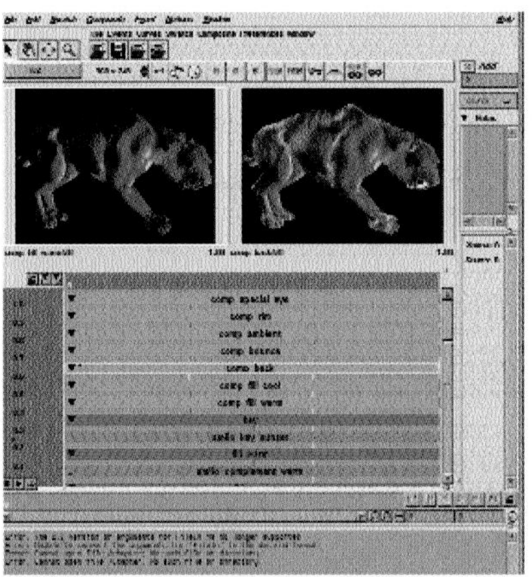

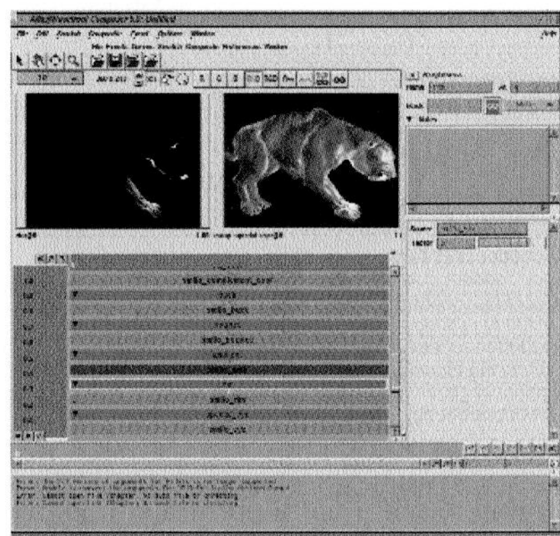

Figures 8.22 and 8.23
Using Composer to adjust lighting levels.

5. After you have adjusted your lighting intensities, note which light layers were changed and reflect these values in Maya. As an example, if a Brightness layer were moved to 2 (from a starting value of 1) in Composer, the associated light intensity value in Maya would change to 4 (from a previous value of 2). Using Brightness effects will result in a good approximation of the actual light effects in Maya: For a perfect preview, use the Photoshop technique described in Exercise 8.2.

Note

You can also use Composer to preview color changes in your lights. First, add an Effect, Color Effect, Adjust HSV event on each Brightness event already set up in your Composer script. Make changes to the colors in your scene by tweaking the Hue sliders. Then, note the + or - change in Hue for each layer and reflect these changes in your Maya light color selections, adding or subtracting according to the + or - Hue values in Composer.

Exercise 8.2: Previewing Maya Lighting in Photoshop

1. In Photoshop, import your Maya-rendered frames as separate layers using the method described in Exercise 8.1. The following figure was generated from a Maya file with three lights: red, blue, and green. Each light was rendered as a pass, and then brought into Photoshop as a separate layer. (When pasting these images as layers, make sure to use the Screen layer type.)

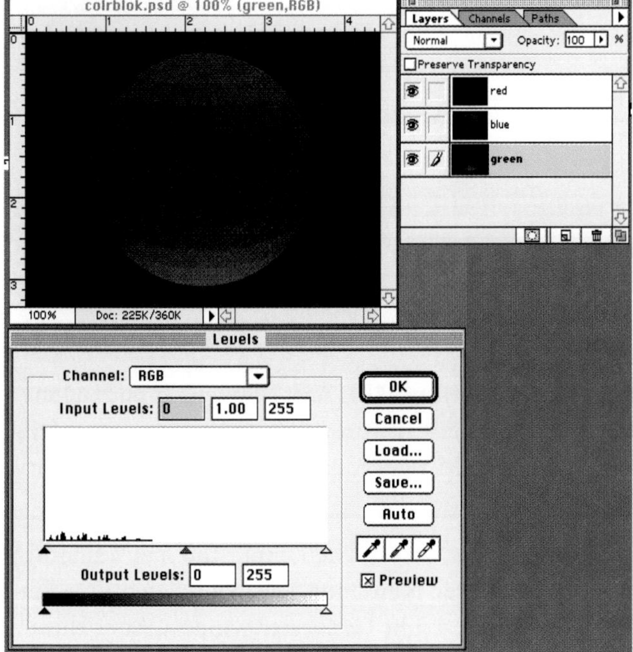

Figure 8.24
A Maya scene with three lights: red, green, and blue.

2. First, change the light intensities. Use the Adjust, Levels window to adjust the light intensity for a specific layer. Moving the white input increases the intensity, and moving the white output decreases it. Be sure to keep one of these two controls at its starting level—don't move both! In the following figure, the white input level is set to 170 (from 255).

Exercise 8.2: continued

3. Next, modify the colors for each lighting pass. Using the Adjust, Hue/Saturation window, you can preview light colors changes. In the following figure on the right, the Hue is set to 23 and Saturation is set to 60.

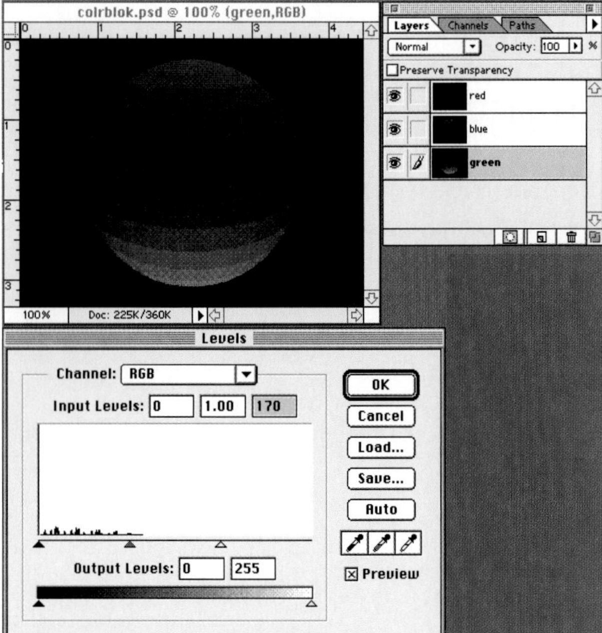

Figure 8.25
Editing lighting levels in Photoshop.

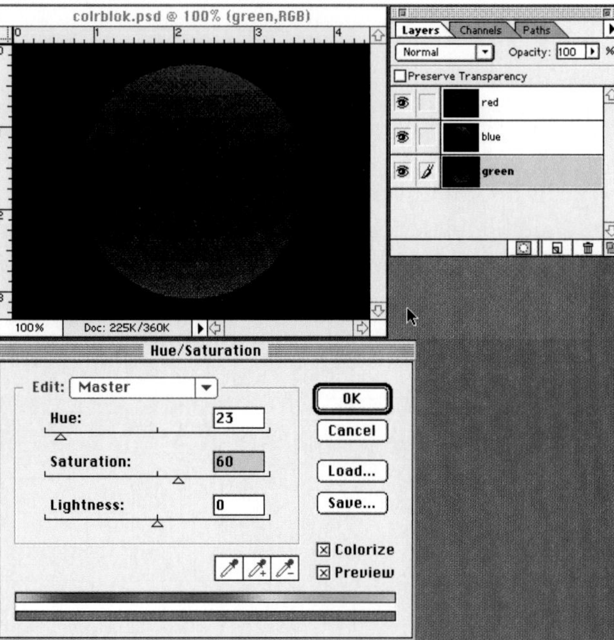

Figure 8.26
Editing lighting colors in Photoshop.

4. Back in Maya, launch the Light Converter MEL script by typing **photoConv** in the script window. Naturally, you must first install the light converter script (photoConv.MEL) from this CD-ROM into your maya/2.0/scripts directory. The light converter window will appear.

5. Select the first Maya light—in this case, green. Then, input the new values for White Input Level, White Output Level, New Hue, and New Saturation from Photoshop. Click the OK boxes in the Light Converter window. After this information is entered, the light converter will set the intensity and color of your light to match the settings you made in Photoshop.

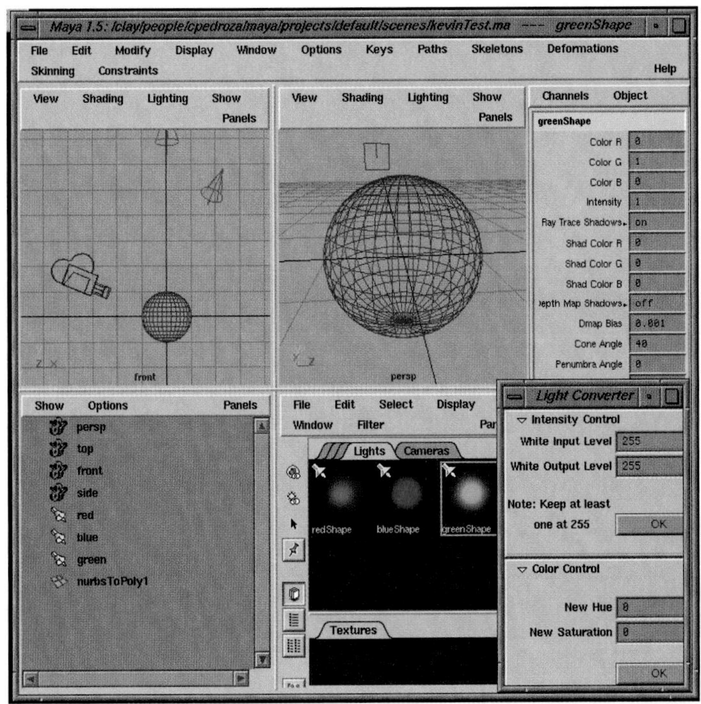

Figure 8.27
Viewing the Light Converter in Maya.

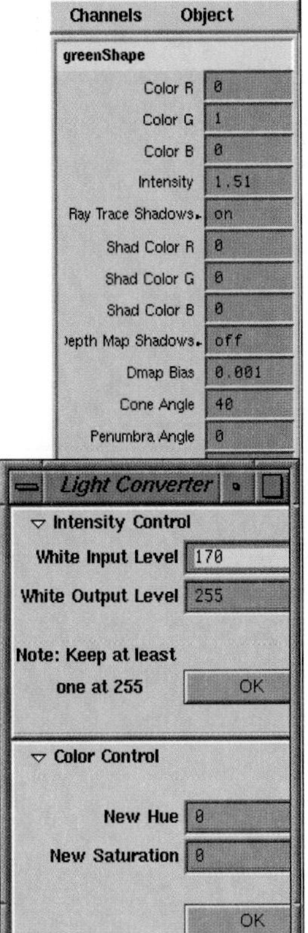

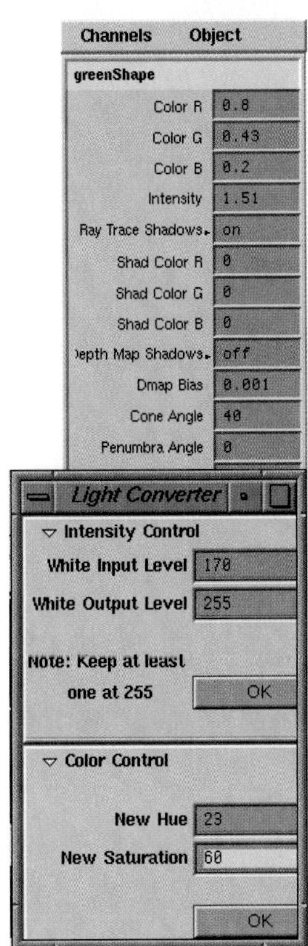

Figures 8.28 and 8.29
Using the Light Converter window, Photoshop light and color changes are applied to Maya lights.

Exercise 8.2: continued

6. Finally, Display, Show, All and render a test frame in Maya. Compare the following rendered frame to the Photoshop image shown in Figure 8.26: They will match. This proves that the lighting changes you made in Photoshop are now reflected in your Maya scene file.

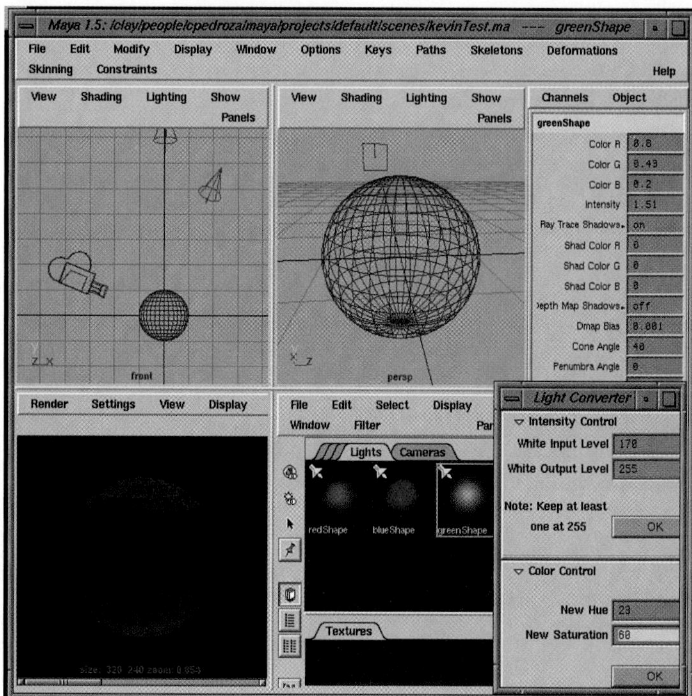

Figure 8.30
A rendered fram from Maya matches the lighting edits made in Photoshop.

Part III

Animating Characters in Maya

Setting Up Your 3D Characters for Animation

Chapter 9

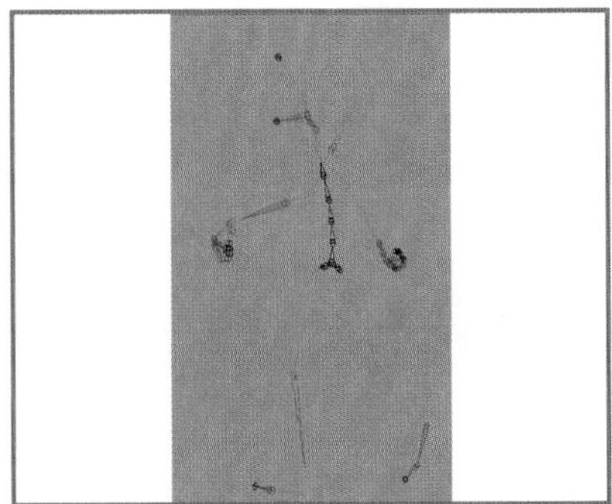

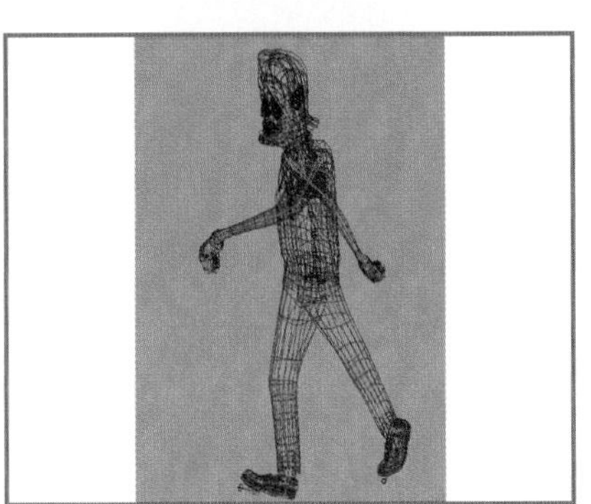

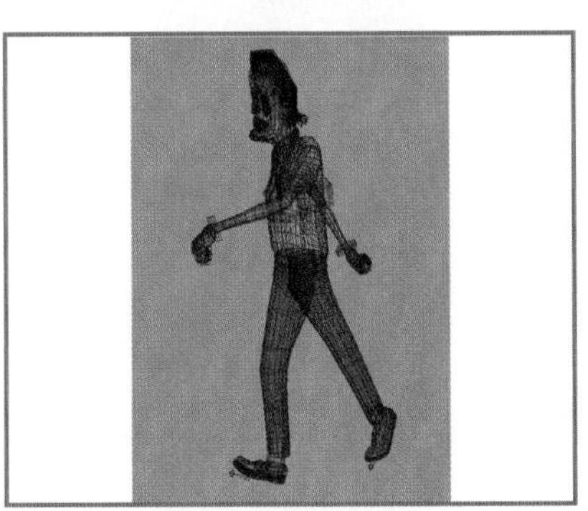

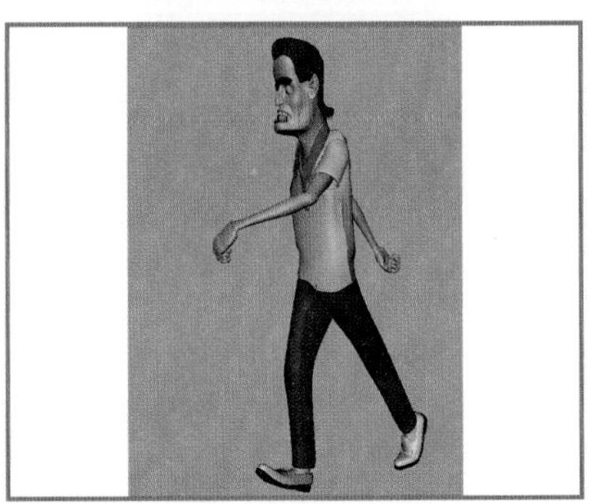

Setting Up Your 3D Characters for Animation

by Tim Coleman

Setting up the skeleton joint structure of a 3D *computer-generated (CG)* character is an extremely important step in the character animation process. A character's skeleton joint structure is much like the skeleton in the human body. It is an underlying structure of joints that the animator uses to pose the 3D character's body in different positions. The character's body is actually animated when the joints are positioned (usually rotated) and keyframed (that is, the positions are saved in time).

When setting up your 3D character, it is essential that you ask yourself some important questions. How will the character move? What ranges of motion can your character achieve? In other words, will your character have two arms or six? Will your character walk on all fours? Is your character flexible or stiff in its movement?

Observing humans and animals in the real world can help you build an understanding of how bodies move. When setting up your characters, look at how your own body moves. When you raise your arm over your head, ask yourself what joints are rotating and where on your body this is taking place. Anatomy books are also very valuable references for setting up the skeletons for CG characters; they can give the animator insight as to where to position joints in the body. Books such as *The Illusion of Life* by Frank Thomas and Ollie Johnston and *The Human Figure in Motion* and *Animals in Motion* by Eadweard Muybridge are great motion references for a variety of body types.

Keep in mind that there is no one way to set up the skeleton for a CG character. In some instances, you may find that particular setups are better for particular types of animation scenarios. For instance, you may set up your character one way for walking and another way for climbing a rope.

In this chapter, we take a look at some of the tools and terminology involved in setting up character skeletons in Maya. We will also provide some exercises that you can follow for setting up different kinds of characters.

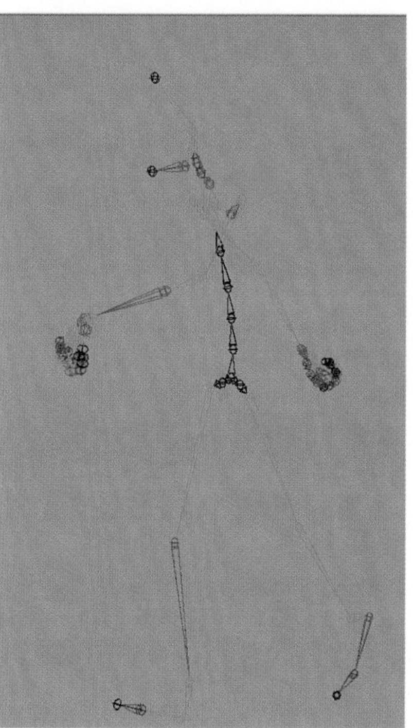
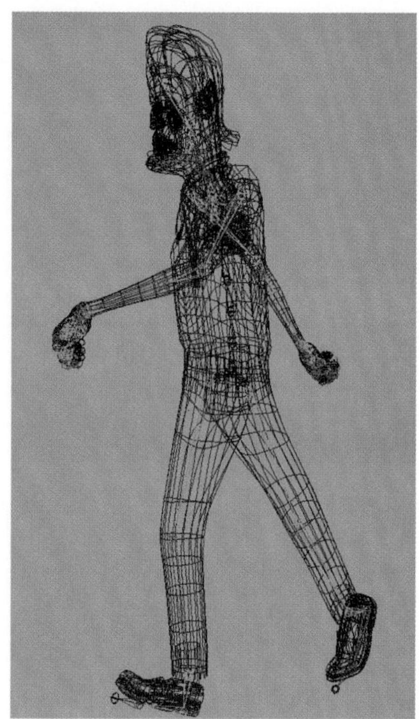
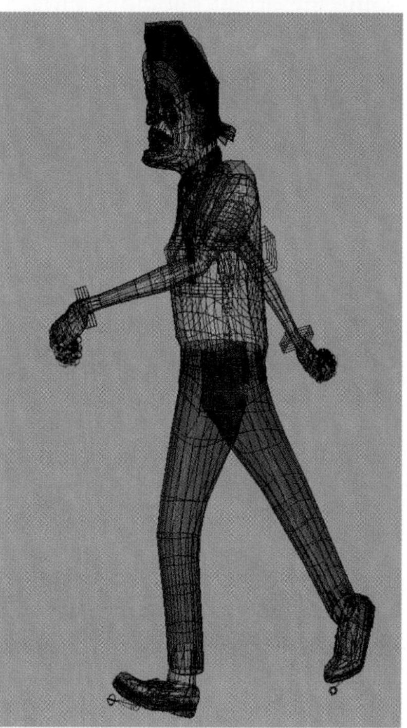
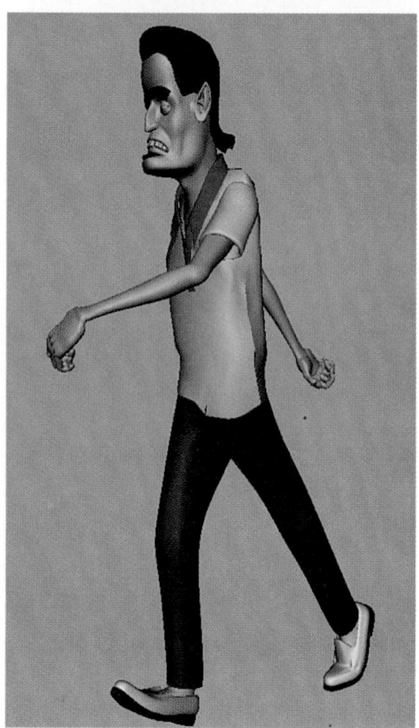

Figure 9.1
A character and its skeleton.

9.1 Concepts and Terminology

A lot of terminology is associated with the setup of skeletons for CG characters, and this can make the process somewhat confusing. Some of the confusion comes from the fact that many times the terminology varies depending on what software package you are using. Here, we will talk about terms used with Maya; however, you can carry many of these terms and concepts into other 3D packages as well.

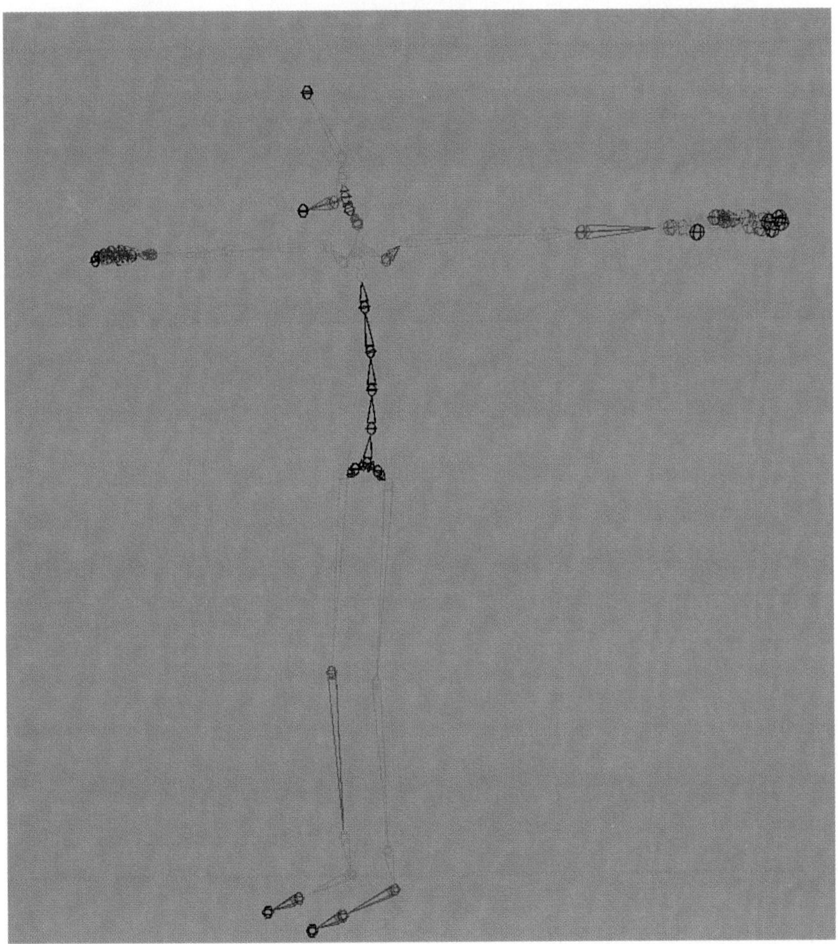

Figure 9.2
A skeleton for a biped character.

9.1.1 Hierarchies

Hierarchies may well be the most important concept for the character animator to understand when manipulating CG skeletons. The skeleton is basically a hierarchy of joints (pivot points, or locations in 3D space). The first joint the Maya animator draws (by going to Skeleton, Joint Tool) is at the top of the hierarchy (also known as the root of the skeleton). Each subsequent joint the animator draws is a child of the joint preceding it.

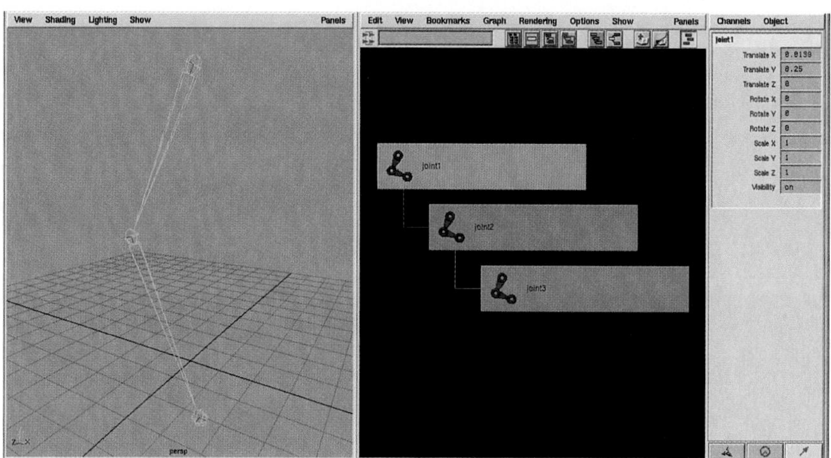

Figure 9.3
A simple joint chain, and Maya's Hypergraph.

When you rotate a joint that is the parent of other joints, all the children inherit the same rotation. A good example of this is seen when you rotate a simple arm, with shoulder, elbow, and wrist joints. When you rotate the shoulder joint, the elbow and wrist joints follow. When the elbow rotates, the wrist joint follows. Notice, however, that the shoulder does not inherit the rotation of the elbow or wrist because it is higher in the hierarchy. Transforming the parent joint affects all the children beneath, but a child never causes the parents above in the hierarchy to inherit its transform.

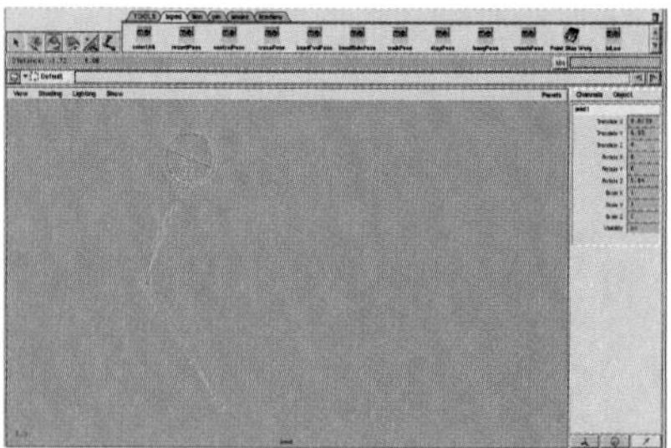

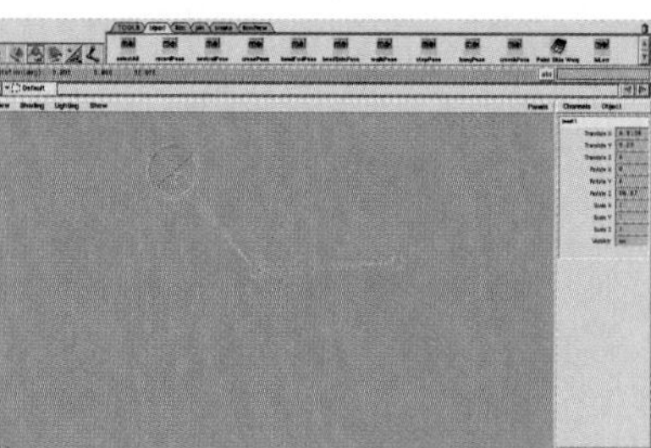

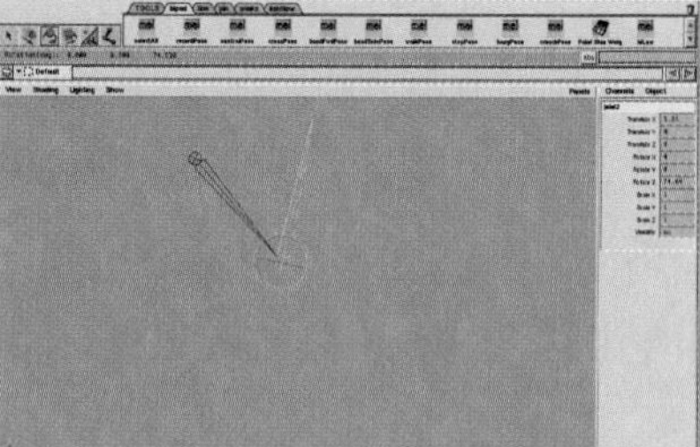

Figure 9.4
A simple arm rotating at the shoulder and elbow.

Understanding hierarchical animation such as this simple example of the arm forms the basis for knowing how to set up and animate your 3D character successfully. Things can start to get more complex with the introduction of multiple chains that branch out from each other, but taking a step back and looking at your character one limb at a time simplifies the task of setting up even the most complex characters.

9.1.2 Bones and Joints

Bones are connected to one another by joints. Bones are represented as long, pyramid-like icons, and joints are displayed as circles between the bones. The wide end of the bone is higher in the hierarchy; the thin end points to the next joint down the hierarchy. The joints themselves are transformed (rotated, translated, scaled) to move the bones in a hierarchical manner.

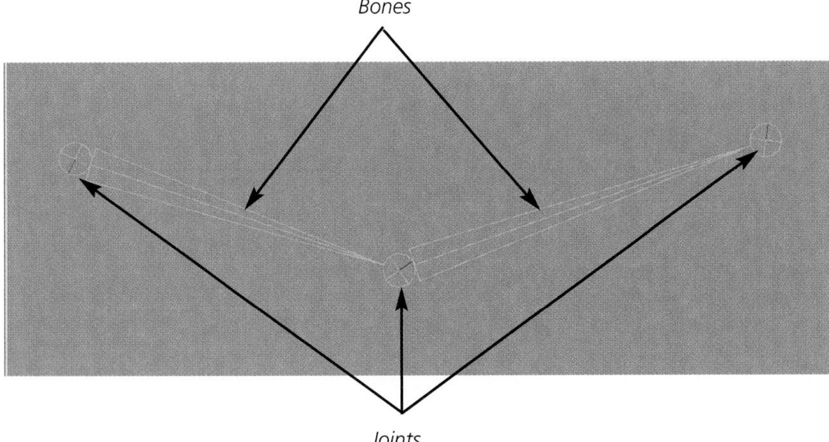

Figure 9.5
A diagram of joints and bones.

9.1.3 Forward Kinematics

The term *forward kinematics* is used when an animator manipulates joints at the top of the hierarchy first, rotating them into position. (Remember that all the child joints will inherit this rotation.) The animator then does the same for the next joint down in the hierarchy, and so on, until that chain is in its pose.

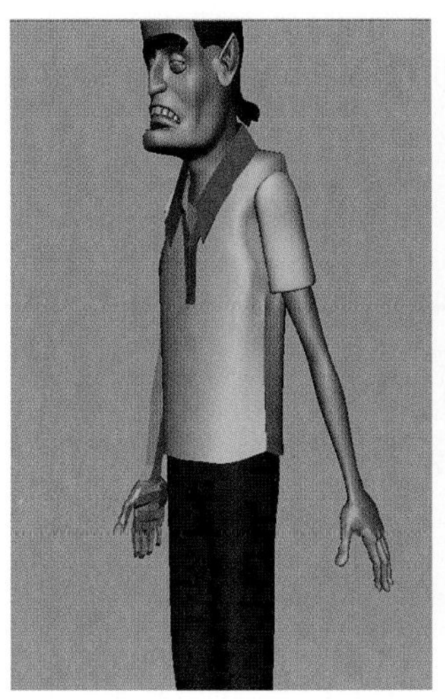
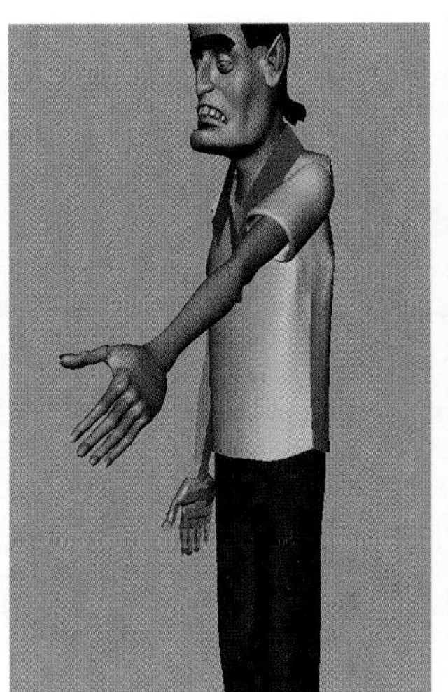
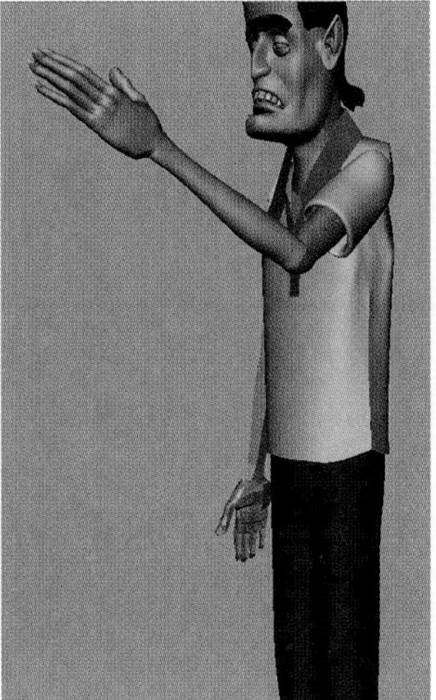
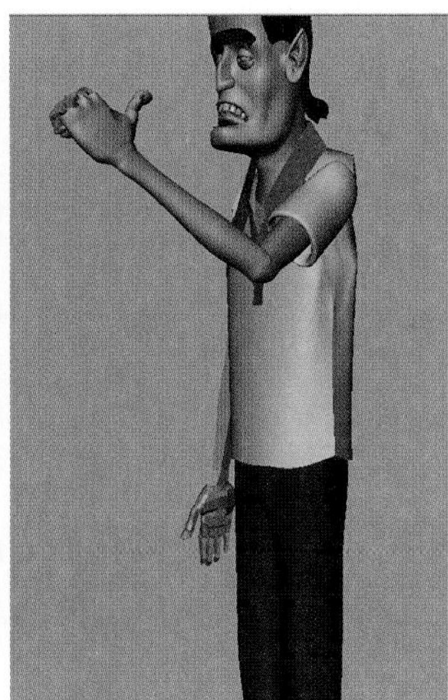

Figure 9.6
Using forward kinematics to move an arm.

This workflow gives the animator direct control over the manipulation of the joints. The animator can animate joints one by one and modify how each is animated. Overlapping and offset motion are very easy to animate when using forward kinematics.

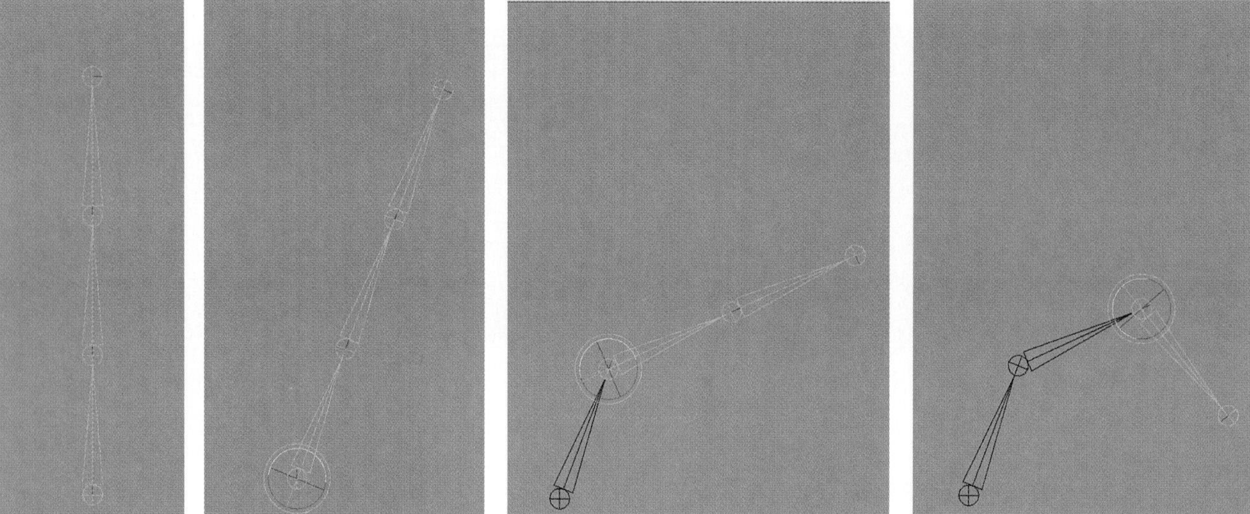

Figure 9.7
An example of overlapping action.

When only a few joints are available to select (for example, an elbow in an arm, or a knee in a leg), the forward kinematics process is very simple and straightforward. However, when you must manipulate multiple joints, this process can get very complex and tedious. We will look at some tools and concepts later that can automate the task of working with many joints and with inverse kinematics, which we will take a look at next.

9.1.4 Inverse Kinematics

Inverse kinematics (IK) solvers are tools in Maya that enable the animator to establish a chain of joints that will be rotated through the translation of an *end effector (IK handle).* Three such solvers exist: Single Chain, Rotate Plane, and IK Spline—and each one gives the animator varying levels of control over the skeleton's joints.

For instance, following the previous arm example, go to Skeleton, IK Handle Tool and select the shoulder joint first, followed by the wrist; an IK handle is created at the wrist. By selecting the IK handle, the animator can then translate this IK handle in 3D space. The IK solver will rotate the elbow and shoulder joints back up the chain (thus, the name "inverse" kinematics). In this example, the animator has to deal with just the IK handle to position the arm, unlike in forward kinematics, where the animator must rotate the shoulder and then the elbow. IK solvers can really streamline the animation of complex skeleton joint hierarchies. There will also be fewer animation curves in your scene; all the animation for the skeleton chain will be on the IK handle and not each individual joint.

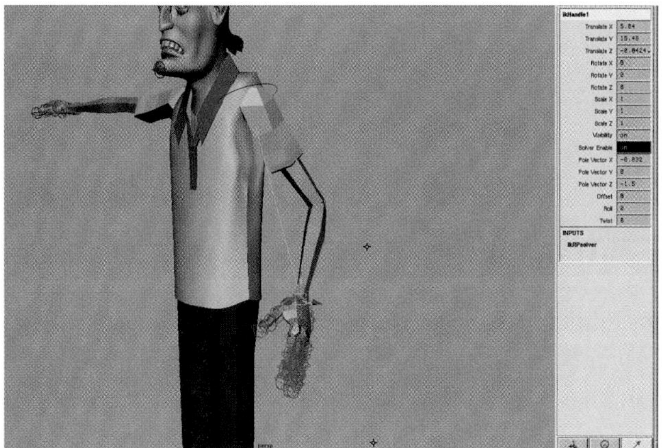
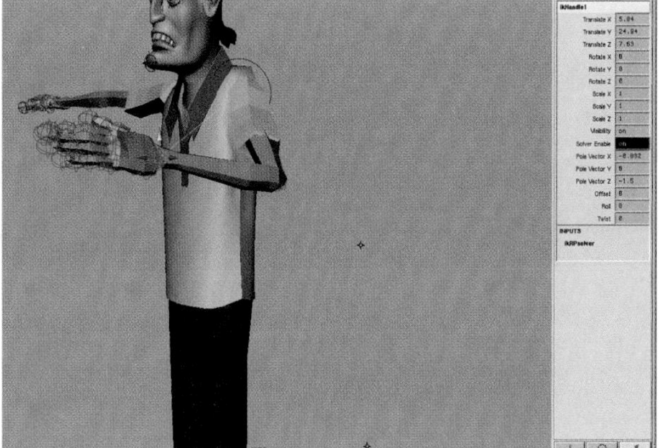

Figure 9.8
An IK handle at the wrist, and the effect of translating the IK handle in 3D space.

Using IK solvers also allows the animator to lock skeleton chains in place. This is great when you are trying to keep feet pinned to the ground or hands constrained to an object.

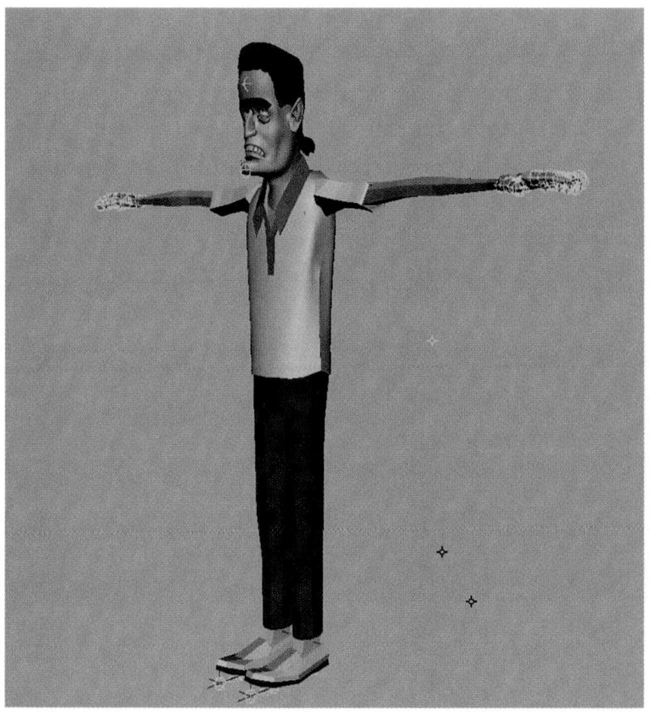
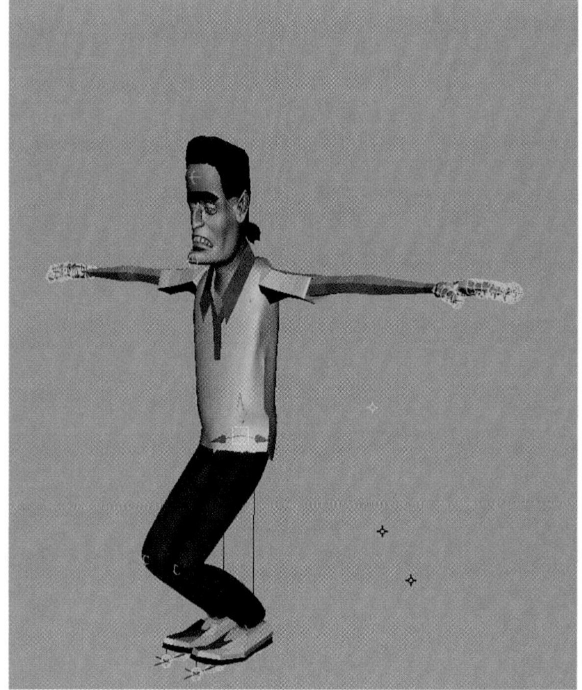

Figure 9.9
IK handles keep feet planted to the ground. Without IK handles, feet leave the ground.

However, using IK solvers has its drawbacks. Animating offset and overlapping motion through a chain of joints can be difficult to achieve because when the IK handle is translated, all the joints rotate at the same time. It is difficult to move the joints at different times (for example, to rotate the shoulder joint, and then six frames later to rotate the elbow joint). Some animators dislike IK solvers because they lose direct control over the joint (in other words, animators can no longer pick a joint and rotate it). With IK solvers, you have indirect control over the joint; the solver is essentially rotating the joints for you.

9.1.5 Types of CG Characters

3D characters can be categorized into three body types: segmented (robots, mechanical structures, insectoid/exoskeleton); skinned single-surface (humans, animals, creatures); and cartoon (squash-and-stretch characters, possibly anthropomorphic). Of course, exceptions exist, but these three body types are most prevalent.

In addition, these body types may vary in terms of how they articulate. Some may walk upright, while others walk on all fours. Other may have no legs at all! The approach you take when animating your characters depends on all these factors.

9.1.6 Getting Started: The Neutral Pose

It is important when setting up your 3D character that you start with the character modeled in a neutral position. For example, many times, animators model their 3D characters standing straight up with the arms out to the side (in a cross position), or with the arms straight down along the side of the body. This allows you to draw the skeleton joints into your character much more easily. For instance, it is much easier to create a skeleton chain for a 3D hand when all fingers are parallel to one another than when the 3D hand is modeled in a tight fist. In addition, this neutral pose will make it much easier for you to test deformations in your character when you start attaching the geometry to the skeleton.

Figure 9.10
A character in a cross pose.

9.1.6.1 Exercises

Before getting started on the exercises in this chapter, you will need to copy the project files from the CD-ROM to your project directory.

On the CD-ROM, go into the chapter 9 directory. If you're on an SGI machine, copy the chapter 9 projects directory to *HOME DIRECTORY*/maya/projects/. If you're on an NT machine, copy the directory to *DRIVE*:\WinNT\Profiles\UserName\maya\projects\.

9.1.6.2 Prepare for Exercises: Loading Shelf Files

Before starting the exercises for this chapter, be sure to copy a set of shelf files to your Maya preferences. The shelves contain tools and scripts that will aid you throughout the exercises.

Go to the Chapter 9 directory on the CD-ROM that comes with this book and open the chp9Shelves directory. If you're on an SGI machine, copy the shelf files to *HOME DIRECTORY*/maya/2.5/prefs/shelves. If you're using an NT machines, copy the files to *DRIVE*:\WinNT\Profiles\UserName\maya\2.5\prefs\shelves. (Replace 2.5 depending on the version you use.)

After you have copied the shelf files into your shelf preferences directory, you will have to re-launch Maya if it is currently running. The shelf files and other preferences are loaded on startup; therefore, you must restart Maya for the new shelf files to be read. After you have launched Maya again, you should see several new shelf files along your shelf. We will refer to these different shelves throughout this chapter.

9.2 Parenting Objects to Skeleton Chains

A good way to begin to understand how to use skeleton joint hierarchies is to simply parent objects to joints. This is the basic approach an animator uses in setting up a robot or a knight in shining armor.

We will start with a basic example: a robot arm. The robot arm is made up of rigid objects (cubes and cylinders) that we will simply parent to corresponding joints in the hierarchy. Use the robotArm.ma file on the accompanying CD as reference.

The round cylinders of the robot arm signify its points of rotation. We basically have a base, a joint at the base, an elbow, a wrist, and a claw that needs to pinch its "fingers" together. We will use the cylinders to help determine where we will place joints in the robot arm.

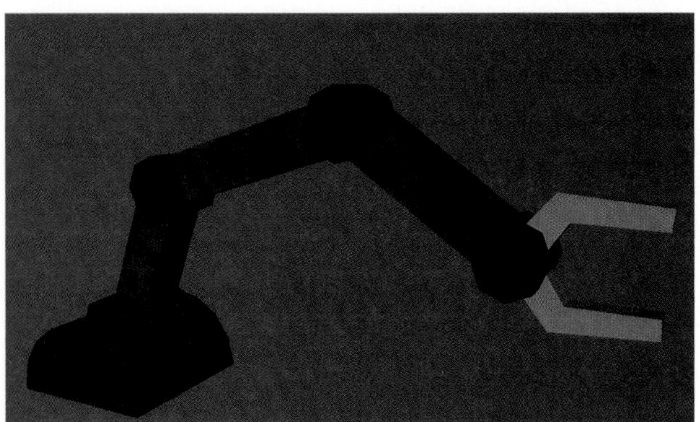

Figure 9.11
A robot arm in modeling views.

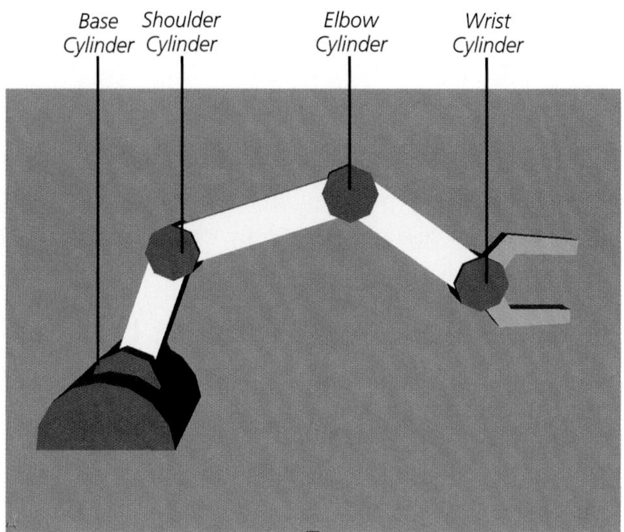

Figure 9.12
Different parts of the robot arm, along with cylinders labeled for joint placement.

Exercise 9.1: Creating Skeleton Joints for the Robot Arm

1. Go to File, Project, Set and set the current Maya project to robotArm. Go to File, Open Scene and open the scene file robotArm.mb. Draw a skeleton chain for the arm, making sure you are in the Animation menu set, and choose the menu Skeleton, Joint Tool.
2. Click once at each of the following locations, in this order: the middle of the base cylinder, the shoulder cylinder, the elbow cylinder, the wrist cylinder, the base of the top claw, and the tip of the claw.
3. Press the up arrow twice on the keyboard to step back up the hierarchies two levels.
4. Click the base of the bottom claw and the tip of the bottom claw and press Enter to finish drawing the skeleton chain.
5. On the main menu bar, Choose Window, Outliner. In the Outliner, double-click on joint1, and name it **baseJoint**.
6. Shift+LMB+click the small white arrow next to baseJoint to expand the hierarchy of joints. A listing of the joints you have drawn for the robot arm appears, indented to show their respective level in the hierarchy.
7. Name the remaining joints: **shoulderJoint**, **elbowJoint**, **wristJoint**, **topClawBaseJoint**, **topClawTipJoint**, **bottomClawBaseJoint**, and **bottomClawTipJoint**.

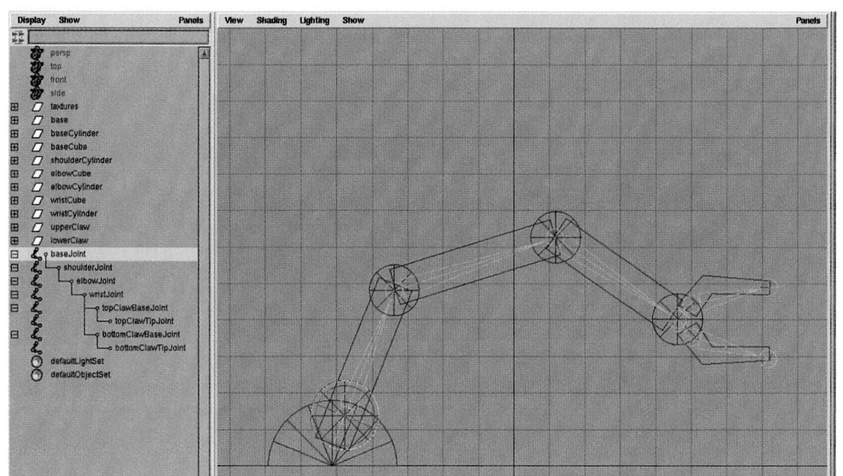

Figure 9.13
The Outliner with the renamed joints.

Exercise 9.2: Testing the Rotation of the Robot Arm Joints

1. Make sure you are in Object mode, with skeleton joints on in the Pick Mask, and then select across the baseJoint bone. Selecting the bone picks the joint on the high end of the hierarchy.
2. Press E to get to the Rotate Manipulator Tool and use it to rotate the baseJoint. All the other joints should follow when the baseJoint is rotated.
3. Press Z to Undo.
4. Use the down arrow to select the elbowJoint. Rotate this joint and note which bones rotate and which do not.
5. Test the other joints in the arm.
6. Undo your rotations after you have tested the joints.

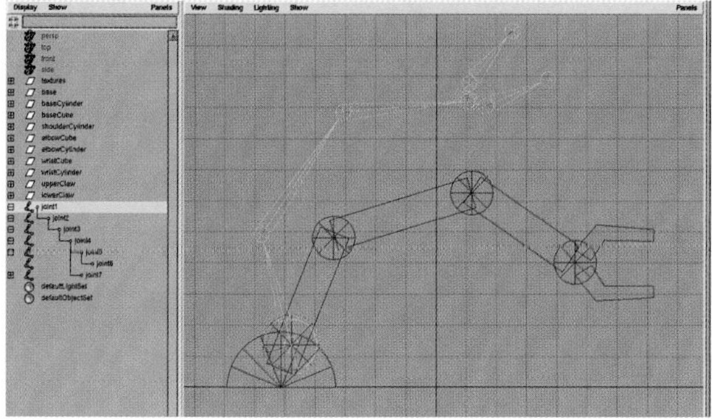

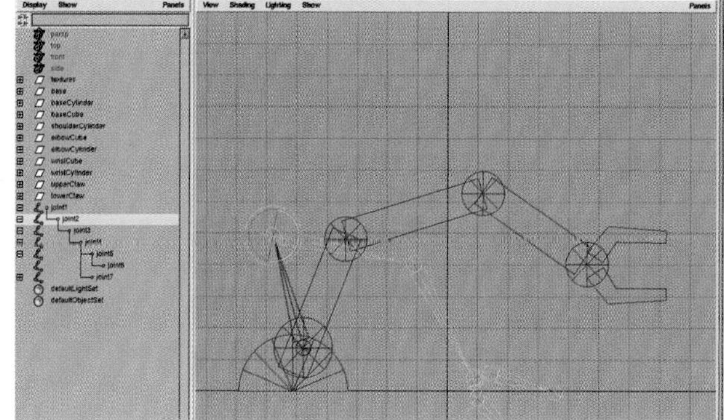

Figure 9.14
The joints rotated on the arm.

Exercise 9.3: Parenting Objects to the Robot Arm Joints

1. Go to the Outliner again and notice the nodes that represent the geometry that makes up the robot arm.
2. In the Outliner, select the baseCylinder. Ctrl+select the baseCube and the baseJoint and choose Edit, Parent (or press P) to make the first two selected objects children of the baseJoint.
3. Select the baseJoint and rotate it. The baseCylinder and baseCube inherit the rotation of their parent.
4. Select the elbowCylinder, Shift+select the elbowCube and the elbowJoint, and choose Edit, Parent.
5. Select the wristCylinder and the wristJoint, and choose Edit, Parent.
6. Select the upperClaw and the upperClawJoint, and choose Edit, Parent.
7. Select the lowerClaw and lowerClawJoint, and choose Edit, Parent.
8. Test the rotation of the joints on the robot arm skeleton chain. Be sure to Undo after you finish testing the rotation.

Tip

When you're setting up parenting, remember to select all children first and the node you want to be the parent last, and then choose Edit, Parent.

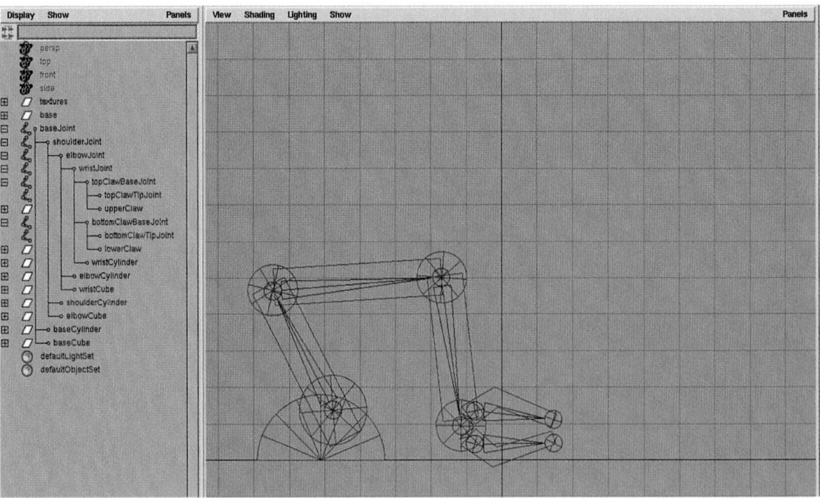

Figure 9.15
The robot arm in different positions after parenting.

Exercise 9.4: Adding an IK Handle to the Robot Arm

1. Go to Skeletons, IK Handle Tool, choose ikSCsolver in the Current Solver option box, and close the option box. The ikSCsolver is the Single Chain solver. This solver rotates joint along a two-dimensional plane. The Single Chain solver is ideal for joints that rotate in a hinge-like fashion (such as the elbow or knee).

2. Select the baseJoint and then the wristJoint. Maya displays an L-shaped icon at the wristJoint and a line drawn from the baseJoint to the wristJoint. The L icon is the IK handle, and the line represents the range of joints affected by the IK solver you just created.

3. Select the IK handle, press W to translate, and middle-mouse-button-drag the IK handle around in the Side View.

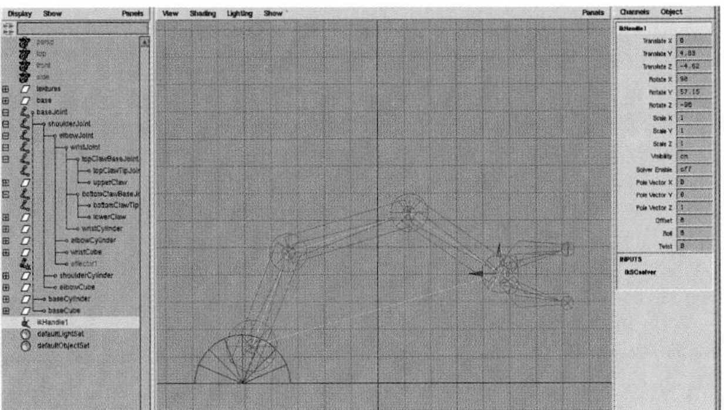

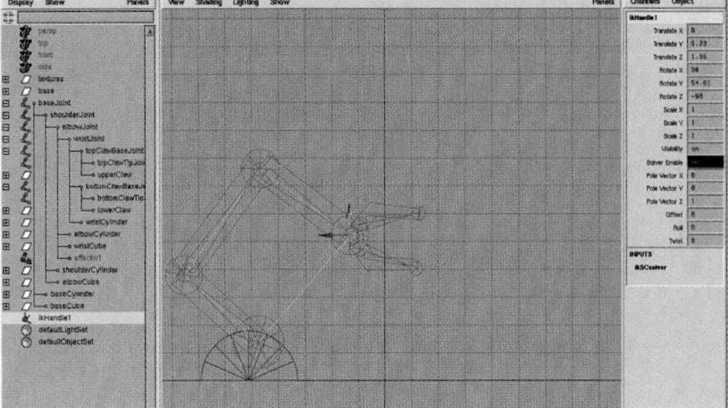

Figure 9.16
An IK handle, and an IK handle translated.

Exercise 9.5: Animating the IK Handle and Joints on the Claw

1. Go to frame 1 in the time slider, select the IK handle, and press Shift+W to keyframe the translation channels of the IK Handle at its current position.

2. Go to frame 30, translate the IK handle to a different position, and press Shift+W to keyframe the translation of the ikHandle once again.

3. Click the Play button and review your animation.

4. Rewind the current time to frame 1, select the lowerClawJoint and Shift+select the upperClawJoint, and press Shift+E to keyframe only the rotation channels of the two joints.

5. Go to frame 30 and rotate each of the claws into an open position. (You may have to rotate the claws individually.)

6. Select both the lowerClawJoint and the upperClawJoint, and press Shift+E to keyframe them.

7. Play back the 30-frame animation.

Exercise 9.5: continued

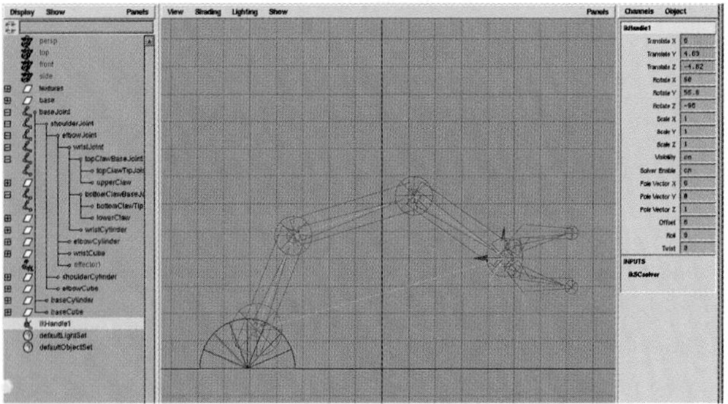

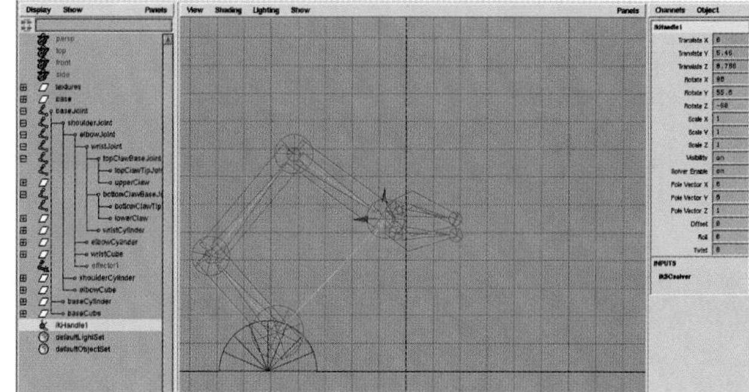

Figure 9.17
The robot arm at frames 1 and 30.

Congratulations! You've set up and animated your first character in Maya. Notice the importance of understanding the hierarchy of joints and other objects that make up the character. This plays an important part in determining how your character will move. Now let's take a look at some other skeleton setups.

9.3 Setting Up a Biped Skeleton

The most common skeleton setup situation is the bipedal skeleton (the character that walks on two legs). You will cover many concepts in this section that you will be able to carry over to other types of characters. This particular setup will best suit a character that needs to be able to walk, run, sit, gesture, and so on. We will use the bipedSetup.ma file from the accompanying CD as reference.

Figure 9.18
A biped character in different poses.

Exercise 9.6: Templating the Character Surfaces

1. Go to File, Project, Set and set the current Maya project to biped. Go to File, Open Scene and open the file bipedModel.mb.
2. If the layer bar is not showing, in your interface go to Options and turn on the layer bar. The layer bar appears horizontally across the upper half of the interface. Click the New Layer button on the far left of the layer bar, and a new layer appears on the Layer Bar named layer 1. Double-click layer 1 and name the layer **`characterGeom`**.
3. In Object Mode, select all the geometry for the character. Place the cursor over the characterGeom layer bar and hold down the right mouse button. Select Assign Selected in the pop-up menu. Then hold the right mouse button down again and select Template from the characterGeom layer pop-up menu.
4. Maya grays out the surfaces to indicate that they are no longer selectable. If you wanted to make the surfaces selectable again, set the characterGeom menu to Standard (but do not do this at this point in the exercise). The reference option in the layer menu keeps the surfaces visible but not selectable.

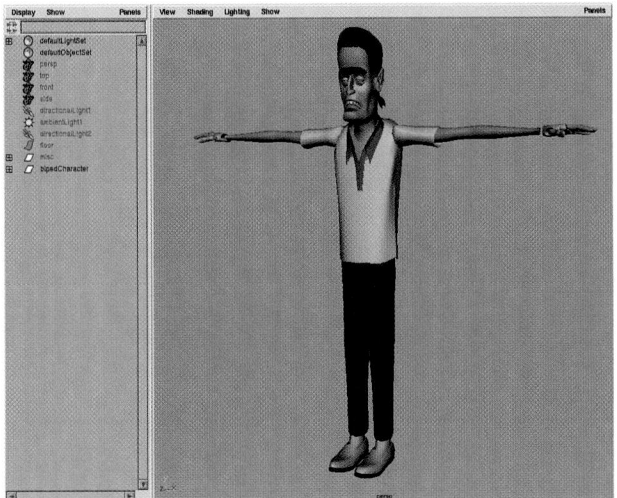

Figure 9.19
A biped character.

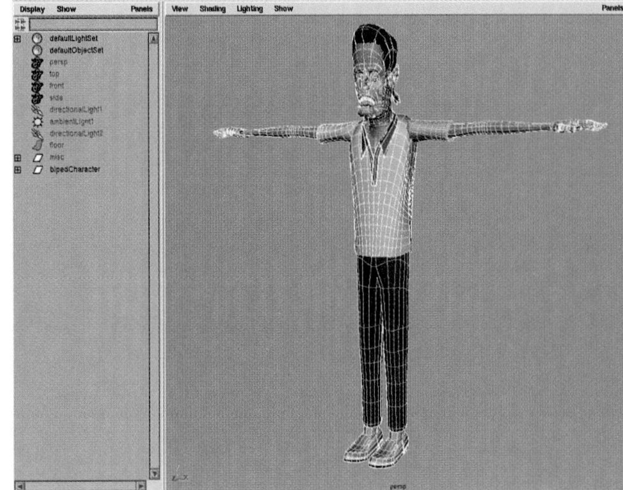

Figure 9.20
The layer bar.

Exercise 9.7: Creating Leg Skeletons

1. Choose Skeleton, Joint Tool, and set Auto Joint Orient to XYZ. Auto Joint Orient lets Maya orient the XYZ axis for the joints as you create them. In this case, with Auto Joint Orient set to XYZ, the X axis of each joint will

Exercise 9.7: continued

point down the bone. This allows for consistent orientation of all the axes of the joint chain that you create. If Auto Joint Orient is set to None, the joint axis will be aligned along the world axis.

2. In the Side View, draw a skeleton chain starting from the hip, to the knee, to the ankle, to the heel, to the ball of the foot, and finally to the tip of the shoe. Name the joints according to their location in the leg: **hip**, **knee**, **ankle**, **heel**, **ball**, **toe**.
3. In the Front View, select the hip joint and move to the left side of the character. When joints are created, Maya always places them along the 0 axis of the plane in which they are drawn. Make sure the leg skeleton is in the center of the leg geometry.
4. Select the hip joint, go to Skeleton, Mirror Joint, and set Mirror Across to YZ. Click the Mirror button and close this option box.

   The Mirror Joint duplicates the entire joint hierarchy across the Y and Z axes of the character. (Look at the red, green, and blue world axis in the lower-left corner of the Perspective View; the YZ plane splits through the center of the character.)
5. Select the hip joint for the left leg. Choose Modify, Prefix Hierarchy Names and type **left_** in the text field.
6. Repeat step 4 with the hip joint for the right leg and use **right_** as the prefix.

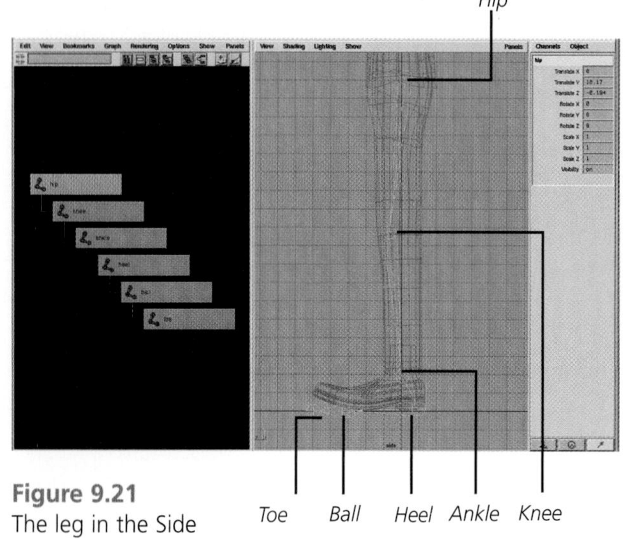

Figure 9.21
The leg in the Side view with joints labeled.

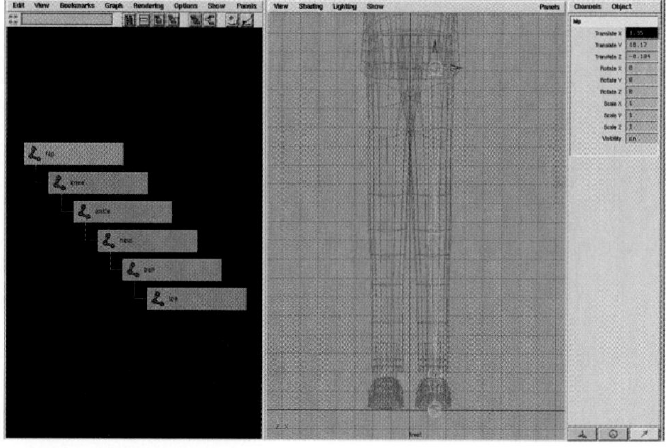

Figure 9.22
The leg moved over in the Front view.

7. Select both the right_hip and the left_hip joints in the Perspective View and open Hypergraph, Window, HyperGraph. Press F while your cursor is in the Hypergraph window to frame the view of the selected objects. Observe the hierarchy of joints that make up the skeleton for each leg. Note how the hip joints are at the highest level, then the knee, and so on.
8. Select one of the hip joints and rotate it. Take note of how all the joints below it in the hierarchy follow along. Test the rotations of other joints in the hierarchy. Be sure to undo any rotations you have placed on any of the leg joints. You can also pick the hip joints and go to Skeleton, Assume Preferred Angle to reset the position of the joints.

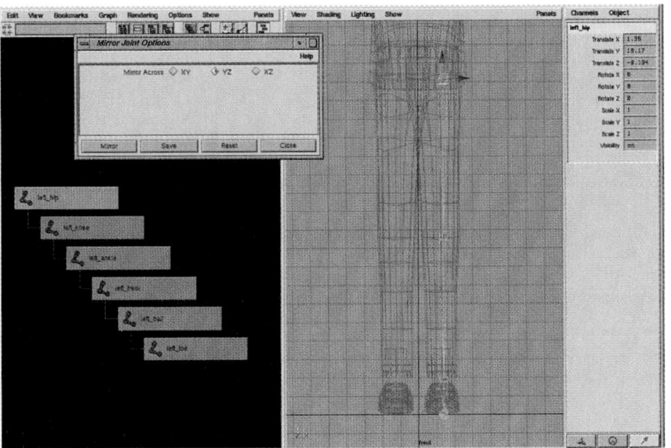

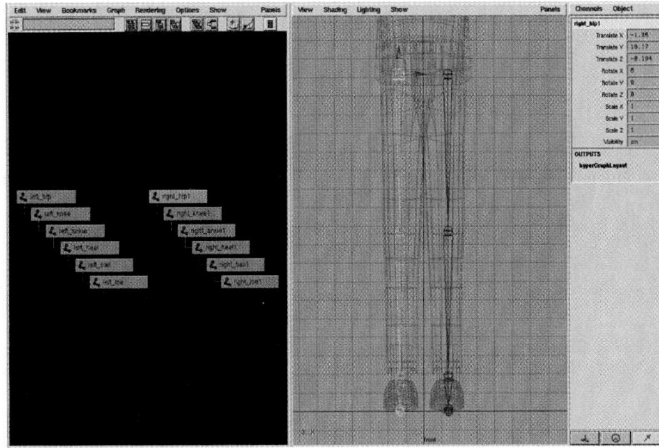

Figure 9.23
The YZ Axis Perspective view.

Exercise 9.8: Creating the Back, Neck, and Head Joints

1. Choose Skeleton, Joint Tool (it should still be set to Auto Orient Joint XYZ), and draw 6 to 12 joints from the waist to the base of the neck. Use the diagram in the following figure for the general positioning of the joints in the back.
2. Next create two more joints at the middle and top of the neck. With the Joint Tool still active, click the next joint at the middle of the neck, then another joint at the top of the neck. Again, see the following figure for joint placement.
3. Add two more joints up into the head and one more joint to the forehead of the character.
4. Use the up arrow to step up the joint hierarchy two levels, draw joints for the jaw hinge and the chin, and press Enter to finish drawing the skeleton chain.
5. Name the joints according to the diagram.

Exercise 9.8: continued

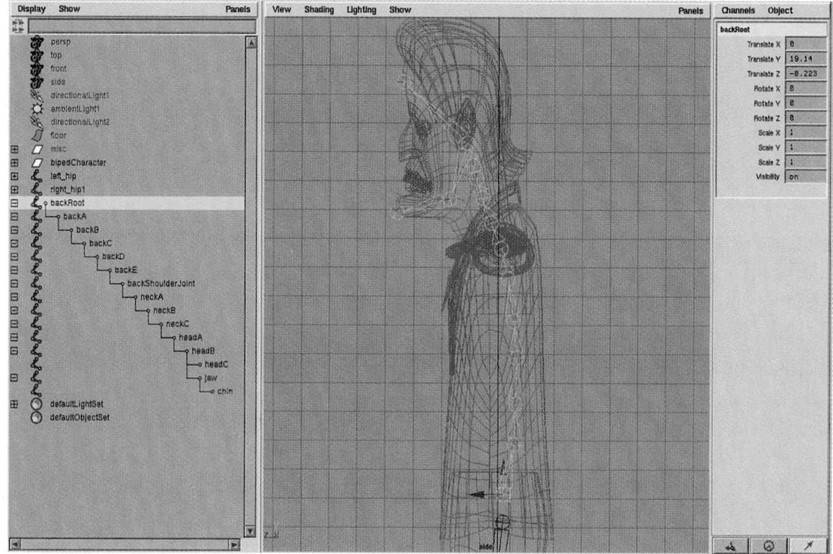

Figure 9.24
The upper body in the Side view, with the joints labeled.

Tip

If you draw a skeleton joint in the wrong location, select the joint, press W to translate, and then press Insert to set the pivot. Move the joint into a new location. The selected joint moves to a new position. When you reset the location of the pivot point for the joint, the joint (which is essentially the pivot point of the bone) follows. Press Insert again to get out of setting the pivot point. You can also select a joint and simply translate it. The difference with translating is that the selected joint and its children will be moved.

Exercise 9.9: Creating Pelvis Joints and Parent Chains Together

1. In the Front view, create a single skeleton joint between the left_hip joint and the back_root joint, and name this joint **`pelvis`**.
2. With the pelvis joint selected, go to the Side View and move the pelvis joint in line with the left_hip and back_root joints. To do this, use the Translate tool, or set the pivot of the joint.
3. Choose Skeletons, Mirror Joint Tool to mirror a duplicate of the pelvis joint across the YZ axis to the other side, naming these joints **`left_pelvis`** and **`right_pelvis`**, respectively.
4. Parent the left_hip joint to the left_pelvis joint by first selecting the child (left_hip) and then the parent (left_pelvis). Then choose Edit, Parent. Maya creates a bone that connects the left_hip and left_pelvis together.

5. Select the left_pelvis joint and the back_root joint, and then choose Edit, Parent.
6. Repeat steps 4 and 5 to create the same parenting on the rightside.
7. Select the backRoot joint and choose Edit, Group to group it to itself, creating an extra pivot point above backRoot. Name this node **`bipedRoot`**.
8. Save your work, naming your scene **`myBipedSkeleton.mb`**.

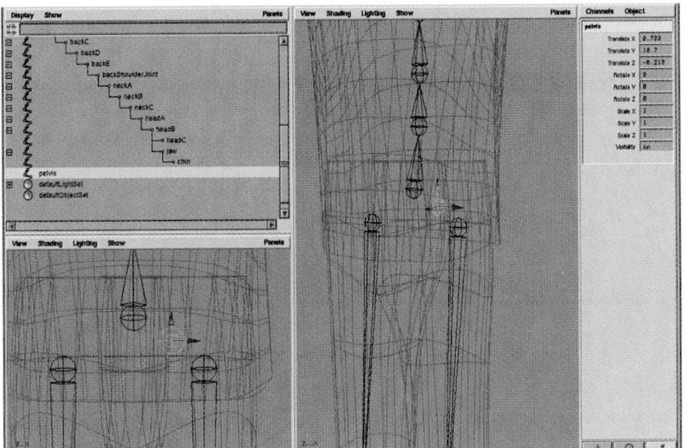

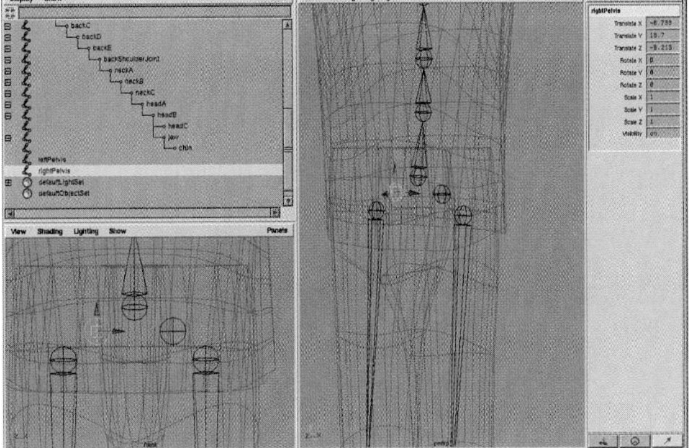

Figure 9.25
The pelvis joint before the mirror joint step.

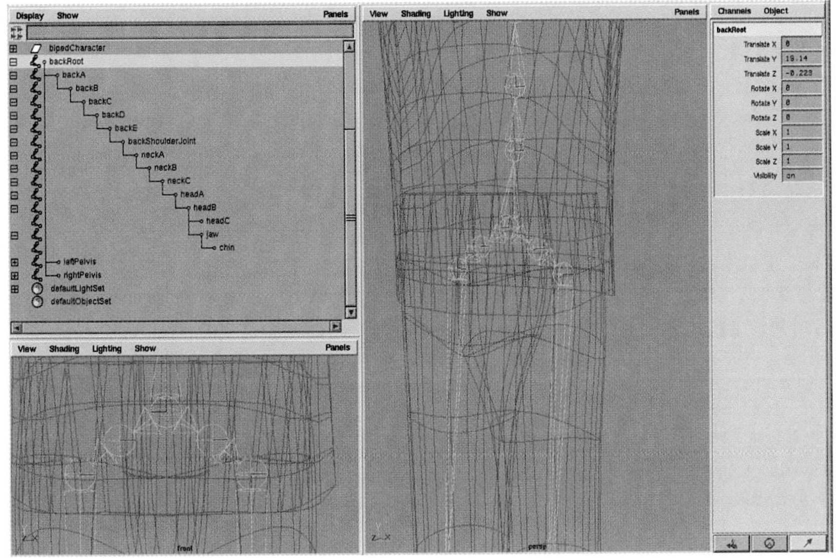

Figure 9.26
The leg after the joints have been parented to the spine.

Exercise 9.10: Testing the Joints You Have Created

1. Select the joints and rotate them. Pay close attention to which parts of the skeleton move when you rotate the joints.
2. Select the knees and rotate them.
3. Select the joints in the spine and rotate them.
4. Select the back_root joint and try rotating and translating it. When you're finished, select the back_root joint and choose Skeleton, Assume Preferred Angle to reset all the joints back to their original position.

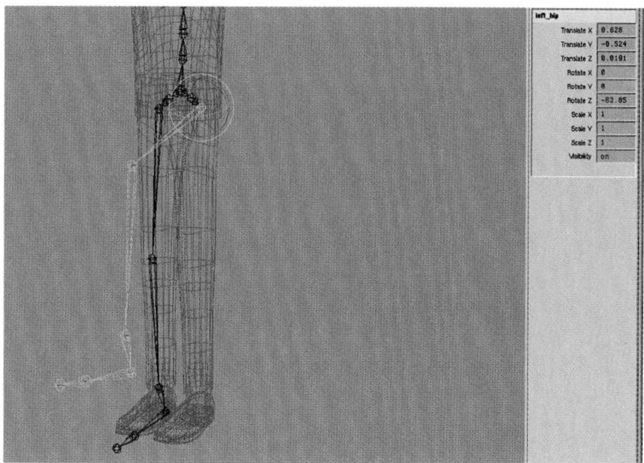

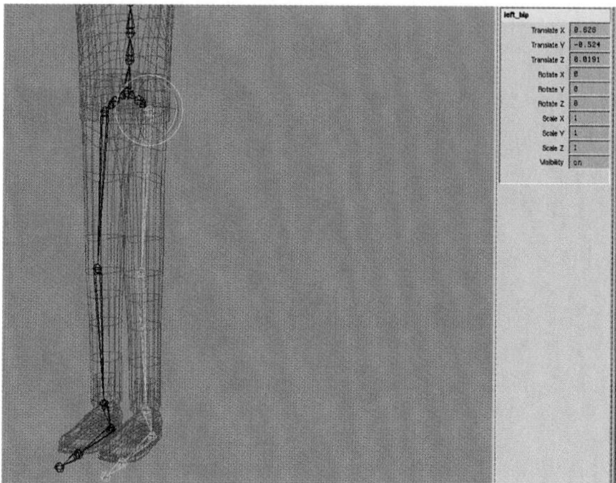

Figure 9.27
Testing joint rotations and assuming Preferred Angle.

You now see how parenting joints together allows you to connect different skeleton chains. Drawing your skeleton chains individually and then connecting them with parenting makes creating the skeleton much simpler. Approach your character one limb at a time and then connect all the limbs to the spine.

9.3.1 Creating a Single Chain IK Solver for the Legs

We want our character to walk and run, so we will use Single Chain (SC) IK solvers to help us animate the legs. The SC IK solver will enable us to simply animate the IK handles to rotate all the joints in the legs, and keep the feet pinned down in place (this occurs when keyframes are set on IK handles). Using forward kinematics to keep the feet pinned in place is difficult (but not impossible!) and requires many more keyframes on the joints in the legs. We will also look at how grouping our IK handles into hierarchies can help in controlling the joints in the legs. Grouping the IK handles will give us additional pivot points that we can use to move the feet in specific ways.

Exercise 9.11: Creating IK Handles for the Skeleton Legs

1. Choose Skeleton, IK Handle Tool, and set Current Solver to SC solver.
2. Click the left_hip joint, then the left_ankle joint. An IK handle will be created at the left_ankle joint.
3. With this IK handle selected, name it **leftAnkleIK**.
4. Repeat steps 1 through 3 for the right leg, and name the IK handle **rightAnkleIK**.
5. Select leftAnkleIK, press W to translate, and use the mouse to move the leg. Notice that translating the leftAnkleIK rotates the knee and hip joints.

Exercise 9.12: Creating IK Handles for the Skeleton Feet

1. Choose Skeleton, IK Handle Tool (or press Y to return to the last tool used).
2. Click the left_ankle joint, and then the left_ball joint. Name the IK handle **leftBallIK**.
3. Press Y to return to the IK Handle tool, and select the left_ball joint and then the left_toe joint. Name this IK handle **leftToeIK**.
4. Repeat steps 1 through 3 for the right foot, naming the IK handles appropriately.

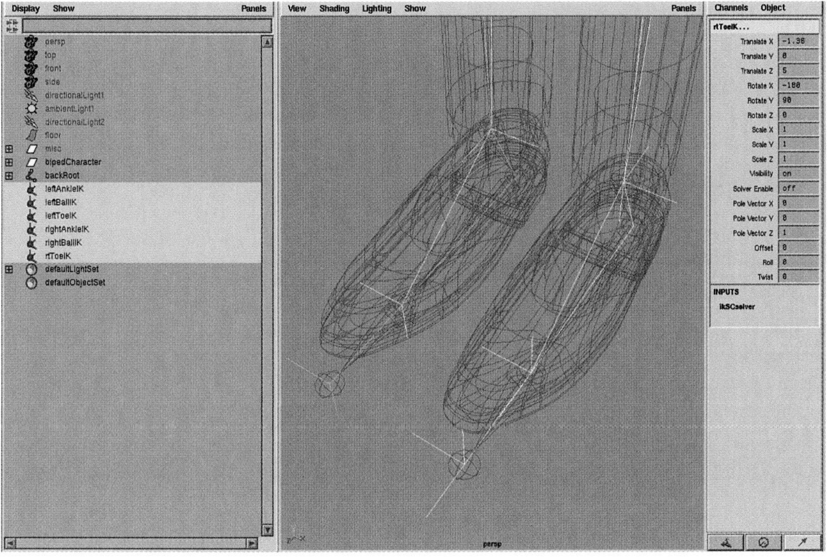

Figure 9.28
The feet with labeled IK handles.

Exercise 9.12: continued

The heel joints for both feet are used as a visual reference for the bottom of the feet. Look at your own foot and notice how it rotates at the ankle. We will turn the degrees of freedom for the heel joint off for the X, Y, and Z axes to ensure that the heel does not rotate when the IK handles in the foot are moved.

Exercise 9.13: Restricting the Heel Joint Rotations

1. Select the left_heel joint and open the Attributes Editor by choosing Window, Attributes Editor or pressing Ctrl+A.
2. In the Limit Information, Rotation section of the Attributes Editor, turn off the X, Y, and Z degrees of freedom (un-check the check boxes).
3. Repeat steps 1 and 2 for the right_heel joint. Now both heel joints will not rotate.

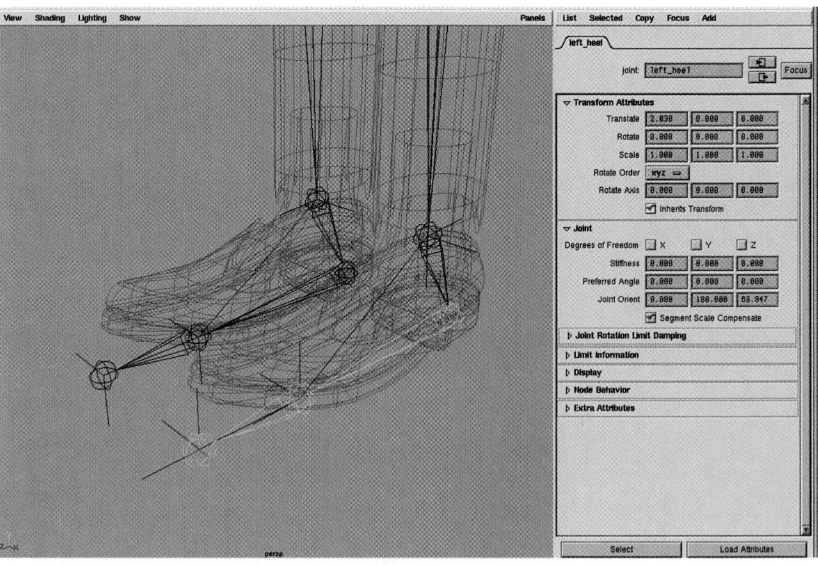

Figure 9.29
Degrees of freedom turned off in the Attributes Editor.

We will group the IK handles to give ourselves extra nodes (pivot points) that we can use to rotate the foot in a rolling motion from heel to toe. Each time you group an object, you create a new node in the scene. This new group node is now the parent of the object you had selected when you went to Edit, Group.

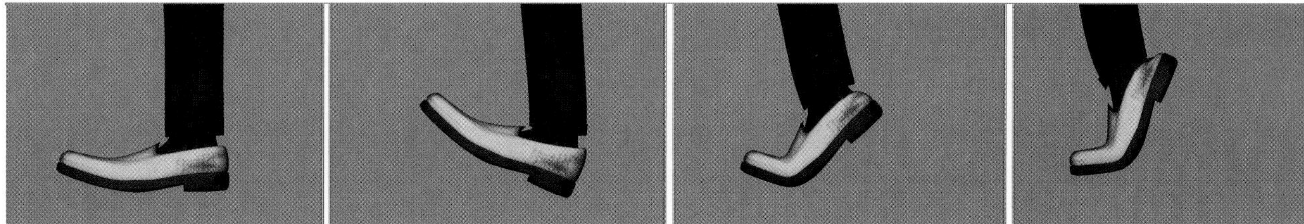

Figure 9.30
The foot in different roll positions.

In the case of making the foot roll, each new group node (pivot point) you create will serve to rotate the foot in different ways. For instance, when we create the leftToePivot group, that group node will rotate the foot from the toe of the foot. The leftHeelPivot group node will rotate the foot from the heel of the foot, and so on. The new group nodes you will be creating will be higher in the hierarchy than the IK handles; therefore, the IK handles will follow when you rotate one of the new group nodes.

Exercise 9.14: Grouping the IK Handles to Create Extra Nodes

1. Select leftToeIK, and choose Edit, Group to place a new node above the leftToeIK node. Name this new node **`leftToe`**.
2. With leftToe still selected, press Insert to set the pivot point of leftToe, hold down the V key to activate point snapping, and then use the middle mouse button to drag the pivot to the center of left_ball joint. Release the middle-mouse button and the V key when you see the pivot point snap to the left_ball joint. Toggle the Insert key to get out of setting the pivot point location.
3. Repeat steps 1 and 2 with the leftAnkleIK, grouping it and snapping the new pivot point to the left_ball joint. Name this new node **`leftAnkle`**. Do not follow these steps for the leftBallIK.
4. Repeat steps 1 through 3 for the ankle and toe IK handles on the right foot and name the nodes appropriately.
5. Select the leftToe, leftBallIK, and leftAnkle nodes and choose Edit, Group. Name this new group node **`leftToePivot`**, and set the pivot point for leftToePivot to the left_toe joint (using the same point snapping technique used before).
6. Select leftToePivot and choose Edit, Group to group it to itself. Name this node **`leftHeelPivot`**, and set the pivot point for leftHeelPivot to the left_heel joint.
7. Select the leftHeelPivot and group it to itself. Name this node **`leftFoot`**, and set the pivot point for leftFoot to the left_ball joint.

Exercise 9.14: continued

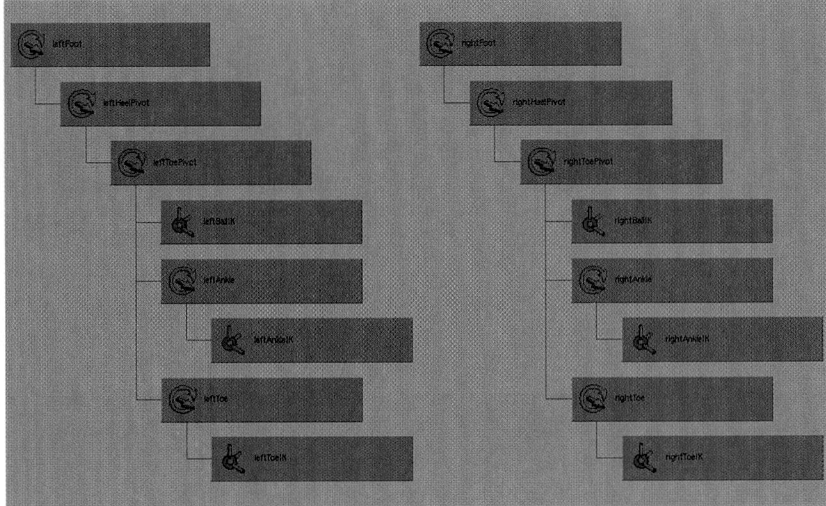

Figure 9.31
The foot control hierarchies in the Hypergraph.

8. Repeat steps 5 through 7 for the right foot IK handle nodes.
9. Save your work as **`myBipedSkeleton1.mb`**.

It may seem somewhat confusing at this point why we are creating all these extra nodes and placing their pivot points in different locations. Remember that these extra nodes will enable us to manipulate the rotations of the foot from different locations. In the following steps, you will see how important these extra group nodes are in allowing us to control the foot in various ways.

9.3.2 Creating Your Own Control Attribute

We will now take a look at one of the most powerful workflows in Maya: creating your own attributes and connecting them to other attributes in your scene. We'll start by creating an attribute called roll, and in the next section we will connect it to the rotations of nodes in the foot hierarchy we just created.

Exercise 9.15: Creating a Control Attribute

1. Select leftFoot and choose Modify, Add Attribute.
2. Name the attribute **`roll`**, set the Data Type to float, the Minimum value to –5, the Maximum value to 10, and the Default value to 0, and click the Add button. In the Channel Box on the right side of the screen, Maya has added the roll attribute to the keyable attributes of leftFoot.

At this point, you have simply added your own attribute named roll. We now need to connect it to attributes on other nodes in the scene.

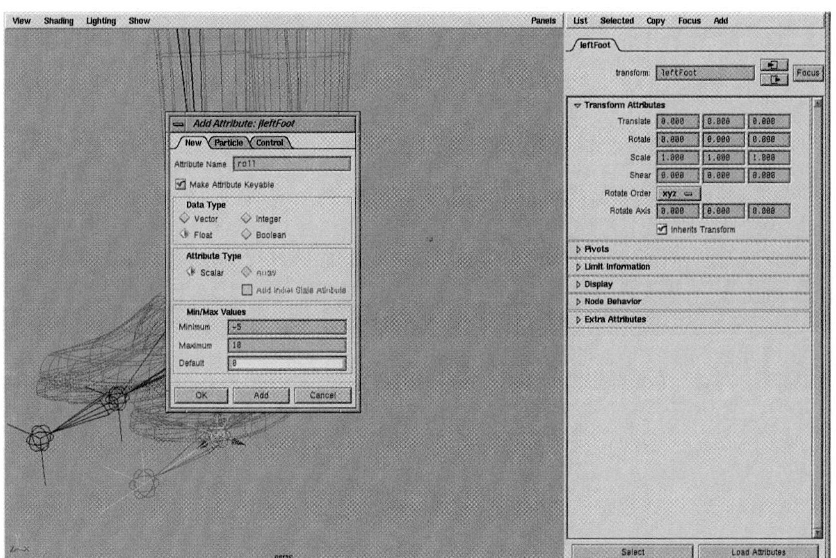

Figure 9.32
The Add Attribute window.

9.3.3 Connecting Your Roll Attribute Using Maya's Set Driven Key

Set Driven Key (SDK) is a very powerful tool that essentially enables you to have a node's attribute(s) drive the attribute(s) of another node(s). After you have set up the Driver/Driven relationship, you can modify how Maya interpolates the attributes with a Set Driven Key Curve.

With our foot set up, we will use the roll attributes we have just added on leftFoot and rightFoot to drive rotation attributes on nodes in the foot control hierarchy we set up earlier. For example, when roll is at a value of 0, the foot will be flat on the ground. At –5, the foot will be rotated back on its heel. At 5, the foot will be rotating around the ball of the foot, and at a value of 10 the foot will rotate off of the toe.

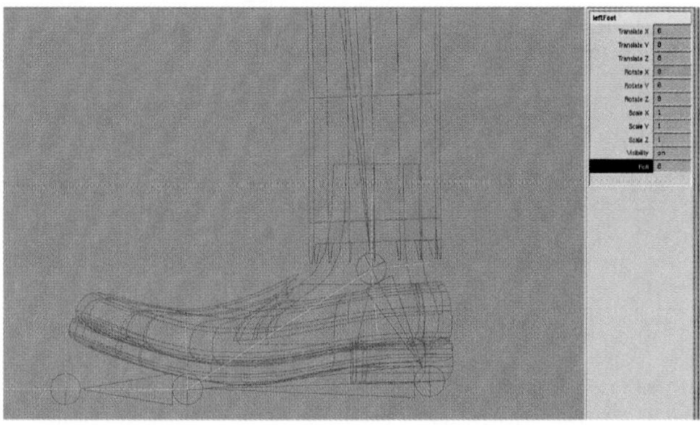

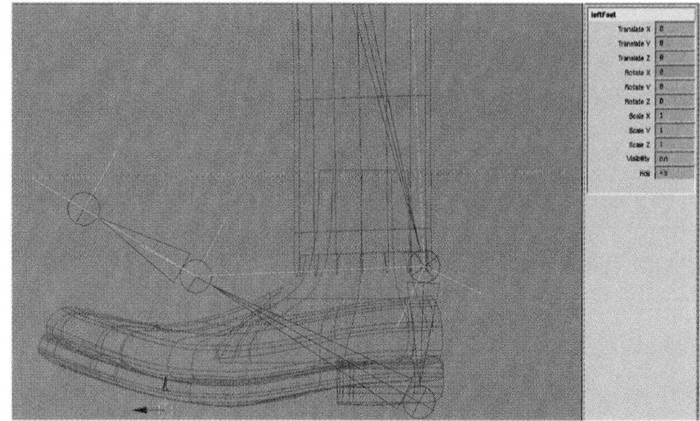

Figure 9.33
Roll values and foot positions.

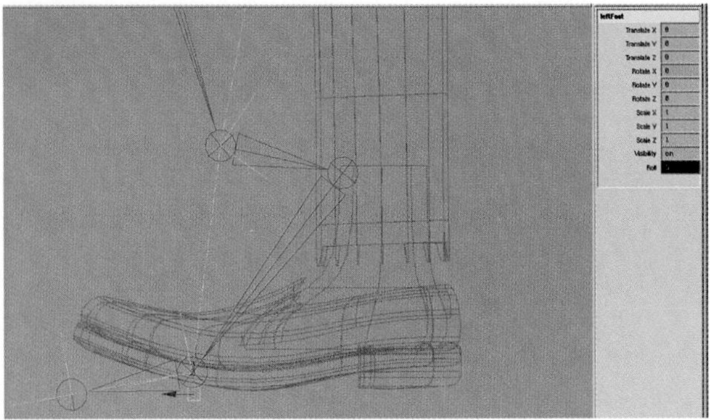

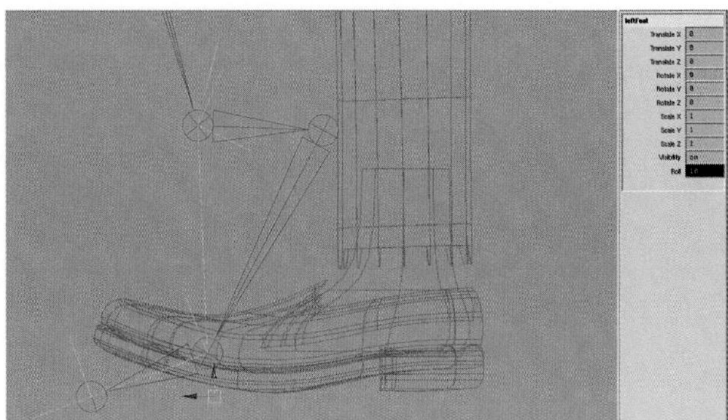

Figure 9.33
Roll values and foot positions, continued.

Exercise 9.16: Loading the Driver and Driven Nodes

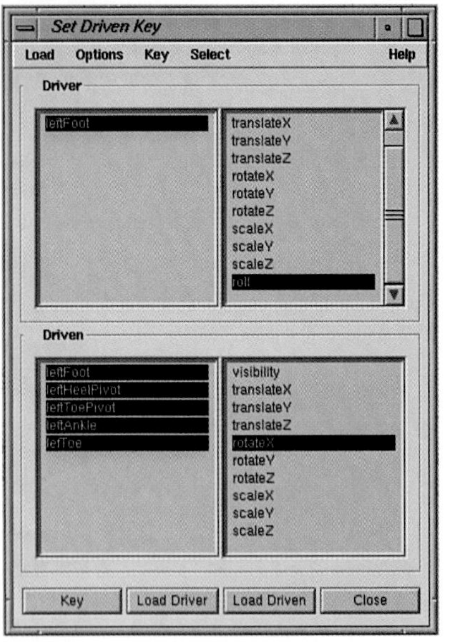

Figure 9.34
The SDK window with foot nodes loaded.

1. In the Animation menu set, go to the Animate menu to Set Driven Key, Set, to open the SDK window. Select the leftFoot, and click the Load Driver button in the SDK window. Maya displays leftFoot in the top-left box and its keyable attributes in the top-right box.
2. Select the leftHeelPivot, the leftToePivot, the leftToe, the leftAnkle, and the leftFoot, and then click the Load Driven button. The objects display in the lower-left box and their keyable attributes in the lower-right box.

We now have the Driver and Driven nodes loaded into the SDK window. Next, we will select which attributes will either drive or be driven. The keys we set here can almost be thought of as poses or extremes for the positions of the foot. Here, roll will be the Driver and rotate X will be the driven attribute of the Driven nodes.

Exercise 9.17: Setting the Initial Set Driven Keys

1. Check that leftFoot's roll value is set to 0.
2. Check that all the Driven object's X rotations are set to 0. Again, you can select the nodes right from the SDK window.

3. Make sure all the names of the Driven nodes are highlighted in the SDK window. Select the roll attribute in the driver attribute section and rotateX in the driven attribute section and click the Key button at the bottom of the Set Driven Key window. This sets the initial keyframes for all the Driven object's X rotations.

Tip

You can click leftFoot in the SDK window to select it so that you don't have to select it in the modeling views.

Exercise 9.18: Settin the Keys for the Foot Rotating back on the Heel

1. Select the leftFoot, and set roll to –5 in the Channel Box.
2. Select leftHeelPivot, and set its X rotate value to –25.

Tip

In your Set Driven Key workflow, you will always want to change the Driver value first and the Driven value second. Remember that the Driver attribute(s) will drive the Driven attribute(s). Therefore, if you were to edit the Driven attributes first, they would change as soon as you changed the value of the Driver attribute, causing incorrect results.

3. Make sure that only leftHeelPivot is highlighted in the Driven area of the SDK window, and click the Key button in the SDK window.

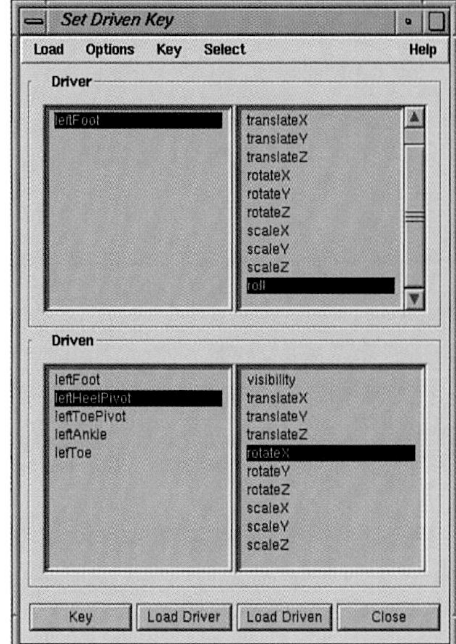

Figure 9.35
The SDK and leftHeelPivot.

Exercise #19: Setting the Keys for the Foot Rotating Around the Ball of the Foot

1. Select the leftFoot, and set its roll attribute to 5.
2. Select leftAnkle, and set its X rotate to 40.
3. Make sure that only leftAnkle is highlighted in the Driven area of the SDK window, and click the Key button in the SDK window.

Exercise 9.19: continued

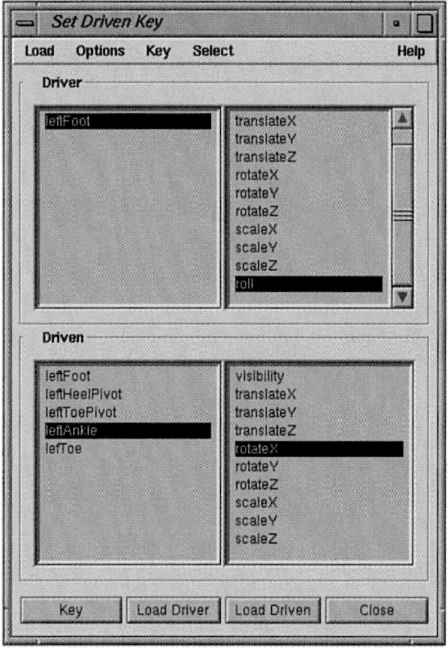

Figure 9.36
The SDK and leftAnkle.

Exercise# 20: Setting the Keys for the Foot Rotating Off of the Toe

1. Select the leftFoot, and set roll to 10.
2. Select leftToePivot, and set its X rotate to 20.
3. Make sure that only leftToePivot is highlighted in the Driven area of the SDK window, and click the Key button in the SDK window.

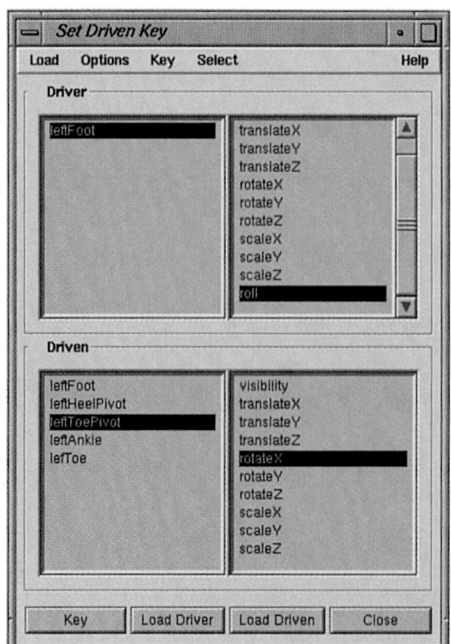

Figure 9.37
The SDK and leftToePivot.

Exercise 9.21: Testing the Foot Roll Attribute

1. Select the leftFoot, and click the attribute named roll in the Channel box.
2. In any of the modeling windows, middle mouse button-drag the mouse horizontally. Maya updates the roll attribute values, and the foot rolls forward and backward.
3. Save your work as **`myBipedSkeleton2.mb`**.

You have now set up the rolling of the foot using your own custom attribute (roll) and Set Driven Key. Now, to animate the foot rolling, you simply have to set keyframes on the roll attribute within the value range of –5 to 10. Keep in mind that you can use Modify, Add Attributes to add your own attributes to virtually any node in Maya. Then you can use Set Driven Key to connect that custom attribute to any other attribute in your scene (including lights and texture attributes).

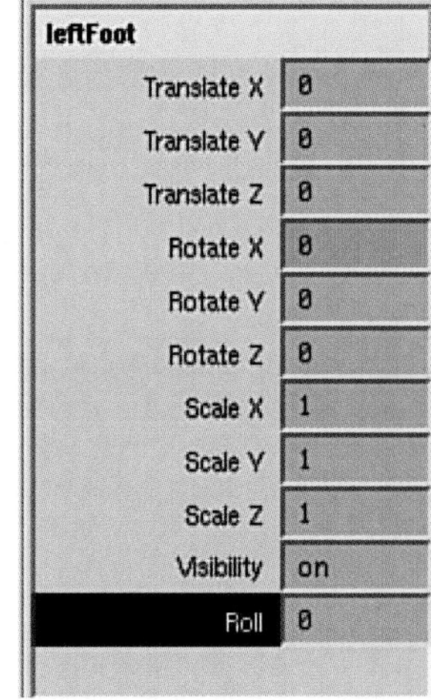

Figure 9.38
Roll selected in the Channel box.

9.3.4 Modifying Set Driven Keys

It is possible for you to edit these Set Driven Key relationships later on. Remember that all the editable SDK curves will occur only on the Driven object's attributes.

Let's increase the amount of rotation on the leftToePivot when the foot rolls forward.

Exercise 9.22: Increasing the Rotation on the leftToePivot

1. Select leftFoot, and set the roll value to 10. The foot rolls forward off of the toe.
2. In the HyperGraph or Outliner, select leftToePivot, the Driven object whose animation we must modify.
3. Choose Window, Animation Editors, Graph Editor. Press the F key while your cursor is in the Graph Editor to frame the animation curves.

   Typically, the Graph Editor displays animation information with Time (in frames) displayed along the horizontal axis and Attribute Values displayed along the vertical axis. In the Graph Editor, a red curve represents the X rotation SDK we set up earlier for the leftToePivot node. With SDK, the information in the Graph Editor is represented differently: The value of the Driver's attribute (roll = –5 to 10) appears along the horizontal axis, and the value of the Driven attribute (leftToePivot = 0 to 20) appears along the vertical axis. Time is not graphed when you are using SDK; however, you would animate the roll attribute to Time when you start animating your character.

Exercise 9.22: continued

The Driver and Driven object names appear in the lower right and upper left of the Graph Editor.

4. Take a look at one of the keyframes we set using SDK for the leftToePivot. Along the horizontal (X) axis in the Graph Editor, go right to 10. Go up along the vertical (Y) axis in a positive direction to 20.

   This point represents the keyframe for the leftToePivot at 20 degrees of X rotation when roll is at a value of 10.

5. Select the keyframe at 10,20, press the W key to translate, hold down the Shift key (to constrain the direction we will move the key), and MMB drag the key to 25. (Toggle on Value Snap if you want to snap to whole number values.) The leftToePivot will now rotate 25 degrees in X when roll is at 10.

6. Repeat steps 1 through 5 for the rightFoot and rightToePivot nodes. Set the roll values for both feet back to 0 when you are finished editing the SDK curves for the toePivots.

Tip

You could also select the key and use the stats window in the upper left of the Graph Editor to type the values you want for X and Y.

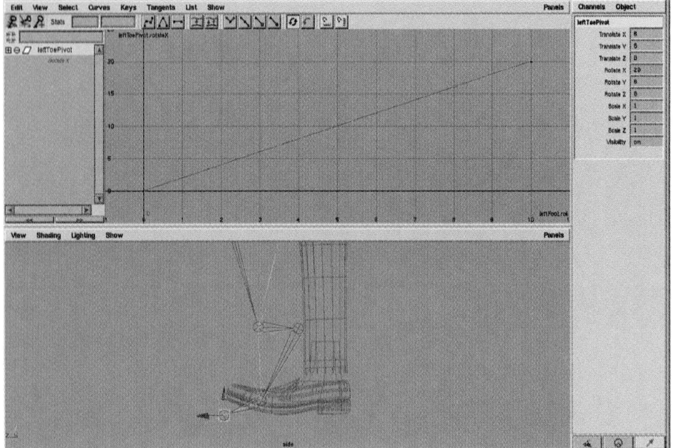

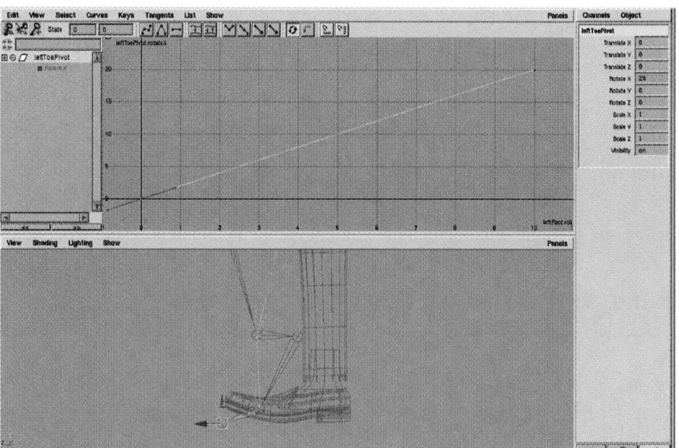

Figure 9.39
The SDK curve in the Graph Editor before editing (left), and after editing (right).

9.3.5 Adding a Selection Handle to the backRoot Joint

Let's use some selection handles to make it easier to pick important nodes of our 3D character. *Selection handles* are icons that can be placed anywhere in the scene. When selected, they select the node they represent. Selection handles are not nodes themselves; they simply enable you to pick nodes more conveniently.

Exercise 9.23: Display a Selection Handle to the backRoot

1. Select the backRoot joint in the HyperGraph or Outliner, and choose Display, Object Components, Selection Handles. It may be hard to see, but right on top of the backRoot joint is a black crosshair icon.
2. Go to Component Mode (you can do this quickly by using the F8 hotkey), turn off all component categories, and then turn on the Selection Handle pick mask. In Component Mode, you affect the selection handle only and not the object it selects.
3. Select the selection handle, press W, and drag the handle straight behind the character. Make sure that only the Selection Handle moves. If the skeleton moves, undo the translation and verify that you are in Component Mode.

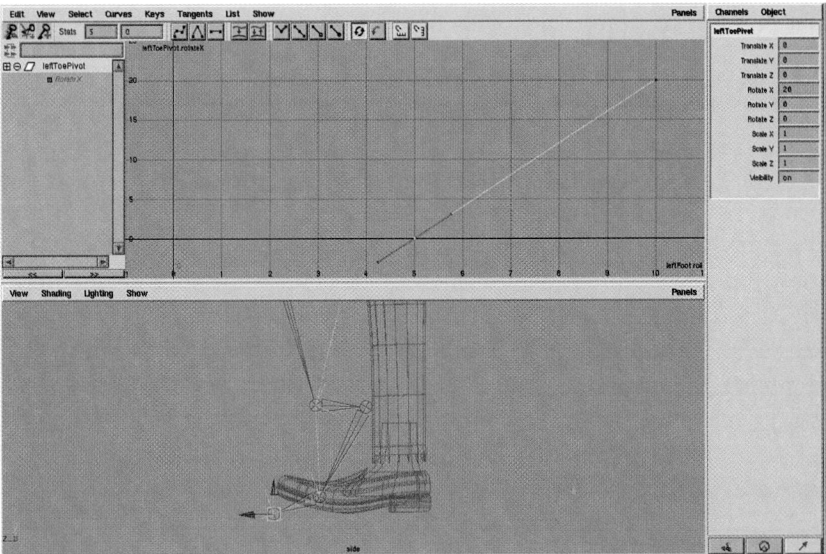

Figure 9.40
The selection handle behind the character.

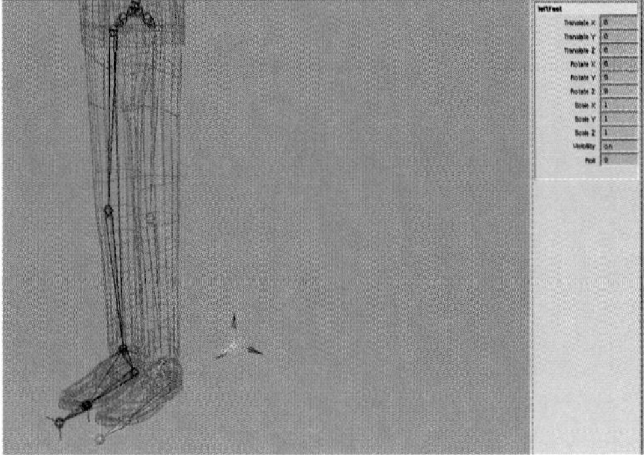

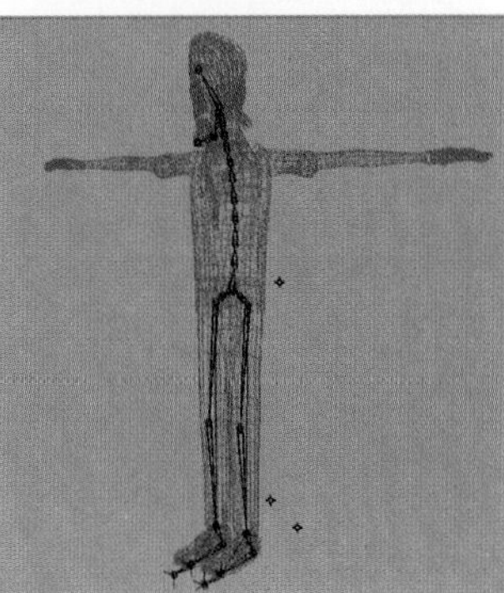

Figure 9.41
The backRoot and feet selection handles translated behind the skeleton, continued.

Exercise 9.23: continued

4. Go to Object Mode (press F8 on the keyboard again; this hotkey toggles between Object and Component Modes). Select the selection handle you've just created behind the character (if it is not already selected). Note that the backRoot is selected in the Channel Box. Translate the backRoot to test the movement of the skeleton.

   This time, the skeleton moves.

5. Display selection handles for the leftFoot and rightFoot nodes and move the handles behind the character. You may want to raise one of the foot handles higher than the other so that when you are in the Side view, the selection handles won't be on top of each other. Move the handles somewhere away from the character, but not too far. Moving the handles slightly behind the character prevents us from accidentally picking the wrong object.

6. Save your work as **`myBipedSkeleton3.mb`**.

9.3.6 Using the Rotate Plane (RP) IK Solver

In addition to the Single Chain IK solver that we used in the feet and legs of our biped character, you can use the Rotate Plane IK solver for some additional control. In the leg, for example, the Single Chain solver rotates the knee and hip joint around a fixed 2D plane (invisible to the viewer). In animating the biped's leg, the only way to point the knee in a particular direction is to rotate the entire foot hierarchy. This, however, rotates the foot along with the leg. If we wanted to keep the foot in place while pointing the knee, we could use the Rotate Plane solver. Let's take a brief look at how we can use the Rotate Plane Solver in the leg.

Figure 9.42
Pointing the knees with the Single Chain Solver and with the Rotate Plane Solver.

Exercise 9.24: Using the Rotate Plane Solver

1. Select the IK handle on the left ankle (leftAnkleIK), and open the Attribute Editor for the leftAnkleIK Ik handle.
2. Go to the IK Solver Attributes section and switch ikSCsolver to ikRPsolver. A circular icon appears at the left hip joint with a pointer inside. The pointer represents the position of the rotate plane.
3. With the leftAnkleIK IK handle still selected, set the Twist attribute in the Channel Box to –30 and note how the direction of the left knee changes and how the foot stays in place even though the leg is rotated.
4. Set the twist attribute back to 0.
5. Repeat steps 1 through 4 to change the Single Chain IK handle on the right ankle to the Rotate Plane Solver.

The Rotate Plane Solver is good to use in situations where you may need more control over the joints in your character.

9.3.7 Creating Joints for the Arms

Here again is one way of setting up arms for a 3D character. Of course, Maya has a multitude of ways to set up the joints for animating a character's arms. In this chapter, we'll set up the arms so that they can be used with forward kinematics animation techniques.

Exercise 9.25: Creating Five Joints for the Arms

1. Choose Skeleton, Joint Tool, Option, and verify that Auto Joint Orient is still set to XYZ.
2. Create a collar bone, shoulder, elbow, forearm (yes, forearm!), and wrist joints, and name the joints accordingly. Make sure that the joints you create for the arms lie inside the templated geometry of the character. Remember that you can position joints after creating them using Translate or Set Pivot.
3. Select the collarbone joint once again and use the Mirror Joint tool to mirror the arm joint across the YZ axis.
4. Parent the arm skeleton to the spine joints. Select the collarbone joint, Shift+select the backShoulderJoint, and choose Edit, Parent. Maya creates a bone connecting the two joints.x

Exercise 9.25: continued

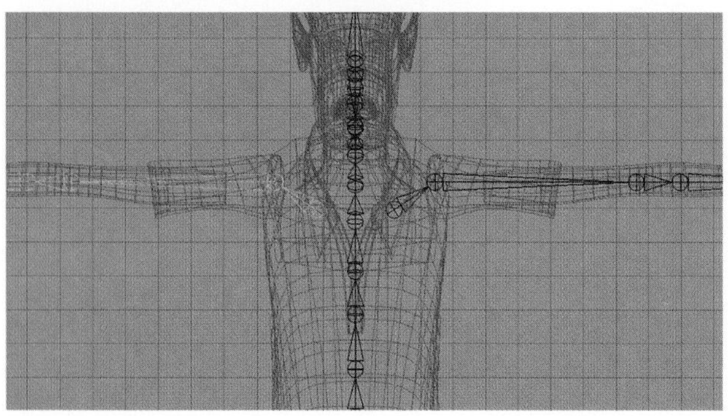

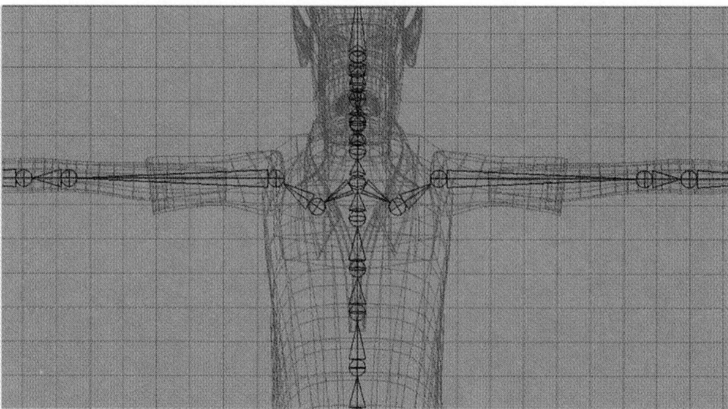

Figure 9.43
The arm mirrored to the right side.

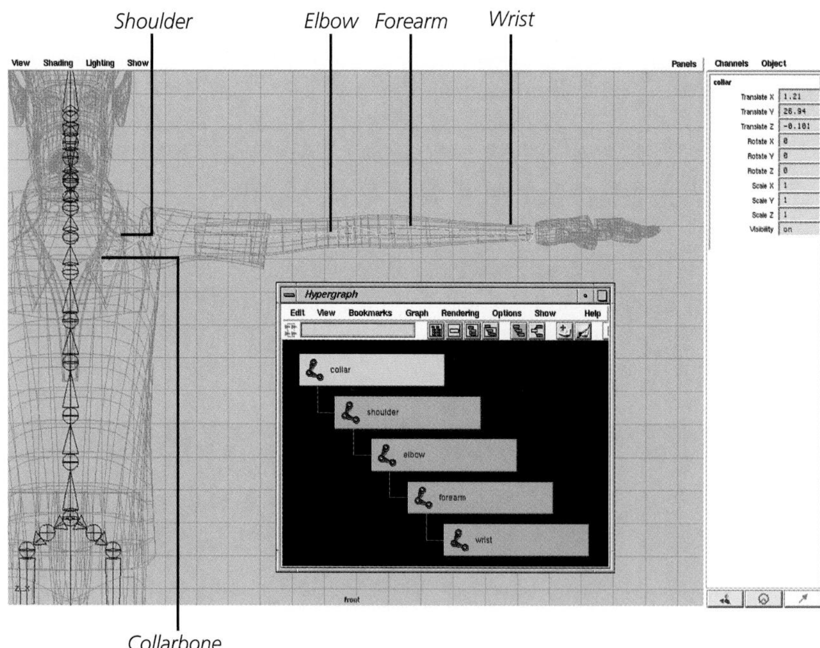

Figure 9.44
Diagram of arm joints, and parented to spine.

Exercise 9.26: Laying Out the Basic Skeleton Joint for the Hand

1. Use the previous methods to create hand joints for the character. Use the following figures as a guide for the placement and number of joints to use for the hand.
2. Use Set Pivot and Translate to move joints into the proper position.

3. Flip the local rotation axis of the joints so that the joints throughout the hand are consistent with one another. (Note in the diagrams the extra joints used in the thumb and pinky.)
4. Open bipedSkeletonFinish.mb for an example of this and compare your work with the hand in that file.

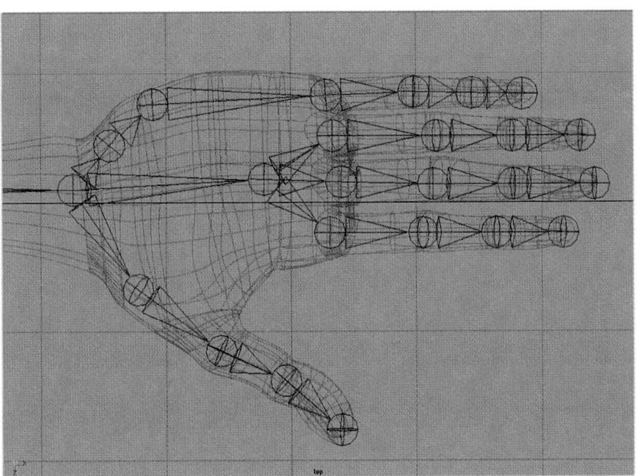
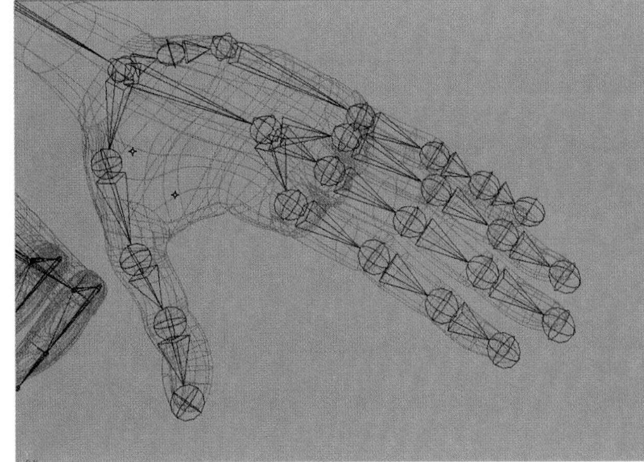

Figure 9.45
Hand joints, and Perspective view of hand joints.

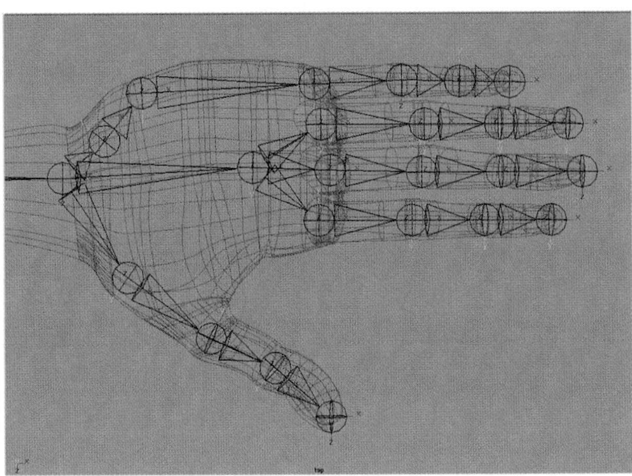
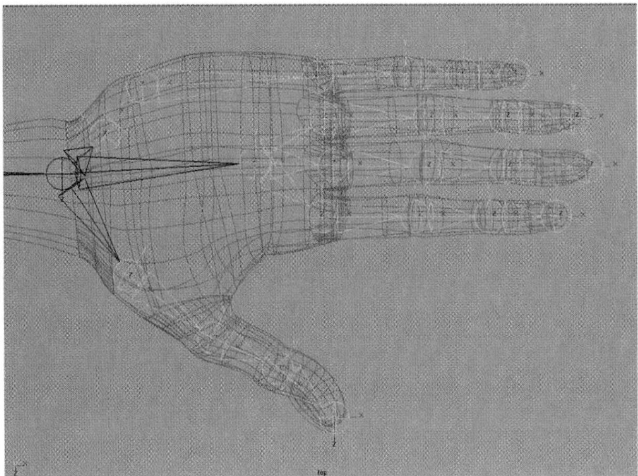

Figure 9.46
Hand joints before and after Local Rotation Axes are flipped.

9.3.8 Binding Character Geometry to Skeleton Joints

Binding the geometry or skin of a character is the process of assigning points on the character's surface geometry to bones on the character's skeleton. Therefore, when the skeleton is moved (rotated), the surfaces will deform (bend, fold, and bulge) to follow the joint position. You can think of it as wrapping a layer of clay around a metal wire. When you bend the wire, the clay bends and folds to follow. This occurs with your arm as well. As your elbow joint rotates, your skin moves and stretches to follow.

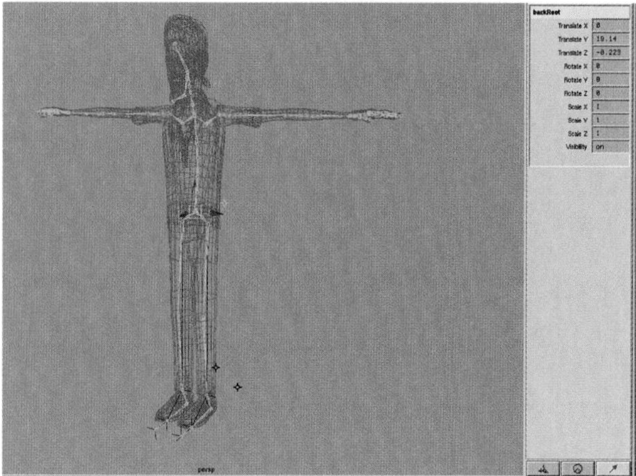

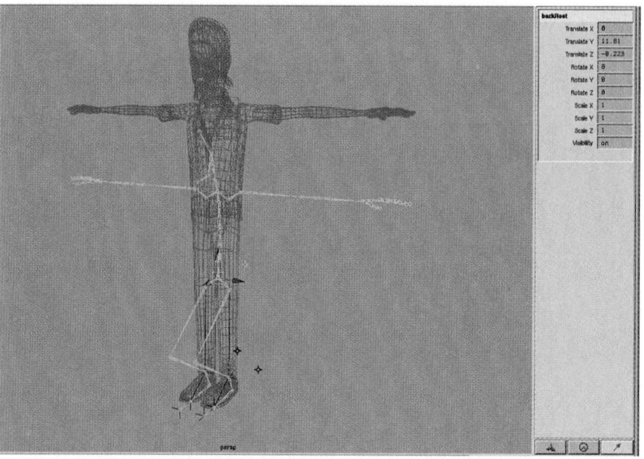

Figure 9.47
Biped skeleton in neutral position, crouched position, and with skeleton and geometry bound together.

The animator can bind the skin of the character to the skeleton joints directly and/or indirectly. There are two direct methods of "skinning": rigid skinning (Skin, Bind Skin, Rigid Skin) and smooth skinning (Skin, Bind Skin, Smooth Skin). Both rigid and smooth skinning involve the assignment of points on the surface geometry to joints in the skeleton. When joints in the skeleton are transformed, or rotated, they will influence the geometry to bend and deform. The points can then be weighted by the user to refine the way the skinned geometry bends and folds around the influencing joints. Smooth skinning adds preset weights to surface points by default, whereas Rigid skinning does not.

Indirect skinning can be achieved in two ways as well: through the use of lattice and/or wrap deformers. The animator can use a lattice or wrap to deform a surface (or surfaces), and then bind the lattice or wrap deformer to the skeleton. With indirect skinning, the skeleton deforms the lattice or wrap deformer, which is deforming the surface geometry. We will take a look at examples of binding a character's surface geometry directly, indirectly, and using a combination of both.

To bind skin, the animator must first choose a skeleton joint hierarchy and surfaces (or deformers) to bind, then choose Skin, Bind Skin, Rigid Skin (or Smooth Skin). Maya will determine which surface points are closest to which bones, group those points into one or more clusters (special types of groupings of points that allow for weighted effects on individual points), and make each cluster a child of that joint. The animator can then edit the membership (which points belong to which joint) and weights of points to fine-tune the bind skin of the character.

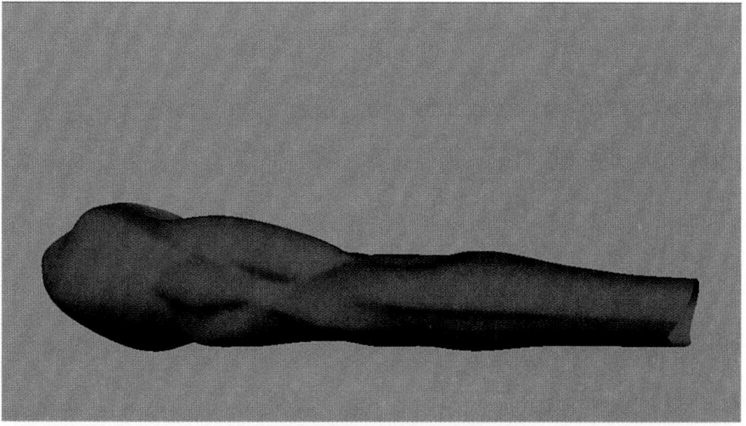

Figure 9.48
The surface of the arm.

Exercise 9.27: Direct Skinning: Binding a Simple Arm to Skeleton Joints Using Rigid Skin

1. Go to File, Project, Set and set the current Maya project to armSkinning. Go to File, Open Scene and open the file arm.mb (or, you can build your own arm or use a cylinder for this task). A detailed arm model will appear in your scene.
2. In the Side View, using the Joint tool, draw the shoulder, elbow, and wrist joints inside the arm geometry.

Exercise 9.27: continued

3. In the Outliner, select the arm group node (containing all the objects that make up the arm) and Shift+select the shoulder joint (the top of the skeleton hierarchy).
4. Go to the Skin, Bind Skin, Rigid Skin, Option, and click the Color Joints check box to turn it on. Color Joints will color surface points the same color as the joint of which the points are members, making membership editing much more intuitive.
5. Select the elbow joint and rotate it on the Z axis so that the arm surfaces (from the elbow down to the wrist) are now following the rotation of the elbow.
6. Go to Skin, Bind Pose. This returns the joints and the geometry to the positions they were in when the Bind Skin operation was performed.
7. Press the up arrow on the keyboard to get to the shoulder joint and rotate it on the Z axis as well. All the points follow the shoulder because it is at the top of the hierarchy.
8. Go to Skin, Bind Pose to return the arm joints to their original position when the bind operation was performed.

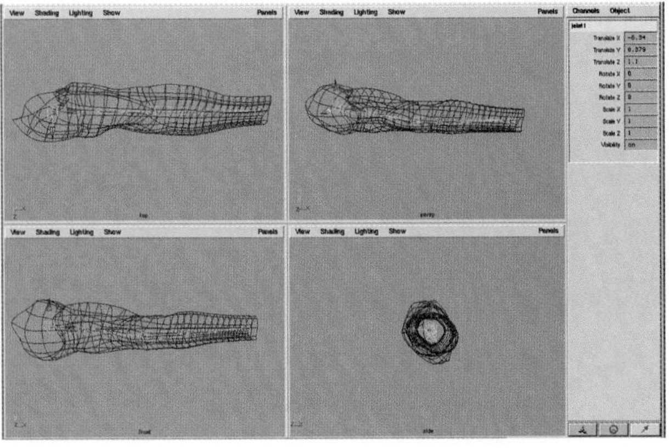
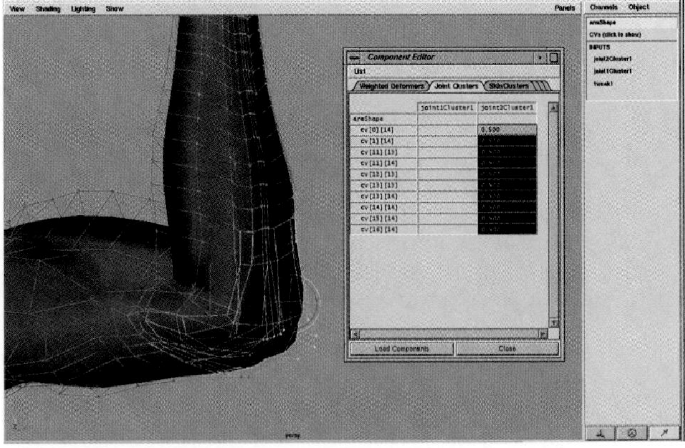

Figure 9.49
Joints in the arm, and weighting points at the elbow on the attached arm in the biceps pose.

When rotating the elbow joint, you'll notice that the surfaces around the elbow crimp and stretch unnaturally. Let's look at how we can weight the points around the elbow to get a more natural deformation.

Exercise 9.28: Weighting the Points in the Elbow

1. Select the elbow joint, and rotate it as if the arm were flexing its biceps.
2. In the Side View, select points at the bottom of the elbow that are stretching too far, and choose Windows, General Editors, Component Editor. In the Component Editor select the Joint Clusters tab. Maya lists information in columns for the points you have selected.
3. In the Component Editor, find the column with the weight values for the selected points. In this example, it's the joint2Cluster1 column.
4. Click one of the fields that has a cluster weight value and then click the name of the cluster at the top of the column in the Component Editor. Type **.5** in the number field. You will see that all the weight values for the selected CVs update. In the 3D views you will see the selected CVs jump into a new location because you have reduced their weight from 1(100%) to .5(50%).
5. Continue selecting points around the elbow and weight their values in the Set Editor. You will only see weight adjustments on points that move when the elbow is rotated. The points from the elbow to the wrist are members of the elbow joint.

When weighting points in clusters, you are specifying that points inherit a percentage of the transformation from the joint they are members of. For example, if the elbow rotates 90 degrees, points that are members of the elbow joint with a weight of .5 (50%) would rotate only 45 degrees, whereas points with a weight of 1 (100%) would rotate the full 90 degrees. (You can weight cluster points above 1 and below 0.)

9.3.9 Using Lattice Flexors

You can see that weighting points manually can get pretty tedious, especially on a more complex character. However, Maya offers other tools that speed up your workflow when working with deforming surfaces.

Lattice joint flexors offer built-in attributes to aid you in smoothing deformations around joints. A *lattice* is a type of deformer that places a bounding box of points around selected items (surfaces and/or points). The lattice points influence points on the geometry. When lattice points are transformed, they pull on the surface points. The lattice deformation of the surface geometry occurs in the bounding box area of the lattice. A lattice joint flexor influences points around a specified joint.

Exercise 9.29: Creating a Lattice Flexor for the Arm

1. Repeat the steps in the preceding exercise with the arm.mb file. Create a simple arm skeleton (shoulder, elbow, wrist), and bind the arm geometry to the skeleton. This time, however, don't weight any of the CVs with the Component Editor.

2. Select the elbow joint and go to Skin, Edit Rigid Skin, Create Flexor. In the Create Flexor option box set Flexor Type to Lattice, turn on the At Selected Joints option, and set the S to 2, T to 5, and U to 2. The divisions specify how the lattice you've created is divided in X, Y, and Z space.

3. Turn on the Position the Flexor option, and click the Create button. With the Position Flexor on, Maya selects the Lattice Group node when you have finished creating the Lattice Flexor.

   Maya creates a Lattice around the elbow joint. Notice that the Lattice Group node is selected in the Channel box.

4. Scale the lattice so that it fully encompasses the arm on both sides of the elbow.

5. Select and rotate the elbow joint so that the arm looks as if it is flexing its biceps muscle. Lattice flexors are automatically parented to the selected joints, so they will follow the character joints when they are moved.

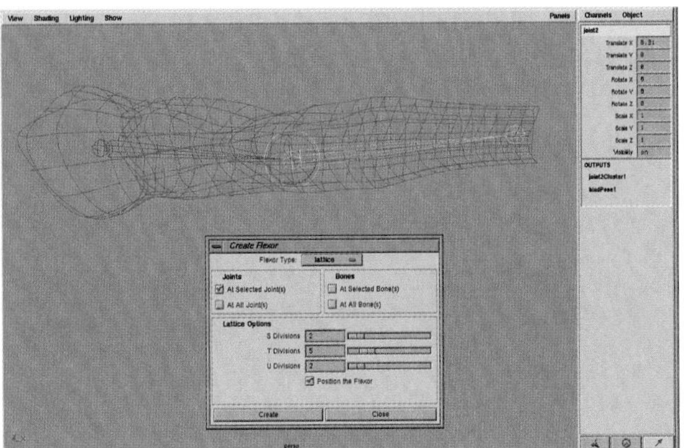

Figure 9.50
The Create Flexor window.

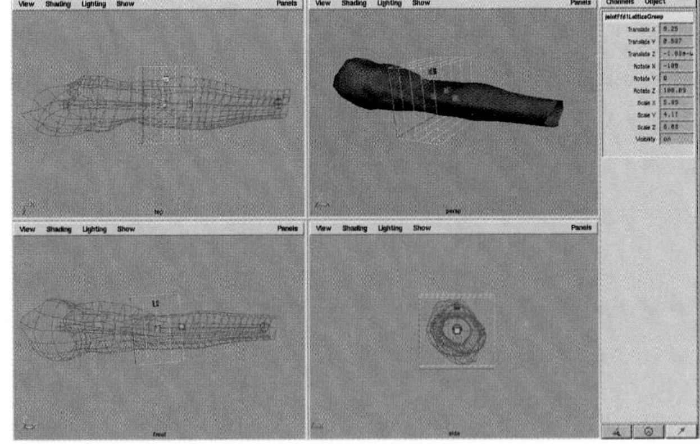

Figure 9.51
The arm with lattice flexor resized.

Warning

Make sure when you scale the lattice that you have the Lattice Group node selected and not just the lattice. The lattice object will actually deform the surfaces when moved. If the arm geometry moves when you try to reposition the lattice, Undo and select the Lattice Group node one level above the lattice. You can quickly get to the Lattice Group by selecting the lattice and then pressing the up arrow on the keyboard to step up one level in the hierarchy to the Lattice Group. Use the Lattice Group node to reposition or resize the lattice around the joint.

6. With the left mouse button, select the lattice and look over at the Channel box. In the Shape section, Maya displays attributes unique to lattice flexors.
7. With the elbow still rotated, adjust the values for Creasing, Rounding, Length In, and Length Out until the surfaces around the elbow deform more cleanly.

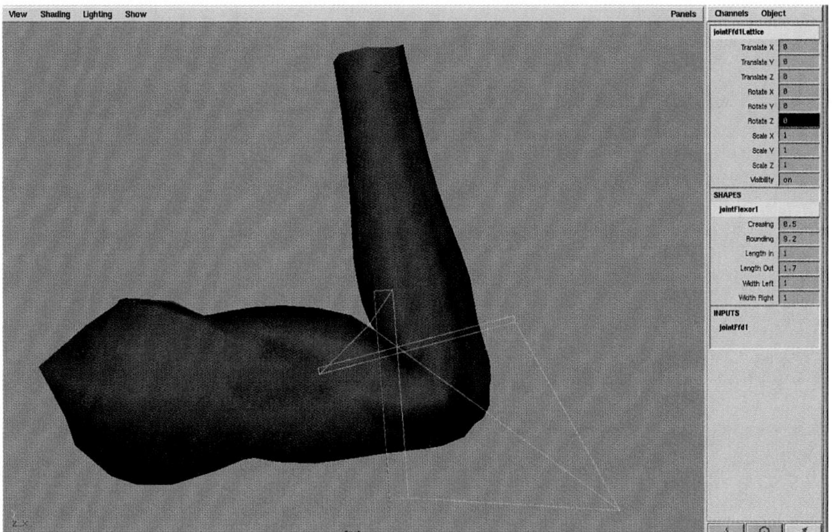

Figure 9.52
The arm rotated at the elbow with Flexor Attributes adjusted.

Note

You can think of lattice flexors as having their own Set Driven Keys built into themselves. When the joint is at its Bind Pose, the lattice flexor is unaffected. You can modify the lattice shape when the joint rotates by going to the shape section of the Channel box of the lattice flexor and modifying the Creasing, Rounding, Length In, Length Out, Width Left, and Width Right attributes. These Lattice Flexor attributes affect how the lattice deforms the surface area it is surrounding. You can use these attributes to create bulge effects. After you have modified these attributes, every time the elbow is rotated the lattice changes shape automatically to the attributes you specified.

8. Adjust the Creasing, Rounding, Length In, Length Out, Width Left, and Width Right for those attributes and look at how Maya modifies the deformation at the elbow.
9. Select the shoulder joint and create a lattice flexor, this time using the At Selected Bone option. This option works similarly to the lattice flexor at the joint; however, the lattice is placed along the length of the bone and not around the joint. The lattice flexor at the bone has built-in attributes for quickly creating biceps and triceps deformations.
10. Adjust the biceps and triceps attributes and note how the deformation on the arm changes.

Exercise 9.29: continued

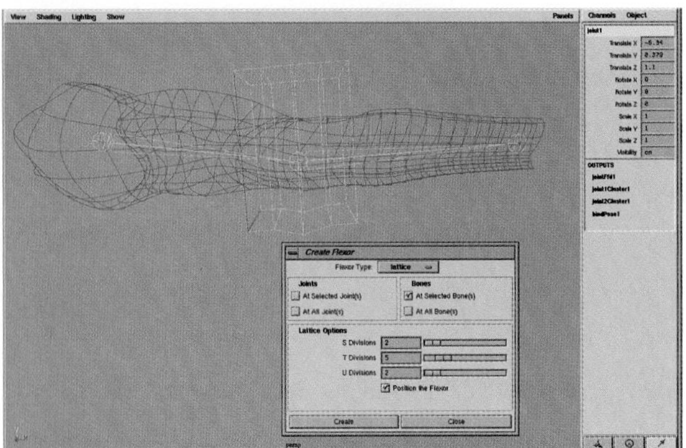

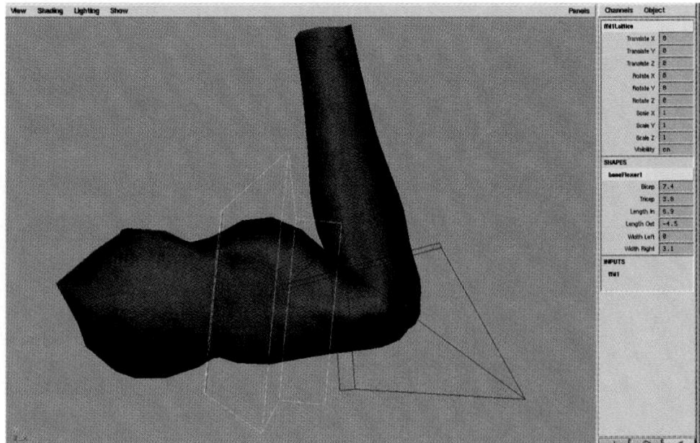

Figure 9.53
The Channel box with bone lattice attributes adjusted.

Note

Using lattice flexors allows you to quickly fix deforming areas of your character. This method is much quicker than having to weight individual cluster points. However, in some situations where you may need very detailed deformations (such as for realistic muscles or tendon movement), you may need point-by-point control over deforming areas. Lattices generally smooth out points in deforming areas; this may be undesirable at times when more intricate detail is needed.

9.3.10 Indirect Skinning: Binding a Lattice to Skeleton Joints

Maya also allows you to bind lattices to skeleton joints. Lattice deformers are slightly different from lattice flexors. Lattice flexors work around joints and bones, whereas lattice deformers in general can be used to deform entire surfaces, parts of surfaces, or groupings of surfaces. Lattices are effective in situations where you may have many surface points or multiple surfaces. A few lattice points can be used to influence hundreds of surface points.

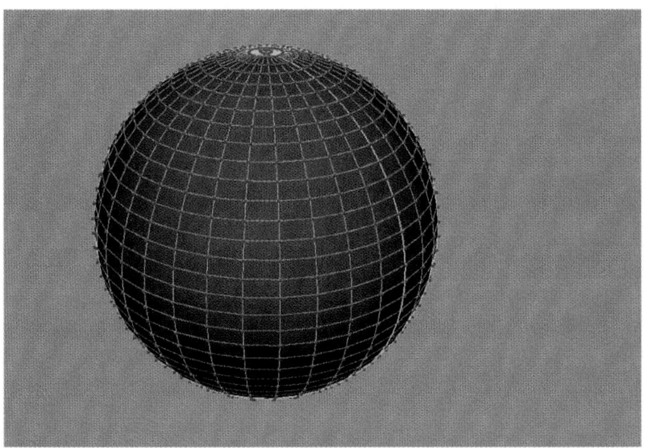

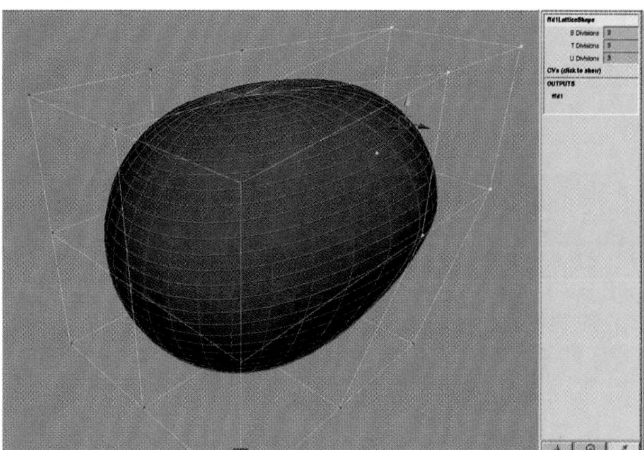

Figure 9.54
Lattice points affecting surface geometry.

Exercise 9.30: Binding a Lattice to the Arm Joints

1. Open the file armSkeleton.mb. This file already has the skeleton joints created for you.
2. Select the arm geometry.
3. Choose the Deform, Create Lattice, Option, set the S division to 10, T to 4, and U to 2, turn on Center Around Selection and Group Base and Lattice Together, and click the Create button. Maya creates a lattice that encompasses all the arm geometry.
4. With the lattice selected, go to Component Mode and turn on lattice points in the Pick Mask. You can also place the cursor over the lattice and hold down the right mouse button to bring up the marking menu for showing the lattice points. Select a few lattice points, translate them, and take a look at their effect on the arm geometry. Undo all your lattice point translations.
5. Go back to Object Mode, select the lattice and the joint at the top of the skeleton (shoulder), and choose Skin, Bind Skin, Rigid Skin. Be sure that you do not have the arm surface selected when performing the Rigid Skin.
6. Select the elbow joint and rotate it. Watch how this rotation affects the lattice.

Tip

Using the Component Editor, you can weight lattice points being deformed by the skeleton the same way we weighted the CVs in the arm.

Exercise 9.30: continued

Note

In this situation, Maya assigns the lattice points to the joints instead of the surface CVs of the arm. Remember that the CVs of the arm geometry are being deformed by the lattice which, in turn, is being deformed by the skeleton. Binding the lattice results in much fewer points that need to be modified in the deformation of the arm. The lattice also offers a nicely organized box of points to select, as opposed to a complex surface with hundreds of points on top of one another.

7. Select the lattice points you need to weight, go to the Component Editor, and weight the selected points highlighted in the joint cluster list.

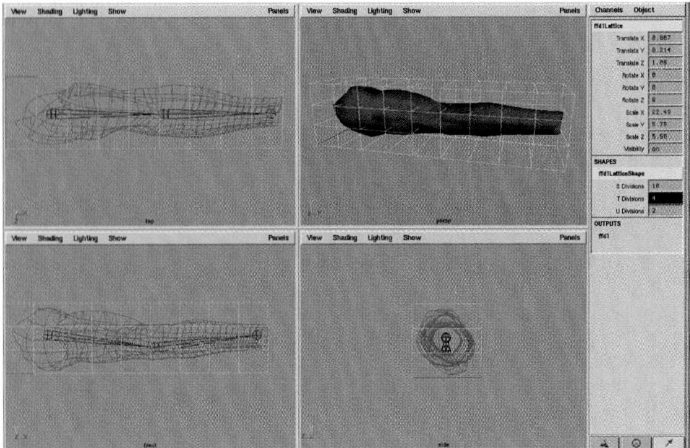

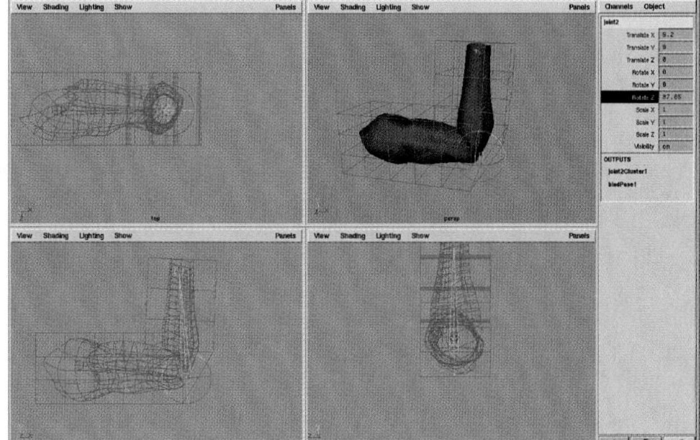

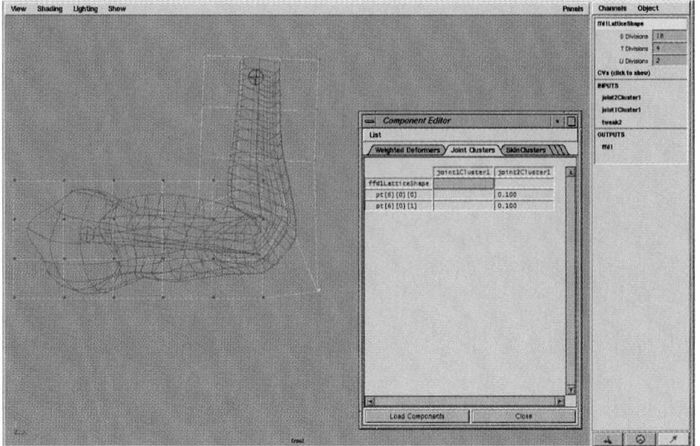

Figure 9.55
A lattice around the arm, a lattice bound to the skeleton, and lattice points weighted.

Exercise 9.31: Binding a Wrap Deformer to the Arm Joints

Let's look at another way to bind the arm geometry to the skeleton indirectly, this time using a wrap deformer. A wrap deformer enables you to deform a surface with another surface. What you will do here is create a low-resolution version of the arm to wrap-deform our original arm. We will then deform the low-resolution arm with a skeleton.

1. Open the file armSkeleton.mb once again.
2. Select the arm surface geometry and go to the Modeling menu set on the main menu bar or in the hotbox. Go to Edit Surfaces, Rebuild Surfaces, Option.
3. In the Rebuild Surfaces Option box, click the Reset button at the bottom of the window. In Number of Spans U and V, set each to 10 and turn on (check) the Keep Original option. Click the Rebuild button at the bottom of the window. A lower-resolution copy of the original arm surface is created. Close the Rebuild Option window.

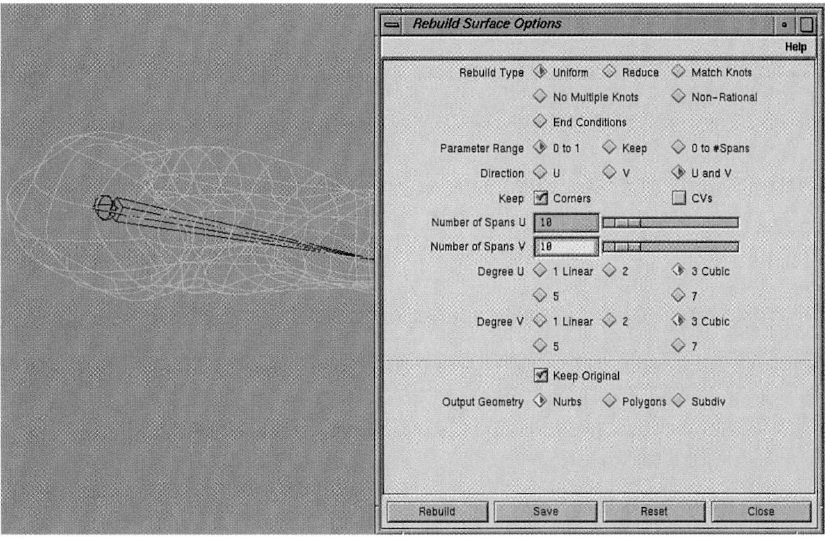

Figure 9.56
Rebuild Surfaces Options window settings for rebuilding the arm surface.

4. With the rebuilt arm surface still selected, go to Edit, Delete by Type, History to remove the construction history for the rebuilt arm.
5. In the Channel box, rename the rebuilt arm surface to armLow.
6. Open the Outliner window and select the arm node, and then Ctrl+select the armLow node. Go to the Animation menu set, then select Deform, Create Wrap. Now the arm geometry will follow the shape of the armLow geometry. When creating wrap deformations, always select the object that will deform the other surfaces last. In the outliner you will also see an additional armLowBase node created. This is the intermediate object or reference to the armLow surface before any deformation has been applied.

Exercise 9.31: continued

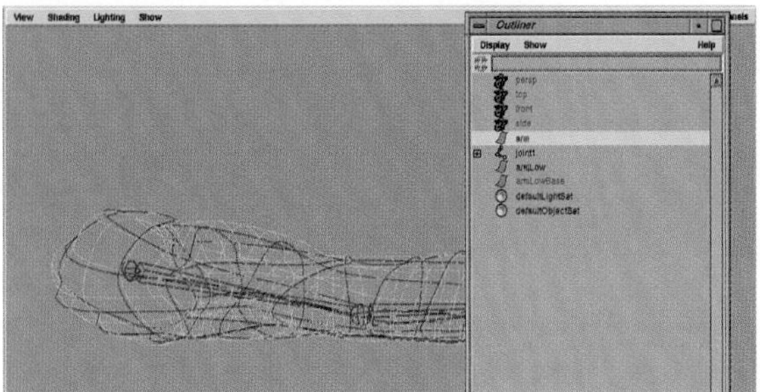

Figure 9.57
Rebuilt copy of original arm surface.

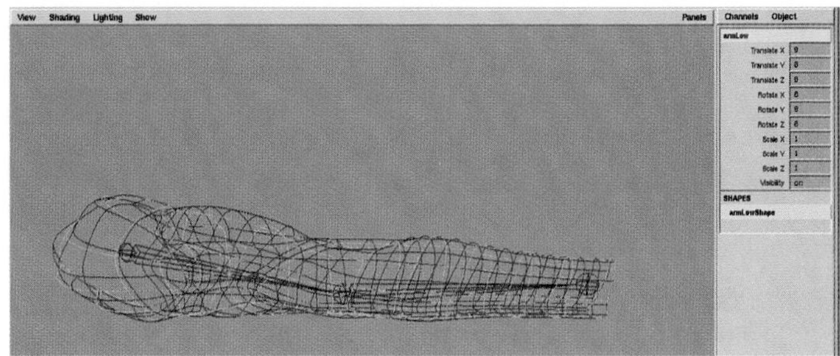

Figure 9.58
After creating the wrap deformer, an armLowBase object will also be created.

7. Select armLow in the Outliner and then Ctrl+select joint1. Go to Skin, Bind Skin, Rigid Bind and use the default settings to bind armLow to the skeleton.
8. Select the elbow joint (joint2). Rotate the elbow into a flexed position. If you look closely, both the arm and armLow geometry are deforming to the skeleton.
9. Select armLow and go to Display, Hide, Hide Selection. Turn on hardware shading by pressing the 5 key on the keyboard.

Any deformations you now apply to the armLow geometry will be carried through to the arm geometry. For instance, you could add a lattice flexor to the elbow of the arm skeleton (which is deforming armLow) and the deformation would carry through to the higher-resolution arm surface. You could use this technique for binding low-resolution geometry to a skeleton that is a wrap deformer for high-resolution geometry. The low-resolution geometry results in less data for you to have to deal with when you're animating.

Figure 9.59
Arm geometry "indirectly" bound to the skeleton using a wrap deformer.

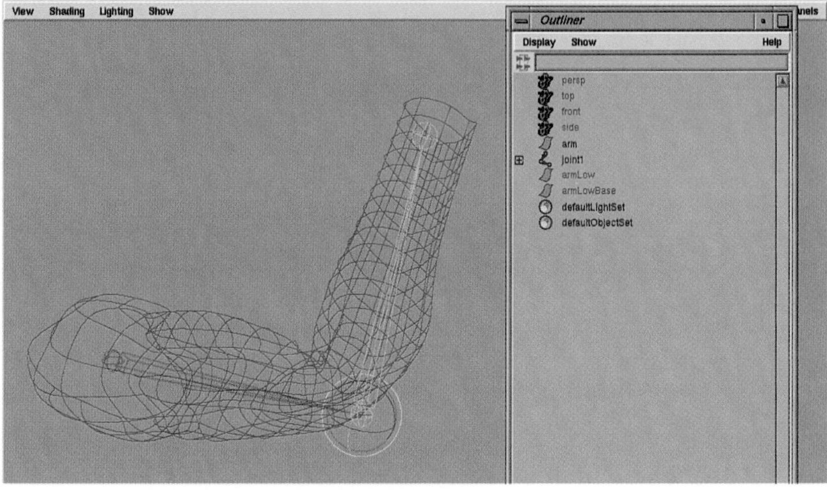

Exercise 9.32: Adding an Influence Object to the Wrap Deformer

Let's look at an approach to binding our biped character geometry to the skeleton we have created. We'll use a combination of direct skinning techniques through the use of Rigid Bind Skin and a new technique found in Maya 2.0: Smooth Skin Bind. We will utilize Smooth Skin Bind in areas of the character that are typically problematic to bind to the skeleton: where the arms and legs connect to the torso. The Smooth Skin Bind will help us to maintain smooth deformations in these areas where there is a wide range of motion. Smooth Skin Bind will apply weighted deformations to the surfaces by default. This makes the weighting of surface points much quicker. We will look at how you can change the weights of the Smooth Skin Bind using the Paint Skin Weights tool.

1. Go to File, Project, Set and set the current Maya project to biped. Go to File, Open Scene and open your previous file (myBipedSkeleton3.mb) or the file bipedSkeletonPreBind.mb (which has already been completed for you).
2. Open the Outliner window and click the + symbol inside the square next to the bipedCharacter node to expand the hierarchy. Select rtSock, lftSock, lftShoe, rtShoe, lftArmGroup, rtArmGroup, and head. Go to Edit, Group. Name this group **`rigidSkin`**.
3. Select rtLeg, lftLeg, shirt, rtSleeve, and lftSleeve and group these nodes. Name the new group node **`smoothSkin`**.
4. In the Outliner, click the + symbol inside the square to the left of the bipedRoot node to expand its hierarchy. Select the backRoot joint node.

Figure 9.60
Biped with skeleton ready for "smooth skin" bind.

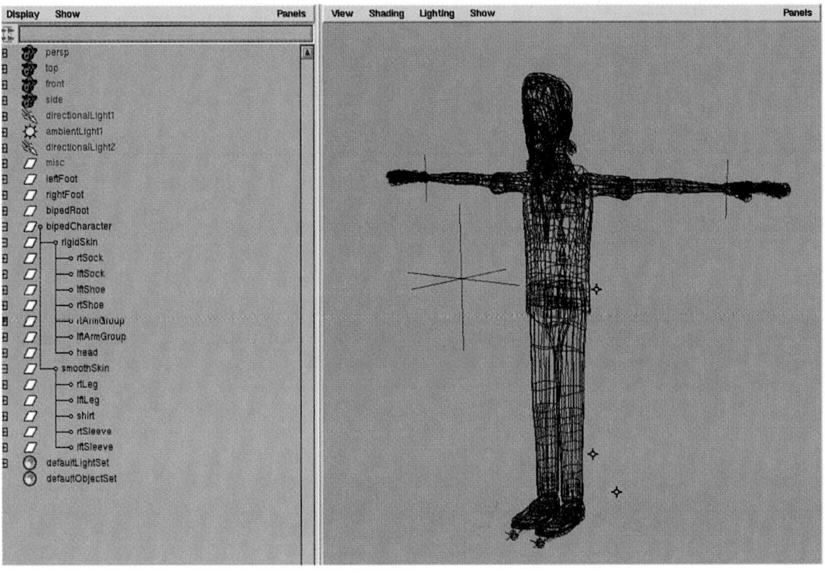

Figure 9.61
Grouping objects for rigid and smooth skinning.

Exercise 9.32: continued

5. Shift+select the rigidSkin node and go to Skin, Bind Skin, Rigid Bind, Option. In the Rigid Skin Option window, click the Reset button and then the Bind button. Maya binds the selected geometry to the skeleton using the Rigid Bind method.

Tip

A good way to test the Bind Skin of a character is to move the character and look for any problems, such as parts of the character's surface geometry not following the skeleton correctly.

6. Deselect everything and then select the smoothSkin node and the backRoot joint. Go to Skin, Bind Skin, Smooth Skin-Option. In the Smooth Skin Option window, click the Reset button and then the Bind button. Maya attaches the selected geometry to the skeleton using the Smooth Skin method.

7. Select the selection handles for the bipedRoot, leftFoot, and rightFoot.

8. Press W to translate, and drag the three nodes back along the Z axis.

   If the entire character moves backward without coming apart, the Bind Skin was successful. Undo the translation of the character.

   If parts of the character are left behind or move ahead of the character, you may have had some Bind Skin problems. Undo the translation, check your steps, and then re-bind the character.

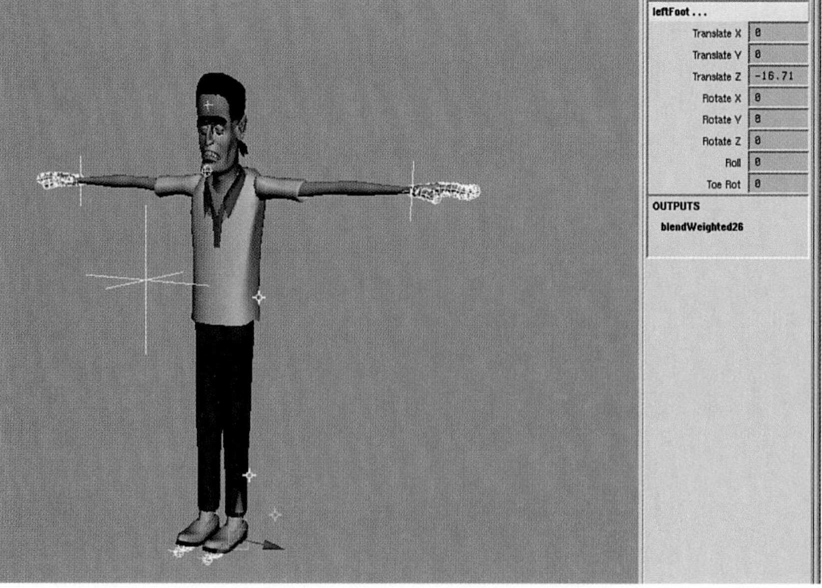

Figure 9.62
Testing the rigid and smooth skin bind by translating feet and bipedRoot nodes.

Exercise 9.33: Refining the Smooth Bind Skin Weights Using the Paint Skin Weights Tool

Next, we'll look at refining the weights of the Smooth Skin bind at a few key areas of the character. You will use the Paint Smooth Skin Weights tool (Skin, Edit Smooth Skin, Paint Skin Weights, Option) to paint weight values onto the different surfaces of the character.

1. Turn on Hardware Shading by pressing the 5 key. Go to the biped shelf and click the CrossPose shelf button. This will position the arms and legs of the character into an X position, allowing you to identify areas that need to be refined. The problem areas include the waist, between the legs, and the shoulder areas. In the crossPose, we can see that the geometry in these areas is either bulging or stretching unnaturally. To place the character back to its original pose, click the Neutral shelf button.
2. Select the left pant leg surface of the character. Open the Paint Skin Weights tool by going to Skin, Edit Smooth Skin, Paint Skin Weights Tool, Option(there is also a shelf button in the biped shelf that will open the tool for you).

   When you select the Paint Weights Tool Option window, the left pant leg surface turns black. When painting skin weights, black represents a weight of 0, white represents a weight of 1, and gray values represent the weights between 0 and 1. With the Paint Weights tool, we will select different joints of the skeleton and adjust their influence on different parts of the surface by painting weight values.
3. In the Tool Settings window for the Paint Skin Weights tool, go to the Influence section. In the text field is a list of all the joints of the character. By selecting the joint names in the influence list, you are instructing Maya to display the weights of influence on the selected surface. In the Stamp Profile section, change the Radius U and L to 1.0.
4. Select rightHip in the Influence list and look at the left pant leg surface in the perspective window. You will see the color values on the left pant leg change. There are lighter values near the top of the pant leg where the two pant legs come together. The right hip joint has a slight influence on the left pant leg as shown by the gray values on the left pant leg surface. You need to reduce the amount of influence the right hip joint has on the left pant leg.

Figure 9.63
Place character in a cross pose using the CrossPose button in the biped shelf.

Exercise 9.33: continued

5. In the Tool Settings window, go to the Operation section and change the mode to Smooth. In Smooth mode we will blend the dark and light areas of the weight map on the left pant leg. When you bring the mouse cursor over to the Perspective, the cursor has turned into a small paintbrush. Place the paintbrush cursor over the left pant leg. A circular red icon representing the paintbrush flows along the surface.
6. Paint the upper area of the left pant leg until you notice the left pant leg changing shape. The effect should be subtle; continue to move the paintbrush across the surface until the left pant leg deformation is smoother.
7. Repeat the same process on the right pant leg surface, but this time select the leftHip joint in the influence list of the Paint Weights Tool Settings window.
8. Now select the shirt body surface. Set the Operation section of the Paint Skin Weights tool to Replace. In the Stamp Profiles section, set the Value to 0. With the Value at 0 you will be painting black onto the shirt surface.

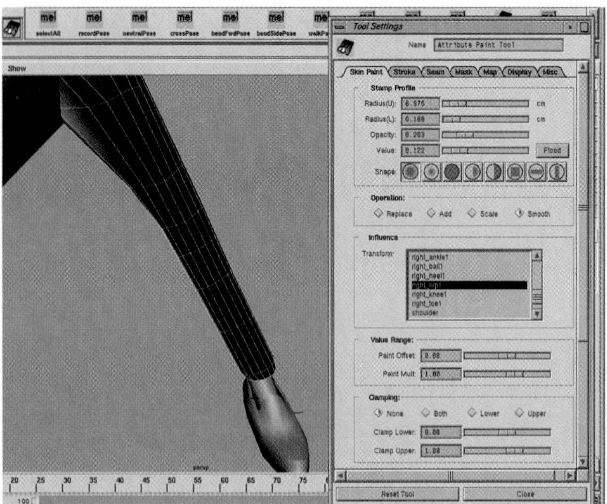

Figure 9.64
Paint Skin Weights tool with left pant leg selected.

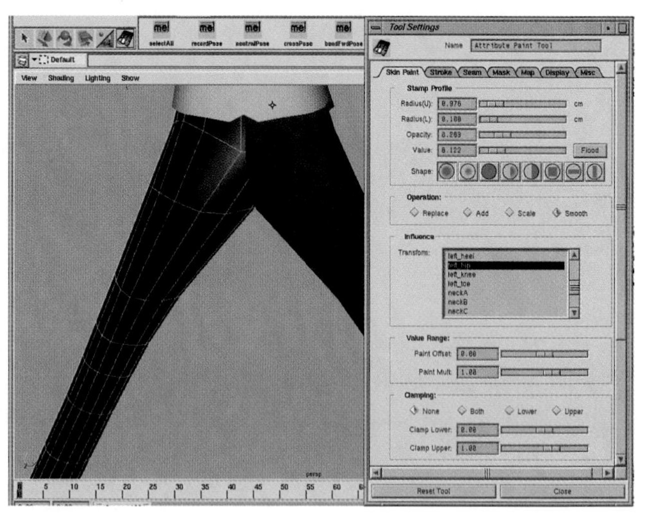

Figure 9.65
Both pant legs smoothed using the Paint Weights tool.

9. In the Influence section of the Paint Skin Weights tool, select the leftShoulder joint. The shirt surface will display the weight values of influence—lighter areas near the shoulder and black over the rest of the shirt.
10. Paint the area of the shirt under the arm (armpit area) to give the left shoulder joint less influence on the shirt below the arm.
11. Repeat the process for the influence of the right shoulder joint on the shirt surface.

12. Adjust the joint influence of the left and right hip joints on the bottom of the shirt (at the character's waist) by using the Smooth Operation with the Paint Skin Weights tool. Smooth the area where the black and gray areas meet to fix the unnatural bulge on the shirt.

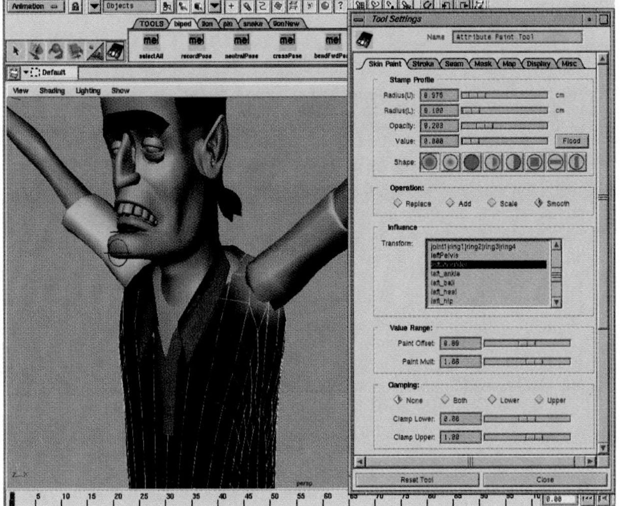

Figure 9.66
Shoulder influences for shoulder joints adjusted on shirt.

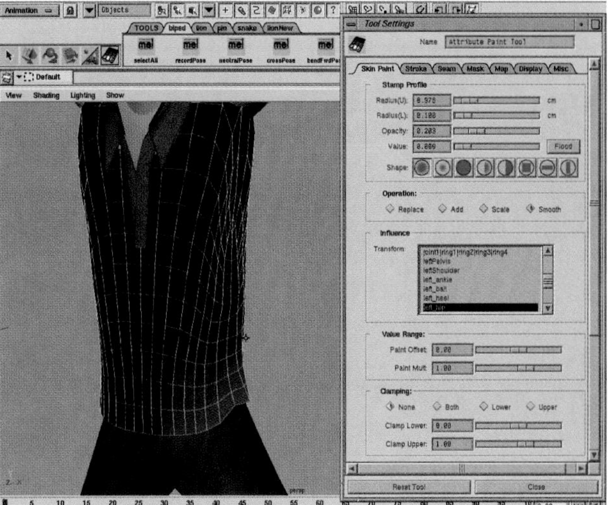

Figure 9.67
Shirt weights adjusted in hip area.

Exercise 9.34: Refining the Smooth Bind Skin Weights Using Tweak Nodes and Set Driven Key

Next we'll take advantage of Maya's tweak nodes to adjust the deformations occurring on the knees of our character. When deformations (such as skeletons, lattices, clusters, and so on) are applied to a polygonal or NURBS surface, Maya creates tweak nodes. These *tweak nodes* allow you to adjust the deformation of the surface through the manipulation of surface points. You are able to push and pull points on the surface to essentially adjust the deformation. Maya puts the tweak node first in the deformation order of the surface so that the point adjustments are calculated before the other deformers. Later on in this chapter we will look more at deformation ordering.

1. Set the biped character back to its initial position by clicking the NeutralPose shelf button in the biped shelf.

2. Select the bipedRoot node (the selection handle behind the pelvis of the character). Set the Translate Y attribute to –9 in the Channel box. Take notice of the deformation taking place in the knees of the character. The deformation appears flat, much like a bent straw. You need to adjust the deformation in the knee so that it is more rounded and less flat.

Exercise 9.34: continued

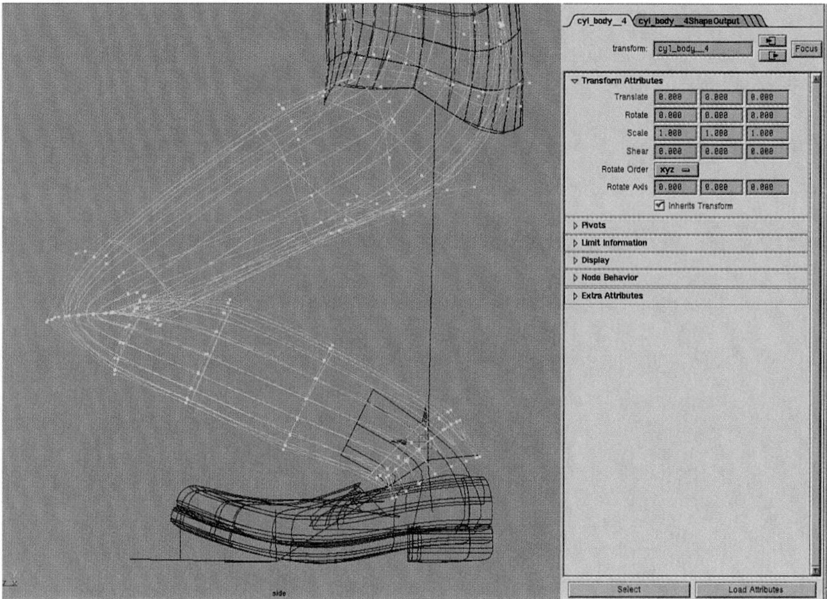

Figure 9.68
Deformation in knees appears too flat, like a bent straw.

3. Set the bipedRoot node's Translate Y back to 0.
4. Select the left pant leg surface and click the tweak node in the input section of the Channel box. Open the Set Driven Key window, Animate, Set Driven Key, Set, Option. You will see both the left pant leg surface and the tweak node loaded as the Driven objects. Click the tweak node in the right side of the SDK window, and then select Envelope as the driven attribute in the right side of the window.
5. Select the left knee joint and click the Load Driver button in the SDK window. Select rotateZ as the driving attribute in the right side of the window.
6. Select the left pant surface once again and select the tweak node in the input area of the Channel box. Set the envelope attribute to 0. Make sure the bipedRoot translate Y set to 0, and then click the Key button in the SDK window.
7. Select the bipedRoot node and set the translate Y to –9.
8. Select the tweak node in the SDK window and then set the envelope attribute to 1 in the Channel box. Click the Key button once again in the SDK window. You haven't had to rotate the left knee joint. By translating the bipedRoot down the Y axis, the knee is bending automatically.

   Now that we have set up the SDK relationship, we can edit the points around the knee in the bent position to make the knee appear more rounded.

9. Select the left pant leg and display its CVs by going to Display, NURBS Components, CVs. Use the move tool to select CVs on the front and back of the knees and move them into position. Use the figure as a guide to how the knee should appear after point tweaking.

10. After you finish with point tweaking, test the knee bend deformation by translating the bipedRoot node up and down the Y axis. You can always go back and tweak the CVs on the knees more by simply translating the bipedRoot node to –9 on the Y axis and moving points into new positions.

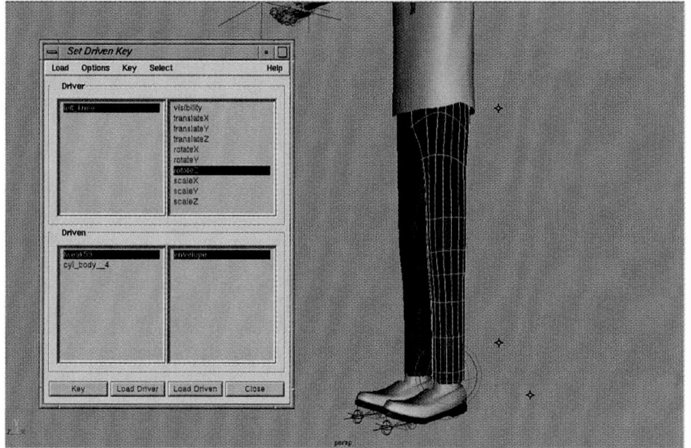

Figure 9.69
Tweak node and left knee joint loaded into Set Driven Key Window.

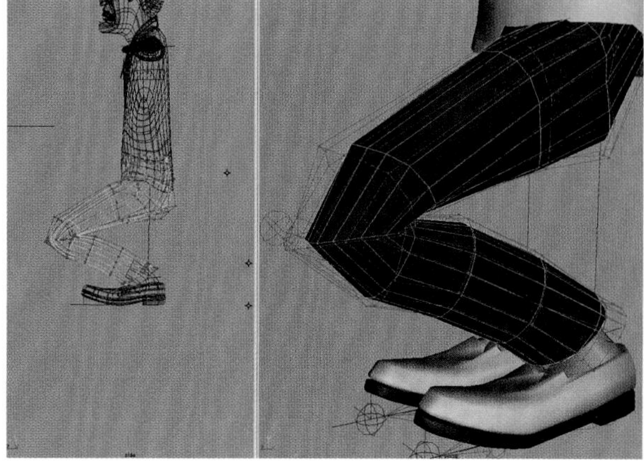

Figure 9.70
Make knee deformation "rounder" by pushing and pulling points on the pant leg surface.

You can see that point tweaking deformed surfaces is a very quick method for fine-tuning deformations on your character.

Later, in Chapter 11, "Facial Animation," we will look at how we can use the Artisan tool to paint the weights of CVs. Painting weights allows you to quickly and interactively edit cluster weights, even on very complex characters with many points.

9.3.11 Using Lattice Flexors on Deforming Joints

Yet another way to weight points at areas of deformations on your character is to use joint flexors. On our character, we will add a type of joint flexor called a lattice flexor in several areas to improve the deformations. The lattices help to smooth the deformations and prevent the surfaces from intersecting.

Exercise 9.35: Adding Lattice Flexors to the Skeleton

1. Select the rightElbow joint, go to the Top View, and rotate the rightElbow joint 100 degrees on the Z axis. The surface above and below the elbow now intersects.
2. Set the rightElbow joint back to 0 degrees in the Z axis.
3. With the rightElbow joint still selected, go to Skin, Edit Rigid Skin, Create Flexor.
4. In the Create Flexor window, set Flexor Type to Lattice, select the At Selected Joints option in the Joints section, use the default S, T, U divisions, and select the Position the Flexor option. Click the Create button. Maya creates a lattice surrounding the rightElbow joint.
5. Use the Translate and Scale Transform tools to resize the lattice flexor so that it is closer to the elbow.

Note

Remember, you can transform the lattice flexor without deforming the surface only when the Lattice Group (the parent node of the lattice and the lattice base node) is selected. If you transform the lattice, the arm geometry deforms incorrectly.

6. Rotate the rightElbow joint again 100 degrees on the Z axis. The deformation is much smoother.
7. Use the lattice flexor controls under the Shapes section in the Channel box to modify the lattice flexor further.
8. You can quickly copy the lattice flexor on the right elbow joint using the Copy Flexor tool. Select the lattice flexor on the right elbow joint. Shift+select the left elbow joint. Go to Skin, Edit Rigid Skin, Copy Flexor. Maya will duplicate the lattice and all the attribute settings you have set over to the left elbow joint. This allows you to quickly copy flexors from one side of the character to the other with the exact same flexor attribute settings.
9. Repeat steps 1 through 7 to create lattice flexors for other joints on the character. Use lattice flexors on the ball and ankle joints in the foot (this improves deformations in the shoe when using the Roll attribute), the knee joints, the wrist joints, and the forearm joints. With the forearm joint, resize the lattice flexor so that it encompasses the entire forearm to simulate the way the forearm flesh twists when rotating your wrist.

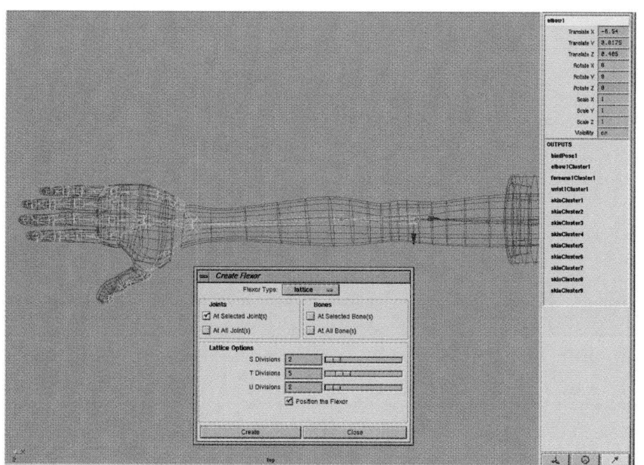

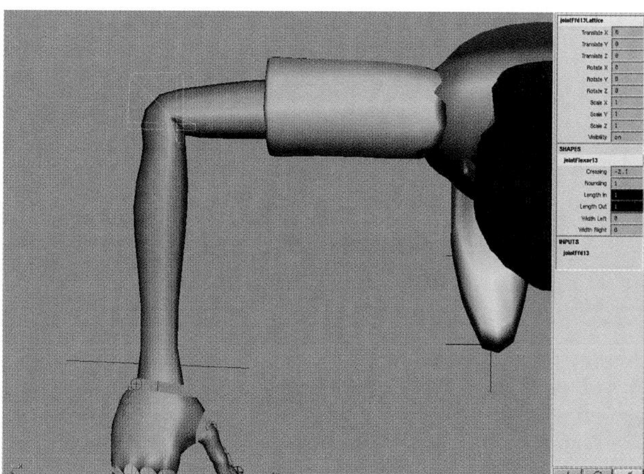

Figure 9.71
The lattice flexor on the elbow, and the elbow joint rotated.

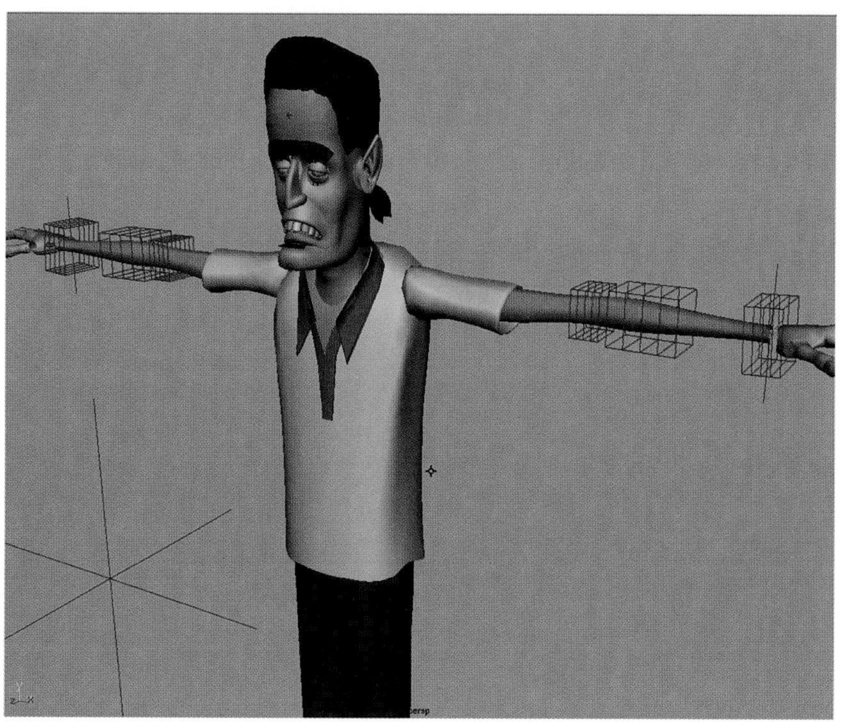

Figure 9.72
Additional lattice flexors on the biped.

9.3.12 Using the Blendshape Tool to Create Custom Deformations

One of the most powerful ways to control the deformation of your character is to use the BlendShape tool to morph your surface(s) into different shapes. Used in combination with Set Driven Key, this tool enables you to have any attribute drive the morph of your surface.

In this example, we will set up a Set Driven Key to have the rightShoulder joint drive the morph of the right sleeve surface. As the shoulder rotates the arm toward the body, we will morph the sleeve to simulate the way the cloth would press against the body. Then, when the arm raises above the head, we will morph the sleeve to simulate how the bottom of the sleeve would drop away from the arm.

Exercise 9.36: Setting Up a Set Driven Key for the Right Shoulder Joint

1. In the Front View, rotate the rightShoulder joint 80 degrees on the Z axis. Notice how the geometry of the right sleeve intersects with the body. Undo the rotation.
2. Select the rtSleeve surface and choose Edit, Duplicate, Option. In the Duplicate option box, click the Reset button and then click the Duplicate button.
3. You now need to unlock the keyable attributes of the duplicated sleeve geometry. With the duplicate sleeve still selected, go to the Channel box and highlight the attribute names. Hold down the right mouse button for the Channel box pop-up menu and go to Unlock. Move the duplicate of the rtSleeve upward and to the side of the character's head. Name this duplicate **`rtSleevePressed`**.
4. Duplicate rtSleevePressed, and name it **`rtSleevDrop`**. Move rtSleevDrop to the side of rtSleevePressed.
5. Select rtSleevePressed, Shift+select rtSleeveDrop and rtSleeve, and choose the Deform, Create BlendShape, Option. In the BlendShape Node text field, type the name **`rightSleeveFIX`** and click the Create button.

Tip

When using the BlendShape tool, select the morph targets first and the base object last.

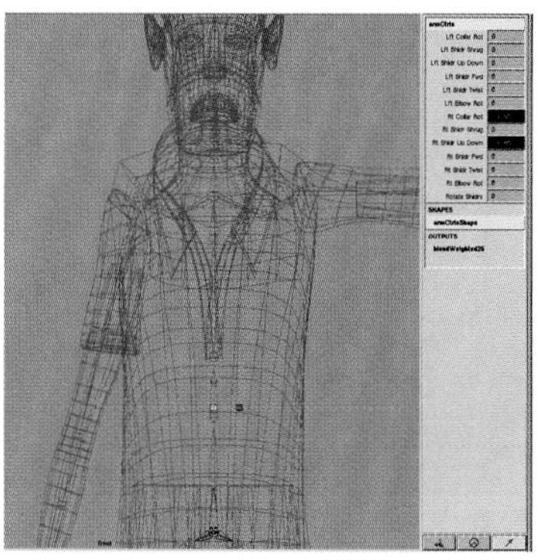

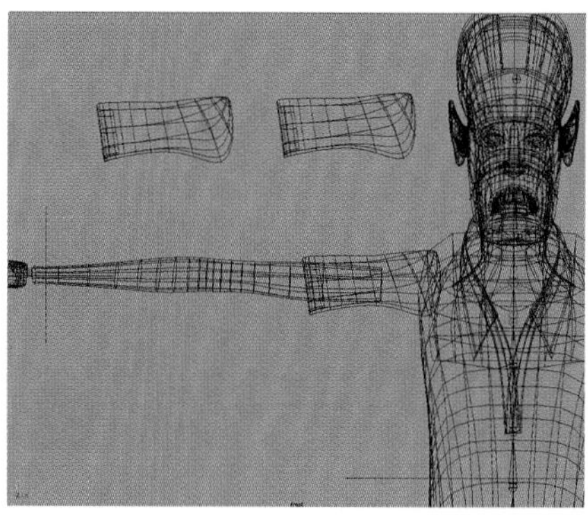

Figure 9.73
The sleeve intersecting with the torso (left). Duplicates of the right sleeve surface (middle and right).

Now we will set up our Set Driven Key to connect the rotation of the rtShoulder joint to the morph of the BlendShape. You may have noticed that the duplicates of the sleeve are identical to rtSleeve. We will reshape them after the SDK is set up.

Exercise 9.37: Connecting the Shoulder Joint Rotation to the Morph

1. Open the SDK window and load rtShoulder joint as the Driver. Select Z rotate as the Driving attribute.
2. Choose Windows, Animation Editors, BlendShape. A window with the rightSleeveFIX blendshape node displays with two vertical slider bars (one for rtSleevePressed and one for rtSleeveDrop) which drive the morph from 0 (no morph) to 1 (full morph).
3. Click the Select button to select the rightSleeveFIX blendshape node.

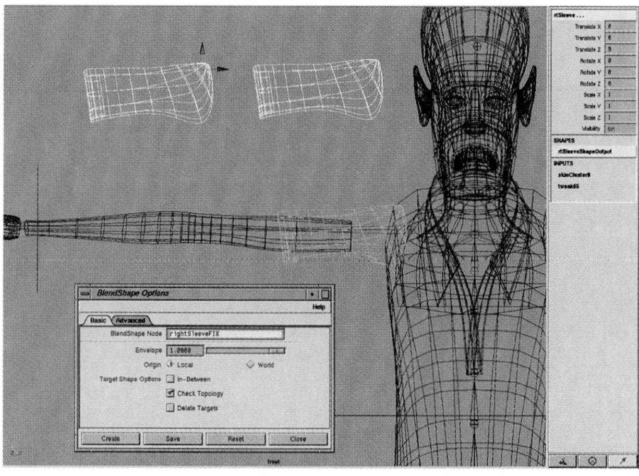
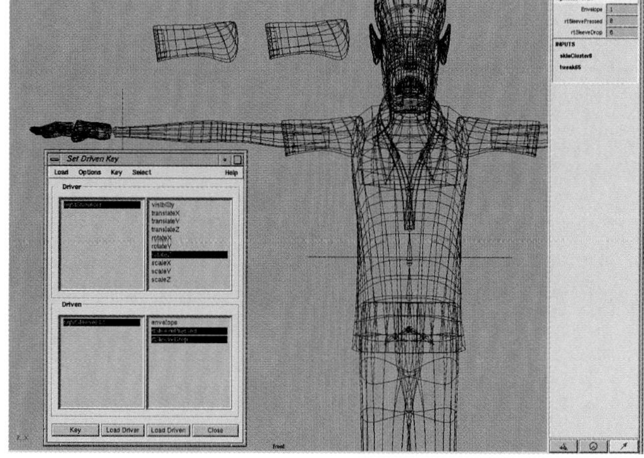

Figure 9.74
The BlendShape Editor, and the SDK window with the right shoulder joint and rightSleeveFIX blendshape.

Exercise 9.37: continued

4. Go back to the SDK window, click the Load Driven button, and select rtSleevePressed and rtSleeveDrop as the Driven attributes.
5. Select the rtShoulder joint, rotate it 65 degrees on the Z axis, and click the Key button in the SDK window. This sets the initial key for the Set Driven Key relationship between the blendshape and the shoulder joint rotation.
6. Rotate the rtShoulder joint (Driver) 80 degrees on the Z axis to rotate the arm close to the body. Edit the Driver attribute first and then the Driven attribute.
7. Select rightSleeveFIX (Driven) in the SDK window, set the rtSleevePressed attribute in the Channel box to 1, and click the Key button. The rtSleeve surface jumps away from the skeleton, back to its initial location. When there is more than one deformation acting on a node, Maya orders the deformations in the order in which they are created. In this case we have the sleeve geometry (rtSleeve) being deformed by the skeleton (with the smooth bind we performed earlier) and the blendshape deformation we just created.

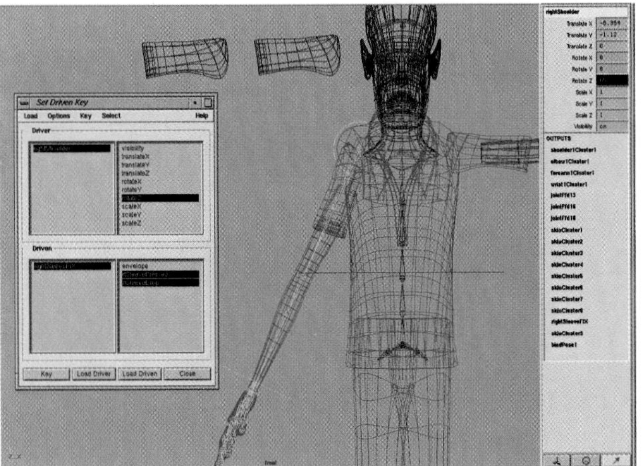

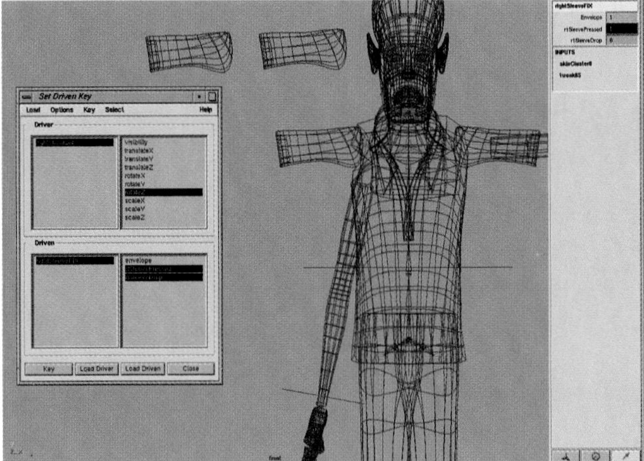

Figure 9.75
The right shoulder rotated 65 and 80 degrees on the Z axis.

Maya also created another deformation node when the Smooth Skin bind was performed—a tweak node. Because we created the skeleton deformation first and the blendshape second, the deformation order performs the tweak deformation first, followed by the smooth skin deformation and then the blendshape deformation. You can also think of deformation ordering as layering deformers. Here, the skeleton deformation of the sleeve geometry is being deformed by the blendshape. We need to switch this order so that the blendshape deformation occurs before the skeleton deformation.

8. Place your mouse cursor over the right sleeve geometry in one of your modeling views and hold down the right-mouse button. A pop-up marking menu appears. Go down the menu to Inputs, Complete List.

   A window with the history list appears for the rtSleeve object. The history list shows all the input nodes acting upon the right sleeve geometry. In this list we can rearrange the order of these input nodes. Maya evaluates each of the input nodes from the bottom of the list to the top.

9. Holding down the middle mouse button, drag the entry on the list for the BlendShape (rtSleeveFIX) on top of the Cluster node (the Smooth Skin cluster). Make sure that the tweak node remains at the bottom of the input history list. The cluster node represents the Smooth Skin deformation on the sleeve, which occurs at the bottom of the list because it was created first.

   The sleeve jumps into the proper location. Now the blendshape deformation occurs first on the sleeve and then on the skeleton deformations, essentially layering all the deformations on the sleeve geometry.

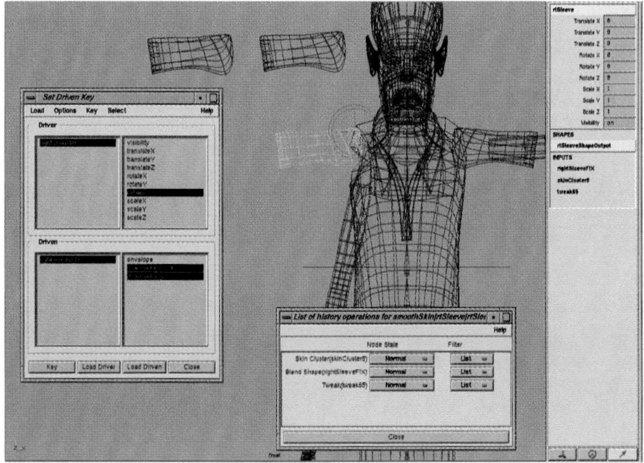
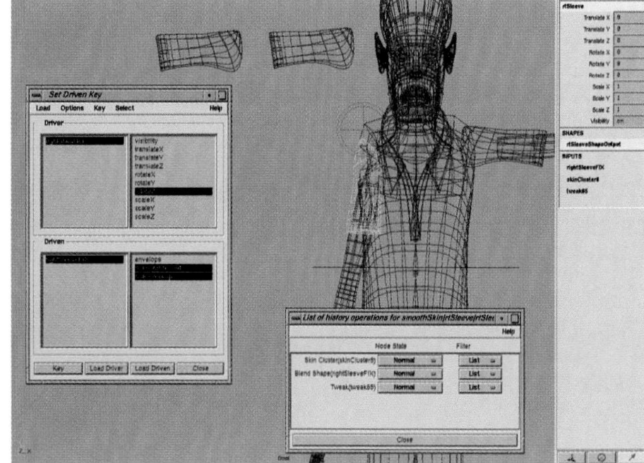

Figure 9.76
Reorder deformations.

Now we need to change the shape of the rtSleevePressed geometry.

10. Let's use a lattice to quickly model the rtSleevePressed geometry. Select the rtSleevePressed surface, choose Deformations, Lattice, and change the S division to 5, T to 5, and U to 2.

11. Turn on the lattice points, move the lattice points at the bottom of rtSleevePressed, and shape the rtSleevePressed so that the sleeve appears to rest against the side of the body. As you change the shape of rtSleevePressed, you should see the rtSleeve geometry on the character update. Remember that the Set Driven Key relationship we created earlier morphs rtSleeve into

Exercise 9.37: continued

rtSleevePressed when the shoulder joint is rotated down toward the torso of the character.

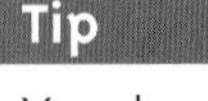

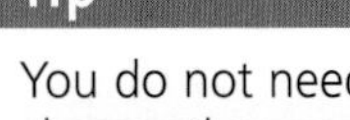

Tip

You do not need to change the Deformation Ordering again; it just needs to be done once.

12. Rotate the rtShoulder joint to –30 degrees on the Z axis to rotate the arm into the up position.
13. Repeat steps 10 and 11, but this time with the rtSleeveDrop surface. Shape the rtSleeveDrop surface so that it appears that the sleeve sags when the arm is raised. You do not need to change the Deformation Ordering again. It just needs to be done once. You can then repeat the same steps for the left sleeve.

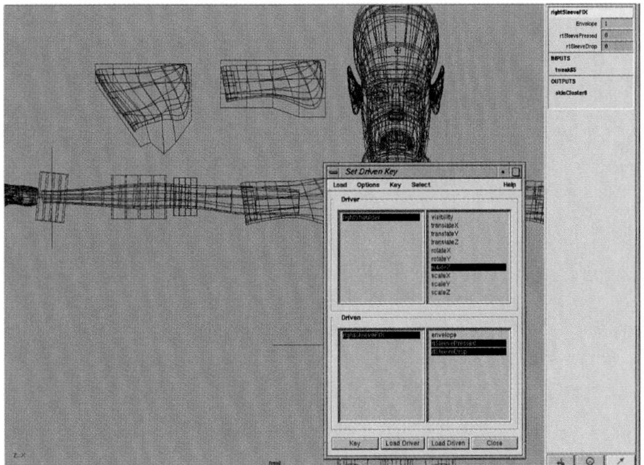

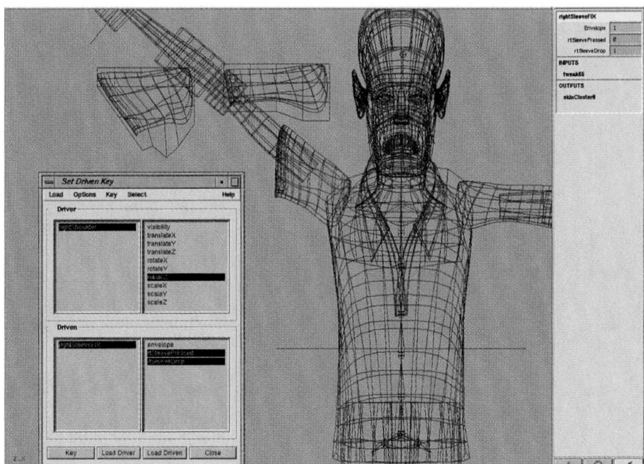

Figure 9.77
Model both blendshape targets (left). At right, we see the shoulder rotated up and the sleeve sagging.

After you are finished, it is a good idea to keep all your morph targets in case you need to edit them later on (you can edit your morph targets at any time). You can hide the geometry and lattices for the morph targets so that they are not in the way and, more importantly, so that they do not render. Using Set Driven Key and blendshape can be a very powerful combination for controlling how your character's geometry deforms. This combination can be used not only for simple clothing (such as in the sleeve example above), but also for simulating bone/muscle movement, skin stretching, squash and stretch effects, and correcting deformations. Blendshape allows you to model your deformation exactly the way you want it. Set Driven Key allows you to instruct Maya when you want the deformation to occur.

9.3.13 Centralizing Character Controls

Maya offers many ways to optimize the way you animate and control your 3D character. Another powerful concept in Maya is to centralize the controls of your character. The most tedious aspect of animating a jointed character is that you have to keep track of so many joints throughout the body. Imagine the amount of time it would take to simply pose the hand into a fist. You would have to deal with at least 15 joints. With custom attributes (remember the roll attribute we set up for the feet), Set Driven Key, and the Connection Editor, we can create nodes in the scene with attributes connected to anything we want in the scene. With this, the animator just needs to pick the control node and animate custom attributes that control joints throughout the character.

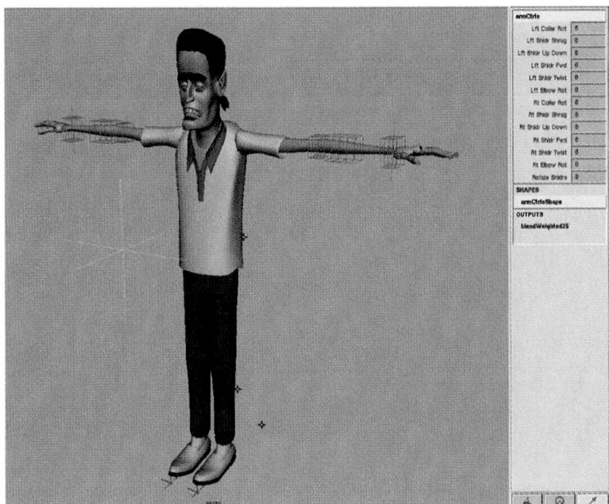

Figure 9.78
Biped with custom attributes on Locator.

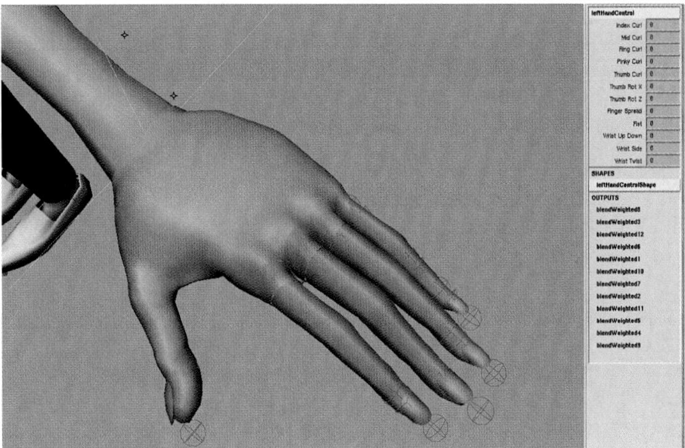

Figure 9.79
Hand controls on Locator parented to wrist.

We will accomplish this control by placing custom attributes on a Locator (Create, Locator). The Locator is good to use because it marks a point in space (with what looks like a crosshair), and it will not render. The Locator also has the same transformation channels as other objects in the scene. Let's create a Locator, add a custom attribute, and connect that attribute to joints in the body with Set Driven Key.

Exercise 9.38: Creating a Locator for the Skeleton Joints

1. Create a Locator and name it **`bodyControl`**.
2. With the bodyControl Locator selected, move it so that it is in front of and about chest level with the character.
3. Choose Modify, Add Attribute, and name the attribute **`leftElbowRot`**. Use a data type of float, set min to 0, max to 10, and default value to 0, and click OK. A leftElbowRot attribute appears in the Channel box.
4. Open the SDK window, and click the Load Driver button. The bodyControl appears in the Driver area. Click leftElbowRot as the Driving attribute.
5. Select the left elbow joint, click the Load Driven button in the SDK window, and click Rotate Z as the Driven attribute.
6. With both objects loaded in the SDK, click the Key button.

Exercise 9.38: continued

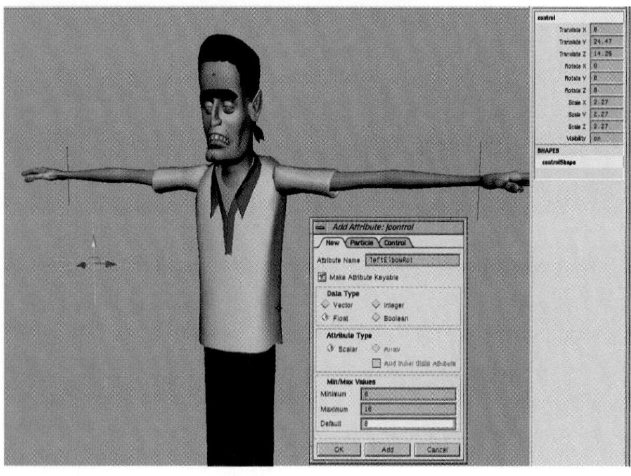

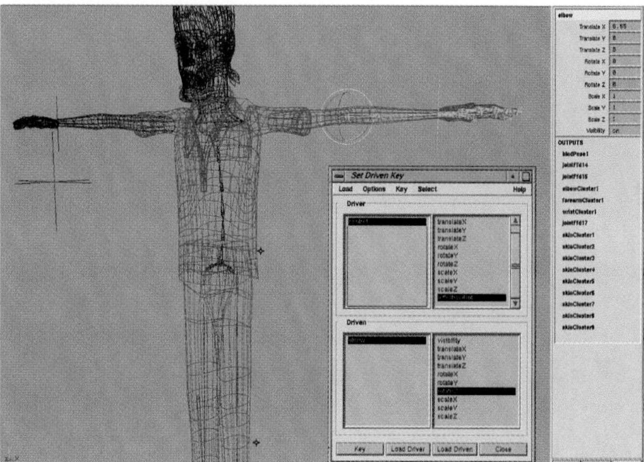

Figure 9.80
Add attributes to Locator (left). SDK window with driver and driven objects (right).

Tip

Using SDK with your own attributes builds in a minimum and maximum amount of rotation for the elbow. Keep in mind that if you ever need to get the elbow back into its starting position, you simply need to set the leftElbowRot attribute to 0.

7. Select bodyControl and set leftElbowRot to 10.
8. Select the left elbow joint and rotate it 120 degrees on the Z axis.
9. Click the Key button in the SDK window.
10. Select bodyControl in the SDK window once again and select the leftElbowRot attribute in the Channel box.
11. In any of the Modeling windows, middle mouse button-drag back and forth horizontally. This updates the value of the leftElbowRot attribute, which, in turn, is driving the rotation of the left elbow joint. You should see the left elbow of the character bending.
12. Set the leftElbowRot attribute back to 0.

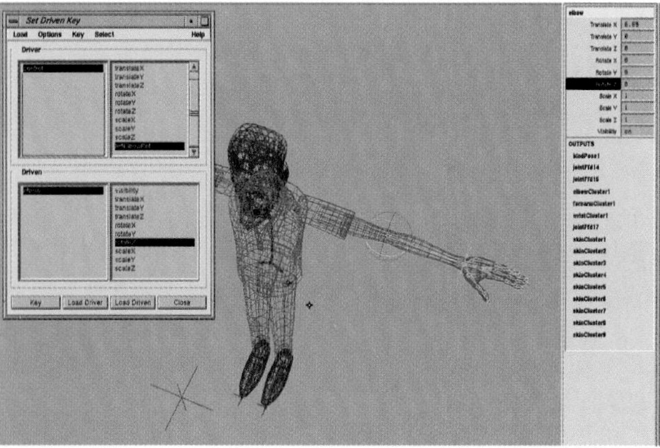

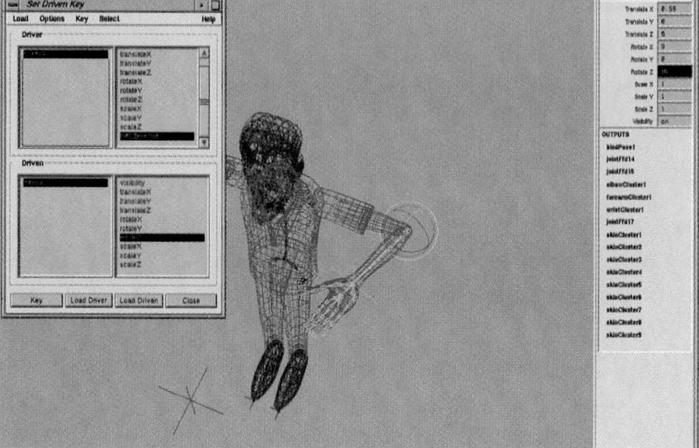

Figure 9.81
leftElbowRot attribute set to 0 and 10.

Exercise 9.39: Creating a Character Set for the Biped Character

Character sets enable you to quickly access the attributes you will use to animate the character. A *character set* is a collection of objects and their attributes that you can organize and quickly access through the Character Set menu.

1. If you have not had the chance to add your own custom attributes to the biped character, you can open the bipedFinished.mb file or use your current file.
2. In the Animation Menu set, go to Character, Create, Option. In the name field, type **biped**. This creates a Character Set named biped in the current scene.
3. Open the Relationship Editor for Character Sets by going to Window, Relationship Editors, Character Sets. The Relationship Editor window will open with the biped character set loaded on the left column. Click the name biped in the left column.
4. In the right column of the Relationship Editor, click the + symbol inside the square next to the bipedRoot node. Select the bodyCtrl node. Click the + symbol inside the square next to bodyCtrl and then select armCtrl. You have added the bodyCtrl and armCtrl nodes to the biped character set. This means that whenever you select the character set biped, Maya will select both the bodyCtrl and armCtrl nodes.
5. Close the Relationship Editor and then go to the main menu bar once again to Character, Select Character, biped. Make sure the Channel box is showing in the interface. You will see that Maya has loaded all the attributes for the bodyCtrl and armCtrl nodes into the Channel box.
6. Let's now add the two feet nodes to the biped character set. Open the Relationship Editor for Character Sets once again. Select biped in the left column. In the right column, select rightFoot and leftFoot (there is no need to Shift+select items in the Relationship Editor; they are either selected or deselected).
7. In the left column, click the + symbol inside the square next to biped. You will see a list of all the attributes associated with the biped character set. Highlight the XYZ translate attributes for both the left and right feet.
8. In the Relationship Editor, go to Edit, Remove Selected from Character. This action removes these attributes from the biped Character Set, allowing you to remove unwanted attributes from your Character Set.

Continue experimenting with creating character sets. If you have many attributes for your character, you can also split up the controls over several character sets. For example, you could have a character set bipedLowerBody that contains attributes

Exercise 9.39: continued

for the lower body, and a character set bipedUpperBody with attributes for the upper body. This would lessen the amount of attributes that would be listed, making it easier for you to locate specific attributes faster.

Tip

You can add attributes into a Character Set by selecting an object in the right column and then expanding its attribute list by clicking on the + symbol in the circle. There you can select or deselect attributes to the current character set.

You have set up your own character control attribute on the bodyControl Locator. From here, you can probably imagine other character controls you can set up. Create attributes that control back rotations (try multiple joints controlled by a single attribute), finger rotations, neck rotations, arm rotations, or anything else you may need.

The bipedFinished.mb file is a finished version of the biped character you have been setting up. In this file, many custom attributes are already set up on Locators (one parented to each wrist for the hand controls and one in front of the body for body controls). You can parent these control locators to joints; the locators will then follow the animated character. You can use Windows, General Editors, Channel Control to remove keyable attributes from Locators to make room for your custom attributes.

armCtrls	
Lft Collar Rot	0
Lft Shldr Shrug	0
Lft Shldr Up Down	0
Lft Shldr Fwd	0
Lft Shldr Twist	0
Lft Elbow Rot	0
Rt Collar Rot	0
Rt Shldr Shrug	0
Rt Shldr Up Down	0
Rt Shldr Fwd	0
Rt Shldr Twist	0
Rt Elbow Rot	0
Rotate Shldrs	0

bodyCtrl	
Back Fwd	0
Back Side	0
Back Twist	0
Neck Fwd	0
Neck Side	0
Neck Twist	0
Rt Pelvis Up Down	0
Rt Pelvis Fwd Bwd	0
Lft Pelvis Up Down	0
Lft Pelvis Fwd Bwd	0

leftHandControl	
Index Curl	0
Mid Curl	0
Ring Curl	0
Pinky Curl	0
Thumb Curl	0
Thumb Rot X	0
Thumb Rot Z	0
Finger Spread	0
Fist	0
Wrist Up Down	0
Wrist Side	0
Wrist Twist	0

Figure 9.82
Custom control attributes on Locators for biped.

9.4 Setting Up a Snake Using the IK Spline Handle Tool

Let's take a look at another way to control skeleton joints through the use of the IK Spline Handle tool. The IK Spline Handle tool (Skeleton, IK Spline Handle Tool) creates a NURBS curve on which the skeleton joints align themselves. Pulling CVs on the spline curve moves the joints to keep them aligned on the curve. Just a few CVs on the spline curve can affect many joints. The animator just needs to animate the spline CVs, and the skeleton joints will update their location throughout the animation. This tool is great for serpentine-type effects on skeleton joints—tails, antennae, tentacles, snakes, long necks, and other body parts. In addition to characters, the IK spline is great for rope-like effects.

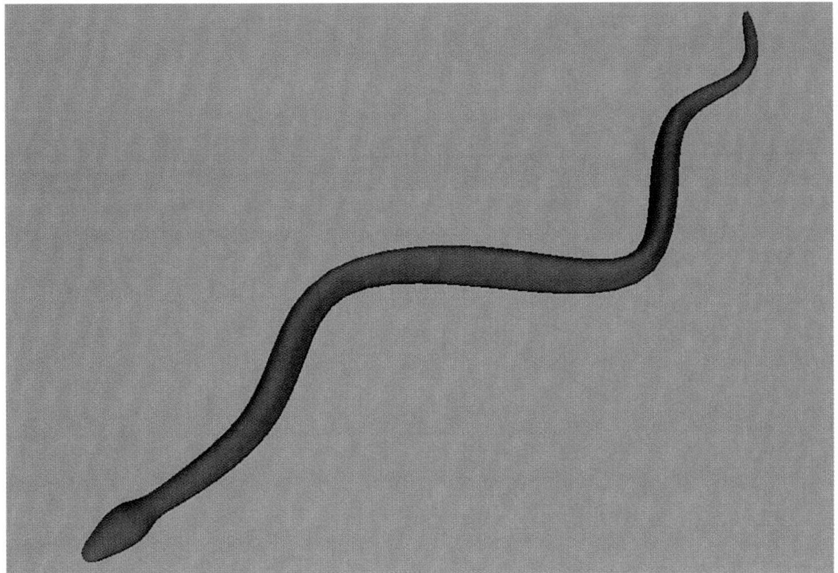

Figure 9.83
Snake model with IK Spline.

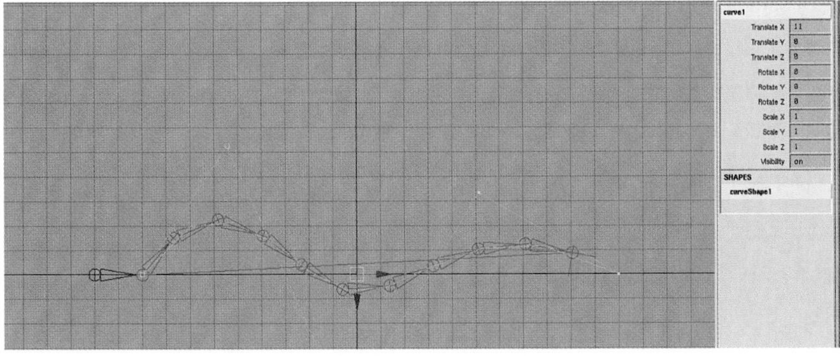

Figure 9.84
Skeleton joints controlled with an IK Spline.

We will take a quick look at using the IK Spline Handle tool for controlling the joints of a snake.

Exercise 9.40: Using the IK Spline Handle Tool to Control the Snake Joints

1. Set the current Maya project you are using to "snake." Open the snakeModel.mb file, which contains a model of a snake that we will use with the IK Spline Solver.
2. In the Outliner, select the top of the snake model geometry hierarchy "snake."

Exercise 9.40: continued

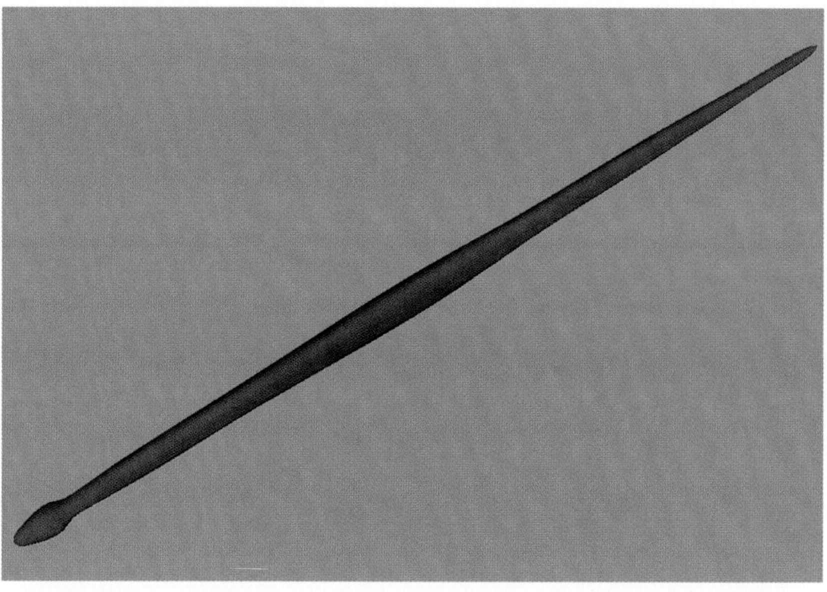

Figure 9.85
Default snake model.

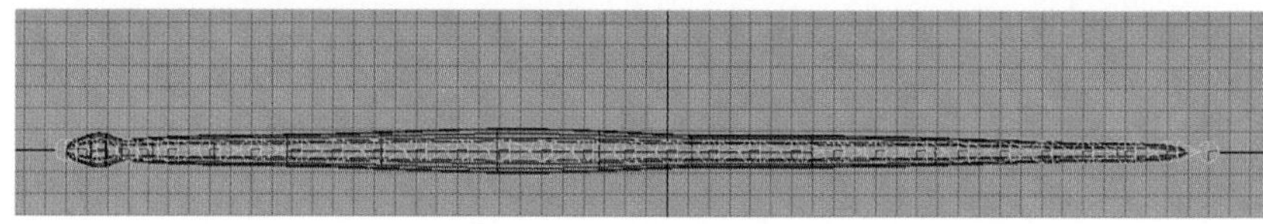

Figure 9.86
Skeleton joints in snake geometry.

3. Create a skeleton joint hierarchy from the base of the neck to the tip of the tail, using approximately 10 to 15 joints. After you finish drawing the skeleton, press Enter. Name the first joint (the top joint of the hierarchy) **`snakeRoot`**. Again, to get smooth deformations and movement, we need many joints in the tail.

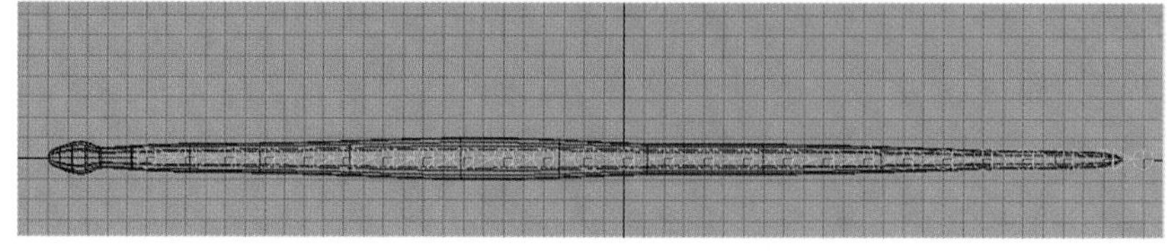

Figure 9.87
Skeleton joints in the snake tail.

4. Create another skeleton hierarchy from the base of the neck to the tip of the snake's nose, placing the first joint next to the first tail joint you created. Use four to six joints to do this. Name the top joint of this hierarchy **`neckBase`**.

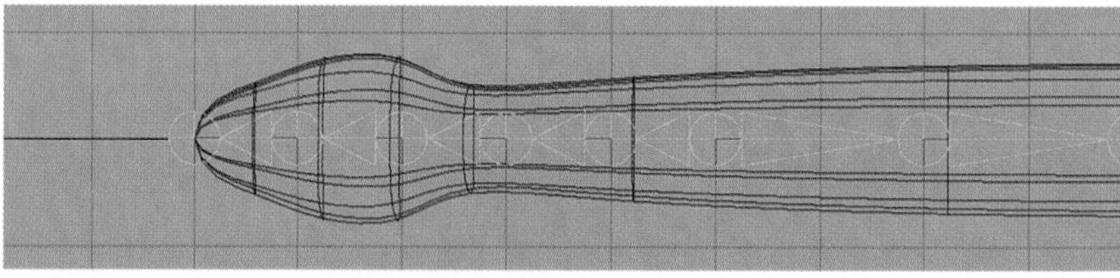

Figure 9.88
Skeleton joints in the snake head.

5. Parent the neckBase joint to the snakeRoot joint. Select neckBase, Shift+select snakeRoot, and choose Edit, Parent to parent the neckBase joint to the snakeRoot joint. Maya will create a bone that connects the two joints.

6. Go to Skeletons, IK Spline Handle Tool to create an IK spline handle. Using the default settings in the option box (click the Reset button if you need to), click the second joint (joint2) in the tail from the snakeRoot joint, and then click the joint at the tip of the tail.

   Maya creates an IK handle at the tip of the tail. It may be hard to see, but right down the middle of the joints in the tail is a NURBS curve.

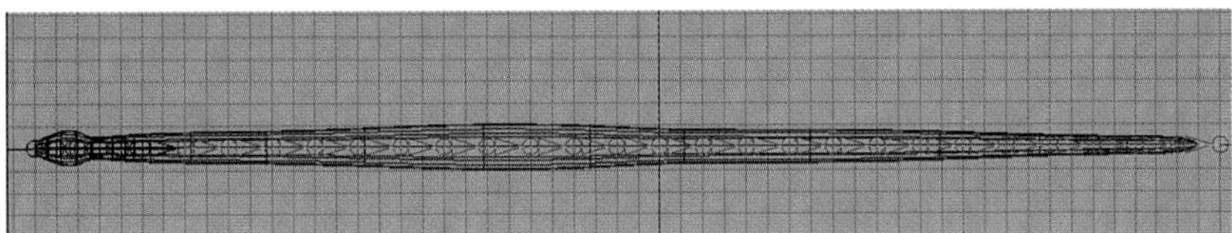

Figure 9.89
The IK Spline in the snake tail.

7. In Object Mode, set your Pick Mask to Curves Only. Select the curve in the middle of the snake's tail.

8. Go to Display, NURBS Components, CVs to turn on the CVs for the curve.

9. Select a CV on the curve and translate it to see how the joints align themselves to the shape of the curve. Undo any CV manipulation.

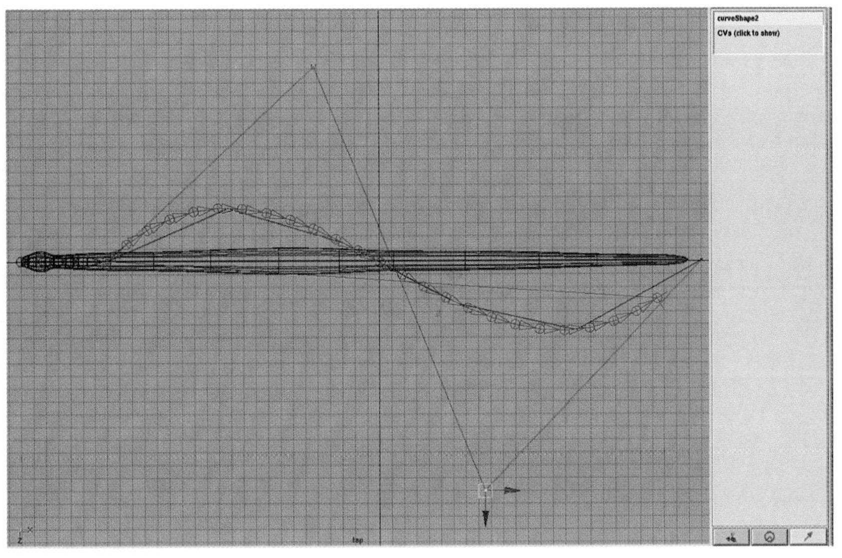

Figure 9.90
Modifying spline CVs moves joints in the tail.

10. Select the curve in the tail and name it **`tailSpline`**.

Exercise 9.40: continued

11. Right now this curve has only four CVs. We need to add more CVs so that we have more control over the shape of the tail. In the Modeling Main Menu Set, choose Edit Curves, Rebuild Curve, Option, set the number of spans to 8, and click the Rebuild button.

    Maya has rebuilt the tailSpline curve so that we now have 11 CVs on the curve that we can manipulate.

12. To create the IK Spline for the head skeleton, repeat the same process used for the tail, or you may choose to use forward kinematics (manual rotation of joints) for the neck and head.

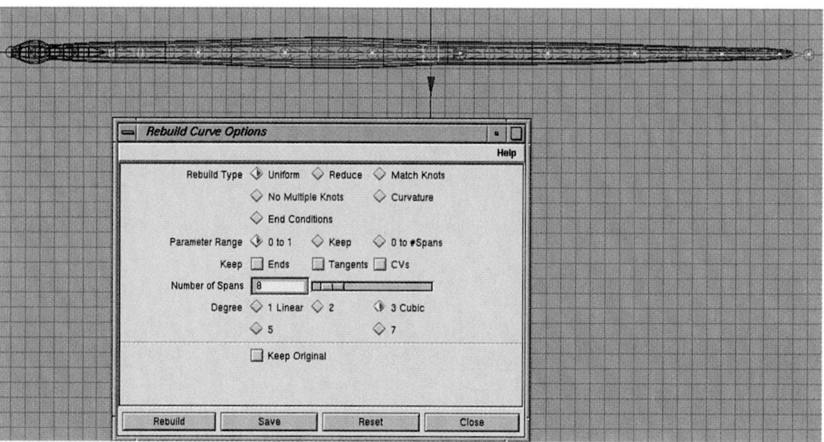

Figure 9.91
Rebuilt spline in the snake tail.

Tip

When creating IK splines in skeleton joint hierarchies, it is a good idea to not have the spline going through the root joint (top of the skeleton hierarchy). Doing so may cause undesirable rotations and translations on the root joint, potentially causing the entire character to flip or move unpredictably.

Clusters are another type of deformer you can use to control groups of points. We have already seen clusters in action on the skeleton (Bind Skin groups points into clusters and then parents the cluster of points under a joint). Clusters have their own transform channels that we can animate. We are using Relative mode here so that we can parent the clusters to the snake's skeleton and not have any double transformations.

Exercise 9.41: Creating Clusters for the Snake

1. In the Top View, select the second CV in the tailSpline from the left near the head, and choose Deform, Create Cluster, Option. Make sure Relative is on in the option box and hit the Create button to create a cluster for this selected CV.

   A C icon, representing the cluster, appears on top of the point. You will notice that when you select a CV there are no keyable attributes in the Channel box for you to animate. Creating a cluster gives us a transform node that we can use to animate the position of the CV.

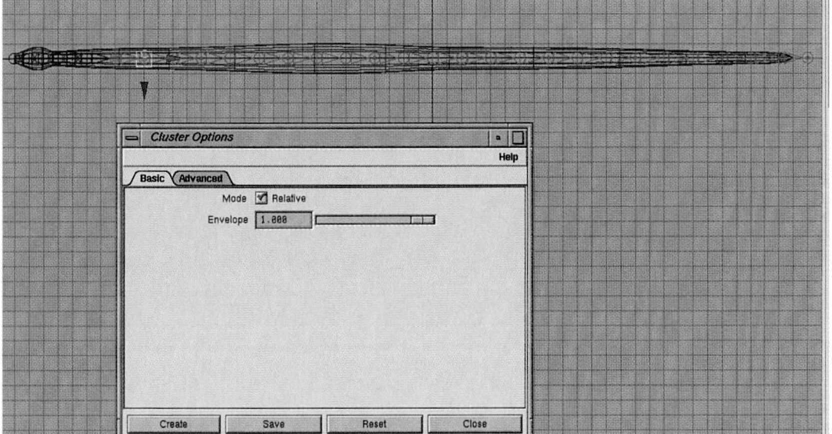

Figure 9.92
Cluster created on second CV in tailSpline.

2. Repeat the creation of clusters for each individual CV on the tailSpline curve.

   It is not necessary to create a cluster for the first CV (closest to the head) on the tailSpline. You should have a total of 10 clusters in the tail of the snake. Each cluster can now be animated independently of the others.

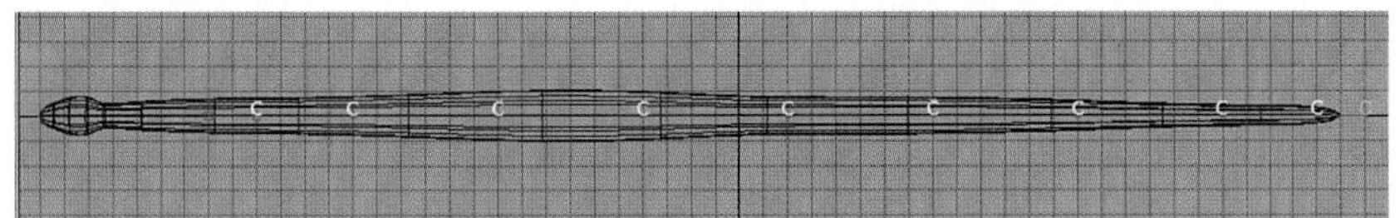

Figure 9.93
Clusters in the tail.

3. Select the snake surface and Shift+select the snakeRoot joint. In the Animation Main Menu Set, go to Skin, Bind Skin, Smooth Skin (make sure you are using the default settings for the Smooth Skin Bind; click the Reset button in the option window if you are not sure). Maya binds the snake surface to the skeleton using Smooth Skin Bind.

4. Test the Bind Skin by translating clusters in the tail. You should see the snake surface update. Undo any cluster transforms.

Exercise 9.41: continued

5. Group all the clusters together and parent them under the snakeRoot joint. The clusters will then move with the snake when you translate the snakeRoot joint (the C icons will follow the snake wherever it is placed in the 3D scene). You can animate the entire snake translation using the snakeRoot joint. Then you can animate the bending serpentine motion by keyframing the translation of the individual clusters.

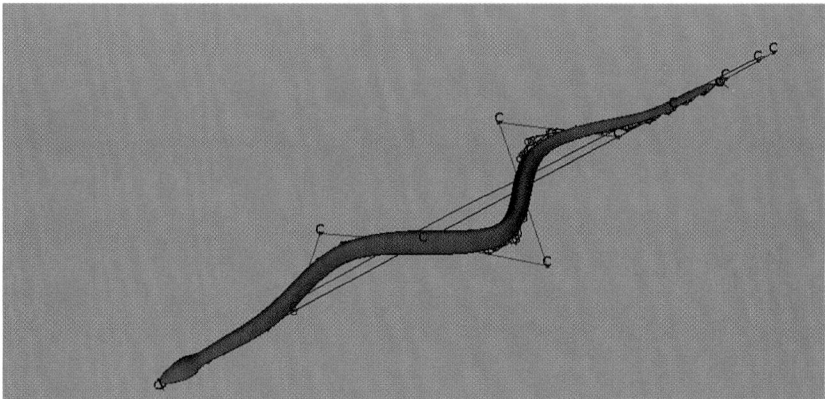

Figure 9.94
Test the Snake deformation.

6. Use joint rotations in the neck joints to position the head and neck.
7. Open snakeFinished.mb if you want to compare your work with a finished version of this exercise.

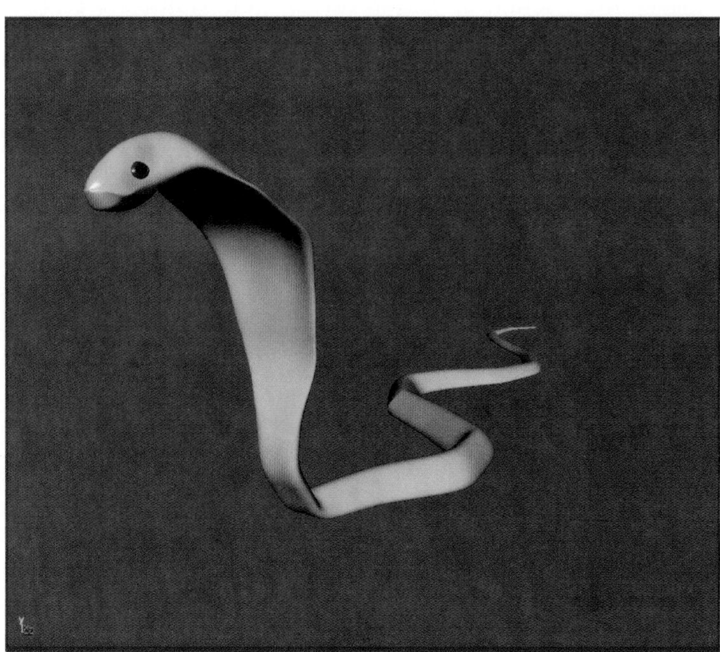

Figure 9.95
Snake in the final pose.

Animating Characters and Keyframing

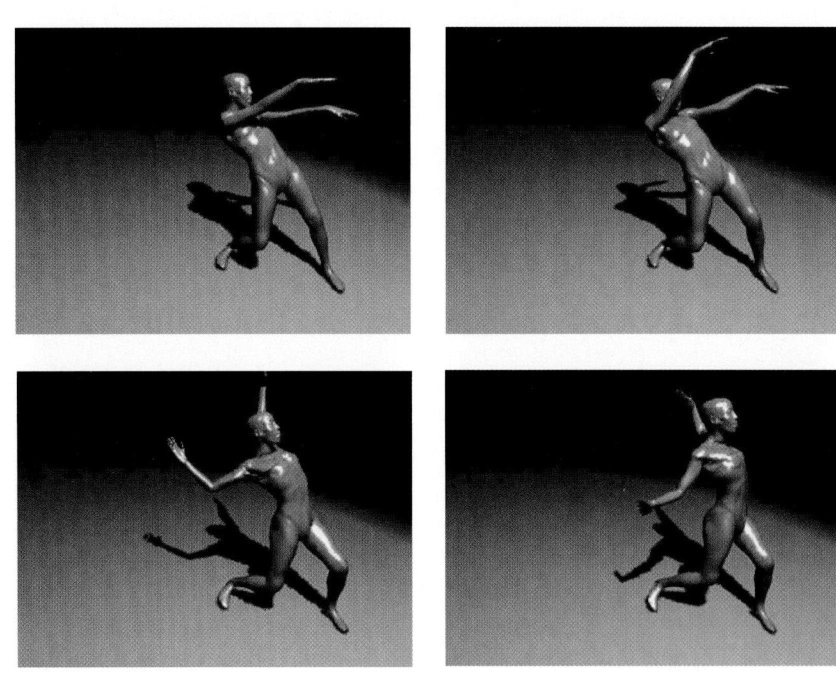

CHAPTER 10

Animating Characters and Keyframing

by Tim Coleman

Now that you have set up your skeleton, we will look at the process of animating your 3D characters. The basic idea is to pose your character at different frames in time. You record each pose in time as a keyframe. Maya then interpolates how your character gets from one keyframe to the next, with help from you, the animator. The most challenging aspects of the character animation process are achieving strong poses with your character and achieving the correct timing between these poses. Being able to pose your character properly is highly dependent on how you modeled and set up your character with its skeleton. The way you time your character between poses depends upon your understanding of motion and the way things move.

One of the best ways to start understanding motion is by observing the world around you. Observe people walking down the street. Watch a dog walking or running in the park. Observe yourself moving and doing different things. Use a stopwatch to get a feel for how much time it takes for something to move.

Videotaping is another great way to study how things move. Taping enables you to record the motion of objects, and then play it back as many times as you want at any speed. You can even play the video back frame by frame, which is essential for breaking down complex or subtle motions. For stylized or exaggerated motion, such as what you see in cartoon animation, studying cartoons and comics from the traditional masters is a great way to examine timing and animation techniques.

Let's not forget, however, that just starting to animate is important as well. It is best to start with simple and straightforward animation situations. You will see many beginning animators starting with a bouncing ball. This may sound too easy, but it is a great exercise in timing. Later on, you can work up to more complex animations. It can be a very frustrating experience to start animating a very complex scene if you haven't experimented with and discovered some of the basic rules of animation. Over time you will develop your own instincts and methods for animating 3D characters. You will quickly be able to see when motion appears correctly or incorrectly.

This chapter covers different methods for animating your 3D character in Maya. Along the way we will look at some basic animation rules to keep in mind as you animate. We will also look at methods for improving your animation workflow in Maya. Finally we will look at some special animation scenarios and how we would solve them in Maya.

10.1 Methods of Animating 3D Characters

It is always a good idea to start with some sort of road map of what you want to animate. This could be a storyboard or series of sketches of what you want to do with your character. It is good to try to previsualize your animation as much as possible. Even if it is in your head, play it over and over in your mind and get a good sense of what the character needs to do. Try to "be" your character, challenge yourself to get out of your chair and act out the animation of your character. Use a stopwatch and time yourself as you act out the actions for your character. Physically getting into the character and getting a feeling for the timing and how your character moves from point A to point B is really important for the animator. Don't be shy—you will find some of the best animators jumping around like crazy people trying to flush out the timing and movements for their characters.

After you have your animation idea down on paper, you need to determine how to go about animating your 3D character. There are numerous methods that an animator can use. Some methods lend themselves to specific animation situations. In other situations you may use a combination of methods to animate your character. There may be methods that you may prefer over others. Over time you may develop your own personal methods to approaching the animation of 3D characters.

Let's take a look at several different methods for animating 3D characters. We'll note the strengths and weaknesses of each method. Again, in some situations, you may use a combination of methods.

10.1.1 Pose-to-Pose

The *Pose-to-Pose* method is great for characters that must be animated through a series of very distinct poses. Each pose of the character conveys an idea or action. Each pose can be thought of as the traditional 2D "keypose." After the poses are established, the animator must determine the speed and timing between the poses and how the character moves from pose to pose.

In the following example, we have a lion jumping through a flaming hoop, from one platform to another. The keyposes for the lion are standing, crouching (ready to jump), begin jump (front legs leave platform), jump in the air, landing, and once again standing on the platform. Using photographs, sketches, and video for reference, the lion was posed into these seven positions. The lion jumping in midair is one of the most distinct poses that needs to be achieved. The lion is posed in each position and then keyframed every 10 frames.

Figure 10.1
The lion jumping from platform to platform, using pose-to-pose animation.

Spacing the poses every 10 frames is arbitrary; spacing the poses makes it easier to see how the body of the lion changes between each pose. After the poses are keyframed, you must start working with the timing between the poses. You have to determine how quickly or slowly the lion progresses between its poses. For instance, the timing of the lion going into its crouch pose would be slower than when the lion goes into its leaping pose because leaping is a quicker movement.

Figure 10.2
Frame 1: The lion is standing.

Figure 10.3
Frame 9: The lion crouches down.

Figure 10.4
Frame 20: The lion begins to jump.

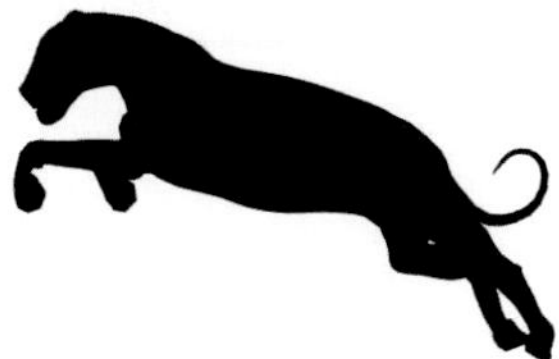

Figure 10.5
Frame 30: The lion jumps in the air.

Figure 10.6
Frame 40: The lion's landing.

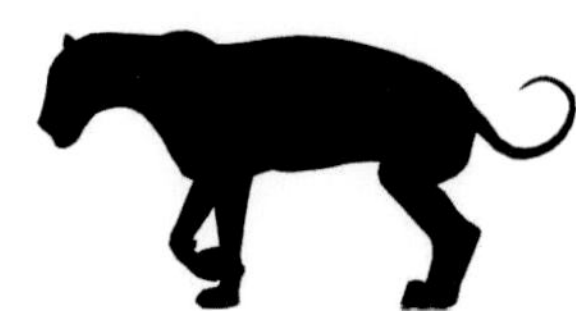

Figure 10.7
Frame 50: The lion standing on the platform.

Prepare for Exercises: Loading Shelf Files

Before starting the exercises for Chapter 10, be sure to copy a set of Shelf files to your Maya preferences. The Shelves contain tools and scripts that will aid you throughout the exercises.

Go to the chapter 10 directory on the CD-ROM that accompanies this book. Open the chp10Shelves directory and copy the Shelf files (all the files with the suffix .mel) to the following directories on your computer.

If you're using an SGI machine, copy the files to *home directory*/maya/2.0/prefs/shelves. If you're running an NT machine, copy them to *drive*:\WinNT\Profiles\<YourUserName>\maya\2.0 prefs shelves.

After you have copied the Shelf files into your Shelf preferences directory, you must re-launch Maya if it is currently running. The Shelf files and other preferences are loaded upon startup; therefore, you must restart Maya for the new Shelf files to be read. After you have launched Maya, you should see several new Shelf files along your Shelf. We will refer to these different Shelves throughout this chapter.

Exercise 10.1: Do Pose-to-Pose Animating with the Leaping ion

1. After loading the Shelf files, you should see a Lion tab on your Shelf. If not, follow the steps above to copy the Shelf files from your CD-ROM to the appropriate location. This lion Shelf contains MEL scripts to help you animate the lion.
2. Copy the lion project directory to your Maya home directory from the accompanying CD-ROM. Go to File, Project, Set and set the current Maya project to lion. Go to File, Open Scene and open the file lionJumpStart.mb.
3. In the Shelf area of the interface, click the Lion tab to make it the current Shelf. In the Shelf you will see several MEL Shelf buttons that will be used throughout this exercise.
4. The seven Shelf buttons in the lion Shelf are labeled lionPose1 through lionPose7. When clicked, each of these seven buttons positions the lion character into a different pose. Click each of the buttons and examine the different poses of the lion. When you are finished examining the different poses, click the lionPose1 Shelf button. The posing of the lion character has been done for you so that you can concentrate on the "timing" of the lion poses. Experiment with recording poses on your own later.

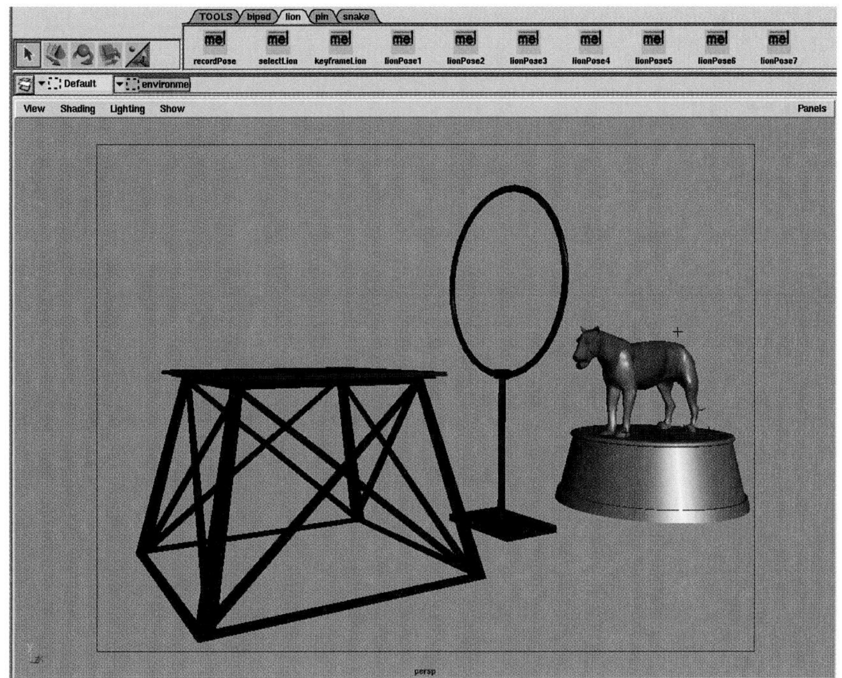

Figure 10.8
The new Lion tab on your Shelf.

Note

These Shelf buttons were created using the recordPose.mel script, which is available in the AnimationScripts directory on the CD-ROM. This script "records" the settings of attributes for the selected objects. In the case of the lion, the feet, back cluster (for the spine), and lionControl locator were selected before the recordPose.mel script was executed. The attribute settings are recorded to the history section of the Script Editor. Highlight the list of setAtttr commands and middle-mouse-drag it to the Shelf. Each time you click that MEL icon, it will set the attributes of those objects.

5. Make sure you are at frame 1 and click the lionPose1 Shelf button on the lion Shelf. Then click the selectLion Shelf button on the lion Shelf. This button is a macro for selecting the different nodes of the lion character, such as the feet and root of the skeleton. Press S on the keyboard; this is a hotkey for setting keyframes on the selected objects.
6. Go to frame 10 using the Time Slider and then click the lionPose2 Shelf button to pose the lion in its crouching pose. Make sure the different objects of the lion are still selected; if they aren't, click the selectLion Shelf button. Press S once again to set keyframes for the lion objects at frame 10.
7. Set the current frame to frame 20 and click the lionPose3 Shelf button. Press S to keyframe the selected items of the lion at the current frame.

Exercise 10.1: continued

8. Set the current frame to 30, and then click the lionPose4 Shelf button. This poses the lion so that the front feet start to leave the platform. Press S to keyframe the selected objects at frame 30.
9. Continue setting keyframes for each of the poses, spacing them 10 frames apart.
10. Next, we will Playblast the animation so that we can examine the timing of the lion animation. Using the middle-mouse button, click in the Perspective window to make it active. Go to Window, Playblast. In the Option window, click the Reset button and then the Playblast button. Maya will take a screenshot of each frame of the animation one by one, which will take a few seconds. When the Playblast is finished, Maya will open a movie file of the animation. If the movie is not playing, click the Play button on the movie control window. Upon examination of the animation, you will notice that the timing of the lion between each of its poses is somewhat unnatural and robotic. You will need to adjust the timing between some of the poses to make the animation more convincing. We'll focus on adjusting the timing of the takeoff and landing of the lion so that it occurs more quickly.
11. With the objects of the lion still selected (if they aren't, click the selectLion Shelf button), go to Window, Animation Editors, Dope Sheet. The *Dope Sheet* is an ideal editor for adjusting the timing between keyframes in your scene.
12. In the Dope Sheet, we will use the Dopesheet summary line running horizontally near the top of the window to adjust the keyframes of all the selected objects. On the left side of the Dope Sheet, you will see a vertical listing of the selected objects. Running horizontally in the Dope Sheet are a series of small squares, each representing a frame in time. Keyframes are represented as brown squares in the Dopesheet. Several keyframes should be showing in the Dopesheet, spaced 10 frames apart from frame 1 to 70. Horizontally on the Dopesheet summary line, use the left mouse button to select the keyframes from frame 20 to frame 70 (when selected, the keyframes will turn yellow).
13. With the keyframes selected in the Dopesheet, press W on the keyboard to translate the keyframes. In the Dopesheet, use the left mouse button to move the selected keyframes five frames to the left. This makes the timing of the lion crouching to beginning to take off occur five frames sooner.
14. Select all the keyframes in the Dopesheet summary line from frame 25 to frame 65 and translate them five frames to the left. This again makes the motion of the lion taking off five frames faster.

15. Select the keyframes from frame 30 to frame 60 and move them six frames to the left. Now the motion of the lion leaving the platform to mid-air is six frames quicker.
16. Select the keys from frames 34 to 54 and move them four frames to the left.
17. Select the final two frames (from frame 40 to 50) and move them six frames to the left. This makes the timing of the lion landing on the high platform quicker.
18. Playblast the animation once again to examine the changes in timing we have performed using the Dopesheet.
19. You can continue to adjust the timing of the lion by translating the keyframes in the Dopesheet. You can also select the different parts of the lion, such as the feet and root, using the selection handles and edit their timing individually in the Dopesheet. Open the scene lionJumpFinish.mb to view a completed version of the lion jump animation.

The strong points of the Pose-to-Pose approach are that it is very intuitive and methodical. It follows the traditional workflow of 2D animation. You start with strong, easily identified character poses, then set the time between those poses. The quality of your animation relies heavily on how well the character poses relay a message or idea to the audience. If one of the poses is incorrect, it could throw off the entire animation.

One of the weaknesses of the Pose-to-Pose method can be managing all of the keyframed objects of your character. For instance, all four feet, the back, the neck, and head need to be keyframed to hold the lion in each of its keyposes. You will have to keep track of all of these objects when adjusting poses and timing between poses.

10.1.2 Rotoscoping

Rotoscoping is the process of matching the movement of your 3D character to a sequence of 2D images. For instance, shoot some video of someone throwing a ball, digitize the video into your computer (using a video capture card), and save the digitized video as a sequence of image files or as a movie file. These files can then be loaded into the background 3D views, where you can then pose your character using the background images as a reference. The following steps demonstrate how to load a movie file into the perspective window to use for rotoscoping.

Exercise 10.2: Do Rotoscoping Animating by Throwing a Baseball

1. Set the current project to rotoscope by going to File, Project, Set and select the rotoscope directory. Then go to File, Open Scene and select the file pinRoto.mb.
2. With the middle mouse button, click in the perspective window to make it the current window. Then on the panel menu bar for the perspective window, go to View, Select Camera. Open the Attribute Editor for the perspective camera by going to Window, Attribute Editor.
3. In the Attribute Editor for the perspective camera, go to the Environment section and click the Create button next to Image Plane. This creates a default Image Plane node for the perspective camera.
4. In the image plane attributes, browse using the folder icon in the Image name field. Find the images in the bballRoto directory on the CD-ROM. You can select any frame in the sequence to load it into the image plane. In the Placement section of the Attribute Editor for the Image Plane, change the Fit from Best to Fill.

Figure 10.9
Loading the sequence you will use for the rotoscoping exercise.

5. Set the Time Slider to frame 1 if it is not set already.
6. In the Image Plane Attributes section of the Attributes Editor for the Image Plane, click in the box next to Use Frame Extension. This tells Maya that you are loading an image sequence into the Image Plane. When you click on Use

Frame Extension you will see the Frame Extension area underneath become editable. Make sure that Frame Extension is set to 1; then, over the word Frame Extension, right-mouse-click for the pop-up menu, then go to Set Key.

7. Set the Time Slider to the frame 150. Go back to the Attributes Editor and set the Frame Extension to 150, then set a keyframe for the Frame Extension at frame 150 using the right mouse button pop-up menu.

8. You have now keyed the Frame Extension of the Image Plane from frame 1 to 150. As you now update the time on the Time Slider, you will also see the image sequence in the Image Plane Update.

   You now need to match the view of the character with the images of the Image Plane. The Image Plane stays locked to the character, so wherever you move the camera, the Image Plane will always stay locked right in front of the view of the camera. You can tumble, pan, zoom, translate or rotate the camera to line up the Image Plane and character.

9. Go to frame 1 in the Time Slider. Tumble, track, and zoom the camera to line up the Image Plane with the camera. It won't be a perfect match because of the proportion differences between the actor on the Image Plane and our cartoony character. Just try to get the angle correct and line up the feet of the actor on the image plane with the feet of the 3D character. Zoom the camera in or out to get the size of the actor on the Image Plane close to the size of the 3D character.

10. After lining up the 3D character, you should set a keyframe for the perspective camera's translate and rotate attributes. Make the perspective camera the current camera by clicking in the perspective view with the middle mouse button. In the panel menu bar for the perspective window go to View, Select Camera. In the Channel box, highlight all the translate and rotate attributes, and hold down the right mouse button and the Key Selected button. You have set a keyframe for the camera in its current position. Now if you accidentally move the camera and lose your alignment with the Image Plane, you can rewind to frame 1, which will reset the camera to the keyframes you set.

11. Now you can start animating the character to the images on the Image Plane. Select both feet and the pelvis using the three selection handles behind the character, and set keyframes for the translate attributes in the Channel box. You can highlight the translate attributes in the Channel box for the feet and pelvis, and then click Key Selected.

12. Go to frame 10 and use the selection handles to translate the feet and pelvis to match the actor in the Image Plane. After you have positioned the feet and pelvis, set a keyframe for the translate channels at frame 10. Now, continue animating the feet and the pelvis every 5 or 10 frames one time through to frame 150.

Tip

You can also "lock" the camera attributes (select the camera, highlight the transform attributes (translate and rotate) in the Channel Box, right-click in the Channel Box for the pop-up menu, and click Lock Selected to lock attributes.

Exercise 10.2: continued

Figure 10.10
Image Plane loaded into Perspective window panel.

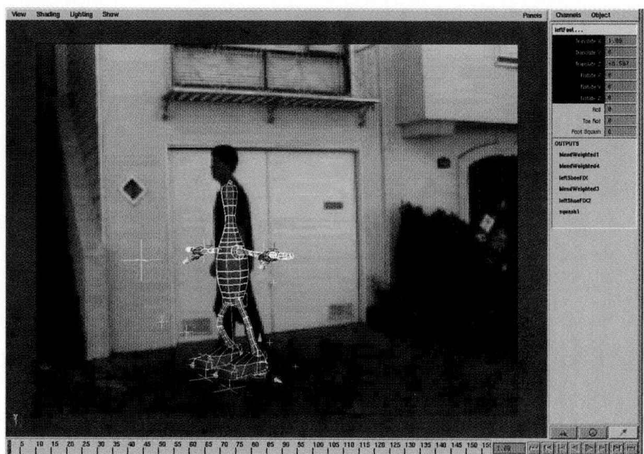

Figure 10.11
Lining up feet and pelvis of character with Image Plane.

13. After keyframing the translation of the feet and pelvis every 5 to 10 frames from frame 1 to 150, you can animate the rotation channels every 5 to 10 frames. Again, use the images on the Image Plane as a reference. Your character doesn't have to match the Image Plane exactly, but try to match the limb positions as closely as possible.

Figure 10.12
Moving character's feet and pelvis to Image Plane sequence.

14. After you have blocked in the animation of the feet and pelvis, select the locator in front of the pin character to get to the controls of the arms. Use the attributes in the Channel box to position and keyframe the arms every 5 to 10 frames. Highlight the attributes in the Channel box and use the right mouse button to access the pop-up menu and click Keyframe Selected.

This is your "blocking pass." You will rotoscope your character in many passes. You start by roughly blocking in the poses of the character to match the Image Plane over the entire length of the animation. Then rewind back to frame 1, and fine-tune the motion until it matches the Image Plane images more exactly. It is a good idea to also start with the main body parts first, like the feet and pelvis. Work with your animation from general body parts (feet, pelvis, arms, back, and neck) to detailed body parts (fingers, ears, eyes, facial parts and so on).

This method is great for animating subtle and complex motion for your 3D character. By using the footage as a guide for your animation, you will be able to pick up on the subtle nuances of motion and timing. This method is also great method for learning about how things move. It is very similar to the Pose-to-Pose method with the additional image references. In addition, rotoscoping is essential for matching live action footage. If you are animating a character that will later be composited into video or film, it is essential that you have footage to which you can match timing and camera angles.

One weakness of this method is the difficulty of obtaining the proper footage for your animation situation—especially when animating imaginary characters. Imagine trying to find video for an 18-legged beast! However, you could always grab a video camera and record yourself hopping around like an 18-legged beast. Another problem can be lining up your 2D rotoscoping imagery with your 3D character. It can be difficult to rotoscope to footage in perspective. Using video shot from the side or front can be more straightforward. Reference footage shot in multiple angles, such as from the side and front, can be helpful too.

10.1.3 Layering

The process of animation can also be approached in layers. *Layering* the different aspects of your animation can be compared to painting or drawing. The general idea or design is sketched in; then the fine detail is gradually added. The detail is added only after the basic framework has been established.

The same method can be used when animating a 3D character. You want to start with general, overall movements, then add detail later on. For example, if you wanted to animate a character walking from one end of the room to the other to pick up an object on a table, you would start by animating the feet and pelvis of the character getting to the table. Then you would animate the arms and hands picking up the object on the table.

Try to develop a methodology or an approach to animating your 3D character. Animate the legs and feet first, then the pelvis, then the back, then the shoulders, then the elbows, and so on. Treat each animated aspect of your character as a "layer." Only go on to the next layer when the current is nearly complete. The sum of all your layers is the entire character animated. Using a methodology like this also makes it straightforward to edit and fix animation in your character.

Exercise 10.3: Walk Your Character Across the Room with the Layering Animation Method

1. Open the bipedWalkStart.mb file from the accompanying CD-ROM. This file is the finished biped you worked on in Chapter 9, "Setting Up Your 3D Characters for Animation."

   We will be using the bipedRoot, leftFoot, and rightFoot nodes to animate the pelvis and each of the feet to start. You can access these objects by clicking the selection handles that have been placed behind the character. For the arms and other parts of the body we will use custom attributes that have been added to the locator in front of the character (as covered in Chapter 9).

2. First, we need to change the default keyframe interpolation and the frame rate for our animated walk cycle. Choose Options, General Preferences. Click on the Animation tab. In the Keys section of the Animation Preferences, set the Default In Tangent and the Default Out Tangent to Flat. This interpolation type will give us a nice "ease in" and "ease out" of keyframes that we set.

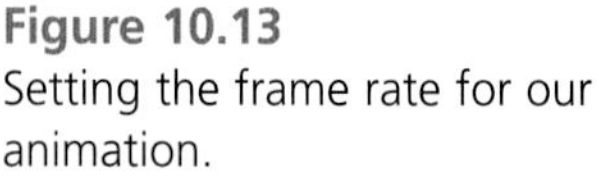

Figure 10.13
Setting the frame rate for our animation.

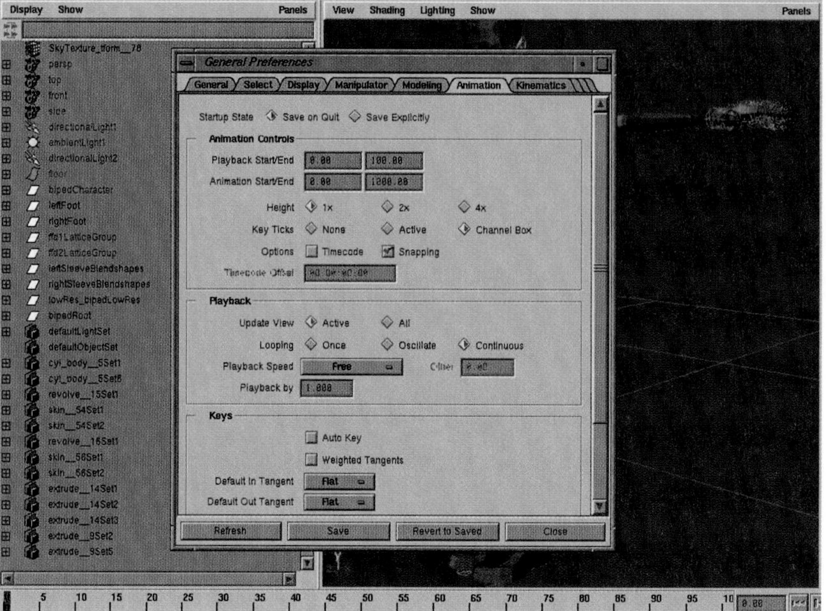

3. In the General Preferences window, go to the Units tab if it is not already showing (you may have to click the overlaying tabs to see a drop-down list with the units selection). Set the time to NTSC (30fps). We will animate the walk of our character at 30 frames per second of animation. This frame rate works well for animation that will be put to videotape. Close the General Preferences window and set the time range slider to frame 1 for the start and frame 100 to end.
4. Position the character in the starting position for the walk. Select the rightFoot selection handle and move the right foot back slightly.
5. Select the leftFoot selection handle and move the left foot slightly forward.

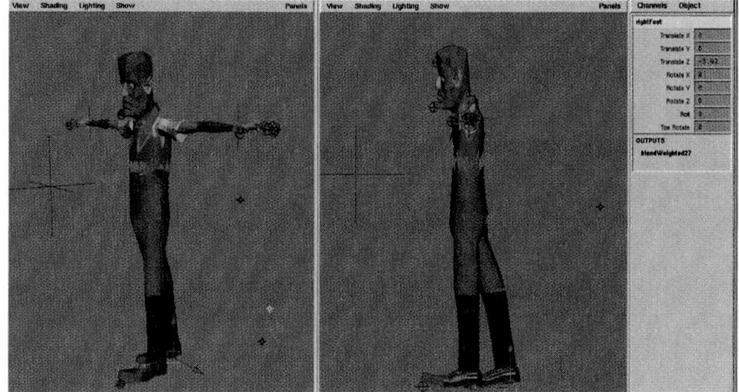

Figure 10.14
Moving the rightFoot selection handle to the starting position for the walk.

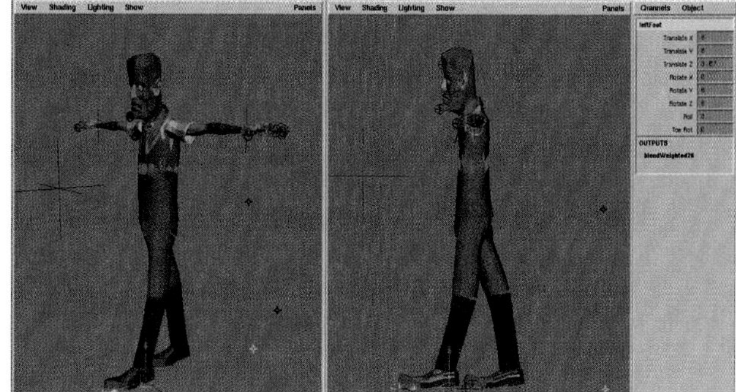

Figure 10.15
Moving the leftFoot selection handle to the starting position for the walk.

6. Select the bipedRoot selection handle and move the pelvis down slightly to add a slight bend in the legs. We want our character to have a natural bend in the legs at the start of the walk.
7. Select the leftFoot selection handle and Shift+select the rightFoot and bipedRoot selection handles.
8. In the Channel box, highlight the translate X, Y, and Z attributes. In the Channel box, hold down the right mouse button to get the Channel box pop-up menu and go to Key Selected to set a keyframe at frame 1.

   Later on in this chapter we will discuss how to create macros using the MEL scripting language to automate the keyframing process.
9. Set the current time to frame 20 using the Time Slider. Select the rightFoot and move it forward along the Z axis, ahead of the left foot. You will notice that the pelvis is left behind; we will fix this in the next step.
10. Select the bipedRoot node and move it forward along the Z axis so it lies between the two feet. Select the selection handles for the two feet and the pelvis and set a keyframe for translate x,y,z at frame 20.

Exercise 10.3: continued

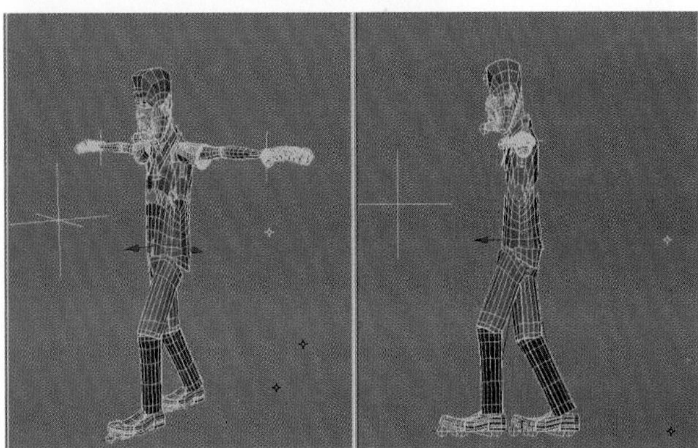

Figure 10.16
Adding a natural bend at the knees to our character.

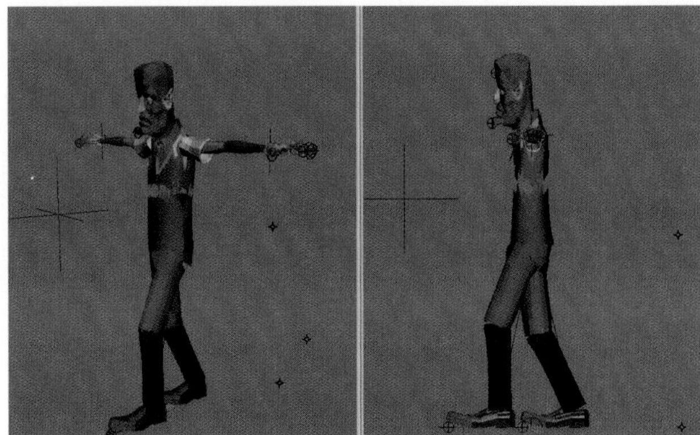

Figure 10.17
Biped character with left foot forward and knees bent.

You can scrub through the animation you have just keyframed between frames 1 and 20 by dragging the Time Slider bar back and forth along the time line. You will see how Maya is interpolating the motion of the pelvis and two feet between frames 1 and 20. Although the left foot has not moved from frame 1 to 20, we still set a keyframe for it to ensure that it remains in place until the right foot has stopped moving.

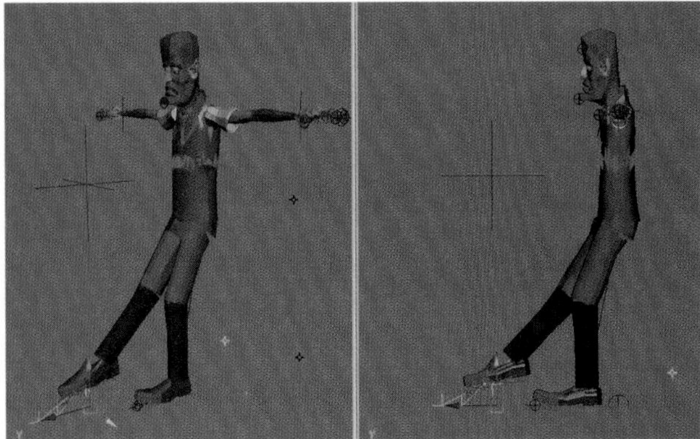

Figure 10.18
Notice how the pelvis is left behind when you move the right foot forward.

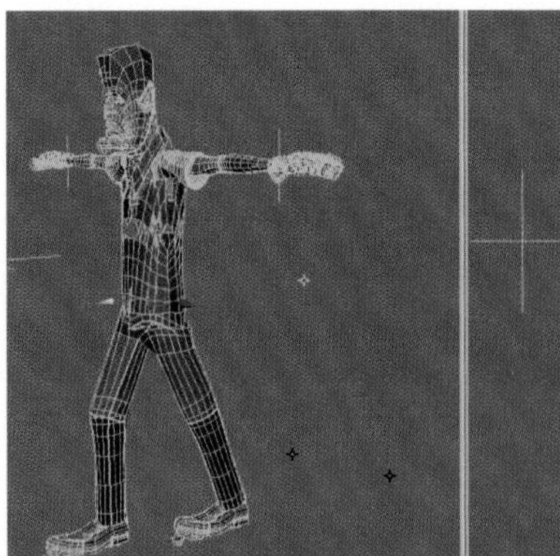

Figure 10.19
Right foot and pelvis forward on Z axis at frame 20.

11. Set the Time Slider to frame 40. Select the leftFoot selection handle and translate it along the Z axis so it is ahead of the right foot. Again, the pelvis lags behind. Select the bipedRoot and move it so it is centered between the left and right feet. Set a keyframe for both feet and the pelvis translation in the Channel box at frame 40.

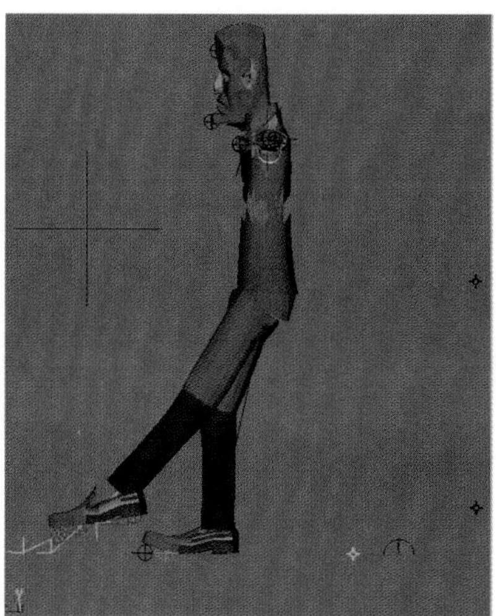

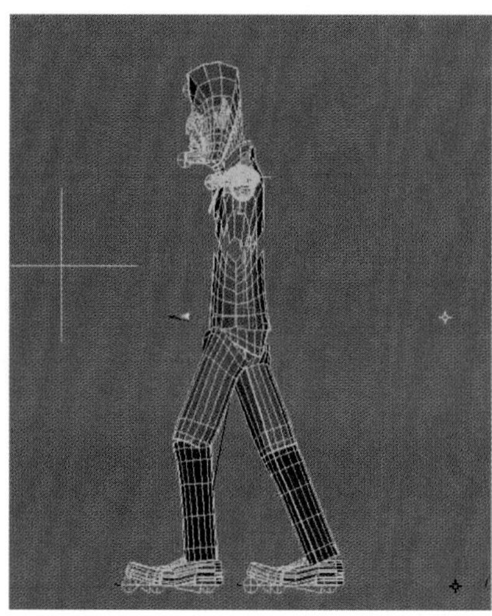

Figure 10.20
The left image shows the pelvis being left behind as the foot is dragged forward. The right image shows the corrected pelvis.

12. Repeat the alternating translation of the right and left feet while keeping the pelvis centered for the rest of the time range (up to frame 100). Set keyframes for both feet and the pelvis at frames 60, 80, and 100. At frames 60 and 100, the right foot should be ahead of the left foot. At frame 80, the left foot will be ahead of the right foot. Continue to keyframe the translation of the pelvis at frames 60, 80 and 100 to keep it centered between the feet.
13. Rewind the animation and click the Play button. The biped character should be "sliding" its feet forward for about five steps.

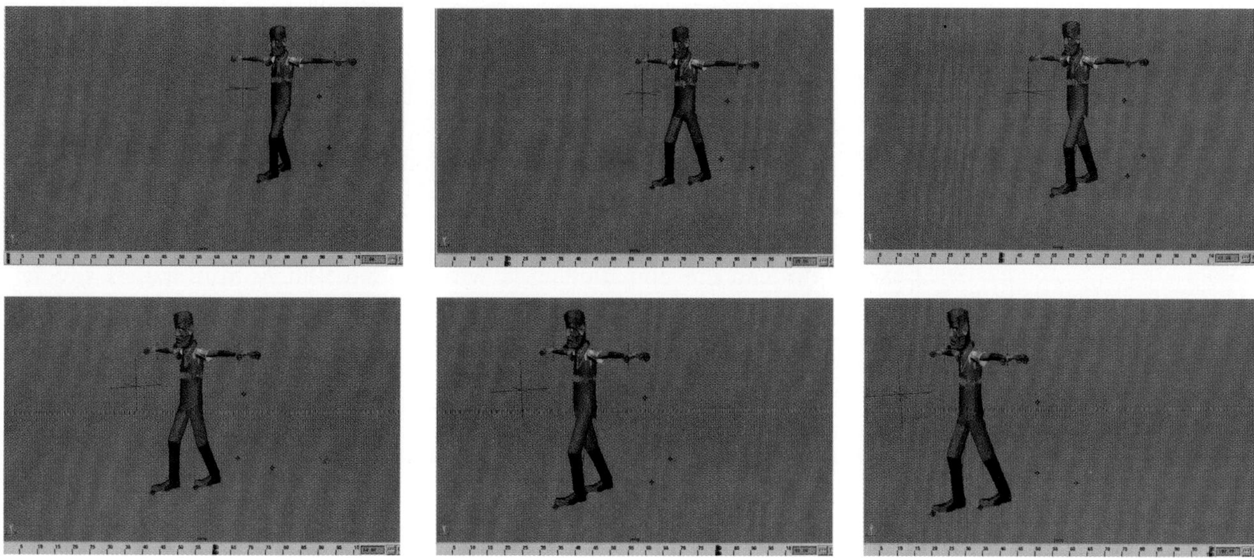

Figure 10.21
Viewing the animation you have created to this point.

Exercise 10.3: continued

At this point the animation looks pretty stiff and unconvincing. However, this is just our first pass at animating the character. With each pass through the 100 frames of animation, we will refine and add new motion to the character.

Exercise 10.4: Refine Your Character's Walk Animation

The first task we need to do to begin refining our animation is to lift the feet off the ground at the middle of each step.

1. Select the rightFoot and set the Time Slider to frame 10. This is the "in-between" step for the right foot. Translate the rightFoot on the Y axis so it is raised above the ground. Highlight only the Translate Y channel of the rightFoot in the Channel box and click Key Selected in the Channel box pop-up menu.

Figure 10.22
Right foot translated up on the Y axis at frame 10.

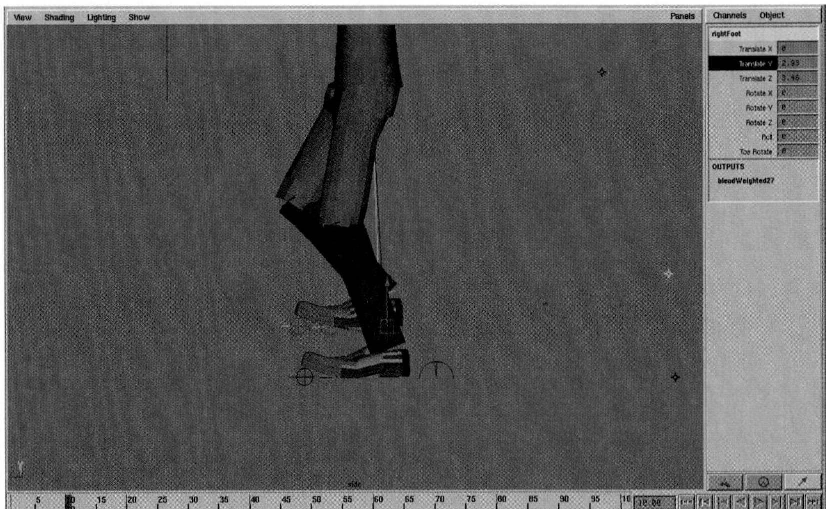

2. Scrub through the animation again with the Time Slider from frames 1 to 20. Watch how the right foot lifts off the ground at frame 10 then lands back on the ground at frame 20. Repeat the keyframing of the right foot lifting off the ground at frames 50 and 90.

   We will keyframe the Y translation of the leftFoot using Copy and Paste in the Time Slider.

3. Select the leftFoot and set the current time to frame 30. This is the first "in-between" step of the leftFoot. Translate the leftFoot up along the Y axis and set a keyframe for translate Y in the Channel box.

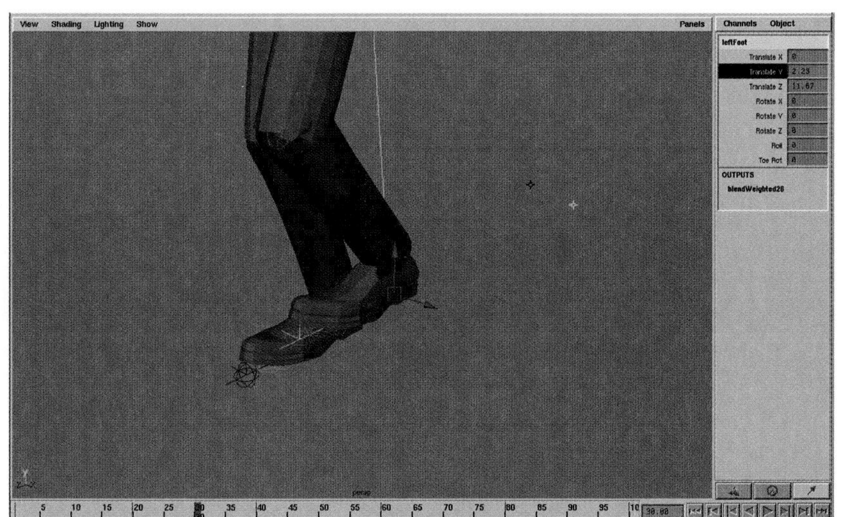

Figure 10.23
Left foot translated on the Y axis at frame 30.

Take a close look at the Time Slider. You will notice thin, red tick marks. These represent keyframes for the selected objects—in this case, the leftFoot. Right now, the red keyframe tick marks represent keyframes for all the animated channels on the currently selected objects—in this case, XYZ translate. Let's change the Animation preferences so that we can display the keyframes for the selected attributes in the Channel box of the selected object.

4. Go to the Main Menu bar and choose Options, General Preferences. Click the Animation tab in the General Preferences window. In the Animation Controls section of the Animation Preferences, set Key Ticks to Channel Box. With this setting, you select an object and highlight channel names in the Channel box; the red keyframe tick marks represent the keys for just the selected channels. In the following steps we need to copy and paste just the Y translate keys and not the X or Z translate channels.

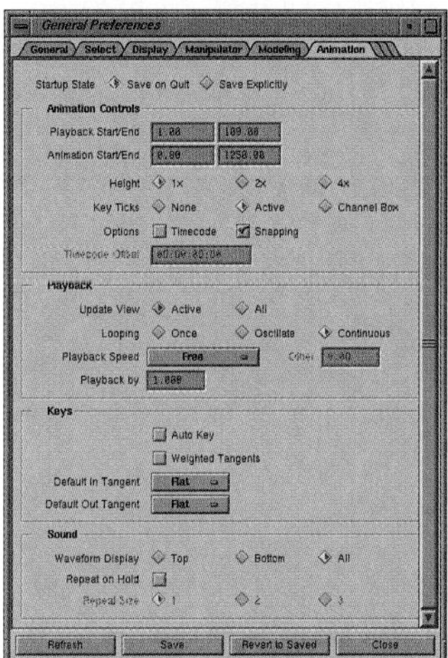

Figure 10.24
Animation Preferences with Key Ticks set to Channel Box.

Exercise 10.4: continued

5. While at frame 30, position your mouse in the Time Slider and right-mouse-click to get the Time Slider pop-up menu and go to Copy. You have just copied the Y translate keyframe for the leftFoot.

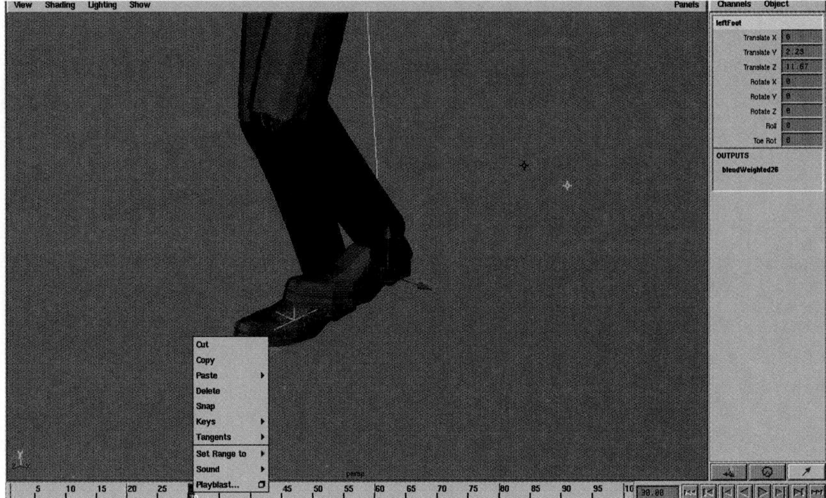

Figure 10.25
Pop-up menu in the Time Slider for copying keyframes.

6. Set the Time Slider to frame 70, which is the next frame where the left foot should be off the ground. In the Time Slider, Shift+left-click at frame 70; a red bar will appear at frame 70. Again, in the Time Slider, hold down the right mouse button and go to Paste to paste the copied keyframe (from frame 30) at frame 70.

When you Shift+click at a frame in the Time Slider then paste the keyframe(s), you are inserting keyframes without moving other keyframes back. You can also paste keyframes without Shift+clicking on the frame you want to paste. Doing this will paste the keyframes and "push" keyframes to the right. Pushing the keyframes over would be good if you had to insert a range of keyframes in the middle of an animation and push the existing animation later in time.

Exercise 10.5: Animate the Pelvis Moving Up and Down and Side to Side

When the feet of the character are at their widest stride, the pelvis should be somewhat lower than when the feet are in mid-stride (feet being side by side). As the feet go through mid-stride, the pelvis will lift up slightly higher along the Y axis. We will use Auto Key to help automate the keyframing of the bipedRoot node.

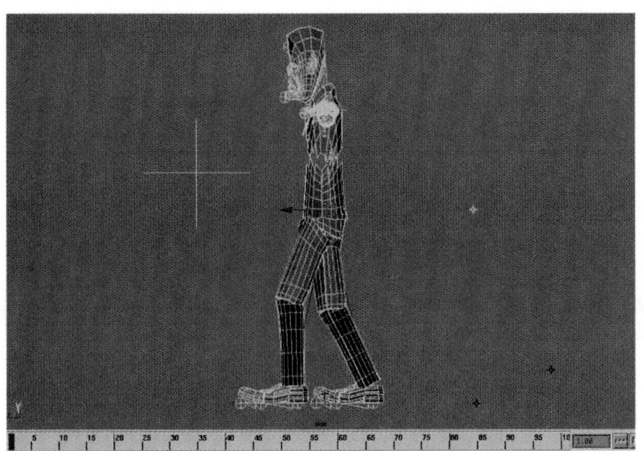

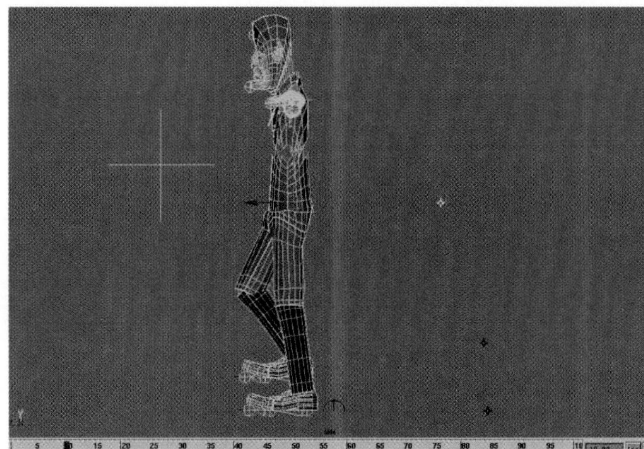

Figure 10.26
Pelvis lower at widest stride and higher in mid-stride.

1. Activate Auto Key by clicking on the Key icon in the lower-right corner of the interface. Notice that we have existing keyframes on the Y translate of the bipedRoot already. With Auto Key activated, anytime you change the value of the Y translate at a particular frame in time, Maya will automatically insert a keyframe at the current frame.
2. Go to frame 10 and translate the bipedRoot node up slightly on the Y axis to raise the pelvis where the left and right leg are in mid-stride. You will see that Maya automatically adds a keyframe for the Y translate at frame 10.

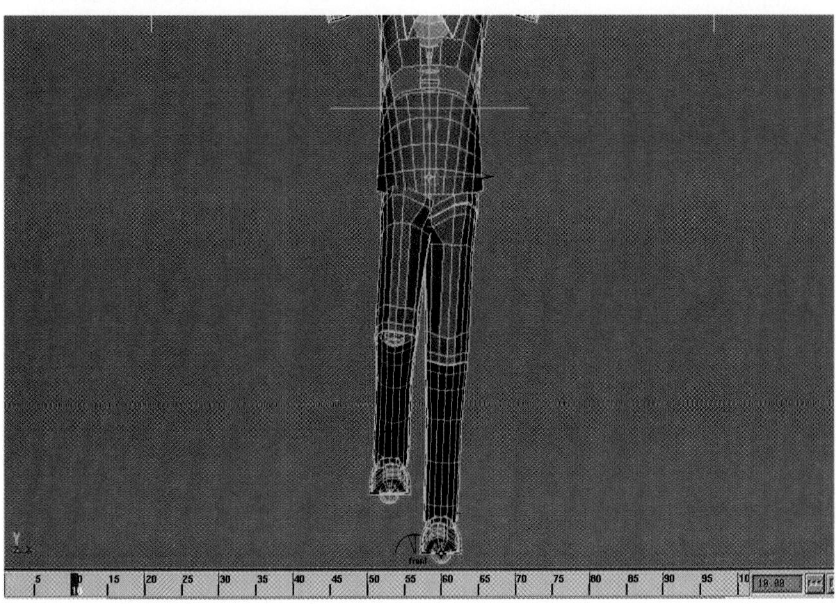

Figure 10.27
Move the pelvis up slightly along the Y axis.

Exercise 10.5: continued

Repeat the previous steps to animate the pelvis moving side to side on the X axis. The pelvis area will move over the leg that is planted on the ground. When that leg begins to lift off the ground, the pelvis will shift over the opposite leg as it is planting on the ground. We have some keyframes already on the X translate channel, so let's delete them and reanimate the X translate of the bipedRoot.

3. Select the bipedRoot selection handle and highlight Translate X in the Channel box. Hold down the right mouse button in the Channel box and go to Delete Selected.

   Make sure Auto Key is still on and set a keyframe at frame 1 for the Translate X of the bipedRoot. This sets an initial keyframe for Translate X, now each time you change the Translate X of the bipedRoot from its initial keyframe, Maya will add a keyframe for you.

4. Go to frame 10, where the right foot lifts off the ground, and shift the bipedRoot over the left leg. A keyframe is added automatically at frame 10.
5. Go to frame 30 where the left foot lifts off the ground and shift the bipedRoot over the right leg.

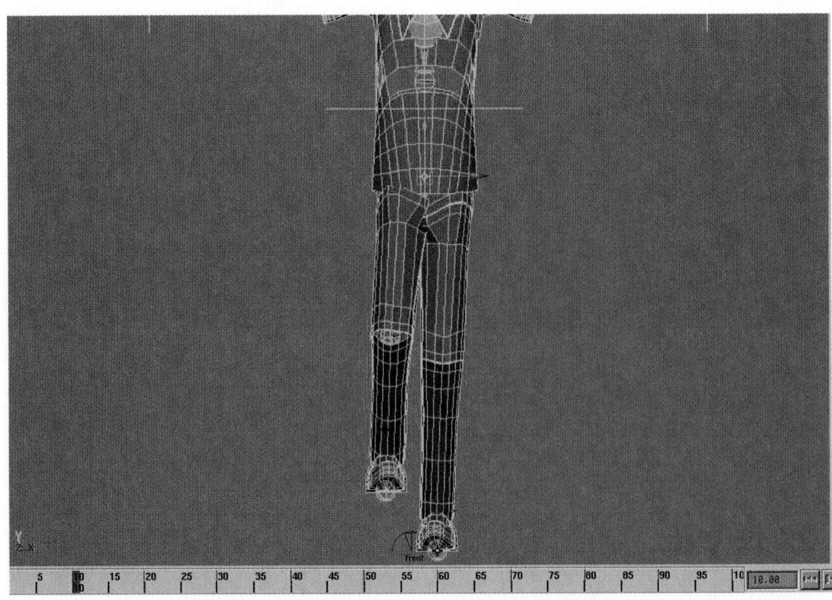

Figure 10.28
Pelvis translated over the left leg when right foot lifts off the floor.

There is no need to add a key for the pelvis in between the two feet. As the bipedRoot is animated on the X axis side to side it will pass through the middle.

6. Repeat the side-to-side motion of the bipedRoot at frames 50, 70, and 90. When you are finished, be sure to click the Key icon in the lower-right corner of the interface to turn Auto Key off.

Exercise 10.6: Animate the Roll of the Feet

Let's take a look at animating the roll attribute we created for the feet from Chapter 9. There will be three main positions of the foot we will need to concern ourselves with when animating the roll attribute. As the foot begins to lift off the ground, the heel will lift off the ground . As the foot begins to land, the heel will come in contact with the ground first. Then as the foot comes to rest on the ground, it will lie flat on the ground.

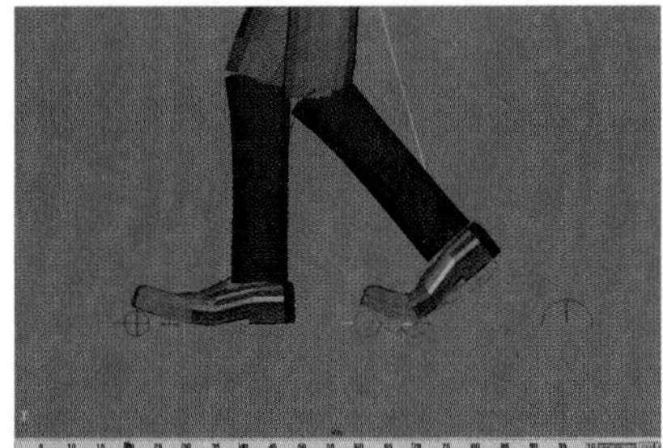

Figure 10.29
The heel lifts off the ground.

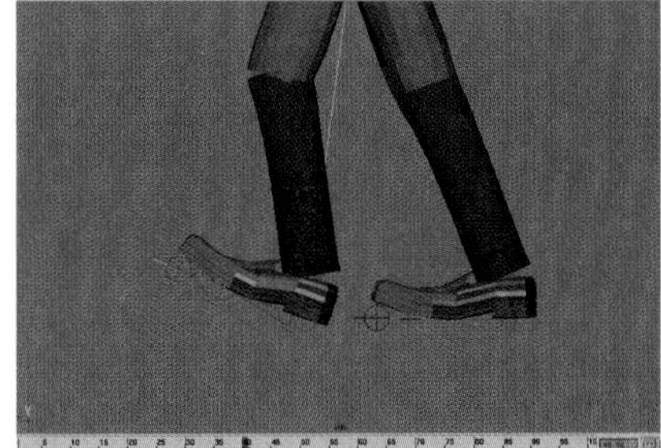

Figure 10.30
The heel lands on the ground first.

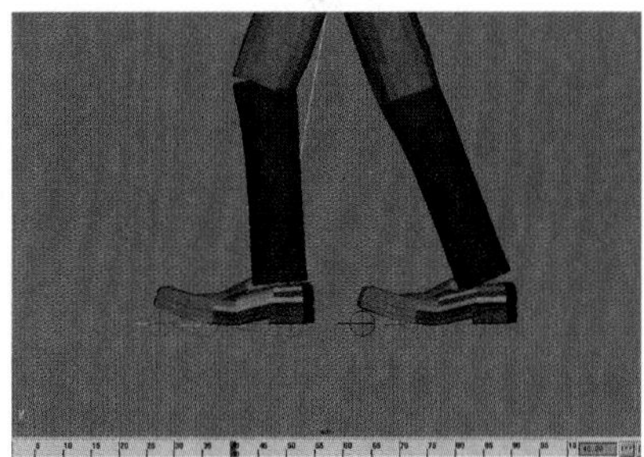

Figure 10.31
The heel lies flat on the ground.

1. Select the rightFoot selection handle and set the Time Slider to frame 1. In the Channel box, set the Roll attribute to 10 and then keyframe the roll attribute.
2. Go to frame 20 and set the roll attribute on the rightFoot to –5 and then set a keyframe.
3. Go to frame 25 and set the rightFoot roll to 0 and then set a keyframe.
4. Repeat the animation of the foot roll for both feet through the rest of the animation. Set keyframes on the foot roll attributes at –5 where the foot is landing and to 10 where the foot is "taking off." Use Auto Key to automate the keyframing process.

Exercise 10.7: Editing the Animation of the "Roll" Attribute for the Feet

At this moment, the roll of the foot is not quite correct. The foot needs to remain flat on the ground for a longer period of time. Right now, the foot rolls too quickly and appears unnatural. Also, we need to make the motion of the foot hitting the ground much faster and snappier. We will be able to adjust the animation curves for the foot roll motion in the Graph Editor. To get the foot to lie flat on the ground longer, we will need to have a flat area in the curve for the roll attribute. To make the motion of the foot hitting the ground faster, we will need to make portions of the animation curve steeper.

Let's first take a look at the curve representing the motion of the Roll attribute in the Graph Editor. We will look at the shape of the curve and interpret what is happening to the foot. Select the selection handle for the rightFoot of our biped character and open the Graph Editor. Highlight the roll attribute of rightFoot on the left side of the Graph Editor. Keyframes with a value of 0 equate to the foot laying flat on the ground. Keyframes with a value of 10 equate to the foot rolling off the ball and toe of the foot. Keyframes with a value of -5 equate to the foot rolling around the heel of the foot.

1. Select all the keyframes for the rightFoot.roll with a value of –5 in the Graph Editor.
2. Click the Break Tangents button in the Graph Editor.
3. Shift+select the red handles on the left of each of the three keyframes you have selected.
4. Press the W key on the keyboard to translate the tangent handles in the Graph Editor.
5. Drag the handles up and to the left so you end up with a steeper region with the curve from –5 to 0.
6. Select all of the keys with a value of 10 that represent the foot rolling forward from the ball to the toe.
7. Click the Break Tangents button in the Graph Editor.
8. Shift+select the blue tangent handles on the left side of each of the selected keyframes.
9. Drag the handle down and to the right, so you end up with steeper regions in the curve where the value of roll goes from 0 to 10.

By creating these flatter areas for the foot roll attribute, the foot remains flat on the ground for a longer period of time. In the areas where the curve is steep, the foot roll will occur much more quickly and appear more lifelike.

When using the layering approach to animating a 3D character, the process is very organized and methodical. You start very generally and refine the motion with each layer and pass over the animation. It is best to start with the primary motion of the pelvis, feet, and arms first, and then work your way to the secondary motion of the hands, head, and face.

One of the drawbacks to using a layered approach is that you are working on only certain parts of the 3D character at a time. The layered approach is not an ideal method for animation where the character must achieve strong poses. A strong pose usually involves positioning several objects on the character into a specific pose. It is much better suited to motion cycles like a walk or run for instance. The Pose-to-Pose method, where you pose the various parts of the character and set keyframes, is a much more intuitive way of animating a character which needs strong keyposes.

10.1.4 Motion Capture

Motion capture is another method for animating your CG characters. This involves recording 3D movement from an object outside the computer and inputting that data into your computer, then interacting with it in 3D software like Maya. Motion capture is most often used to capture human bodies, facial expressions, objects, and camera and light positions.

The process of capturing motion may involve physically placing sensors on an actor, and having the actor perform movements while the computer records the three-dimensional positions of the sensors. This data can then be transferred into 3D software programs like Maya onto skeleton joints or the transformation nodes of other objects.

To observe motion capture animation in action, do the following:

1. Open the file dancer.mb on the CD. This file contains captured motion from a woman dancing.
2. Go to the Perspective window and in the window panel, go to Show, Surfaces to hide the surfaces of the character (this will speed up the playback of the animation on the skeleton). Make sure Show, Joints is on in the Perspective panel.
3. Click the Play button on the Time Slider to play back the animation. You will see the skeleton joints rotating as the figure dances through numerous complex movements. This is a very good example of captured human movement. There are many subtle movements in the motion of the dancer that would be very difficult and time consuming for an animator to reproduce by hand.

Figure 10.32
Stills from the dancer.mb file showing the use of motion capture.

Motion capture can be a very quick method for getting motion into your CG characters. Some motion capture devices and software allow you to get real-time interaction between your real-world actor and CG character. Motion capture also picks up the most subtle nuances of movement that would be very difficult and time consuming to replicate through traditional keyframing techniques. Motion capture is very useful when there is a need for a lot of complex animation in a short amount of time because the turnaround time from captured motion to the CG character can be very short.

Some of the drawbacks to using motion capture are the expenses of the equipment and software needed to actually obtain the captured motion. The cost of the motion capture equipment, most of the time, can run into the thousands of dollars. There are also many technical considerations for getting the motion capture onto your CG character correctly. This includes calibrating the 3D space of the real-world environment with the 3D environment in the computer as well as calibrating the proportions of the real-world actor to the CG character. Motion capture can also be difficult to use when there is a need for highly stylized animation (for example, "cartoony" characters). For instance, how would you capture the motion of a squashing and stretching, 20-foot tall, 12-legged dragon?

10.2 Improving Interaction with Your 3D Characters

Let's look at a few ways that we can accelerate our interaction with our character. As an animator, it is very important to be able to visualize the motion you are creating as interactively as possible. As your characters and environments become more complex, your computer and Maya will take more time to update the visual feedback to your screen.

If you have ever seen a traditional 2D animator working, you might have noticed him penciling in his drawings while flipping through sheets (frames) of his animation. The animator is quickly viewing the progression of his animation, examining, refining, and improving the animation with each pass. Flipping the sheets of drawn animation is done in real time, the sequence of frames being displayed at the speed the audience will be viewing them.

One drawback with 3D animation, especially with complex scenes and characters, is that it takes time for the computer and software to calculate and display the 3D information. A fully modeled character, bound to a skeleton, with a few deformers can slow down the playback rate of your character. In some instances, the animator may have to wait seconds or minutes just to view the next frame in an animation sequence. This is the kiss of death for achieving good animation. It is very difficult for a 3D animator to visualize the motion if they cannot interact with the motion close to real time speed. Even very fast computers will not be able to calculate complex scenes and characters in real time (at least not yet!). The faster you interact with your character, the more iterations you will be able to go through and edit your animation.

We will take a look at a few methods that can help improve the way you interact with even your most complex 3D characters. We will first note some built-in functionality Maya provides to improve interaction. Then we will look at a method for creating low-resolution or stand-in characters and environments to animate with. Using a MEL script, we will be able to switch between a high resolution and low resolution character with the click of a button.

10.2.1 Fast Interaction

This mode is found under Display, Fast Interaction. *Fast Interaction* dynamically simplifies your geometry when you update any of your 3D modeling view panels. When toggled on, you will see your NURBS surfaces appear faceted or polygonalized when you navigate in a 3D window or click Play. After you release the mouse button or click the Stop button, your geometry returns to its original shape. Maya is reducing the amount of geometry data it needs to display to the screen.

Exercise 10.8: Use Fast Interaction to Simplify Your Geometry

1. Open the bipedFinished.ma scene file.
2. Go to Display, Fast Interaction to turn on Fast Interaction in Maya.
3. Navigate the camera (tumble/pan/zoom) with the LMB and the Alt keys. You will see the geometry become simplified as you update the 3D views.

With simple characters, you may not see a significant speed increase in the interaction of your character or scene. But as your characters become more complex, Fast Interaction can speed up screen updates of your scene.

10.2.2 Display Smoothness

Maya also has built-in functionality that allows you to specify the viewing resolution of NURBS surfaces. This is essential when working with complex characters and environments.

Exercise 10.9: Use Display Smoothness to Reduce the Complexity of Displayed NURBS Objects in a Scene

1. Continue using the bipedFinished.ma file. Select the NURBS surfaces of the biped character in the bipedFinished.ma file.
2. Choose Display, NURBS Smoothness, Fine (or press the 3 key on the keyboard). This may take a few seconds.

   This displays the selected NURBS surfaces at their highest screen resolution. The surfaces appear very smooth at the cost of interactive speed. Try navigating the camera in the perspective view: You will see that the update of the view is quite slow.
3. With the surfaces still selected, choose Display, NURBS Smoothness, Hull, and set both the U and V to 2. This tells Maya to display every other hull of the selected NURBS surfaces (a setting of 3 in U and V would display every third hull on the selected surfaces).

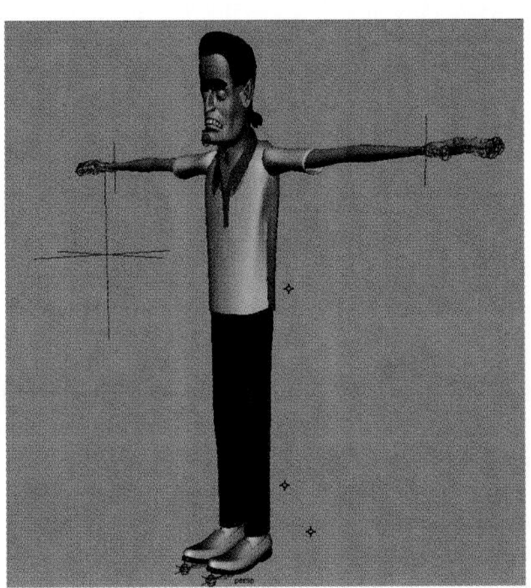

Figure 10.33
Biped character at high resolution, and as reduced complexity display.

This drastically reduces the complexity of the displayed NURBS surfaces in the 3D views. At the cost of smoothly displayed surfaces, your character will update very quickly in the 3D views.

When using Display Smoothness, Hull, you try to balance the surface resolution to be able to recognize important features, with the speed at which you interact with the character.

4. To set the surfaces back to the default resolution, select the surfaces and Display, NURBS Smoothness, Rough.

Keep in mind that the settings of Display Smoothness have no effect on how your surfaces actually render using the software renderer. However, when hardware rendering, Display Smoothness may need to be adjusted for adequately smoothed NURBS surfaces.

10.2.3 Visibility

Another method of improving the speed of interaction that should not be overlooked is the visibility of objects. Maya will not calculate objects that are invisible. You can speed up interaction with your character if you simply hide items that you may not need to use at the time. For instance, if your character has hundreds of teeth, and you are animating a walk cycle, hide the teeth so Maya will not have to calculate them in the 3D views.

10.2.4 Low Resolution "Stand Ins"

An animator can easily create a low-resolution stand-in character to speed up the interaction with even the most complex 3D characters and scenes. The basic idea is to create a copy of your character that is reduced in complexity. You use this simplified character only when animating, for quicker interaction. The simplified character represents the bare minimum amount of geometry you need to represent the character. After you have finished animating, you switch back to the original character model (high resolution) for rendering. All of the animation occurs on one skeleton; you simply hide the high-resolution model and show the low resolution.

Exercise 10.10: Creating a Low-Resolution Stand-In Character

1. Open the scene file bipedFinished.ma.
2. Select only the surfaces that represent the biped character.
3. Show the Layer Bar if it is not showing already by going to Options, Layer Bar. On the Layer Bar click the New Layer button on the far left. Name the new layer **`hiRes`**. Place the cursor over the hiRes layer on the Layer Bar, hold the right mouse button, and go down to Assign Selected.
4. With the surfaces still selected, go to Edit, Duplicate. Click the Reset button, then the Duplicate button.
5. Create another new layer and name it **`lowRes`**. Assign the duplicated geometry, which should still be selected, to the lowRes layer.
6. In the Layer Bar, click with the right mouse button held down on the hiRes layer and select Visible to hide the high resolution geometry.

   The high-resolution geometry is the geometry that is already bound to the skeleton. The low-resolution geometry is a duplicate and has no connections to the skeleton or deformers.

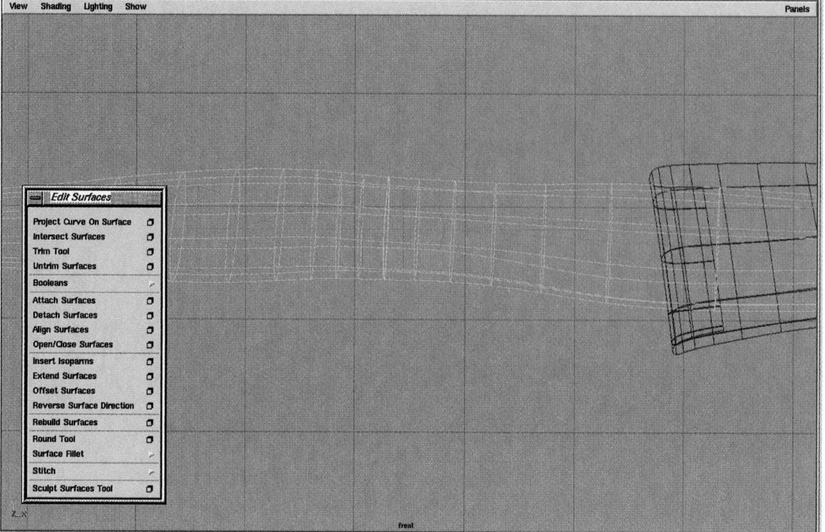

Figure 10.34
Detaching an isoparm at the elbow of the arm surface.

The next step is to slice up or detach our low-resolution character into pieces that we can simply parent to the skeleton. We will detach surfaces at locations on the character where deformations will occur. For instance, on the arm, we will cut the surface of the arm at the elbow. The forearm section will be parented under the elbow and the upper arm will be parented under the shoulder. Surfaces parented to skeleton joints are much faster to calculate then surfaces bound to the skeleton. Maya must calculate all of the points on the surface when bound to a skeleton.

7. Select the left arm surface of the lowRes character geometry.
8. Right-click the left arm surface and select Isoparm from the pop-up menu. This quickly enables you to select isoparms on the left arm surface.
9. Click on one of the vertical isoparms near the elbow.
10. In the Modeling menu, go to Edit Surfaces, Detach Surfaces. The Detach Surfaces tool uses the selected isoparm to cut the single arm surface into two surfaces. We now have the arm surface detached into a lower and upper arm. We detached the arm at the elbow because this is where the deformation will occur.
11. Select the detached upper arm geometry and Shift+select the shoulder joint. Select Edit, Parent. When the shoulder joint moves, the upper arm geometry will follow because it is the child of the shoulder joint.
12. Select the detached lower arm geometry and Shift+select the window elbow joint. Select Edit, Parent.
13. Repeat this process for the rest of the character. Detach surfaces where deformations will take place, such as the elbows, knees, ankles, torso, and so on. Be sure to parent the detached surfaces under the joints you need to control them with.
14. When you're finished, you should be able to animate the skeleton and see the low-resolution geometry follow. Before rendering or to check the high-resolution geometry, you simply need to make the lowRes layer invisible and the hiRes layer visible. Open the file bipedHiLow.mb to view a finished version of this exercise.

You can use other stand-in objects to represent your character geometry as well. For instance, you could use a polygon sphere as a stand-in object for all of the head surfaces. You can also delete items from the duplicated low-resolution geometry. For example, we don't really need the teeth or hair of the character for our low-resolution character. Just pick those unnecessary objects and delete them. Remember, your goal here is to have a stand-in character that represents just enough information for you when animating.

10.3 Characters Interacting with Environments

Often times you may want your characters to appear that they are interacting with the environment around them. Potentially, there are many different ways the character could interact. For instance, your character might hit objects off a table or pick up an object.

Exercise 10.11: Characters Picking Up Objects—Animating the Visibility On and Off

In some situations, your character may need to pick up and interact with objects in the scene. We will achieve this by having two objects for which we will animate the visibility. One of the objects will remain on the table while the other will be parented to the wrist joint of the character's hand. At a particular frame, one object will become invisible as the other becomes visible.

1. Open file pinPickUpObject.mb. This file contains a 150-frame animation of a character picking up a can from a table. The character is not actually picking up the object, but the motion has been animated already for you. It is best to approach your animation this way. Animate the motion of the character first, and then animate the object being picked up; this is what we will do here.
2. Scrub through the animation and determine at which frame the character will need to pick up the object; frame 56 will work great in this example.
3. Go to frame 56 in the Time Slider. Select the can object on the table. Go to Edit, Duplicate. Click the Reset button in the Duplicate option box and then click the Duplicate button.
4. With the duplicate can still active, Shift+select the wrist joint of the right arm. Go to Edit, Parent to make the duplicate can a child of the wrist joint on the right arm.

   Playing back the animation, you will now see that one can remains on the table while the duplicate follows the motion of the right arm and hand.
5. Select the can on the table and set the frame to 55 in the Time Slider. Highlight the Visibility attribute in the Channel box and go to Key Selected.
6. Go to frame 56 and, with the can still selected, change the Visibility attribute to off by typing `0` in the Input field and pressing Enter. You will see the can on the table disappear.
7. Select the can object parented to the wrist joint on the right arm. Keyframe the Visibility on in the Channel box for this can at frame 56.
8. With the can parented to the wrist still selected, go back to frame 55. Keyframe the Visibility attribute off in the Channel box.

Play back the animation. It appears that our character actually picks up the can sitting on the table. But in reality, we are simply animating the visibility of the two cans, which gives the impression that the can is being lifted off of the table.

Exercise 10.12: Character Interacting with Objects Using Rigid Body Dynamics—Pin Character Hitting Cans Off Table

1. Open the file pinHitCans.mb.

   This file contains a "bowling pin" character that has been animated over 45 frames hitting several cans off of the table in front of him. You will notice when playing back the animation that the character's hands pass right through the cans. We will use dynamics to allow us to have the character's hand collide with the cans on the table. It is important to first have the character animated and then go ahead with the rigid body animation.

2. Go to Create, Polygon Primitives, Cube, click the Reset button and then the Create button. This creates a default primitive polygon cube. Name the polygon cube **cubeHand**.

3. With the cube still selected, go to Edit, Delete by Type, History. This gets rid of the extra construction history data for the cube that we no longer need to use.

4. With the cube still selected, go to Bodies, Create Passive Rigid Body with the default settings.

5. Move cubeHand over to the charcter's left hand. Scale and rotate it so the cubeHand cube covers the entire hand. The following transforms work well for this: Translate X=6.5, Translate Y=9, Translate Z=.5, Rotate X=-10, Rotate Y=-12, Rotate Z=35, Scale X=3, Scale Y=5, and Scale Z=3.5.

6. Parent cubeHand to the leftWrist joint. Select the cube object cubeHand and Shift+select the leftWrist joint of the character. Go to Edit, Parent.

7. After parenting the cubeHand object to the leftWrist joint, select cubeHand and go to Solvers, Initial State, Set for Current. This tells the dynamics system of Maya that this is the starting position for the cube.

   Scrub through the animation of the Pin hitting the cans on the table. You should see the cubeHand object following the left hand of the character. You can think of the polygon cube as a stand-in object for the rigid body dynamics animation we are about to do. The cube will calculate much faster than if we actually used the actual geometry of the character's hand. It is important, especially with complex scenes, to try to use stand-ins wherever possible to simplify the amount of data Maya must calculate. Polygon stand-ins are optimal for dynamics; they will calculate much faster than NURBS geometry.

Exercise 10.12: continued

Next, we need to make the cans on the table rigid bodies so they will collide with the cubeHand object when we run the dynamics.

8. Go to frame 1. Select the six polygon cans sitting on the table in front of our character. Go to Bodies, Create Passive Rigid Body. With the six cans still selected, go to the rigid body input (under Shapes in the Channel box). Highlight the Active attribute and Key Selected. Make sure Active is off when you do this.

   We will be animating the active/passive status of the can rigid bodies. At the beginning, we need the cans to stay in place, stacked on the table, so they will begin the animation as passive rigid bodies. As the hand arrives to hit the cans off of the table, we will animate the status to active because we need the cans to collide with the hand and be affected by a gravity field we will add to the scene later.

   Also, notice that the cans are not quite touching each other. This slight spacing will alleviate any inter-penetration errors that might occur upon the intersection of the geometry used in dynamics.

9. Scrub through the animation once again and determine at which frame the character's left hand will arrive right before hitting the cans on the table. This should be at about frame 20. Go to the Channel box and, where it reads Off on the Active attribute of the rigid body Shape node, type **1** or **on**, then click Key Selected. This makes the cans switch from passive to active at frame 20.

10. Go back to frame 1 and then play back the animation. Surprisingly, you will notice that the left hand of the character still passes through the cans on the table.

    The animation of the character's hand hitting the cans occurs so quickly, in just one or two frames, that the collisions are not registered by the dynamics system in Maya. We need to adjust the "step size" of the rigid body solver. Right now the increments that the rigid body solver calculates are too large. We need Maya to calculate finer increments so the collision of the hand and cans is correct.

11. In Dynamics, go to Solvers, Rigid Body Solver. In the Attribute Editor for the rigidSolver, change the Step Size to .001. Play back the animation, and this time you should see the hand colliding with the cans on the table.

12. Adjust the dynamics attributes of the cans. To further optimize the speed of the calculation of the dynamics, select the six cans. In the Channel box under rigid body input node, switch the Stand In attribute from None to Cube. This simplifies how Maya calculates the can surfaces during the collisions. Maya uses a simple Cube to calculate the rigid body dynamics as opposed to the cylindrical polygon shape of the cans that is a bit more complex. You may

want to increase the Mass of the cans to 100 so they don't fly so far away. Decrease the Bounciness to .1 so they don't bounce so much.

13. With the six cans selected, go to Field, Create Gravity. In the Channel box, change the magnitude of the Gravity field to 30. The gravity field, by default, will pull the cans down along the Y axis.

    Adjust the rigid body dynamics attributes for the six cans until you are happy with the animation. You can then bake the dynamics animation into keyframes for the six cans. Right now the rigid body dynamics are being calculated one frame at a time, from beginning to end. If you scrub through the animation at this point you may see some errors in the way that the hand and cans collide. This occurs when the animation is not played back frame by frame. Baking the rigid body dynamics will transfer the motion of the rigid bodies into keyframe animation.

14. Select the six cans and go to Edit, Keys, Bake Simulation. Set the Start/End to 19 and 44 then click the Bake button. You will see the animation play back frame by frame, as Maya writes out keyframe animation for the cans.
15. When Maya is finished baking the simulation, make sure the six cans are selected, rewind the animation to frame 1, and go to Edit, Delete by Type, Rigid Bodies. This deletes all the Rigid Body nodes for the cans, which are no longer needed.
16. Now that the rigid body animation is baked into the cans, you can delete the cube on the hand. Select the cubeHand polygonal cube and then select Edit, Delete.

You have now explored one way of having your character interact with objects in the scene through the use of animating the passive/active status of Rigid Body objects. This method could be used in endless different scenarios where your character needs to collide and interact with objects in your scene.

10.3.1 Keyframe Macro with MEL

Exercise 10.13: Create a Keyframe Macro with MEL

1. Open the file pin.mb.
2. Select the nodes of the character to animate. For instance, select the selection handles for the two feet and the pelvis behind the character.
3. In the Channel box for the selected items, select the channels you want to animate by highlighting the names of the channels in the Channel box. Shift+click if there is more than one channel that you want to key.

Exercise 10.13: continued

4. Open the Script Editor (Window, General Editors, Script Editor). In the Script Editor, go to Edit in the menu bar and select Clear History. This clears the information displayed in the History window of the Script Editor.
5. Right-click in the Channel box for the pop-up menu, then go to Key Selected. You will notice that several setKeyframe MEL commands were reported to the History section of the Script Editor.
6. Highlight the setKeyframe commands reported in the History window of the Script Editor and, using the middle mouse button, drag it to one of your Shelves. A button is created with the label MEL on it.

The MEL button contains the MEL commands you dragged to the Shelf. You now have a macro made of several MEL commands. Every time you click on this MEL button, Maya will execute the setKeyframe MEL commands for the selected objects. You don't even have to select the objects again to setKeyframes. Simply move the objects into position then click on the button. The setKeyframe commands already contain the names of the objects you keyframed originally, so there is no need to select them when you want to keyframe. You can also have as many of these macros as you want. Each one of your macros can keyframe different objects, or you can have one macro that keyframes everything—it is totally up to you. You can delete MEL scripts you drag to your Shelves by middle mouse dragging them to the small trash can on the far right of the screen.

10.4 Editing Animation–Time and Values

Now that you've created your animation, it's time to edit it so that your characters perform the way you want them to.

10.4.1 Scaling Animation Timing with the Dope Sheet

In some instances you may have your character animated in the right poses, but you need to speed up or slow down the animation. A direct and easy way of doing this is through the Dope Sheet. The Dope Sheet can be found under Window, Animation Editors, Dope Sheet. The Dope Sheet represents time in frames and keyframes for selected objects. Frames are represented with boxes and keyframes are represented with brown boxes. The Dope Sheet is great for "sliding" keyframes around in time or for scaling time to make your animation quicker or slower.

Exercise 10.14: Edit the Timing of a Character's Walk Cycle with the Dope Sheet

1. Open the file pinWalk.mb. This file contains a character that already has keyframed walk cycle animation applied to it over 24 frames.
2. To edit the animation, we need to first select the objects that have keyframed animation. In the Outliner, select the rightFoot, leftFoot, pelvis, upperBodyControl, and armControl nodes (the upperBodyControl and armControl nodes are Locators with attributes driving skeleton joints on the character).
3. Open the Dope Sheet (Window, Animation Editors, Dope Sheet). You will see the nodes you have selected appear in a list on the left side of the Dope Sheet. In the rest of Dope Sheet you will see the grid of boxes that represent frames in the scene, and you will see yellow or brown boxes filling some of the frames. The yellow boxes are selected keyframes; the brown boxes are unselected keyframes. Look at one of the nodes on the object list on the left side and read horizontally across the Dope Sheet: These are the keyframes for that object.

   Running horizontally along the top line of the Dope Sheet, you will see the Summary line. The Summary line displays all of the keyframes for all of the objects displayed in the Dope Sheet.
4. Using the left mouse button, select all the keyframes on the summary line. You now have all the keyframes selected for the rightFoot, leftFoot, pelvis, upperBodyControl, and armControl nodes.
5. You can also select keyframes for individual objects by selecting them horizontally across from the name of the object. Shift+select to select keyframes from more than one object.
6. We will now use the Scale tool to scale the whole walk cycle animation for our character. Click the Scale transform tool or press the R key on the keyboard to scale the animation. You will see a white border around all the keys selected in the Dope Sheet.
7. Click the white vertical bar on the right side of the Dope Sheet. Drag it to the right from the current frame 24, to about frame 50. You have now just double the length of the animation from 24 frames to 50 frames. The walk cycle will now take twice as many frames to complete.

In addition to scaling your animation, you can also translate your keys in the Dope Sheet by using the Translate tool or by pressing the W key on the keyboard. This allows you to translate keys left or right to have the animation occur sooner or later.

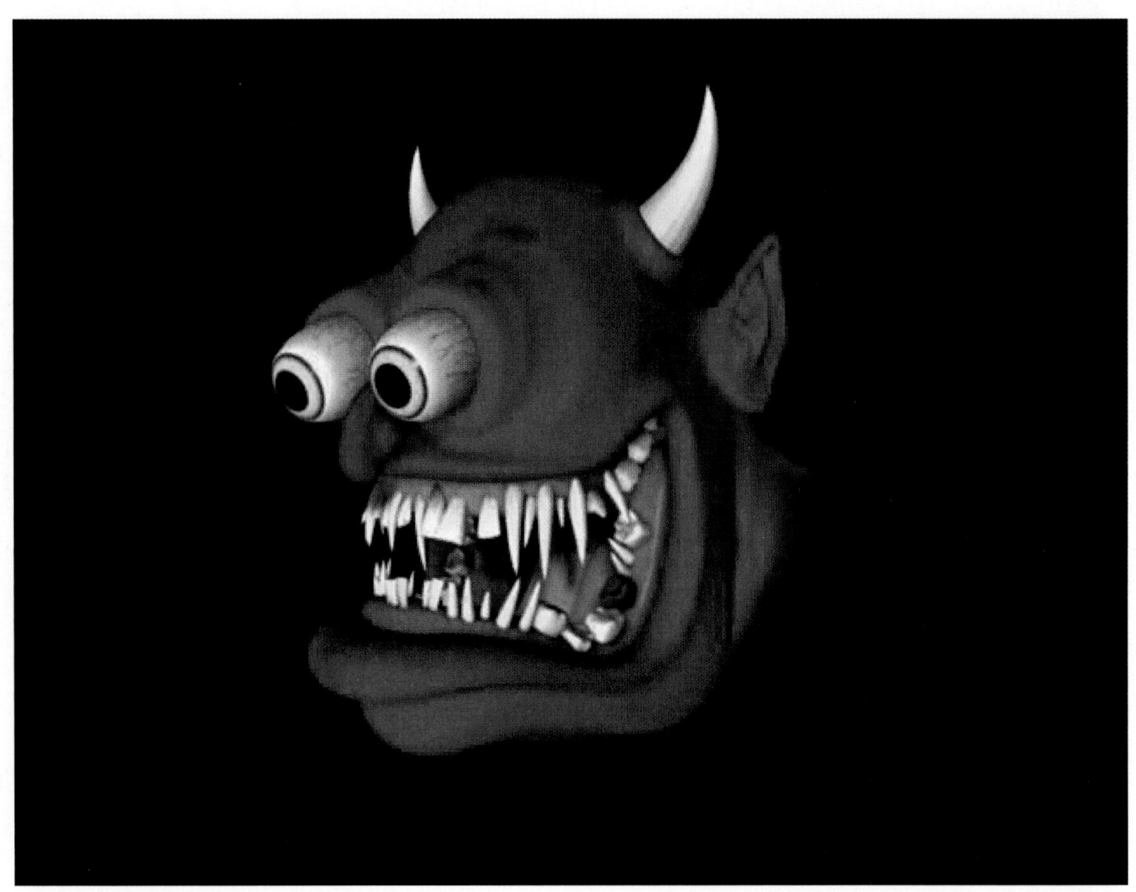

CHAPTER 11

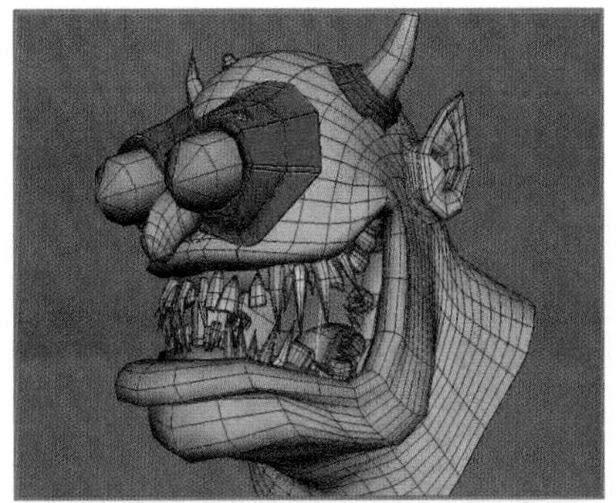

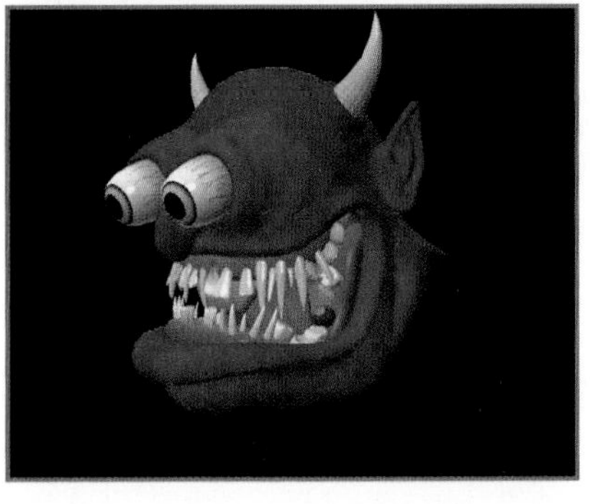

Facial Animation

by Tim Coleman

The ability of your CG character to express emotion and speech with its face and its body motion will help to communicate the actions of the character more thoroughly to the audience. The face can be a very effective way of relaying the thought processes of your character to the viewer. Setting up the head and face for a 3D character can be a difficult and tedious task. It is essential to focus on the creation of effective, identifiable, and contrasting face shapes to bring the face of your character to life.

In this chapter, we will focus on how we can to control the shape of the surfaces that make up the character's head and face. We will look at the process of binding the head surfaces to the skeleton of the character, and the use of the Blend Shape tool to morph the shape of the face between different phonemes and facial positions. To make accessing the controls of the character's face more streamlined, we will centralize the controls of the character's face on a locator with our own custom attributes. Finally, we will examine the workflow of importing sound files into Maya and animating the facial positions of our character to the soundtrack.

11.1 Modeling the Character's Face for Facial Animation

Before we get into the actual animation of your character's head, here are a few important notes about modeling for facial animation that you should keep in mind.

The construction of the character's head and face geometry play a significant role in the facial animation process. Be very careful not to over-model the head. In other words, avoid excess or unnecessary surface detail. Dense geometry, or geometry with an exorbitant amount of surfaces points (CVs), can make the setup and animation process of the character's face very time consuming and difficult. Try to utilize the nature of NURBS surfaces that are inherently smooth with very few spans (isoparms). Space your spans as evenly and uniformly as possible. Add detail for facial features only where necessary.

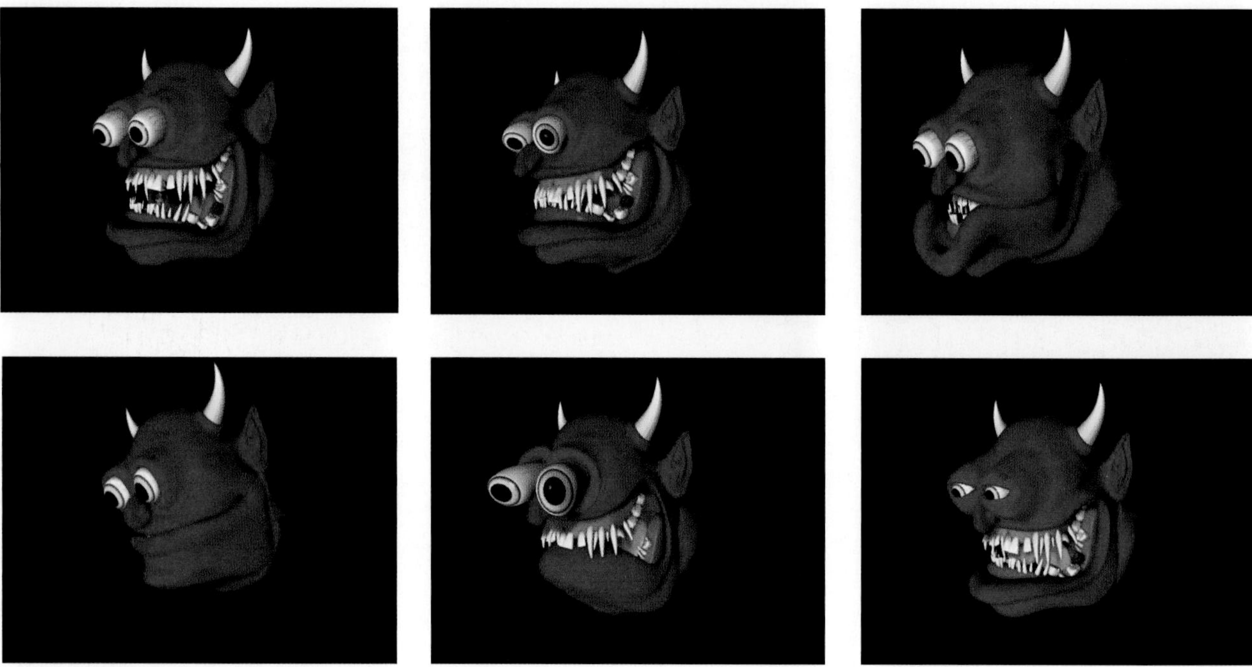

Figure 11.1
Use your character's face to help express emotions such as happiness, derangement, surprise, pride, dementia, and shadiness.

Try not to model the entire head with a single surface. First of all, this can be a very time-consuming task. Secondly, you will most likely end up with a very dense model. Utilize multiple objects located near each other to give the appearance that the head is one contiguous surface. Refer to Chapter 5, "Modeling an Organic Humanoid with NURBS," which describes various modeling techniques.

In addition, modeling the face with spans or isoparms radiating out from the mouth can aid in successfully simulating the natural muscle movements that occur with the mouth. The muscles around the mouth create a circular movement of the flesh around the mouth. Starting with a primitive sphere, orient one of the poles of the sphere as the mouth and then shape the rest of the head around the opening of the sphere. The head of this Devil character is originally derived from a simple sphere.

Other points to note regarding the modeling of a character's head include being sure to add thickness to the mouth. The surfaces of the lip should curl back inside of the mouth giving a thickness to the lips. Nothing is worse than a character with paper-thin lips! Also, add a surface inside of the mouth to simulate the back of the mouth and throat of the character. Using a darker shade on this surface in the mouth will add to the effectiveness of the mouth opening. Don't forget to add a tongue! The tongue plays a very important role in the effective animation of a 3D character's face.

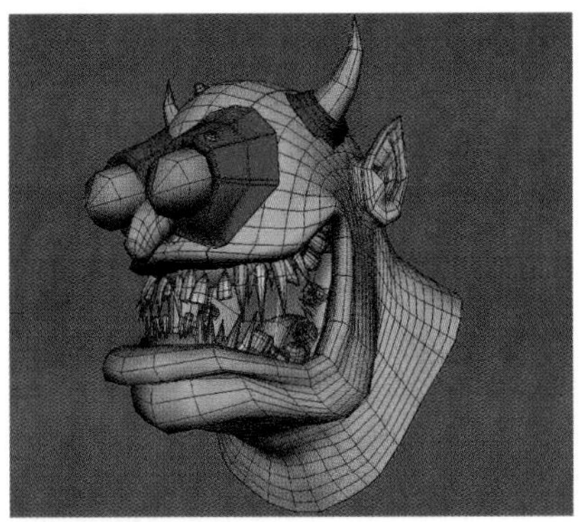

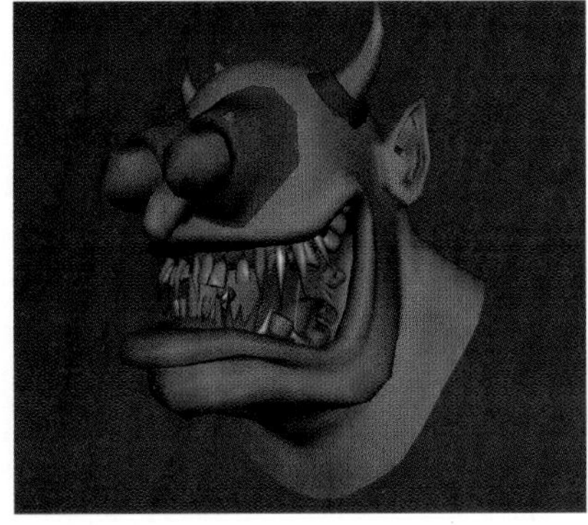

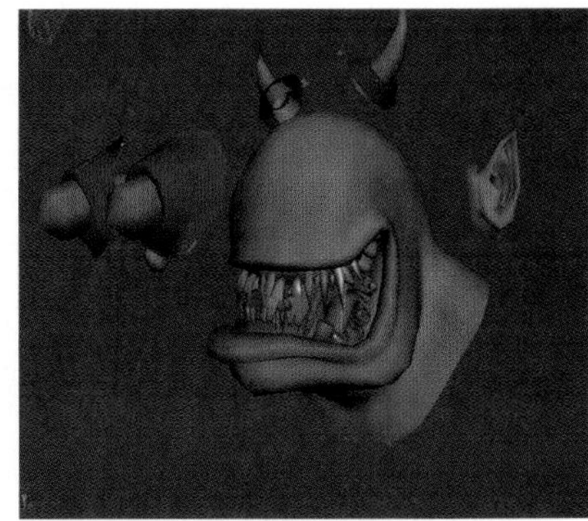

Figure 11.2
The devil's head is comprised of several surfaces laid on top of one another to give the appearance of a continuous, single surface.

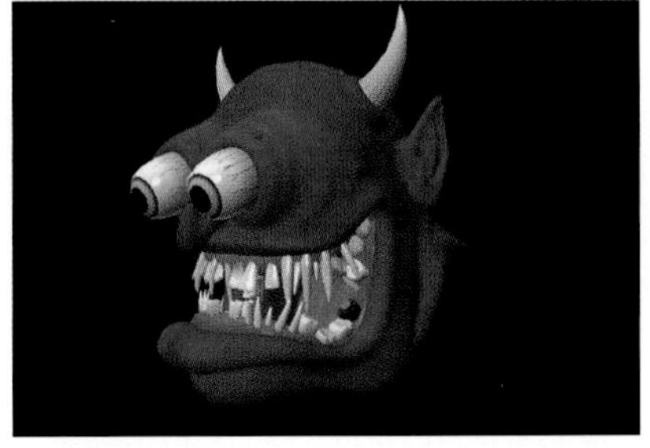

Figure 11.3
Devil's head software-rendered. When using multiple surfaces to model a head, make sure there are no visible seams where the different surfaces meet.

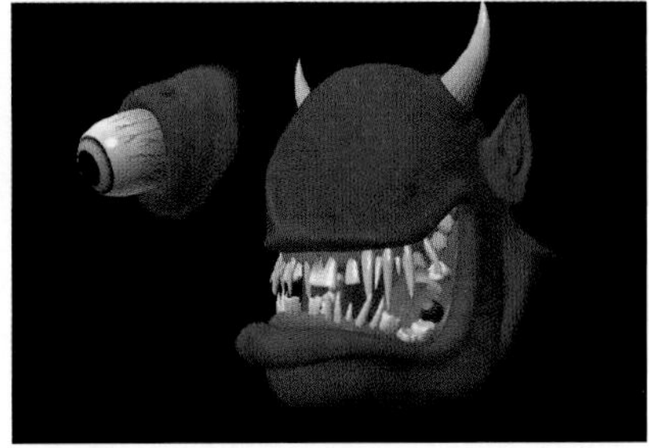

Figure 11.4
Transparency map at the edge of the eye socket surface helps to hide any seams where the eye socket and head meet.

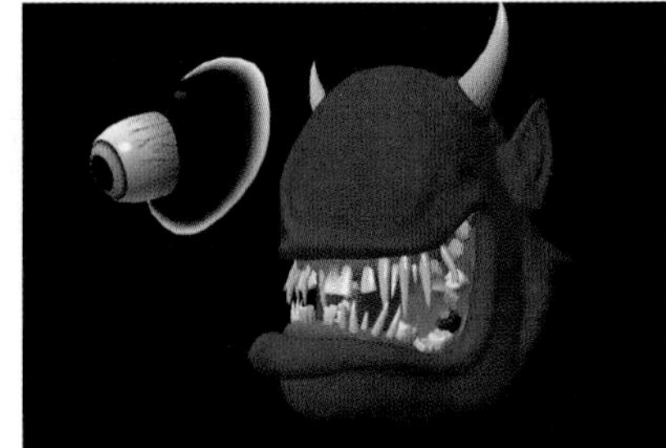

Figure 11.5
A transparency map was used to create a fuzzy, transparent edge on the eye socket. Here the transparency map is mapped in the color channel so you can see the placement and color of the transparency map.

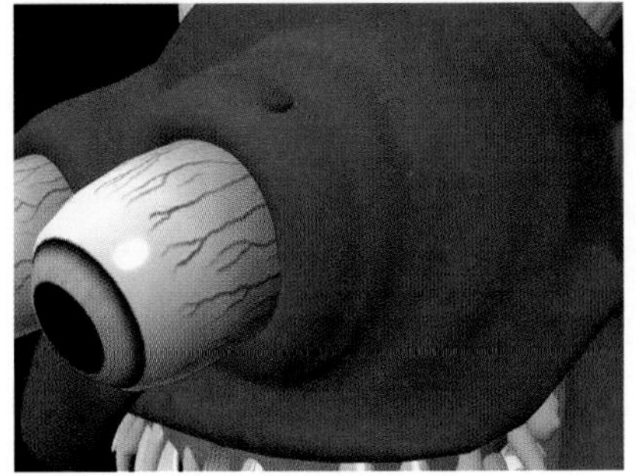

Figure 11.6
Use bump maps to create extra detail, such as this mole, on your characters.

Figure 11.7
The head of the Devil actually started out as a sphere.

11.2 Setting Up the Head

Let's take a look at methods of getting a character's head attached to the body of our character. We will then look at methods that will allow us to animate the head into different expressions and phonetic positions for speech.

Exercise 11.1: Build the Head Skeleton

1. Open the file devilHead.mb from the accompanying CD-ROM.
2. Create joints for the neck and head. There are three joints in the neck (at the base of the neck, mid-neck, and where the head sits on top of the neck). Two more joints extend from the neck up into the head. In the middle of this chain we have two joints that branch out to form the jaw and chin.
3. Name the joints according to the callouts.
4. Select the neckBase joint and Shift+select the neckShoulder joint. Choose Edit, Parent. This basically connects the skeleton of the head to the skeleton of the body. Now, wherever the body goes, the head will follow.

Figure 11.8
The default "devil head" model.

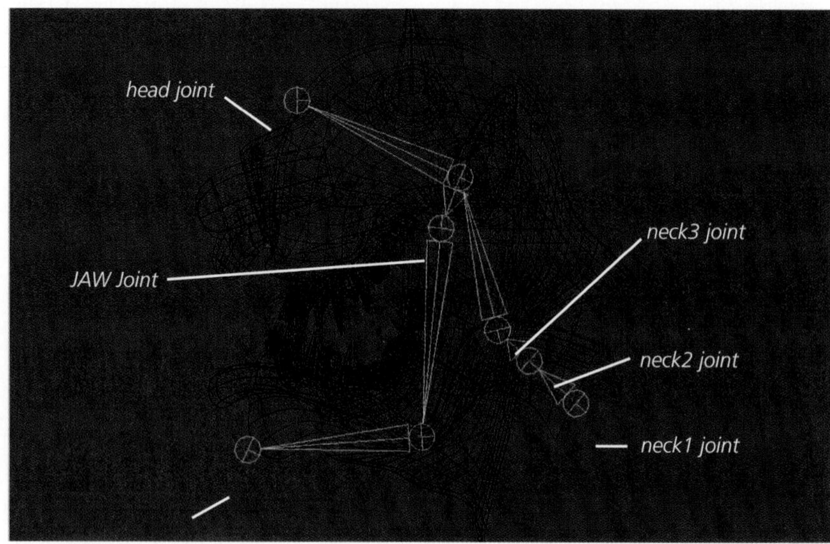

Figure 11.9
Skeleton joint placement in the devil head, side view.

Exercise 11.2: Parent the Teeth to the Skeleton

Some of the objects that make up the head of our character will actually be parented to the skeleton. The rest of the surfaces will be attached to the skeleton using bind skin. Most often you will want to parent objects to the skeleton that will not generally morph or change shape. In this example, we don't necessarily need to have the skeleton joints deform the teeth of the character. In other instances, especially with really cartoony characters, you may need to bind all the surfaces of the head to the skeleton.

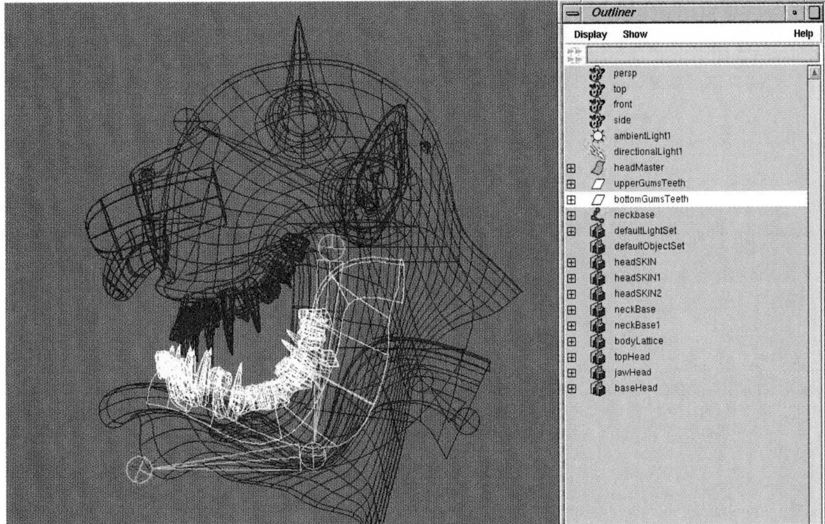

1. Go to Window, Outliner, select the bottomGumsTeeth group, and Shift+select the jawRotate joint. Choose Edit, Parent. We want the lower teeth to follow the rotations of the jaw joint, therefore we have made the lower teeth group a child of the jaw joint).

Figure 11.10
Parenting bottom teeth to jaw joint.

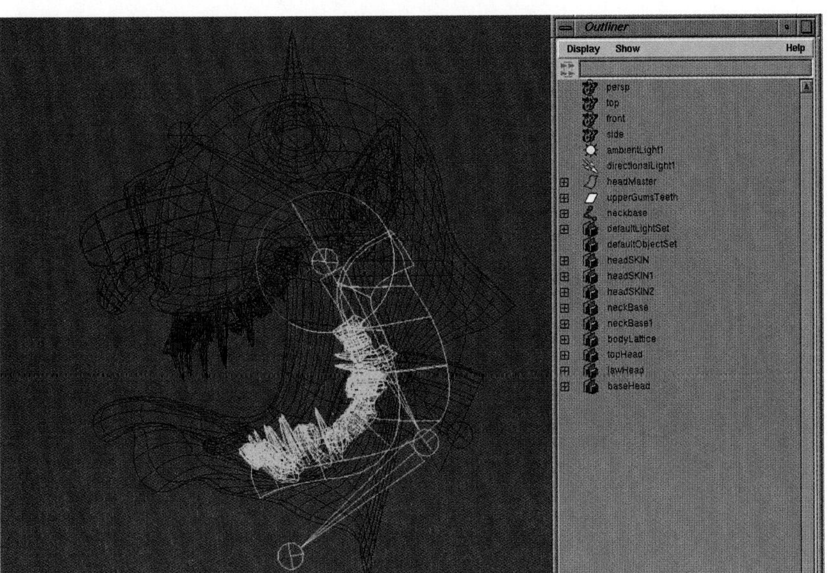

2. Rotate the jaw joint along the Z axis. You should see the lower teeth following the rotation of the jaw joint. Undo any rotations you applied to the jaw joint.

Figure 11.11
Make sure, when the jaw joint is rotated, that the bottom teeth follow.

Exercise 11.2: continued

3. Select the upperTeeth group in the Outliner and Shift+select the head joint. Choose Edit, Parent. We need the upper teeth to follow the head around but we don't want it to inherit the rotations of the jaw joint, therefore we made it a child of the head joint.

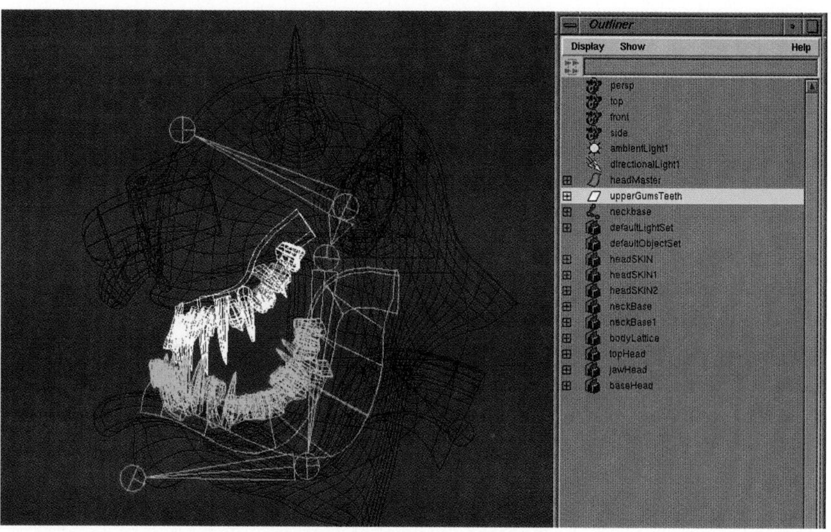

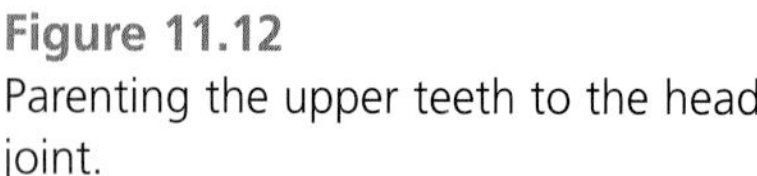

Figure 11.12
Parenting the upper teeth to the head joint.

Exercise 11.3: Bind the Remaining Head Surfaces to the Head Skeleton

Next you will bind the geometry of the head to the skeleton, which assigns the points of the geometry to the different skeleton joints of the skeleton.

1. Select the headSurfaces group.
2. Shift+select the neckBase joint.
3. Choose Skin, Bind Skin, Rigid Bind. This assigns all the surface points of the head geometry to the joints of the head skeleton.

   We are binding the geometry of the head to the skeleton so we will be able to deform the geometry of the head to the skeleton.

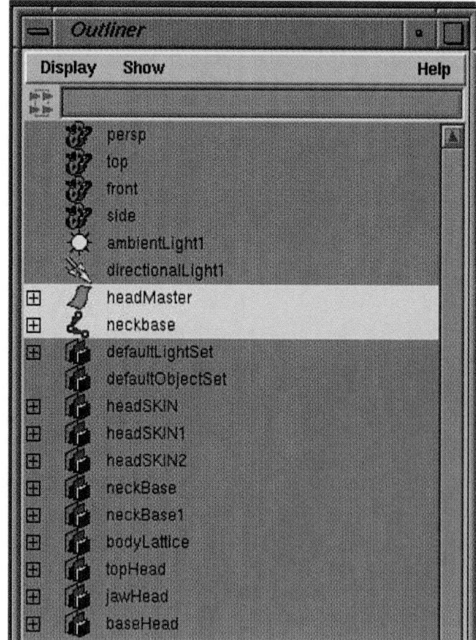

Figure 11.13
Surfaces of head and neckBase joint selected and ready for rigid bind.

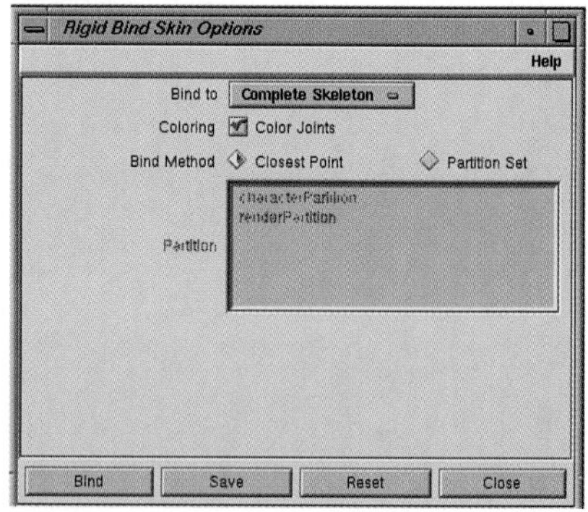

Figure 11.14
Using rigid bind to skin the head surfaces to the skeleton.

Select the jaw joint and rotate it along the Z axis. Notice that some of the upper face follows the rotation of the jaw joint. During the Bind Skin operation, Maya assigned some of the points of the upper face to the jaw joint. We need to correct this problem using the Edit Membership tool.

Figure 11.15
Rotating the jaw joint identifies point membership problems.

Exercise 11.4: Edit the Membership of the Head Surfaces

Next we need to edit the membership of points of the head geometry. We need to re-assign some of the points in the nose area so they are not members of the jaw joint.

1. With the jaw joint still in its rotated position go to Deform, Edit Membership to activate the Edit Membership tool.
2. Select the head joint in the middle of the head. You should see points in the head highlight signifying their membership to the selected head joint.
3. Shift+select the points in the upper face to add them to the membership of the selected head joint.

   As you Shift+select the CVs they should jump back into their original positions. Changing their membership from the jaw joint to the head joint causes the CVs to no longer inherit the rotation of the jaw joint.

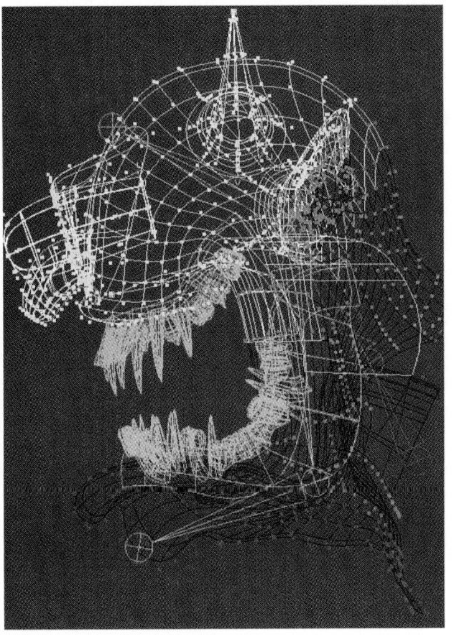

Figure 11.16
Using the Edit Membership tool to add points in the upper face to the head bone.

Exercise 11.4: continued

4. Continue Shift+selecting the CVs of the upper face until you have generally corrected the deformation caused by the jaw joint.
5. Fix the neck by rotating the jaw up so the mouth is in a closed position.

If the jaw is not already rotated, rotate the jaw so the mouth is in a closed position. Notice that the deformations, especially at the corners of the mouth and in the neck, are somewhat unnatural in appearance. Next, you will use Artisan to adjust the weighting in the problem areas of the face.

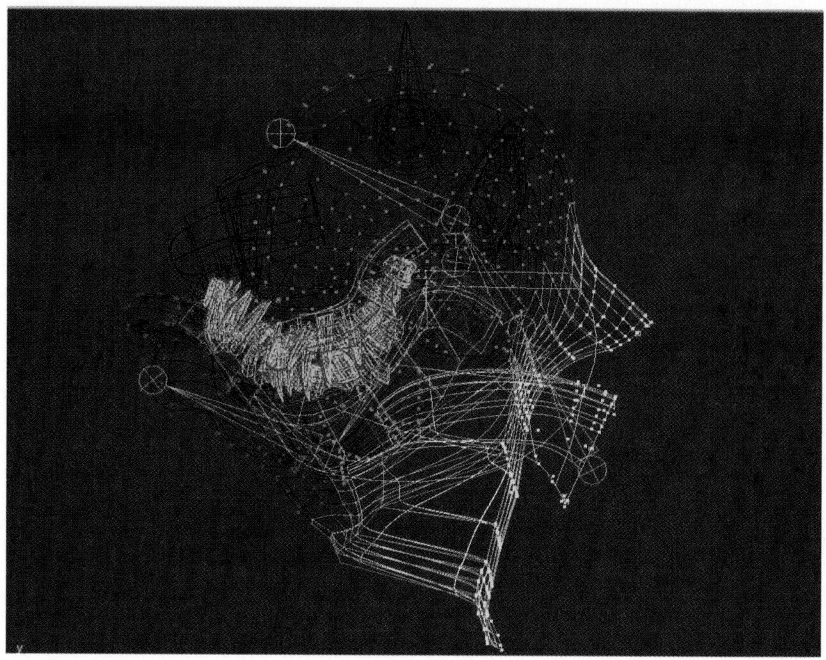

Figure 11.17
Using the Edit Membership tool to fix membership in the neck area.

Tip

You don't have to Shift+select all of the CVs in one shot. You can Shift+select them in small groups or one at a time if you want.

Exercise 11.5: Weight the CVs of the Jaw Joint Using Artisan

1. Select the head surface and shift+select the jaw joint.
2. Choose Deform, Paint Weights Tool, Option. This invokes the Artisan tool, which will allow us to paint the weights of the CVs on the face. In the option window of the Artisan Paint Weights tool, click the Reset button to set the attributes to the default.
3. Go into shaded mode by pressing the number 5 key on the keyboard. You will notice a black and white texture map on the selected head surface geometry. This map represents the cluster weights of the CVs that are members of the jaw joint. The black and white values represent cluster weight values. The white areas represent CVs with a cluster weight of 1 (or 100%) and the black areas represent 0 (or 0%). We can use any level of gray between black and white to get a cluster weight between 0 and 1. This map will make the painting of the CV weights on the head surface much more intuitive. We will use Artisan to smooth the cluster weights between the black and white areas to a level of gray.

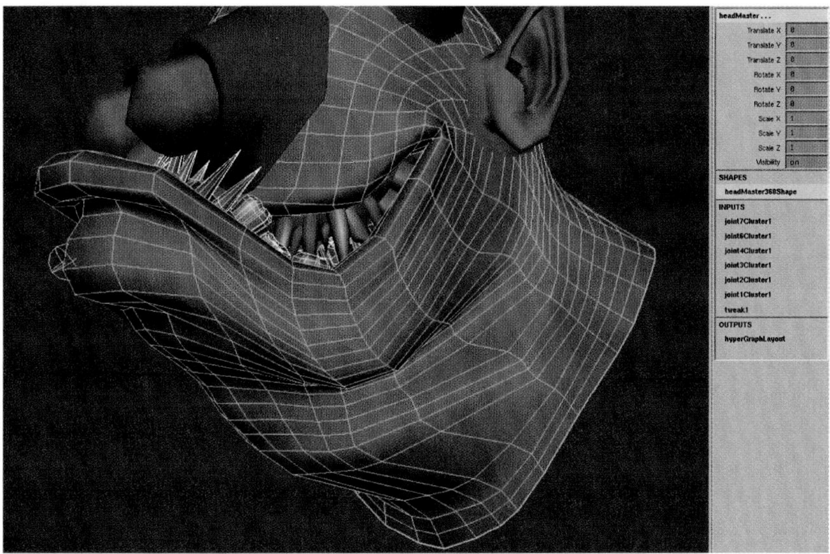

Figure 11.18
Jaw rotated into a closed position to identify areas to fix point weighting.

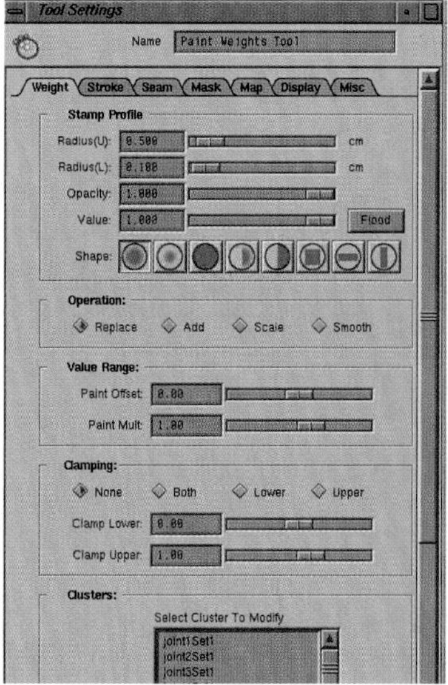

Figure 11.19
Using Artisan's Paint Weights tool to adjust point weighting on the neck and mouth.

4. Switch the Operation in the Artisan interface to Smooth. This setting will smooth the values from black to white.
5. Move the mouse over the head surface in the Perspective View. You will see a circular red icon following along the surface of the head. This is the paintbrush location for the Artisan tool. Paint along the areas where the black and white colors border each other. This is where there is a sudden change in weight from 1 to 0 on the surface. You need to blend the white and black areas together. The goal here is to have a smooth gradient from black to white in the deforming areas. You can paint in a circular motion at the black and white borders. Or you can click the geometry and observe the weight changes after each click of your mouse. You should see the surface update and move to address the new cluster weights you are painting.
6. Change the size of the Artisan paint brush by setting the Radius (U) and Radius (L) options to 5.

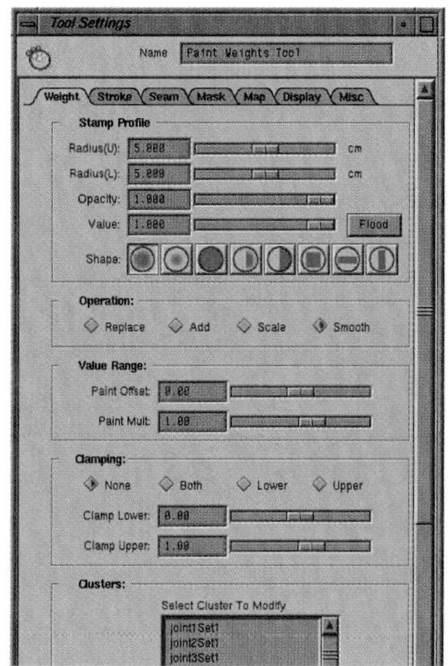

Figure 11.20
Smoothing the white and black areas of the point weight map using the Paint Weights tool.

Exercise 11.5: continued

7. Undo any painting of cluster weights that you may have done.

   You will notice that, as you were painting the cluster weights, you were only affecting one side of the face. We can use the Reflection option in the Artisan tool to paint both sides of the face at the same time so we are weighting the face symmetrically.

8. Go to the Stroke tab in the Artisan Paint Weights tool interface. Click on the Reflection check box to paint on both sides of the face simultaneously. We need to tell Artisan which axis of the surface's U and V coordinates we need to paint. Click the button next to U Dir to reflect paint over the U coordinates of the surfaces. Type **`0.56`** in the number field next to U Dir to offset the paintbrush so the brush is exactly symmetrical on both sides of the head surface.

9. Continue to paint the weights for the jaw until the deformations look smoother. You want to try and make sure that the isoparms or wireframe cross-sections of the geometry are evenly spaced.

10. When finished, select the jaw joint only and set the Z rotation back to 0.

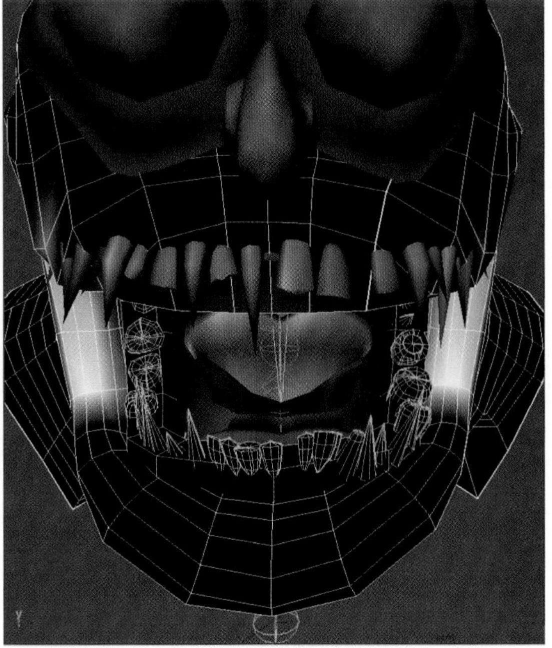

Figure 11.21
Use Reflect Paint to paint weights symmetrically on both sides of the character's face.

When adjusting the cluster weights of CVs, keep in mind that the weight is applied to the CVs only when a transformation has been applied to the joint of which it is a member. If the jaw joint were in its initial position without any rotations applied to it, you would not see any of the weighted effects on the CVs. The cluster weight applied to the CVs is a percentage of the transformation of the joint the CVs are members of. Therefore, always apply a transform to the joint for which you need to adjust the weights of its members. Here, we rotated the jaw into an open position, then painted the weights so we could see them update and determine if the new weights were correct for the mouth.

Painting weights with Artisan can really make the whole process of weighting large numbers of CVs much easier. Artisan even allows you to import and export these point weight maps into and out of Maya. You could export the maps to your favorite 2D paint program for further refinement or apply the maps to other models. The process of binding a character's head geometry to a skeleton involves editing the membership and weights of points. This allows you to make sure the deformations in the face appear natural and smooth.

11.3 Creating Mouth Shapes

The head we have just attached to the skeleton of our character is our master head. We will use a series of target heads (duplicates of the master modeled into new shapes) into which to morph the master head. We will use the Blend Shape tool to accomplish the morphing of the different surfaces of the head.

Creating the target heads involves modeling and editing the master head. This is best accomplished by duplicating the surfaces of the master head group. Make sure the head is in the default position (no skeleton transforms applied to the model) when duplicating.

Let's take a look at a set of phonetic mouth positions for our character. These mouth positions will help convey the speech of our character.

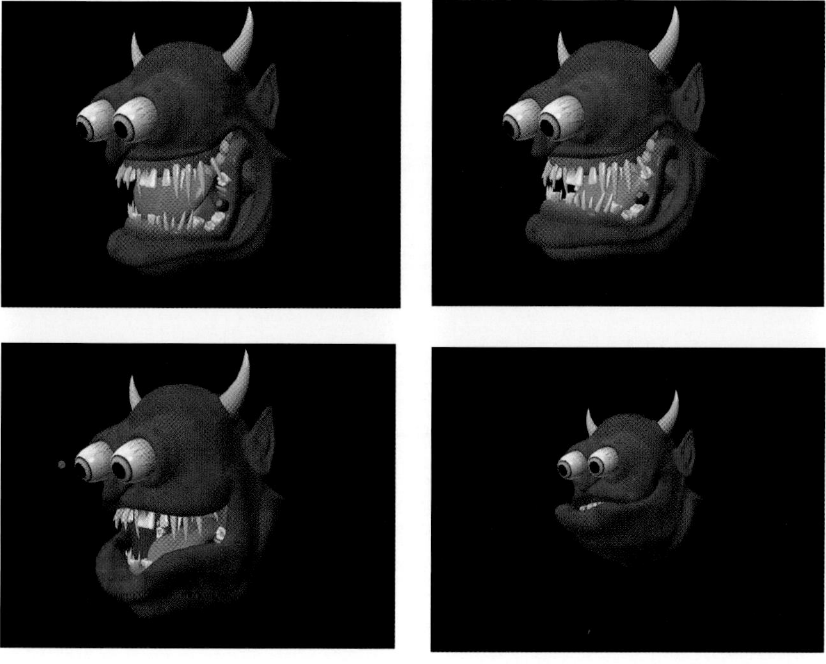

Figure 11.22
The mouth positions for the vowels: A and I, E, O, and U.

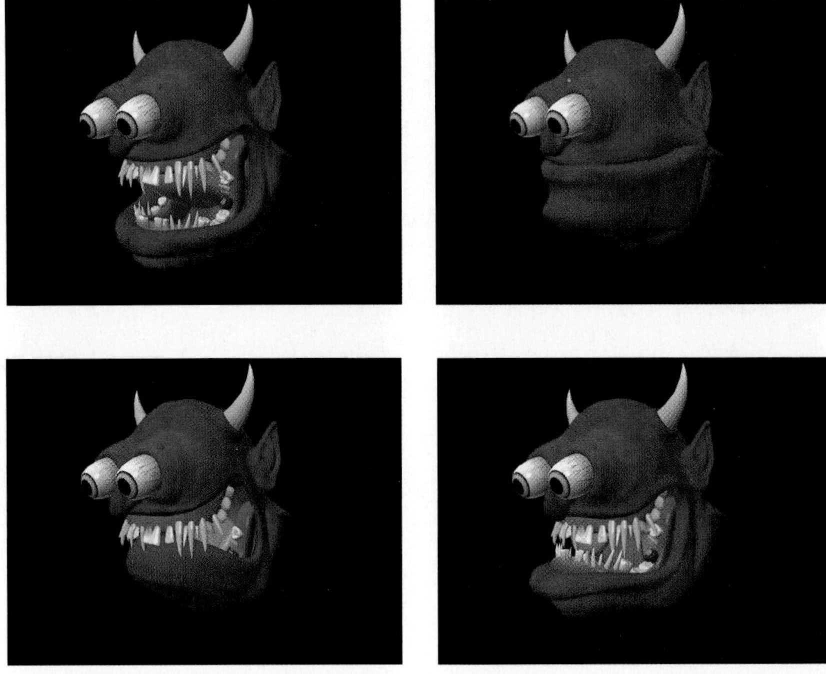

Figure 11.23
The mouth positions for the consonants: L; M, B, and P; F and V; and TH, Y, and Z.

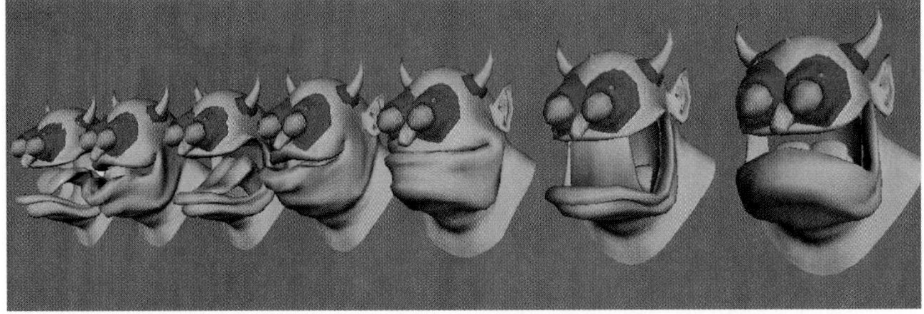

Figure 11.24
Target heads created by re-shaping the original head surfaces.

Each phonetic mouth shape is a remodeled duplicate of the original master head. Using CV pushing and pulling, modify with deformers like a lattice or use the Sculpt Paint with Artisan method to help model these phonetic mouth shapes. The target heads of the Devil character were created using all of those methods.

11.3.1 Morphing Different Mouth Targets

After the mouth shapes have been modified, we can use Blend Shape to morph the master head into the different mouth targets.

Exercise 11.6: Use Blend Shape to Create Phonetic Mouth Shapes

1. Import the file devilHeadPhonemes.mb from the accompanying CD-ROM. This file contains target heads that have already been completed for you.
2. In the Outliner, select the AI, E, O, U, L, W, MBP, F and Cons group nodes. These groups contain an exact duplicate of the hierarchy of surfaces of the master head. They were basically modeled into new mouth positions.
3. Shift+select the headMaster group in the Outliner.

   The targets are always selected first and the master object, or the object you want the morph to be performed on, is selected last. Keep in mind also that we are selecting the top node of the hierarchies here. We are basically telling Maya that we want the entire hierarchy of surfaces to morph into another hierarchy of surfaces. If you ever have odd things happening in the Blend Shape of hierarchies of surfaces, make sure the order of the hierarchies is identical.

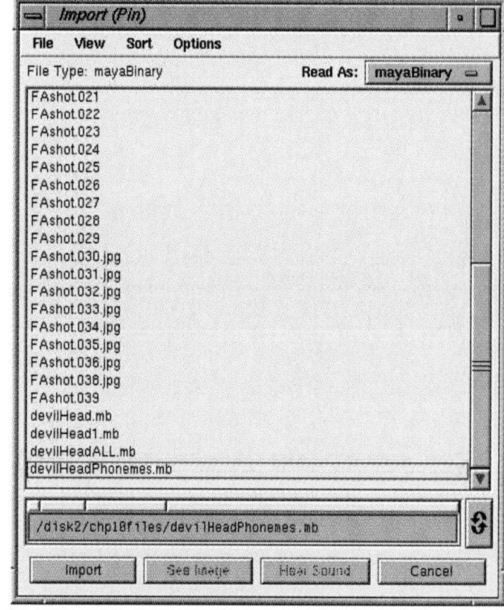

Figure 11.25
Importing the "devilHead Phonemes.mb" file into the scene.

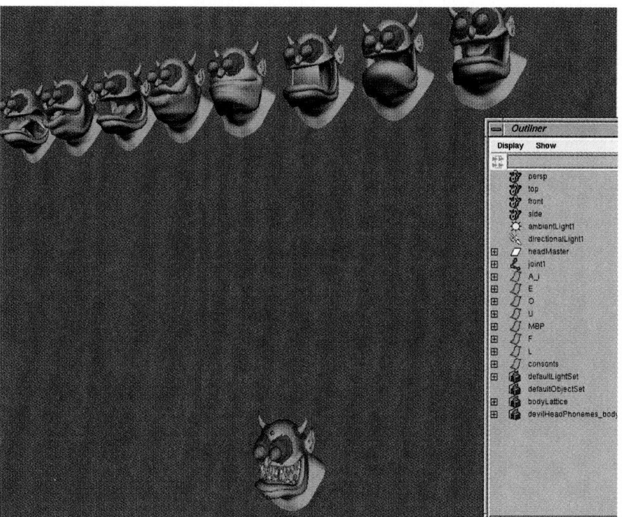

Figure 11.26
Phoneme head targets imported into the scene. The head targets can be placed anywhere in the scene. Here they are placed above the master head.

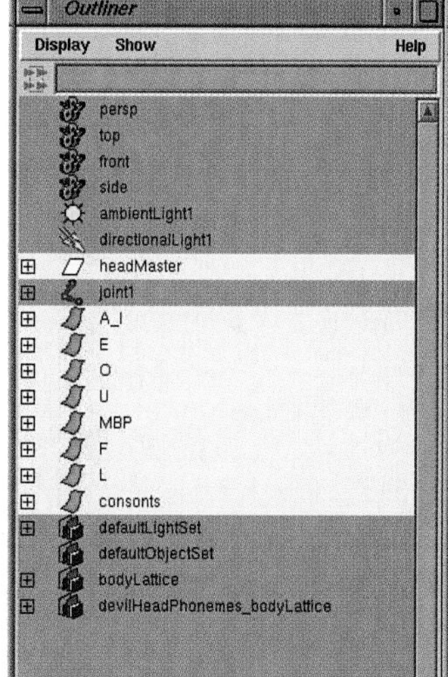

Figure 11.27
Target heads selected first, and then the master head selected, before creating the Blend Shape.

Exercise 11.6: continued

4. Choose Deform, Create Blend Shape Options and name the Blend Shape **phonemes**. Go to the Advanced tab in the option window and change the Deformation Order to Front of Chain. This will ensure that the Blend Shape deformation occurs first in the deformation order of the head surfaces before the skeleton deformation. Click the Create button in the Blend Shape Options window and then click the Close button.
5. Test the phonemes Blend Shape. Choose Window, Animation Editors, Blend Shape. This will bring up the Blend Shape window containing sliders for the Blend Shape targets.
6. Click the vertical slider labeled A and drag it up vertically. You should see the face of the master head morph into the A mouth shape.

   The sliders represent the amount of morph of that particular target that you are applying to the master head, with 0 being no morph and 1 being full morph. This is referred to as the *envelope* of the Blend Shape. You can overdrive the morph as well by taking the envelope value over 1 or below 0. This would be good where you may need exaggeration and anticipation in your facial animation.

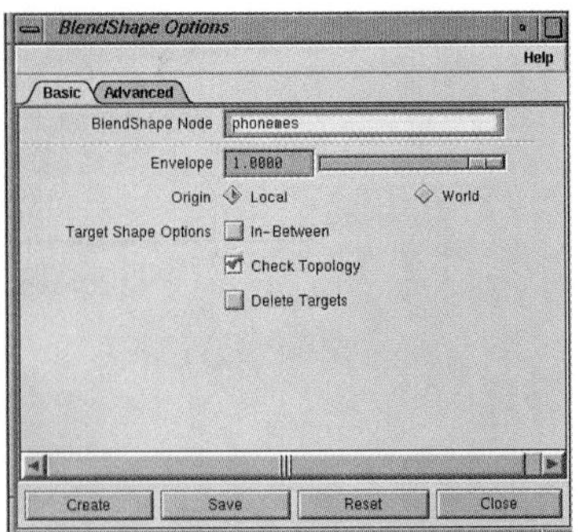

Figure 11.28
Naming the Blend Shape in the Create Blend Shape Options window.

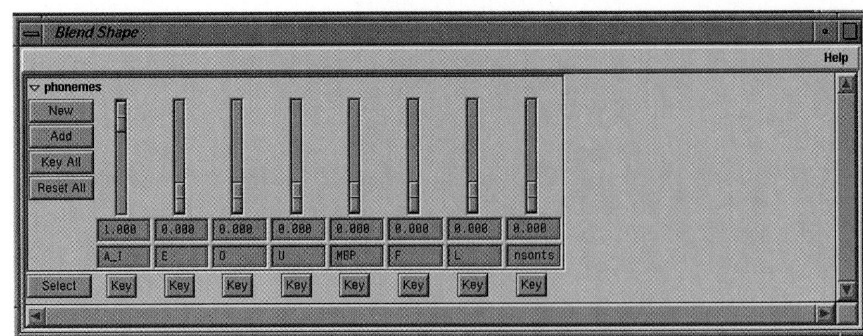

Figure 11.29
Blend Shape window with sliders for each head phoneme target.

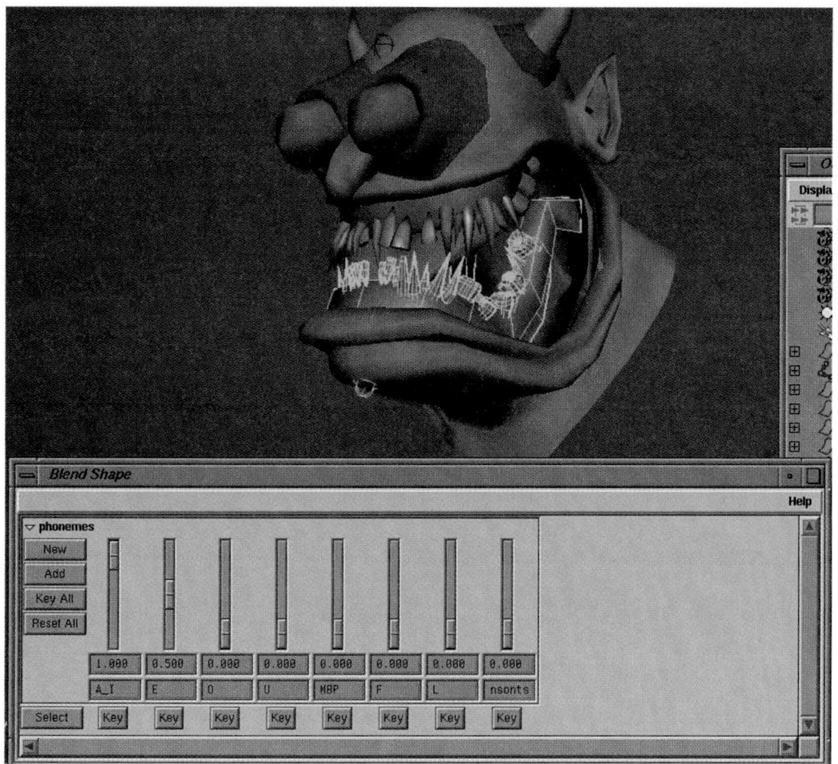

7. Select the next vertical slider, labeled E, and drag the slider up. Notice how the morph of the E mouth shape is mixing with the A mouth shape. This is very powerful—with only a few targets, you can mix together infinite combinations of mouth and face shapes.

8. Set the Blend Shape targets for the A and E targets back to 0 in the Blend Shape window so there is no Blend Shape deformation being applied to the master head.

Figure 11.30
Testing the phonemes' Blend Shape by dragging sliders up and down.

Figure 11.31
Set all the phoneme Blend Shape sliders to 0.

Along with mixing the different Blend Shape targets, don't forget that you can add some rotations to the jaw as well. You can also repeat the above process for the eye positions. Import the devilEyes.mb file into the same scene. Try mixing the eye positions with the mouth shapes.

11.3.2 Creating Centralized Facial Controls

When setting up the face controls for our CG character in Maya, centralizing the controls for the face in one location can make the job for the animator quicker and easier. With skeleton joints controlling some parts of the face and head (the jaw and neck joints and Blend Shapes controlling the shapes of phoneme mouth shapes and eye positions), we need to make access to these controls much more direct. We will create a locator and place it in front of the character's face. On this locator we will create our own attributes and connect them to the various face controls using the Connection Editor and Set Driven Key.

Exercise 11.7: Create Centralized Facial Controls

1. Choose Create, Locator to create a locator. Place the locator somewhere in front of the character's face. Name the Locator headControls. We placed the Locator in front of the character's face to make it very easy to find and pick for the animator.
2. With the headControls locator still selected, choose Window, General Editor, Channel Controls. Highlight all of the Keyable attributes on the left side of the list and click the Move button to move these attributes to the Non-Keyable list. This clears up the attributes for the head-Controls locator in the Channel box. We will need to put our own attributes on headControls and we will not need the XYZ transform attributes.

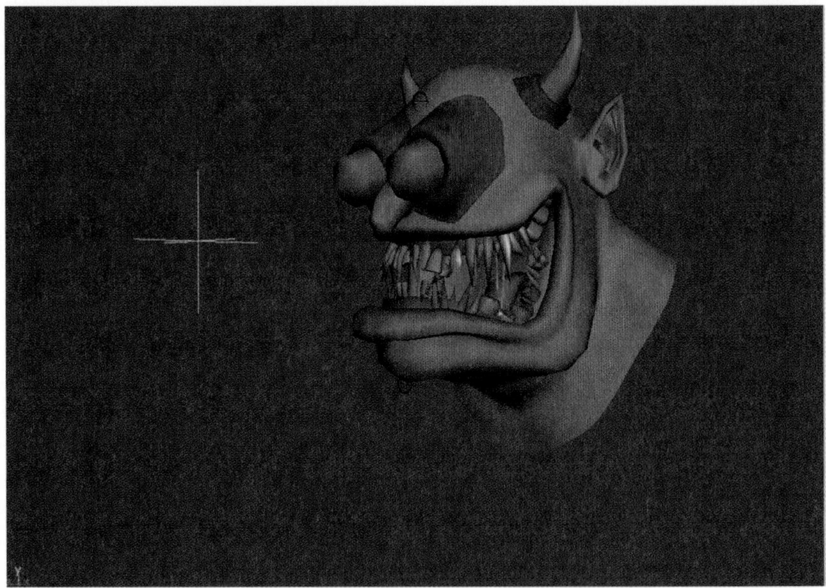

Figure 11.32
A locator placed in front of the character's face will contain all of the control attributes for animating the face/head.

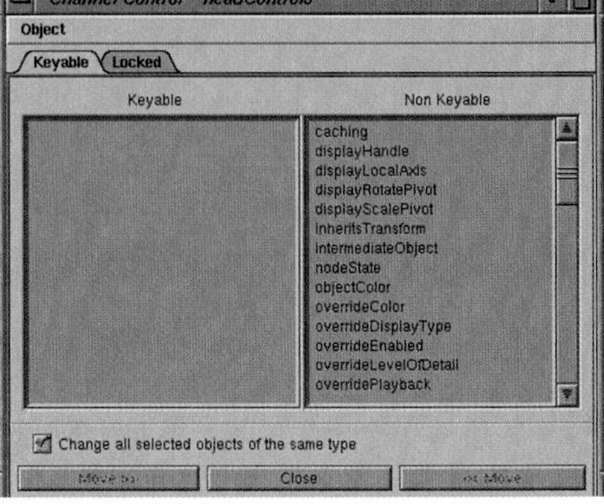

Figure 11.33
Moving all the keyable attributes for the headControls locator to non-keyable using the Channel Control window.

3. With headControls still selected, choose Modify, Add. Use the Default settings and name the attribute **jawRot**. Click the Add button.

4. Open the Connection Editor (Window, General Editors, Connection Editor). In the Options menu of the Connection window, make sure Auto-Connect is on. With headControl selected, click the Load Left button to load it into the Outputs side of the Connection Editor. Select the jaw joint and click the Load Right button to load it into the Inputs side of the Connection Editor. On the left side (headControls), click on the jawRot attribute. On the right side of the Connection Editor, click on the white arrow next to Rotate to expand it then click on Rotate Z. Now the value assigned to jawRot will output directly to the Rotate Z attribute of the jaw joint.

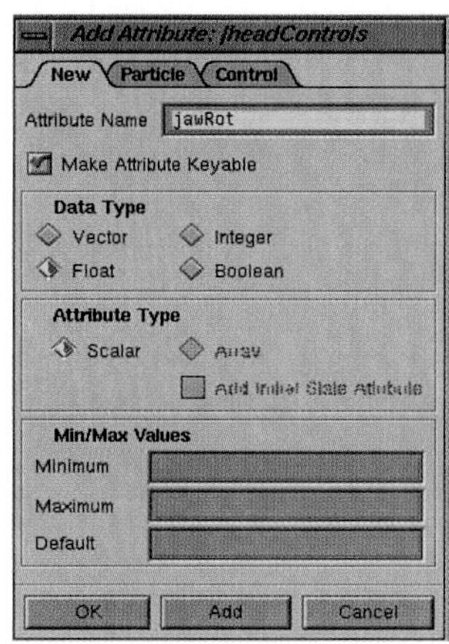

Figure 11.34
Creating the jawRot custom attribute on the headControls locator.

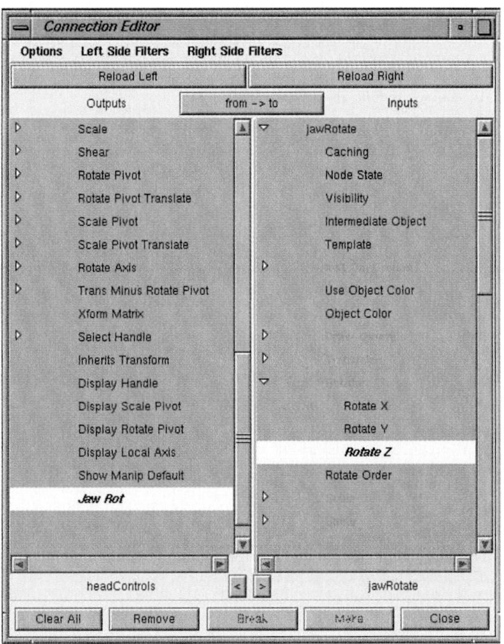

Figure 11.35
Connecting the Z rotation of the jaw joint to the jawRot attribute using the Connection Editor.

5. Select the headControls locator. In the Channel box highlight the jawRot attribute. With the middle mouse button scroll back and forth in the Perspective window. You should see the jaw of the character opening and closing. Set jawRot back to 0 when you are finished testing the jaw.

Exercise 11.7: continued

Figure 11.36
Testing the jawRot attribute. The jawRot is directly connected to the jaw joint's Z rotation.

6. Create an attribute on headControls to connect to the Phonemes Blend Shape. Select headControls and choose Modify, Add Attribute. Name the attribute **A** and give it a min/max of –10 to 10. Click the Add button.

7. Connect the A attribute to the A target of the Phonemes Blend Shape. Select headControls and choose Keys, Set Driven Key. Click the Load Driver button.

8. Select the A attribute as the driving attribute. Choose Window, Animation Editors, Blend Shape. In the Blend Shape window, click the

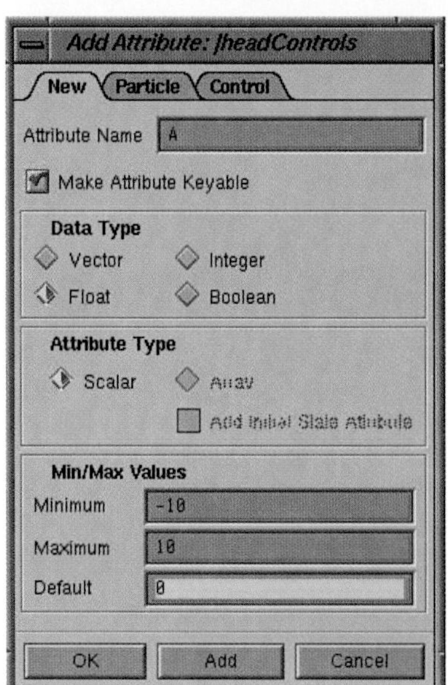

Figure 11.37
Adding the A custom attribute to headControls using the Add Attribute window.

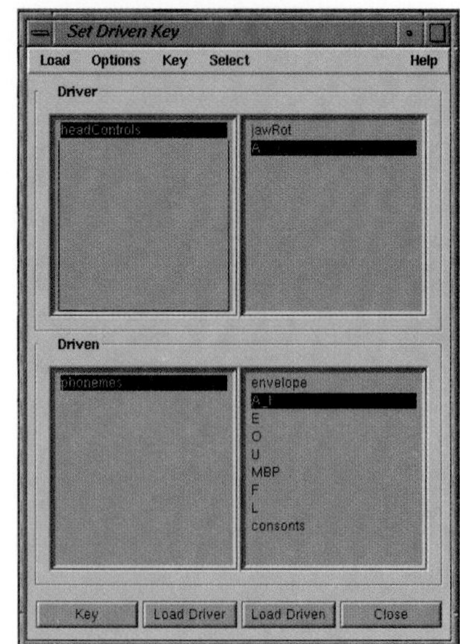

Figure 11.38
Using Set Driven Key to have the A attribute of headControls drive the Blend Shape envelope of the AI phoneme.

Select button for the Phonemes Blend Shape node. In the Set Driven Key window, click the Load Driven button. Select the A attribute of the Phoneme Blend Shape as the driven attribute. Click the Key button to set the initial Set Driven Key.

9. Set the A attribute on the headControls Locator to 10. In the Blend Shape window, click and drag the vertical slider for the A target to 1. Click the Key button.

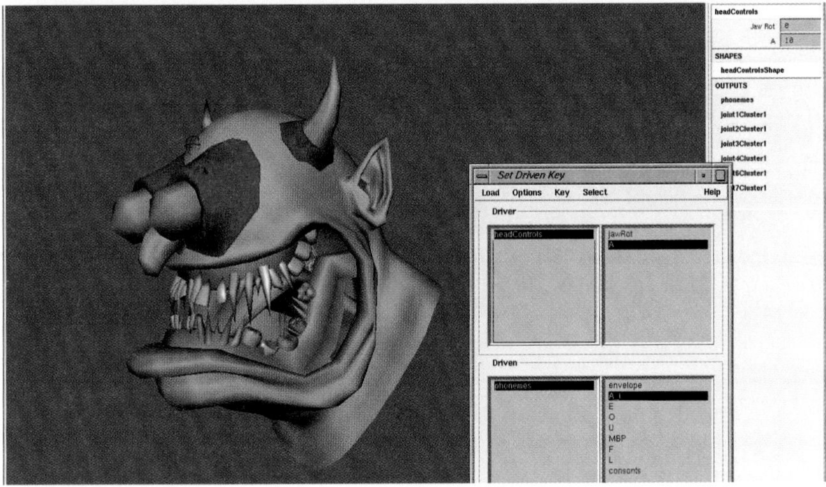

Figure 11.39
Setting Set Driven Key keyframes for the A mouth position.

10. Set the A attribute on headControls to –10. In the Blend Shape window, type **-.25** in the envelope input area. Click the Key button in the Set Driven Key window. Close the Set Driven Key window.

11. Highlight the A attribute on headControls in the Channel box. Update the value of the A attribute and watch the face of the devil morph into the A target Blend Shape. Click on the jawRot attribute to open the mouth and jaw of the devil. Set both attributes back to 0 to set the character's facial controls back to the default.

You can see that adding your own controls to one object can make animating the character's face much more straightforward. You don't need to hunt for joints or the Blend Shape window—simply select the headControls locator and all of your face and head controls are right there. You can continue adding your own attributes to the headControls locator for more practice. Add attributes for the rest of the Blend Shape mouth and eye shapes. Create controls that help with the rotations of the neck joints. You can then parent

Exercise 11.7: continued

the headControls locator to the neck joints so the locator stays in front of the character even if it's running around.

12. Select the headControls locator and Shift+select the neckBase joint. Choose Edit, Parent. You made the headControls locator a child of the neckBase joint, so now it will move with the character.

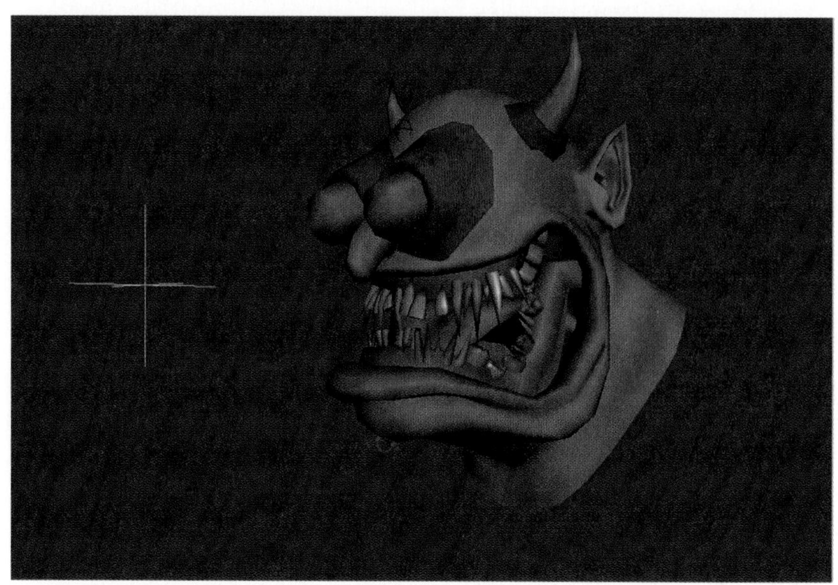

Figure 11.40
Testing the A attribute on the headControls locator.

11.4 Facial Animation

Now that we have the jaw, mouth, and eye shapes working on our Devil character's face, we can take a look at the workflow in animating the face to an imported sound file. We will use the sound file to help us keyframe the appropriate facial positions in time according to the sound track.

Exercise 11.8: Adding Sound

1. Import the .aiff sound file fillerUp.aiff from the included CD-ROM). Choose File Import, By Type, AIFF. This audio contains a voice saying "Fill 'er Up, Buddy!" In the time slider at the bottom of the interface, hold down the right mouse button and go to Sound in the pop-up menu and select the fillerUp sound file. The sound file will appear graphically in the time slider.

   Now we're going to display the sound file in the timeline.

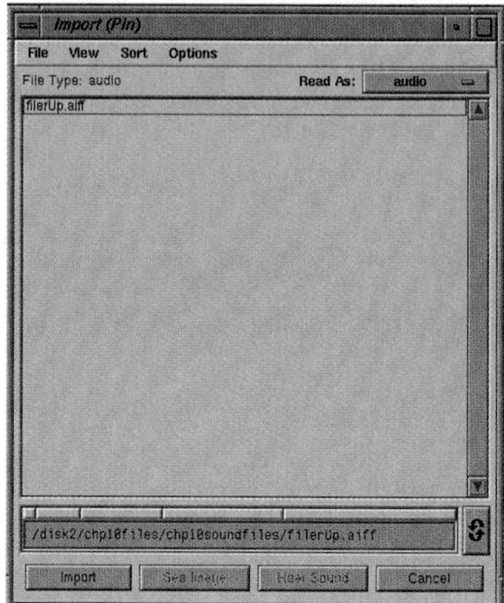

Figure 11.41
Importing the sound file into the scene.

2. In the Animation Preferences (Options, General Preferences, Animation tab) dialog box, set the playback speed to Normal if it is not already set to that speed.

3. Set the current time to frame 1 in the Timeline.

4. Select the headControls Locator, and press the S key on the keyboard to set keyframes for all of the head controls for the character at frame 1.

5. Scrub the Time Slider in the Timeline; you should hear the audio playing back. Make sure the audio on your system is set correctly to enable you to hear the audio. Scrub the audio until you find the start of the F in the words "Fill 'er." Try to locate the approximate frame where you just start to hear the F sound forming.

   When animating the face, you will want to set keyframes for the phonemes just before the actual time the audible sound is heard. The phoneme needs to start forming before the sound is fully audible.

6. With headControls still selected, set the attribute F to a value of 10. Press the S key on the keyboard to set keyframes for all of the attributes on headControls.

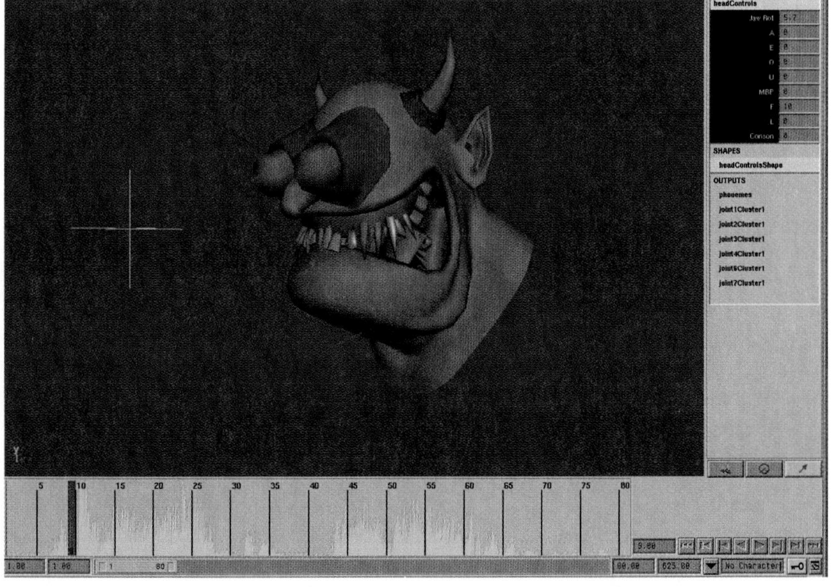

Figure 11.42
Sound file visible in timeline and face set to F position where sound first becomes audible.

7. Scrub over in time a little further until you start to hear the L sound in the words "Fill 'er." Make you sure you find the location in time just slightly before you actually hear the L sound.

8. At the current frame, change the F attribute to 0 and the L attribute to 10. Press the S key on the keyboard to set keyframes for all of the attributes on headControls.

Exercise 10.8: continued

Check the progress of the facial animation by constantly scrubbing the time slider back and forth. Listen for the sound and watch the shapes of the mouth take shape. Edit the location of the keys using the Graph Editor or Dope Sheet to improve the timing of the facial animation.

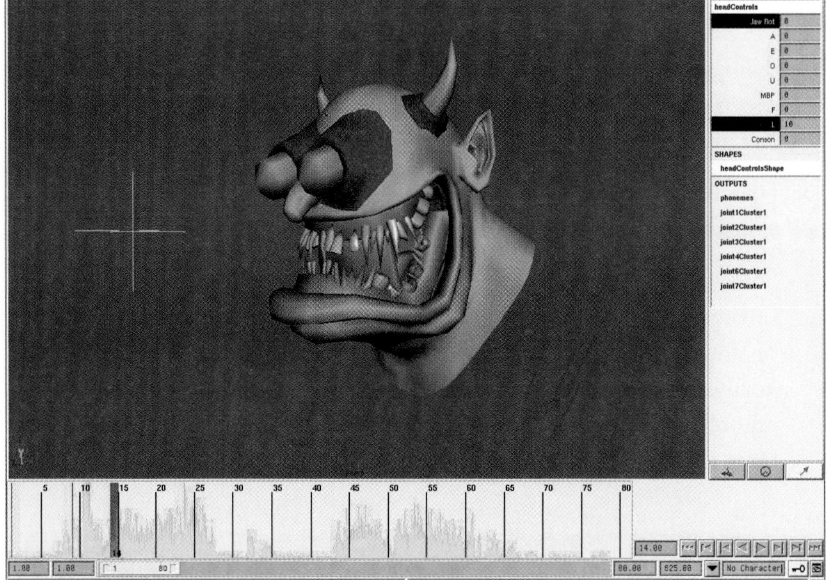

Figure 11.43
Scrubbing through sound file, changing face positions, and setting keyframes on the headControls locator.

If the shaded update of the character is slow, try selecting all of the NURBS surfaces of the face and go to Display, NURBS Smoothness, Hull. This will simplify the amount of data Maya has to calculate to shade and update the model as you scrub through the timeline. Use the Playblast tool under Window, Playblast to view low-resolution tests of your animation. Playblast movies will work best when working with sound in Maya. The movies will play back synched to the sound you have currently loaded in Maya.

Approach the animation of the face in layers. Work on one aspect of the face at a time. You may want to start off by roughly blocking in the phoneme mouth positions according to the dialog. Some animators like to animate the head and neck movements first, and then add the mouth. Many times it very helpful if you start by animating the body of the character first then add the animation of speech and facial expression after that. This allows you to first establish the position of the head and body at different points in time, which will help guide the necessary facial animation.

Index

Symbols

A

B

D

E

Q-R

S

T

U-V

W-Z

New Riders Professional Library

3D Studio MAX 3 Fundamentals
Michael Todd Peterson
0-7357-0049-4

3D Studio MAX 3 Media Animation
John Chismar
0-7357-0050-8

Bert Monroy: Photorealistic Techniques
with Photoshop & Illustrator
Bert Monroy
0-7357-0969-6

CG 101: A Computer Graphics Industry Reference
Terrence Masson
0-7357-0046-X

Click Here
Raymond Pirouz and Lynda Weinman
1-56205-792-8

<coloring web graphics.2>
Lynda Weinman and Bruce Heavin
1-56205-818-5

Creating Killer Web Sites, Second Edition
David Siegel
1-56830-433-1

<creative html design>
Lynda Weinman and William Weinman
1-56205-704-9

<designing web graphics.3>
Lynda Weinman
1-56205-949-1

Designing Web Usability
Jakob Nielsen
1-56205-810-X

[digital] Character Animation 2 Volume 1: Essential Techniques
George Maestri
1-56205-930-0

[digital] Lighting & Rendering
Jeremy Birn
1-56205-954-8 (Publishes in 2000)

[digital] Textures & Painting
Owen Demers
0-7357-0918-1 (Publishes in 2000)

Essentials of Digital Photography
Akari Kasai and Russell Sparkman
1-56205-762-6

Fine Art Photoshop
Michael J. Nolan and Renee LeWinter
1-56205-829-0

Flash 4 Magic
David Emberton and J. Scott Hamlin
0-7357-0949-1

Flash Web Design
Hillman Curtis
0-7357-0896-7

HTML Artistry: More than Code
Ardith Ibañez and Natalie Zee
1-56830-454-4

HTML Web Magic
Raymond Pirouz
1-56830-475-7

Inside 3D Studio MAX 3 Modeling, Materials, and Rendering
Ted Boardman and Jeremy Hubbell
0-7357-0085-0

Inside AutoCAD 2000
David Pitzer and Bill Burchard
0-7357-0851-7

Inside Adobe Photoshop 5, Limited Edition
Gary David Bouton and Barbara Bouton
1-56205-951-3

Inside LightWave 3D
Dan Ablan
1-56205-799-5

Inside SoftImage 3D
Anthony Rossano
1-56205-885-1

Inside trueSpace 4
Frank Rivera
1-56205-957-2

Maya 2 Character Animation
Nathan Vogel, Sherri Sheridan, and Tim Coleman
0-7357-0866-5

Net Results: Web Marketing that Works
USWeb and Rick E. Bruner
1-56830-414-5

Photoshop 5 & 5.5 Artistry
Barry Haynes and Wendy Crumpler
0-7457-0994-7

Photoshop 5 Web Magic
Michael Ninness
1-56205-913-0

Photoshop Channel Chops
David Biedney, Bert Monroy, and Nathan Moody
1-56205-723-5

<preparing web graphics>
Lynda Weinman
1-56205-686-7

Rhino NURBS 3D Modeling
Margaret Becker
0-7357-0925-4

Secrets of Successful Web Sites
David Siegel
1-56830-382-3

Web Concept & Design
Crystal Waters
1-56205-648-4

Web Design Templates Sourcebook
Lisa Schmeiser
1-56205-754-5

What You'll Find on the Maya 2 Character Animation CD

CD-ROM CONTENTS

1. Licensing Agreement
2. Browsing the CD-ROM via the CD-ROM Interface and Installation
3. CD-ROM Contents
4. User Services Information
5. Additional Contact Information

1. LICENSING AGREEMENT

By opening this package, you are agreeing to be bound by the following agreement:

All of the software included with this product is copyrighted, in which case all rights are reserved by the respective copyright holder. You are licensed to use software copyrighted by the publisher and its licensors on a single computer. You may copy and/or modify the software as needed to facilitate your use of it on a single computer. Making copies of the software for any other purpose is a violation of the United States copyright laws. COPYRIGHT © 2000, New Riders Publishing.

This software is sold as is, without warranty of any kind, either express or implied, including but not limited to the implied warranties of merchantability and fitness for a particular purpose. Neither the publisher nor its dealers or distributors assumes any liability for any alleged or actual damages arising from the use of this program. (Some states do not allow for the exclusion of implied warranties, so the exclusion might not apply to you.)

This CD-ROM includes documents in an electronic format. These documents are licensed to you for your individual use on one computer. You may make a single second copy of this electronic version for use by you on a second computer for which you are the primary user (e.g. a laptop). You may not copy this to any network, intranet, the Internet, or any other form of distribution for use by anyone other than yourself. Any other use of this electronic version is a violation of U.S. and international copyright laws. If you would like to purchase a site license to use this material with more than one user, please contact Macmillan at `license@mcp.com`.

2. BROWSING THE CD-ROM VIA THE CD-ROM INTERFACE AND INSTALLATION

**NOTE: This CD-ROM is an SGI- and Windows PC-based hybrid CD-ROM. Some features and content on one operating platform may not be available on the other.*

System Requirements for this New Riders Publishing CD-ROM

Processor:	200Mhz or higher
OS:	Microsoft Windows 98/NT/SGI UNIX
Memory (RAM):	32MB
Monitor:	VGA, 640×480 or higher with 256 color or higher
Storage Space:	1GB minimum (will vary depending on installation)
Other:	Mouse or compatible pointing device
Optional:	Internet connection and Web browser. For your convenience on this disk you will find some Web browsers.

Windows Installation

If you have AUTOPLAY turned on, your computer will automatically run the CD-ROM interface. If AUTOPLAY is turned off, follow these directions:

1. Insert the CD-ROM into your CD-ROM drive.
2. From the Windows desktop, double-click the My Computer icon.
3. Double-click the icon representing your CD-ROM drive.
4. Double-click the icon titled START.EXE to run the interface.

SGI/UNIX Installation

1. Insert the CD-ROM in your CD-ROM drive.
2. Mount the CD-ROM onto your filesystem. Typically, this is done by typing: `mount -tiso9660 /dev/cdrom /cdrom`

NOTE: The mount point must exist before you mount the CD-ROM. If you have trouble mounting the CD-ROM, please consult the main page for the mount command or contact your system administrator.

3. CD-ROM CONTENTS

***APPENDIXES A and B:**

Additional chapters, in electronic format. Please see table of contents for description.

***EXERCISE FOLDER:**

All the files you need to re-create the finished assignments in each chapter.

***EXTRAS:**

Additional Maya 2 files, resources, and tutorials.

***3RDPARTY:**

ADOBE ACROBAT READER 4

We've got the install files for both the Windows and the Macintosh versions of Acrobat Reader 4, so you can browse all the documents that are in PDF format on the companion CD. The free Adobe Acrobat Reader enables you to view, navigate, and browse PDF files seamlessly, either inside a Web browser or in a standalone application. *Adobe Acrobat Reader 4.01 by Adobe Systems, Inc.* `www.adobe.com`

WINZIP 7.0

by Nico Mak Software

WinZip brings the convenience of Windows to the use of Zip files and other archive and compression formats. The optional wizard interface makes unzipping easier than ever. WinZip features built-in support for CAB files and for popular Internet file formats such as TAR, gzip, UUencode, BinHex, and MIME. ARJ, LZH, and ARC files are supported via external programs. WinZip interfaces to most virus scanners.

WS FTP PRO

WS_FTP Pro is the fastest, most powerful Windows FTP client available. With WS_FTP Pro, you can connect to any FTP server, browse through directories and files, and transfer files in either direction—fast. Version 6.01 includes enhancements to both the Classic and Explorer interfaces, including the following: ability to delete folders with subfolders; ability to store FTP site configurations in hierarchical folders; a wizard interface for adding new sites; improved firewall support; automatic detection of the following host types: AMOS TCP/IP, VxWorks, and GCOS 6 HVS AIX; and a graphical interface for connecting to FTP sites (Classic). It also includes a bonus FTP utility pack containing three advanced tools: a Find utility that lets users search an FTP site for files that match the user's criteria, a powerful Scripting utility for automating file transfers and management through a series of WS_FTP commands, and a Synchronize utility for synchronizing or "mirroring" FTP or Web sites. Other features include a Site Manager that stores FTP sites in folders for easy reference, drag and drop support in the WS_FTP Pro Classic interface, Quick Connect for immediate connection to new FTP sites, Recursive Delete, and Auto Re-get.

4. USER SERVICES INFORMATION

Contacting Us for Support

We cannot help you with computer problems, Windows problems, or third-party application problems, but we can assist you with a problem you have with the book or the CD-ROM.

Note: Problems with other companies' programs on the disc must be resolved with the company that produced the program or demo.

If you need assistance with the information provided in this product, please feel free to access our Web site at `http://www.mcp.com/info`. Here you can enter the book's ISBN number, and view a product information page that will include any available downloads or updates. Our User Services department can also be reached at `http://www.mcp.com/support`.

5. ADDITIONAL CONTACT INFORMATION

Please see the front matter of this book for complete info on how to contact us.